AF360728

Eye Movements and Visual Cognition

Eye Movements and Visual Cognition

Special Issue Editors

Raymond M. Klein
Simon P. Liversedge

MDPI • Basel • Beijing • Wuhan • Barcelona • Belgrade • Manchester • Tokyo • Cluj • Tianjin

Special Issue Editors

Raymond M. Klein
Dalhousie University
Canada

Simon P. Liversedge
University of Central Lancashire
UK

Editorial Office
MDPI
St. Alban-Anlage 66
4052 Basel, Switzerland

This is a reprint of articles from the Special Issue published online in the open access journal *Vision* (ISSN 2411-5150) (available at: https://www.mdpi.com/journal/vision/special_issues/Eye_Movements_Visual_Cognition).

For citation purposes, cite each article independently as indicated on the article page online and as indicated below:

LastName, A.A.; LastName, B.B.; LastName, C.C. Article Title. *Journal Name* **Year**, *Article Number*, Page Range.

ISBN 978-3-03936-547-0 (Hbk)
ISBN 978-3-03936-548-7 (PDF)

Cover image courtesy of Jiawei Zhao.

Contents

About the Special Issue Editors

Raymond M. Klein is Professor Emeritus in the Department of Psychology and Neuroscience at Dalhousie University. He has been interested in "how the mind works" since he began graduate school. Working with Michael Posner, he became an expert on the concept of attention, which is at the nexus of this challenging effort, and in the methods of mental chronometry and cognitive neuroscience. The various ways in which attention and eye movements are related has been a continuous preoccupation of his empirical, theoretical, and applied work. Although best known for his work on overt and covert orienting when under exogenous and endogenous control, Dr. Klein has a long-standing interest in applying his knowledge and experience to help solve real-world problems.

Simon P. Liversedge is Professor of Cognitive Psychology at the University of Central Lancashire. He conducted his first eye movement experiment to investigate reading when he was an undergraduate student and has continued to undertake eye movement research in the area of human visual cognition since then. His primary research interest concerns how people read, and much of his work has focused on understanding the visual and linguistic characteristics of text that produce systematicities in the eye movement record. Chinese reading has become a particularly keen interest in recent years because, in its written form, the Chinese language offers the opportunity to investigate theoretical questions that are simply not possible to pursue in alphabetic languages.

Preface to "Eye Movements and Visual Cognition"

This eBook is based on a Special Issue on the topic of "Eye Movements and Visual Cognition" that was published in *Vision*. Our aim in putting together the Special Issue and eBook was to attract a group of high-quality, original, and topical works by leading academic figures in the field of human vision and visual cognition. In so doing, we aimed to stimulate and foster useful intellectual exchange between individuals working primarily on basic theoretical issues, as well as those working on more applied aspects of vision science. From the outset, we were particularly keen to attract papers that review particular topics within this broad field, and we were successful in achieving this. The present volume includes reviews that are narrative (critiquing and summarizing research on a particular topic), tutorial (with a focus on methods and findings), empirical (e.g., meta-analytic), and theoretically synthetic. Indeed, some articles represent a combination of several of these types of reviews. We also included papers with new empirical content when this served to resolve an undecided issue stemming from such a literature review, and we encouraged papers that build bridges between theoretical and applied aspects and between behavior and its neural substrate. All papers were subject to peer review and went through several rounds of revision prior to acceptance.

It is not at all contentious to suggest that eye movement recording is now a well-established method in the field of experimental psychology. Techniques for recording eye movements have changed quite markedly over the last half-century, with recording devices becoming increasingly user-friendly, advanced data acquisition software becoming more flexible and more broadly available, and data analysis software becoming more progressively efficient at dealing with vast data sets that the method generates. Additionally, eye movement methodology has now come to be used very widely to investigate many aspects of human visual and cognitive processes (and more broadly, beyond the realm of visual cognition). Without question, this methodological approach has significantly contributed to the current theoretical understanding of human vision and cognition.

The topics covered in this eBook reflect this breadth, ranging from aspects of relatively low-level vision (unconscious attention, saccadic suppression, inhibition of return) to higher-order aspects of cognition (serialism and parallelism in reading, higher-level comprehension issues, meaning in scenes, and scene interpretation). Developmental themes also feature in relation to performance change with age (children's reading and older adult reading performance), and there is consideration of eye movements in special populations (autism). It is also important to note that the contributions that are represented here vary in the degree to which they focus on "blue skies" theoretical issues (visual salience, inhibition of return, regressions in reading), with some work emphasizing a stronger translational line running through it (medical image interpretation and X-ray baggage screening). Again, this underlines the prevalent extent to which eye movement methodology is employed within the field. In considering the work reflected in the contents of this eBook, what emerges is a clear indication that eye movement research is currently in a very healthy state.

Before closing, it is important to acknowledge the support and assistance that we have received from colleagues at MDPI. In particular, we would like to thank Meirong Duan, who has provided excellent editorial support throughout, and of course, we would like to thank all the authors for taking the time and making the effort to write such high-quality papers for the Special Issue.

Raymond M. Klein, Simon P. Liversedge
Special Issue Editors

Review

Associations and Dissociations between Oculomotor Readiness and Covert attention

Soazig Casteau and Daniel T. Smith *

Department of Psychology, Durham University, Durham DH1 3HP, UK; soazig.casteau@durham.ac.uk
* Correspondence: daniel.smith2@durham.ac.uk

Received: 19 February 2019; Accepted: 25 April 2019; Published: 7 May 2019

Abstract: The idea that covert mental processes such as spatial attention are fundamentally dependent on systems that control overt movements of the eyes has had a profound influence on theoretical models of spatial attention. However, theories such as Klein's Oculomotor Readiness Hypothesis (OMRH) and Rizzolatti's Premotor Theory have not gone unchallenged. We previously argued that although OMRH/Premotor theory is inadequate to explain pre-saccadic attention and endogenous covert orienting, it may still be tenable as a theory of exogenous covert orienting. In this article we briefly reiterate the key lines of argument for and against OMRH/Premotor theory, then evaluate the Oculomotor Readiness account of Exogenous Orienting (OREO) with respect to more recent empirical data. These studies broadly confirm the importance of oculomotor preparation for covert, exogenous attention. We explain this relationship in terms of reciprocal links between parietal 'priority maps' and the midbrain oculomotor centres that translate priority-related activation into potential saccade endpoints. We conclude that the OMRH/Premotor theory hypothesis is false for covert, endogenous orienting but remains tenable as an explanation for covert, exogenous orienting.

Keywords: attention; covert; oculomotor readiness hypothesis; premotor theory; exogenous; endogenous; eye abduction

---

1. Introduction

Covert spatial attention allows us to select important and/or behaviourally relevant visual inputs by enhancing signals arising from attended locations and suppressing signals from unattended locations [1] without actually moving the eyes to that location. Despite many advances in understanding the cognitive processes involved in spatial attentional selection, an enduring issue is the mechanism by which attention is moved from one location to another. It is generally agreed that the orienting of spatial attention can occur in an automatic 'exogenous' mode in response to salient external events (e.g., the flashing lights of an emergency services vehicle) or a controlled 'endogenous' mode in response to the observer's goals (e.g., systematically scanning the road ahead to check for hazards) [2], and that these systems are partially dissociable [3]. It is also widely accepted that eye movements ('overt' shifts of attention) are preceded by a covert shift of attention to the saccade goal, known as 'pre-saccadic attention'. However, there is a long-running debate concerning the relationship between the mental process involved in covert orienting of attention (i.e., attending to things that are not being gazed at), and those involved in overt orienting of attention (i.e., orienting the eye to the stimulus of interest) [4]. One proposal, originally known as the Oculomotor Readiness Hypothesis (OMRH) [5] and later as Premotor Theory (PMT) [6], proposed a complete functional overlap between spatial attention and oculomotor control. OMRH/PMT is often used as shorthand to refer to the general idea that covert attention is, in some way, linked to the oculomotor system. However, this usage does not do full justice to the OMRH/PMT theory, which makes clear and testable predictions about the precise relationship between oculomotor control and covert spatial attention. More specifically, OMRH/PMT holds that the programming of a saccade is both necessary and sufficient for covert orienting of attention [7].

Despite being the original proponents of OMRH, Klein and colleagues concluded that endogenous attention was in fact independent of saccade programming [5,8], although they speculated that OMRH may still hold for exogenous attention. Subsequently, a number of other proposals suggesting differing degrees of overlap between attention and saccade control have been put forward [9–11]. Following the work of Klein and colleagues, we have pursued the idea that the relationship between covert attention and saccade programming may indeed be dependent on the mode of orienting, such that OMRH/PMT was only true when the exogenous mode of orienting was engaged [4]. In this review we outline the main lines of argument for and against OMRH/PMT as a theory of endogenous covert orienting, then explain why we believe that OMRH/PMT is false for endogenous covert orienting, but remains tenable as a theory of exogenous, covert orienting.

2. The Case for OMRH/PMT

The case for OMRH/PMT draws on three main lines of evidence. Firstly, there is clear evidence that saccadic eye movements are preceded by a mandatory 'pre-saccadic' shift of attention [12–18] and a more efficient distractor suppression at non-saccade goals [19]. This pre-saccadic attentional facilitation is clearly tied to the programming of an eye movement, as the effect grows larger with proximity with saccade onset [20,21] and occurs even when the participant expects the probe to appear opposite the saccade goal, implying that programming an eye movement is sufficient to trigger a shift of covert attention [13]. Furthermore, shifts of attention appear to affect the trajectory of saccadic eye movements, consistent with the idea that shifts of attention activate a saccade plan [16,22,23].

Secondly, eye movements and covert shifts of attention appear to activate similar networks of brain areas, including the Frontal Eye Fields (FEF), the Lateral Intraparietal cortex, and the Superior Colliculi (SC) [24–29](see Figure 1), and lesions to these brain areas are associated with deficits of both covert orienting and saccade control [30–36]. Moreover, electrical stimulation of FEF neurons in non-human primates elicited fixed-vector saccadic eye movements, and subthreshold stimulation of the same neurons significantly enhanced perceptual discrimination, even though the monkey was still centrally fixating [37,38]. Using a similar methodology, Moore and colleagues also demonstrated that stimulation of FEF modulated the sensitivity of neurons in V4, an area of the visual cortex that codes for colour, orientation and spatial frequency, and whose visual receptive fields overlap with the motor field [39,40]. The effect of FEF microstimulation on neural responses in V4 was analogous to that observed when the monkey endogenously attended the location [39]. These data suggest a causal role for saccade programming in covert attention, as predicted by OMRH/PMT.

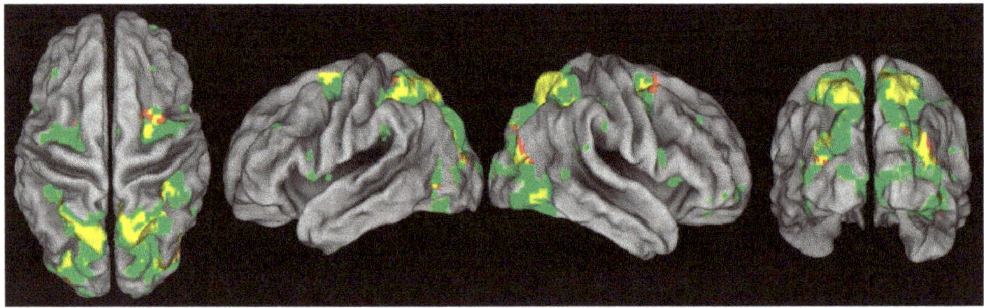

Figure 1. In red are the areas of the brain that are significantly activated in the covert shift of attention task and in green the areas of the brain significantly activated in the overt shift of attention task. In yellow are the areas of the brain activated in both the overt and the covert shift of attention task. Reproduced with permission from [24].

A third line of argument draws on studies in which eye movements are impaired, experimentally restricted, or experimentally modulated. For example, Craighero, Carta and Fadiga [41] observed that patients with a palsy of the VIth cranial nerve were unable to covertly orient attention only when viewing stimuli with their palsied eye, suggesting that the endogenous shift of attention was impaired when viewing with the damaged eye but not when viewing with the intact one. In line with this study, Craighero, Nascimben and Fadiga [42] used an eye abduction paradigm (see Figure 2), where saccadic eye movement programming is disrupted by forcing healthy participants to rotate the eye by 40° into the temporal hemifield. In their experiment participants were presented with a classical Posner cueing task in which a central predictive cue (i.e., a bar attached to the fixation square indicating left or right) indicated in 70% of the cases the accurate position of the upcoming target, which could be either in the nasal hemispace (i.e., at a position that can be reached by a saccadic eye movement) or in the temporal hemispace (i.e., outside a position reachable by a saccadic eye movement). Visual acuity remained unaffected but the attentional benefits typically observed with valid cues were reduced when stimuli were presented in the temporal/eye movement restricted hemispace but not when presented in the nasal hemispace. The authors concluded that, consistent with Premotor theory, covert orienting of attention is subject to the limitations of the saccadic system such that attention cannot be deployed at a location that cannot become the goal of a saccadic eye movement. This led to the proposal that covert attention and saccadic eye movements share the same 'stop limit', which is the range of eye movements, also referred to as Effective Oculomotor Range (EOMR).

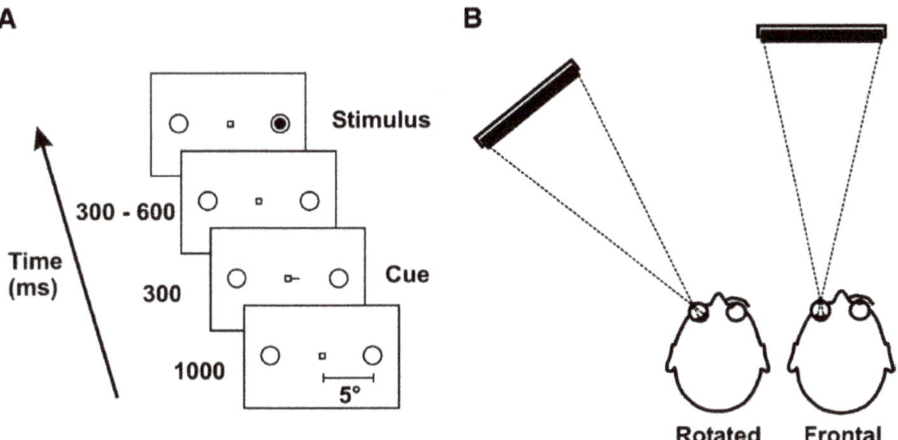

Figure 2. Experimental setup for the eye-abduction paradigm used by Craighero et al. (2004). Reproduced with permission from [42].

Other studies have used the saccadic adaptation technique to dissociate the perceived position of a saccade target from the actual endpoint of the eye movement. In saccadic adaptation tasks the participant makes a saccade to a peripheral stimulus, but during the saccade the stimulus jumps to a new position (double-step task) [43]. At the start of the experiment the participant initially moves to the original stimulus position then, unconsciously, makes corrective eye movements towards the second stimulus position. However, over the course of many trials they adapt the amplitude of the saccade to ensure it lands at the final position of the stimulus rather than its original position (for a review, see [44]). OMRH/PMT predicts that saccadic adaptation should also result in the adaptation of covert shifts of attention, such that the locus of attention should be at the final stimulus position, not the starting position. To test whether attention focus is shifted towards the saccade target or the final eye position, Ditterich et al. [45] asked participants to make a saccade towards a peripheral location and, before the first saccade onset, they briefly flashed a discrimination target at one of four possible locations. The discrimination performances were compared before and after the saccadic adaptation. Prior to adaptation, discrimination performance was best at the goal of the saccade. After adaptation, optimal discrimination performance was still observed at the goal of the first saccade, and not at the endpoint of the adapted saccade. This result is not consistent with OMRH/PMT, and Ditterich et al. concluded that the attentional focus is always directed to the primary target position and not to the saccade landing position [45]. However, Collins and colleagues argued that this conclusion was premature, given that the magnitude of the adaptation effects observed by Ditterich. was somewhat small. In two subsequent studies using more effective adaptation protocols they showed that saccadic adaptation does indeed produce adaptations of pre-saccadic attention [46,47] and that pre-saccadic displacement of attention would be shifted both to the position of the saccadic target and to the landing position of the adapted saccades [48]. In a recent study, Habchi and colleagues claimed that saccadic adaptation leads to changes in the allocation of covert attention, although these changes appear to be due to a more general bias towards the side of adaptation, rather than a modulation of covert orientation per se [49]. Overall, the evidence is consistent with the claim that saccadic adaption is associated with adaptations of pre-saccadic attention, which has been interpreted as evidence for OMRH/PMT.

Further evidence for OMRH/PMT is the finding that covertly attending a location produces a change in the trajectory of saccades, such that they deviate from the intended location [22]. Trajectories of vertical and oblique saccades are never completely straight but curvilinear, even when aiming at an isolated target [50,51], and it has been suggested that saccade curvature is determined by mechanisms situated in the final pathway of eye movement generation [52]. In addition to this natural tendency, other objects presented in the visual scene can influence the magnitude and direction of saccade curvature. Several authors have found that presenting an irrelevant distractor stimulus near a saccade target affects the saccade curvature [22,53–55]. In some instances, saccades can curve towards the irrelevant stimulus, as in visual search tasks [56], when the location of the saccade target is highly unpredictable, or for short-latency saccade [57], but in other cases, there is a tendency to deviate from the position of the distractor, particularly when saccade latencies are long [55], whether the saccade is reflexive or voluntarily triggered [53]. These trajectory deviations are typically attributed to competition between saccade plans associated with the target and distractor, and evidence that covert attention can also cause trajectory deviations [22,58–60] is therefore often cited as evidence for OMRH/PMT.

To briefly summarize, OMRH/PMT argues that covert orienting of attention depends on the activation of a saccade plan. Consistent with this hypothesis, there is a mandatory orientation of attention to saccade goals; covert and overt attention activate overlapping brain areas and damage to these areas causes problems with both overt and covert orienting. For example, ophthalmoplegic patients have deficits of covert attention that seem to mirror their ocular deficit. Moreover, modulating the gain of saccades also modulates the gain of pre-saccadic shifts of attention, and covertly attending a peripheral location affects the metrics of overt saccades, such that their trajectories are deviated away from the attended location. Altogether, these studies seem to offer clear evidence for a tight coupling between attention and oculomotor control.

3. The Case against OMRH/PMT

On first inspection, the evidence for OMRH/PMT seems overwhelming (e.g., [61]). However, we believe there are a number of reasons to be cautious about accepting these lines of evidence as conclusive proof of the claim that saccade programming *is necessary and sufficient* for covert orienting of spatial attention in the absence of an overt eye movement. Firstly, there is evidence that 'pre-saccadic' attention (i.e., the covert shift of attention that precedes an overt eye movement) is qualitatively different to covert attention. Secondly, although neuroimaging and some neuropsychological studies demonstrate associations between attention and oculomotor control, other studies have shown clear evidence of dissociations between saccade programming and covert orienting. Thirdly, behavioural studies that explicitly test the hypothesis that covert, endogenous attentional orienting is caused by saccade programming largely fail to support this hypothesis. Finally, while the evidence of interactions between saccade programming and covert attention suggests a relationship between the two processes, the evidence is not consistent with the claim made by OMRH/PMT, which is that covert orienting of attention is caused by activation of a saccade plan. We expand on these critiques in the following sections.

3.1. Pre-Saccadic Attention Is Not Equivalent to Covert Attention

The intention to make an eye movement produces radical changes in the receptive fields of neurons throughout the visual system, such that they appear to respond to stimuli in their post-saccadic spatial location before the saccade has begun [62]. This neurophysiological mechanism may well underpin the perceptual benefits observed in the moments before a saccade that are typically attributed to covert attention [63]. Critically, however, Duhamel et al. [62] also noted that this 'pre-saccadic remapping' did not occur when attention was deployed *without* a saccade, so cannot be responsible for 'pure' covert orienting (i.e., when the eyes remain fixated). If it is accepted that pre-saccadic remapping underpins pre-saccadic attention, and that pre-saccadic remapping only occurs prior to a saccade, it must also be accepted that pre-saccadic attention and 'pure' covert orienting of attention, which occurs when no saccade is executed, are served by a qualitatively different mechanisms

The proposal that pre-saccadic perceptual enhancements are qualitatively different to covert attention is consistent with neuropsychological evidence of a dissociation between covert attention and pre-saccadic perceptual enhancement. For example, Ladavas [64] asked patients with visual neglect to fixate and report target appearance using a button press response. Targets presented in the neglected field summoned involuntary eye movements on 45% of trials, but only half of these trials were associated with conscious detection of a target. When no saccade was made, only 4% of targets were detected. They concluded that the target could activate the oculomotor system without a concurrent shift of attention. In this case, the amplitude of the eye movements is not reported, so it is not clear whether the saccades that were not associated with target detection actually fixated the target (i.e., they might have fallen short, in which case the shift of attention could also have been hypometric). However, similar results were observed by Benson et al. [65] in a single case study of a patient with hemispatial neglect. In this study, a peripheral cue in the neglected hemifield summoned an eye movement but was not consciously detectible, again suggesting that the programming of eye movements and the orienting of attention can be dissociated. Blangero and colleagues [66] provided evidence of a double dissociation between the two processes. They reported the case of patient O.K., who presented with optic ataxia following a right parietal stroke, but no symptoms of neglect. Patient O.K. could make accurate saccades into the left hemifield and showed the typical pattern of pre-saccadic attentional enhancement at the saccade goal. However, the patient could not covertly attend to the same location when saccades were suppressed, demonstrating a dissociation between pre-saccadic attention and covert attention. Together, these studies suggest that pre-saccadic perceptual enhancements and covert orienting of attention are mediated by different cognitive mechanisms. If this proposal is correct, studies of pre-saccadic perceptual enhancement cannot be taken as evidence that shifts of covert spatial attention that occur in the absence of any overt eye movement rely on saccade programming.

3.2. Association Is Not Causation

The second main line of evidence in favour of OMRH/PMT draws on neurophysiological studies of non-human primates. These studies clearly showed that attention and eye movements share some common neural substrate and elegant work, showing that microstimulation of FEF leads to covert visual selection [37], is often presented as evidence for PMT. However, areas like FEF contain several distinct populations of neurons, some of which are involved in visual selection but not motor control, and others that are involved in saccade control but not visual attention [67–69]. Given that microstimulation of FEF may affect both visual and motor neurons [70], it is impossible to unambiguously attribute the attentional effects of microstimulation to the activation of motor programs. Furthermore, other research has shown that attending a stimulus does not affect the trajectory of microstimulation-evoked saccades [71], and concluded that covert attention is not necessarily associated with activation of a saccade plans, contrary to some of the behavioural findings reported in humans (e.g., [22]). A neurophysiological dissociation between saccade programming and covert orienting has also been observed using EEG in human participants by Weaver and colleagues [72]. The key finding here was that participants could endogenously allocate attention to the target object even on trials where the eyes were involuntarily directed to a salient distractor. This result is hard to reconcile with the claim that saccade preparation is both necessary and sufficient for covert attention. Overall, at best neurophysiological studies demonstrate an association between the brain areas required for saccade programming and those required for covert attention, and the few studies that offer a strong test of the key claim of PMT, which is that endogenous covert orienting is *caused* by saccade programming, seem to argue against this position (e.g., [71,73]).

3.3. Saccade Programming Does Not Necessarily Produce a Shift of Attention

OMRT/PMT argues that saccade programming produces shifts of attention. However, dual task experiments have repeatedly failed to observe attentional benefits at the goal of planned but unexecuted eye movements. In a seminal study by Klein [5] participants were asked to perform a variant of a go-no-go task. In the majority of trials participants were instructed to prepare a saccade to the left or to the right, and execute the prepared saccade when an asterisk was presented at either the left or right location. Participants were faster at executing saccades when the peripheral onset was congruent with the saccade they had prepared. However, in occasional trials they were asked to cancel the saccade and make a manual response instead. The key finding here was that manual detection responses were not faster when probes appeared on the same side as they were instructed to prepare a saccade, suggesting that saccade programming led to shorter saccadic latencies but not a shift of attention. This result is incompatible with the claim that saccade programming is sufficient for covert orienting. A similar result was reported by Remington [74], who found that luminance detection was no better at a saccade goal than at a control location (although saccades were delayed when the luminance change occurred at the control location). Converging evidence for independence was provided by Stelmach and colleagues [75], using a Temporal Order Judgement (TOJ) task whereby two stimuli are sequentially presented with various inter-stimulus intervals, and participants are asked to indicate which stimulus appeared first. In this study endogenous attention to a peripheral location created a prior entry effect, such that the attended stimulus was perceived as appearing before the unattended stimulus. However, consistent with the findings of [5], planning a saccade to a peripheral location did *not* produce prior-entry, suggesting that this programming was not sufficient to orient attention. More recently, Born [76] used a stop-signal paradigm to confirm Klein's claim that a saccade that is programmed but successfully inhibited does not produce a shift of attention.

Other studies have shown that saccades directed towards an intermediate position between two spatially close visual objects presented simultaneously in the periphery, referred to as 'Global Effect' [72,77,78], are not preceded by a shift of attention to the midpoint between stimuli. Rather, there is a subtle attentional enhancement at the location of both objects [73,79,80], even though the eventual eye movement lands at neither location. These observations appear to rule out the mandatory coupling of attention to the saccade landing point (but see Van der Stigchel and de Vries [81] for an alternative interpretation). Thus, the activation of a saccade program alone does not appear sufficient to elicit 'covert' orienting. In a related study, Bedard and Song used a visuomotor adaptation paradigm to dissociate the intended and actual endpoint of ballistic reaching movements [82]. They report that, in the post-adaptation phase, attention was allocated to locations associated with both the intended and the actual endpoint of movements, suggesting that endogenous covert attention can be decoupled from motor programs. In fact, there seems to be very little empirical evidence from human observers that preparing an eye movement is sufficient to produce a shift of attention when no saccade is executed.

Klein [5] conducted a second study to test the idea that attending a location was necessarily associated with the activation of a saccade program targeting the attended location. In this variant of the task, the primary response was a shift of attention, with saccades required on 20% of trials. The data show that attending a peripheral location produced faster manual responses but did not reduce saccade latency. Klein therefore concluded that covert orienting of attention and saccade programming were independent of one another. This conclusion was challenged by several authors, who argue that methodological factors, such as the requirement to make two speeded responses to peripheral events, mean the data are hard to interpret (e.g., [6]), but subsequent studies [8,83] addressed these issues and again found no evidence of attentional facilitation at the saccade goal. However, in a footnote Klein and Pontefract [8] noted that there was a long delay between the onsets of the cue and target, so it remained possible that saccade programming did elicit a shift of attention, just not at the time point measured by [5] or [8]. They speculate that OMRH/PMT might still be tenable for shifting, but not sustaining attention.

The idea that saccade programming could be sufficient for orienting but not for maintenance of attention was explicitly tested by Belopolsky and Theeuwes [84]. They observed that oculomotor priming effects were significantly reduced when a saccadic target is unlikely to appear at a cued location. Furthermore [9,84] demonstrated that participants could sustain attention at a location while simultaneously suppressing saccade programming to that same location. In these experiments, both exogenous and endogenous covert orienting were associated with the activation of a saccade motor plan. However, in the case of endogenous attention, the saccade execution was rapidly suppressed without disrupting the allocation of attention. In a recent study, we also observed that saccadic priming was profoundly affected by the probability that a saccadic response would be required by manipulating the proportion of catch trials in a cueing task. When there were many catch trials (30%), we observed covert orienting without saccadic priming, but when catch trials were removed there was both covert orienting and oculomotor priming [85]. Belopolsky and Theeuwes [84] proposed a revision to OMRH/PMT that they called a 'Shifting and Maintenance (S&M) account of attention'. This revised theory, like that of Klein and Pontefract, retains the core assumption of OMRH/PMT that endogenous orienting depends upon a saccade motor plan but argues that once attention has moved an active saccade plan is not required to sustain attention. However, it is important to note that demonstrating an association between orienting of attention and the activation of a saccade plan is very different to demonstrating that the saccade programming causes orienting of attention. Indeed, this evidence is equally consistent with the idea that attentional selection is a necessary precondition for the programming of accurate saccades, as proposed by [14].

3.4. Impaired Oculomotor Control Disrupts Exogenous but Not Endogenous Covert Attention

Proponents of OMRH/PMT have studied patients with oculomotor problems and used ingenious experimental designs to experimentally constrain saccade programming. For example, Craighero et al. [41] argued that paralysis of the eye due to VIth nerve palsy was associated with deficits of covert, endogenous orienting. However, subsequent studies with both ophthalmoplegic patients and the eye-abduction paradigm reported results in conflict with Craighero and colleagues' [41,42] observations. Smith, Rorden and Jackson [86] reported the case of A.I., who suffered from chronic ophthalmoplegia, a paralysis of the extraocular muscles that made her unable to make any eye movements. They observed a deficit of covert, exogenous attention with intact overt, endogenous orienting. Gabay and colleagues have shown similar effects in patients with Duane's syndrome, a developmental disorder associated with problems making abductive eye movements [87]. The claim that defective oculomotor control is associated with impaired exogenous attention but preserved endogenous attention is consistent with observations in patients suffering from Progressive Supranuclear Palsy (PSP), a disease characterised by vertical paralysis of gaze [88]. For example, Posner et al. [89] examined covert orienting in PSP using a predictive, peripheral cue. When the stimuli were aligned along the horizontal axis, normal exogenous orienting was observed with a cue-target onset asynchrony (CTOA) of 50 ms. However, when the stimuli were aligned along the vertical axis covert orienting was not observed until a CTOA of 1000 ms, indicative of disrupted exogenous attention. Their subsequent study [90] explicitly compared exogenous and endogenous attention using non-predictive peripheral cues to engage exogenous attention and a centrally presented, predictive arrow cue to engage endogenous attention. As in the original study, there was a significant impairment of covert exogenous orienting when stimuli appeared along the vertical axis compared to the horizontal axis, whereas endogenous orienting was largely preserved along both axes. Furthermore, in a recent study we demonstrated that this selective impairment of exogenous orienting in PSP can also be observed in a visual search. More specifically, patients with PSP also suffer visual search deficits when targets appear on the vertical axis, and this deficit was greater for a feature search than a conjunction search [91].

The same dissociation between saccade planning and endogenous covert attention was observed in heathy participants using the eye-abduction paradigm [92]. We subsequently demonstrated that the effect of eye-abduction generalised to visual search, such that feature search was disrupted in the temporal hemispace while conjunction search was preserved [93,94]. Notably, the disruption of saccade programming associated with eye-abduction [95,96] and PSP [91] is also associated with a deficit of short-term spatial memory, which can be at least partly attributed to the failure to attend and encode the relevant locations. On the basis of these results, we concluded that the balance of evidence is more consistent with a weak view of OMRH/PMT that was only true for exogenous orienting.

An important caveat to this conclusion is that the interpretation of eye-abduction data is not entirely straightforward. Firstly, one might argue that participants can still plan eye movements even if they cannot be executed. However, an elegant experiment using eye-abduction demonstrated that the general tendency of saccades to curve away from a distractor location [53] was greatly reduced when the distractor was presented outside the oculomotor range [97]. Given that saccade curvature in a target-distractor paradigm is generally accepted to reflect competition between different saccade plans, this result strongly suggests that eye-abduction leads to impaired saccade programming. Secondly, and more problematically, the pattern of results is rather inconsistent. For example, in a follow-up study to Smith et al. [92], we examined the effect of eye-abduction on social attention (the reflexive shifts of attention triggered by observing an agent change their direction of gaze, also called 'gaze-cueing'), non-predictive arrow cueing and peripheral cueing [98]. As in our previous study, we observed that eye-abduction interfered with covert exogenous orienting. However, in this study the interference effect was observed in the nasal, not the temporal hemifield. Furthermore, we also observed a reduced cueing effect in the nasal hemifield in the arrow cueing task. Interestingly, eye-abduction had no effect on gaze cueing, which was surprising given that gaze cues are known to activate the eye movement system [99,100]. In addition, although not directly relevant to OMRH/PMT, Michalczyk et al. [101] recently observed that eye-abduction disrupted IOR, a result contrary to our 2012 finding. The precise reasons for these disparate findings are not entirely clear. We attributed the nasal-hemifield effect to a reduction in the cost of invalid cues, but as MacLean et al., observed, multiple interpretations are possible, which limits the strength of the conclusions we can draw based on eye-abduction [102]. Given that studies using eye-abduction only use a single Stimulus Onset Asynchrony (SOA), it is also possible that exogenous orienting was delayed by eye-abduction rather than completely abolished in these tasks (as was the case in the studies of patients with PSP [89,90]). A final problem is that eye-abduction creates a very unusual oculoproprioceptive signal, and there is some evidence that oculoproprioception plays an important role in spatial attention (e.g., [103]). It is therefore possible that the impaired attentional orienting observed in ophthalmoplegic patients and studies of eye-abduction was caused by disrupted oculoproprioception, rather than impaired saccade programming per se.

In order to address these issues and provide a more rigorous test of exogenous-only version of OMRH/PMT we developed a new variant of the Posner cueing task in which cues and targets were presented within or beyond the effective oculomotor range (EOMR) [104]. Eye-abduction is thought to induce biased proprioception [105], which could lead to a bias in attention, although a recent study has cast some doubt on this claim [106]; we nevertheless used Presentation in Extreme Periphery (PEP), which has the advantage of presenting stimuli in the far periphery (up to 44° of visual angle) while keeping the participant's eye and trunk in their canonical, natural position. Each participant's Effective Oculomotor Range (EOMR) was computed in order to define the location of the placeholders in the different cueing tasks. In all three experiments reported, the target and placeholders could appear either below or beyond the participants' EOMR. The first experiment examined exogenous, covert orienting using a peripheral cueing task and SOAs of 100, 200, 300 or 500 ms. Consistent with studies with patients [86,87] or with the eye-abduction paradigm [90,92,94], exogenous cueing effects were abolished at all SOAs when stimuli were presented beyond the participant's EOMR, but intact when stimuli appeared within the EOMR. In a second experiment, we tested endogenous covert attention using a predictive, central cue. As with previously reported experiments [86,87,90,92,94], but, contrary to [41,42], there was no deficit in attention when stimuli were presented beyond EOMR. In a third experiment, we tested both exogenous and endogenous attention using a within-participant design. In accordance with the first two experiments, exogenous, covert orienting to peripheral cues was disrupted when targets appeared beyond the EOMR, whereas covert endogenous orienting was preserved (see Figure 3). In a recent experiment we replicated this dissociation using visual search tasks, such that a 'pre-attentive' search for feature singletons (which relies on the same cognitive processes as exogenous attention) was only possible within the effective oculomotor range. When feature search arrays were presented beyond the EOMR, participants had to engage in serial, attentive searching to find the target [107]. These findings rule out the possibility that previous reports of dissociations between endogenous orienting and saccade programming can be explained in terms of abnormal oculoproprioception or in terms of delayed, rather than abolished, covert orienting.

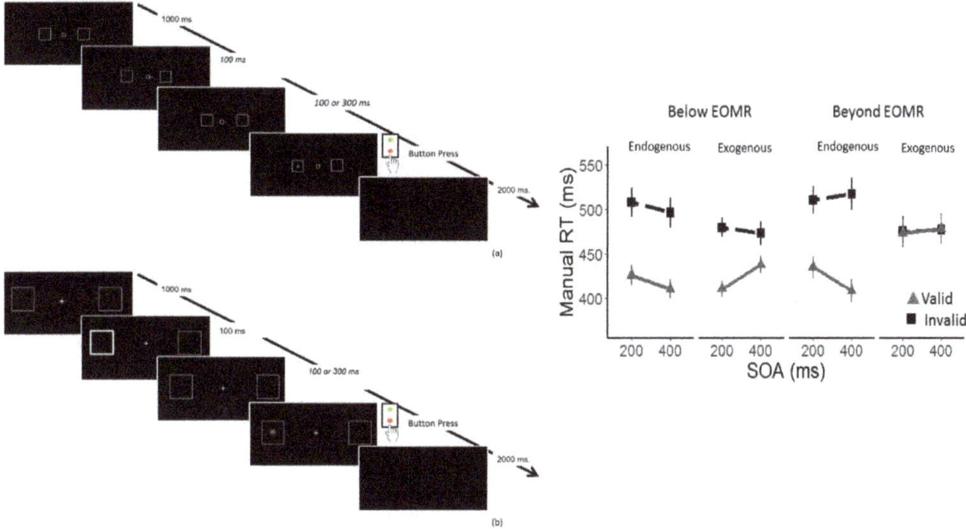

Figure 3. The sequence of events in the endogenous cueing (**a**) and exogenous cueing (**b**) tasks. The right panel shows the mean manual reaction time (RT) in ms as a function of Stimulus Onset Asynchrony (SOA) and cue validity for below and beyond the EOMR separately for the endogenous and exogenous cueing task (Exp. 3). Adapted from [103].

3.5. Saccades Curve away from Attended Locations

The observation that saccade trajectories are affected by covert attention is typically interpreted as evidence in favour of OMRH/PMT (e.g., [59]). However, this interpretation is problematic, because the studies classically report that saccades tended to curve *away* from the cued location [58]. The standard interpretation for this effect is that participants need to inhibit the programmed eye movement to the cued location in order to be able to execute a saccade towards the target location [53,108]. Saccade curvature will depend on the distribution of neuronal activation produced both cue and target. A curvature *away* from the cued location would result from an inhibition of the neurons coding for the irrelevant cued position, allowing the neuronal population coding for the actual target location to take over. This inhibition is thought to come more particularly from projections from the Frontal Eye Field (FEF) and Superior Colliculus (SC) (see [109]). This explanation is consistent with the broad idea that covert attention and motor programming interact. However, it is much harder to reconcile with the specific claim made by OMRH/PMT that shifts of attention are *caused* by motor programs. In fact, the observation that covert, endogenous attention can be allocated to a location that is currently inhibited in the oculomotor system is the *opposite* of what is predicted by OMRH/PMT. Studies of saccade trajectory deviations therefore demonstrate an interaction between covert attention and saccade preparation, but do not provide convincing evidence that covert orienting of attention is caused by the activation of a saccade plan and therefore do not support OMRH/PMT. Furthermore, although the mechanisms underlying curvature towards a distractor are clearly understood [109,110], there is less consensus regarding mechanisms underlying curvature away from distractor. For example, curvatures away from and irrelevant position are observed when participants have a prior knowledge of target position [55], and the direction of the deviation appear to be dependent on response time, such that short latency saccades tend to deviate towards an irrelevant position, whereas slow saccades tend to deviate away [111]. This observation suggests that the oculomotor inhibition operates in the selection process, leaving plenty of time for top-down preparation. Hence, the deviation away observed in the case of covert endogenous shift of attention cannot be explained solely in terms of activation of the oculomotor system.

4. An Oculomotor Readiness Hypothesis of Exogenous Orienting (OREO)

On the basis of these studies, we argue that the data are most consistent with an Oculomotor Readiness Hypothesis that is specific to Exogenous Orienting (OREO). On a theoretical level, the relationship between attention and eye movements can be understood in terms of Biased Competition, such that activation of the motor system exerts a powerful biasing influence on competitive interactions in the visual system [112]. In Biased Competition, the locus of attention arises from a stimulus-driven competition between signals relating to stimulus salience (e.g., their brightness, size, contrast, orientation), which can be biased by goal-driven factors such as the goals of the observer. The competition takes places in a topographic map of space, called a priority map ([113]. The cortical substrates of the priority map are thought to lie in the posterior parietal cortex a region that has dense reciprocal connections with areas known to be directly involved in saccade control such as Frontal Eye Field (FEF) and Superior Colliculus (SC) (for a review, see [114]) When a location is activated in the priority map the activation is passed downstream to oculomotor structures, such as the SC, which represent the prioritized location as the goal of a potential movement. These oculomotor signals are then fed back into the priority map, thus further biasing activity in favour of the activated location [115]. This reciprocal feedback loop will typically produce very rapid selection of a peripherally cued location, which will facilitate target detection, producing the rapidly developing perceptual advantage typically associated with exogenous attention. When the oculomotor system malfunctions, or when targets appear at locations that cannot become the goal of a saccade, the motor system exerts a much weaker influence on the biased competition. If a target is associated with a persistently large salience signal (e.g., in a feature search task in which the search array remains visible until a response is made), the absence of reciprocal reinforcement from the oculomotor system should slow selection of the feature singleton but will not necessarily prevent its selection. This is exactly the pattern we observed, such that placing a salient feature beyond the EOMR delayed, rather than abolished the capture of attention by the singleton [93,94,103]. If salience signal is transient (as in the peripheral cueing task), the absence of reinforcement from the oculomotor system reduces the chance of the cued location 'winning' the competition before the signal decays, and therefore reduces the probability of observing an exogenous shift of attention to the cued location. We can therefore understand the relationship between exogenous attention and saccade programming in terms of oculomotor inputs that bias competition on the priority map in favour of the saccade endpoint. The demotion of the oculomotor system from being the sole arbiter of the locus of attention to being one of many potential influences on the process of biased competition is a key difference between OREO and OMRH/PMT. Importantly it does not deny the possibility that exogenous orienting can be driven by other inputs, such as stimulus salience [116]. Rather, OREO holds that optimally efficient exogenous orienting relies on activation of a saccade plan, and when this activation is disrupted exogenous orienting becomes slower and less reliable.

OREO makes some clear and testable predictions about the interaction between covert, exogenous orienting and saccade programming. Firstly, exogenous orienting should always be associated with the activation of a saccade plan. Secondly, inability to plan a saccade should disrupt exogenous orienting. Thirdly, factors that affect the properties of saccadic eye movements (e.g., their latency, amplitude and direction) should also affect the speed and accuracy of covert exogenous orienting. MacLean et al. [102] tested the first prediction using a variant of the dual task procedure developed by Klein and Pontefract [8]. Contrary to the predictions of OREO, they observed no reduction in saccade latency at peripherally cued locations and concluded that exogenous orienting was not associated with saccade programming. However, this conclusion is premature, as MacLean et al. used a SOA of 250 ms, allowing ample time for the suppression of saccade programming following a shift of attention. Indeed, the authors concede that their results are more similar to those of Belopolsky and Theeuwes [9,84], who previously argued that maintenance of attention was independent of saccade programming. The MacLean study also utilises a very high proportion of 'no-go' trials, where a cue appears but no saccade is permitted, and as we have already noted, a high proportion no-go trials can mask saccadic priming effects caused by peripheral cues [9,107]. We examined the third prediction by using instrumental conditioning of eye movements [117]. If exogenous orienting depends on activation of the oculomotor system, then one might predict that a manipulation that modulates saccade latencies should also affect covert exogenous attention. In our first experiment we found that rewarding eye movements to a specific spatial location reliably reduced saccade latencies to that location, and that this conditioning persisted for 180 trials once rewards were removed. However, in a second experiment this modulation of the oculomotor system had no effect on the magnitude of covert, exogenous orienting or Inhibition of Return. McCoy and Theeuwes [118] report a similar result in a study in which participants learned to made saccades to a location associated with a large reward. As with our study, the high-value location was associated with shorter saccade latencies. However, this oculomotor facilitation did not translate into enhanced performance at the rewarded location in a subsequent task that measured perceptual discrimination at the rewarded location while the eyes remained at fixation. These findings may seem hard to reconcile with the third prediction of OREO, but it is important to note that OREO predicts that reducing the latency of a saccade should lead to a reduction in the rise-time of attention (i.e., the speed at which attention is oriented to the cued location) rather than the absolute magnitude of the cueing effect. Thus, in our view, none of these studies offers a strong test of the predictions of OREO. In contrast, McFadden, Khan and Wallman [119] reported that it was possible to elicit adaptation of exogenous, covert orienting, which was accompanied by an adaptation of subsequent eye movements, suggesting that the adaptation of exogenous attention relied on changes in the oculomotor plans elicited by the peripheral onset. It is not known whether endogenous, covert attention can be adapted in the same way, but such a study would provide a good test of OMRH/PMT and OREO, and the former theories predict an effect of adaptation of endogenous attention on saccade amplitude, whereas OREO does not.

5. Summary and Conclusions

To briefly summarize, OMRH/PMT argues that planning an eye movement is both necessary and sufficient for covert, endogenous orienting of attention. Many studies suggest that there is an association between covert attention and oculomotor control, but none of this evidence demonstrates a causal relationship between saccade programming and covert, endogenous spatial attention. Studies of pre-saccadic attention are problematic because they conflate peri-saccadic perceptual changes ('remapping') with covert attention, and the results are equally consistent with the view that orienting attention is a necessary precondition for saccade programming (e.g., [120]). A single neuropsychological study argues for an association between endogenous orienting and saccade programming, but there are many other examples of double dissociations between oculomotor control and endogenous, covert attention. Studies of healthy participants show no evidence that shifts of attention can be achieved by programming an eye movement, and the weight of evidence from eye-abduction and other manipulations suggests that endogenous covert orienting can be achieved in the absence of saccade programming. Overall, there is surprisingly little evidence from human participants that saccade programming is either necessary or sufficient for covert spatial attention. However, there is a growing body of neuropsychological and experimental evidence that exogenous covert orienting is dependent on the ability to plan and execute eye movements. Neuropsychological patients with paralysis of the eyes reliably present with deficits of exogenous, covert attention and disrupting saccade programming in healthy participants interferes with covert, exogenous orienting. In our view, these findings are powerful and conclusive evidence against the central tenet of OMRH/PMT, which is that saccade programming is necessary and sufficient for endogenous, covert orienting, and thus we should reject OMRH/PMT as a theory of covert, endogenous attention. However, the data are consistent with OREO, which holds that saccade preparation or 'oculomotor readiness' plays a fundamental role in covert, exogenous orienting of attention.

Funding: This research was supported by ESRC grant ES/N018842/1 to DS.

Conflicts of Interest: The authors declare no conflict of interest.

References

1. Carrasco, M. Visual attention: The past 25 years. *Vis. Res.* **2011**, *51*, 1484–1525. [CrossRef]
2. Posner, M.I.; Cohen, Y. Attention and the Control of Movements. *Tutor. Mot. Behav.* **1980**, *1*, 243–258. [CrossRef]
3. Chica, A.B.; Bartolomeo, P.; Lupianez, J. Two cognitive and neural systems for endogenous and exogenous spatial attention. *Behav. Brain Res.* **2013**, *237*, 107–123. [CrossRef]
4. Smith, D.T.; Schenk, T. The Premotor theory of attention: Time to move on? *Neuropsychologia* **2012**, *50*, 1104–1114. [CrossRef]
5. Klein, R.M. Does oculomotor readiness mediate cognitive control of visual attention? *Atten. Perform.* **1980**, *8*, 259–276.
6. Rizzolatti, G.; Riggio, L.; Dascola, I.; Umilta, C. Reorienting attention across the horizontal and vertical meridians: Evidence in favor of a premotor theory of attention. *Neuropsychologia* **1987**, *25*, 31–40. [CrossRef]
7. Rizzolatti, G.; Riggio, L.; Sheliga, B.M. Space and Selective Attention. In *Attention and Performance Series. Attention and Performance 15: Conscious and Nonconscious Information Processing*; Umiltà, C., Moscovitch, M., Eds.; The MIT Press: Cambridge, MA, USA, 1994; Vol. 15, pp. 231–265.
8. Klein, R.M.; Pontefract, A. Does Oculomotor Readiness Mediate Cognitive Control of Visual Attention? Revisited! In *Attention and performance XV: Conscious and Nonconscious Information Processing*; Umiltà, C., Moscovitch, M., Eds.; The MIT Press: Cambrige, MA, USA, 1994; pp. 333–350.
9. Belopolsky, A.V.; Theeuwes, J. Updating the premotor theory: The allocation of attention is not always accompanied by saccade preparation. *J. Exp. Psychol: Hum. Percept. Perform.* **2012**, *38*, 902. [CrossRef]
10. Schneider, W.X.; Deubel, H. Selection-for-perception and selection-for-spatial-motor-action are coupled by visual attention: A review of recent findings and new evidence from stimulus-driven saccade control. *Atten. Perform. XIX* **2002**, *19*, 609–627.
11. Corbetta, M.; Shulman, G.L. Control of goal-directed and stimulus-driven attention in the brain. *Nat. Rev. Neurosci.* **2002**, *3*, 201–215. [CrossRef]
12. Hoffman, J.E.; Subramaniam, B. The role of visual attention in saccadic eye movements. *Percept. Psychophys.* **1995**, *57*, 787–795. [CrossRef] [PubMed]
13. Shepherd, M.; Findlay, J.M.; Hockey, R.J. The relationship between eye movements and spatial attention. *Q. J. Exp. Psychol. A* **1986**, *38*, 475–491. [CrossRef] [PubMed]

14. Deubel, H.; Schneider, W.X. Saccade target selection and object recognition: Evidence for a common attentional mechanism. *Vis. Res.* **1996**, *36*, 1827–1837. [CrossRef]
15. Kowler, E.; Anderson, E.; Dosher, B.; Blaser, E. The Role of Attention in the Programming of Saccades. *Vis. Res.* **1995**, *35*, 1897–1916. [CrossRef]
16. Van der Stigchel, S.; Theeuwes, J. The influence of attending to multiple locations on eye movements. *Vis. Res.* **2005**, *45*, 1921–1927. [CrossRef]
17. Crovitz, H.F.; Daves, W. Tendencies to eye movement and perceptual accuracy. *J. Exp. Psychol.* **1962**, *63*, 495–498. [CrossRef]
18. Bryden, M.P. The Role of Post-Exposural Eye-Movements in Tachistoscopic Perception. *Can. J. Psychol.* **1961**, *15*, 220–225. [CrossRef]
19. Khan, A.Z.; Blohm, G.; Pisella, L.; Munoz, D.P. Saccade execution suppresses discrimination at distractor locations rather than enhancing the saccade goal location. *Eur. J. Neurosci.* **2015**, *41*, 1624–1634. [CrossRef]
20. Deubel, H. The time course of presaccadic attention shifts. *Psychol. Res.* **2008**, *72*, 630–640. [CrossRef] [PubMed]
21. Castet, E.; Jeanjean, S.; Montagnini, A.; Laugier, D.; Masson, G.S. Dynamics of attentional deployment during saccadic programming. *J. Vis.* **2006**, *6*, 196–212. [CrossRef]
22. Sheliga, B.M.; Riggio, L.; Rizzolatti, G. Orienting of attention and eye movements. *Exp. Brain Res.* **1994**, *98*, 507–522. [CrossRef] [PubMed]
23. Moehler, T.; Fiehler, K. Effects of spatial congruency on saccade and visual discrimination performance in a dual-task paradigm. *Vis. Res.* **2014**, *105*, 100–111. [CrossRef]
24. de Haan, B.; Morgan, P.S.; Rorden, C. Covert orienting of attention and overt eye movements activate identical brain regions. *Brain Res.* **2008**, *1204*, 102–111. [CrossRef]
25. Perry, R.J.; Zeki, S. The neurology of saccades and covert shifts in spatial attention: An event-related fMRI study. *Brain* **2000**, *123*, 2273–2288. [CrossRef] [PubMed]
26. Beauchamp, M.S.; Petit, L.; Ellmore, T.M.; Ingeholm, J.; Haxby, J.V. A parametric fMRI study of overt and covert shifts of visuospatial attention. *Neuroimage* **2001**, *14*, 310–321. [CrossRef] [PubMed]
27. Nobre, A.C.; Gitelman, D.R.; Dias, E.C.; Mesulam, M.M. Covert visual spatial orienting and saccades: Overlapping neural systems. *Neuroimage* **2000**, *11*, 210–216. [CrossRef] [PubMed]
28. Corbetta, M.; Akbudak, E.; Conturo, T.E.; Snyder, A.Z.; Ollinger, J.M.; Drury, H.A.; Linenweber, M.R.; Petersen, S.E.; Raichle, M.E.; Van Essen, D.C.; et al. A common network of functional areas for attention and eye movements. *Neuron* **1998**, *21*, 761–773. [CrossRef]
29. Andersen, R.A. Visual and eye movement functions of the posterior parietal cortex. *Annu. Rev. Neurosci.* **1989**, *12*, 377–403. [CrossRef] [PubMed]
30. Thickbroom, G.W.; Stell, R.; Mastaglia, F.L. Transcranial magnetic stimulation of the human frontal eye field. *J. Neurol. Sci.* **1996**, *144*, 114–118. [CrossRef]
31. Grosbras, M.H.; Paus, T. Transcranial magnetic stimulation of the human frontal eye field: Effects on visual perception and attention. *J. Cogn. Neurosci.* **2002**, *14*, 1109–1120. [CrossRef] [PubMed]
32. Muggleton, N.G.; Juan, C.H.; Cowey, A.; Walsh, V. Human frontal eye fields and visual search. *J. Neurophysiol.* **2003**, *89*, 3340–3343. [CrossRef] [PubMed]
33. Muri, R.M.; Hess, C.W.; Meienberg, O. Transcranial Stimulation of the Human Frontal Eye Field by Magnetic Pulses. *Exp. Brain Res.* **1991**, *86*, 219–223. [CrossRef]
34. Muri, R.M.; Vermersch, A.I.; Rivaud, S.; Gaymard, B.; Pierrot-Deseilligny, C. Effects of single-pulse transcranial magnetic stimulation over the prefrontal and posterior parietal cortices during memory-guided saccades in humans. *J. Neurophysiol.* **1996**, *76*, 2102–2106. [CrossRef]
35. Smith, D.T.; Jackson, S.R.; Rorden, C. Transcranial magnetic stimulation of the left human frontal eye fields eliminates the cost of invalid endogenous cues. *Neuropsychologia* **2005**, *43*, 1288–1296. [CrossRef]
36. Smith, D.T.; Jackson, S.R.; Rorden, C. An intact eye-movement system is not required to generate Inhibition of Return. *J. Neuropsychol.* **2009**, *3*, 267–271. [CrossRef]
37. Moore, T.; Armstrong, K.M.; Fallah, M. Visuomotor origins of covert spatial attention. *Neuron* **2003**, *40*, 671–683. [CrossRef]
38. Moore, T.; Fallah, M. Control of eye movements and spatial attention. *Proc. Natl. Acad. Sci. USA* **2001**, *98*, 1273–1276. [CrossRef]

39. Armstrong, K.M.; Fitzgerald, J.K.; Moore, T. Changes in visual receptive fields with microstimulation of frontal cortex. *Neuron* **2006**, *50*, 791–798. [CrossRef]

40. Moore, T.; Armstrong, K.M. Selective gating of visual signals by microstimulation of frontal cortex. *Nature* **2003**, *421*, 370–373. [CrossRef]

41. Craighero, L.; Carta, A.; Fadiga, L. Peripheral oculomotor palsy affects orienting of visuospatial attention. *Neuroreport* **2001**, *12*, 3283–3286. [CrossRef]

42. Craighero, L.; Nascimben, M.; Fadiga, L. Eye position affects orienting of visuospatial attention. *Curr. Biol.* **2004**, *14*, 331–333. [CrossRef]

43. Mclaughlin, S.C. Parametric Adjustment in Saccadic Eye Movements. *Percep. Psychophys.* **1967**, *2*, 359–362. [CrossRef]

44. Pelisson, D.; Alahyane, N.; Panouilleres, M.; Tilikete, C. Sensorimotor adaptation of saccadic eye movements. *Neurosci. Biobehav. Rev.* **2010**, *34*, 1103–1120. [CrossRef] [PubMed]

45. Ditterich, J.; Eggert, T.; Straube, A. Relation Between the Metrics of the Presaccadic Attention Shift and of the Saccade Before and After Saccadic Adaptation. *J. Neurophysiol.* **2000**, *84*, 1809–1813. [CrossRef] [PubMed]

46. Collins, T.; Dore-Mazars, K. Eye movement signals influence perception: Evidence from the adaptation of reactive and volitional saccades. *Vis. Res.* **2006**, *46*, 3659–3673. [CrossRef] [PubMed]

47. Dore-Mazars, K.; Collins, T. Saccadic adaptation shifts the pre-saccadic attention focus. *Exp. Brain Res.* **2005**, *162*, 537–542. [CrossRef] [PubMed]

48. Collins, T.; Heed, T.; Röder, B. Visual target selection and motor planning define attentional enhancement at perceptual processing stages. *Front. Hum. Neurosci.* **2010**, *4*, 14. [CrossRef] [PubMed]

49. Habchi, O.; Rey, E.; Mathieu, R.; Urquizar, C.; Farne, A.; Pelisson, D. Deployment of spatial attention without moving the eyes is boosted by oculomotor adaptation. *Front. Hum. Neurosci.* **2015**, *9*, 426. [CrossRef] [PubMed]

50. Yarbus, A.L. Eye movements during perception of complex objects. In *Eye Movements and Vision*; Springer: Boston, MA, USA, 1967; pp. 171–211.

51. Viviani, P.; Berthoz, A.; Tracey, D. The curvature of oblique saccades. *Vis. Res.* **1977**, *17*, 661–664. [CrossRef]

52. Smit, A.C.; Van Gisbergen, J.A.M. An analysis of curvature in fast and slow human saccades. *Exp. Brain Res.* **1990**, *81*, 335–345. [CrossRef]

53. Doyle, M.; Walker, R. Curved saccade trajectories: Voluntary and reflexive saccades curve away from irrelevant distractors. *Exp. Brain Res.* **2001**, *139*, 333–344. [CrossRef] [PubMed]

54. Van der Stigchel, S.; Theeuwes, J. Our eyes deviate away from a location where a distractor is expected to appear. *Exp. Brain Res.* **2006**, *169*, 338–349. [CrossRef] [PubMed]

55. Walker, R.; McSorley, E.; Haggard, P. The control of saccade trajectories: Direction of curvature depends on prior knowledge of target location and saccade latency. *Percept. Psychophys.* **2006**, *68*, 129–138. [CrossRef] [PubMed]

56. McPeek, R.M.; Skavenski, A.A.; Nakayama, K. Concurrent processing of saccades in visual search. *Vis. Res.* **2000**, *40*, 2499–2516. [CrossRef]

57. Walker, R.; McSorley, E. The parallel programming of voluntary and reflexive saccades. *Vis. Res.* **2006**, *46*, 2082–2093. [CrossRef]

58. Sheliga, B.M.; Riggio, L.; Craighero, L.; Rizzolatti, G. Spatial attention-determined modifications in saccade trajectories. *Neuroreport* **1995**, *6*, 585–588. [CrossRef] [PubMed]

59. Sheliga, B.M.; Riggio, L.; Rizzolatti, G. Spatial attention and eye movements. *Exp. Brain Res.* **1995**, *105*, 261–275. [CrossRef] [PubMed]

60. Van der Stigchel, S.; Theeuwes, J. The relationship between covert and overt attention in endogenous cuing. *Percept. Psychophys.* **2007**, *69*, 719–731. [CrossRef]

61. Craighero, L.; Rizzolatti, G. The premotor theory of attention. In *Neurobiology of Attention*; Itti, L., Rees, G., Tsotsos, J.K., Eds.; Academic Press: Burlington, MA. USA, 2005; pp. 181–186.

62. Duhamel, J.R.; Colby, C.L.; Goldberg, M.E. The updating of the representation of visual space in parietal cortex by intended eye movements. *Science* **1992**, *255*, 90–92. [CrossRef]

63. Colby, C.L. A neurophysiological distinction between attention and intention. In *Attention and Performance XVI: Information Integration in Perception and Communication*; Inui, T., McClelland, J.L., Eds.; MIT Press: Cambridge, MA, USA, 1996; pp. 157–177.

64. Ladavas, E.; Zeloni, G.; Zaccara, G.; Gangemi, P. Eye movements and orienting of attention in patients with visual neglect. *J. Cogn. Neurosci.* **1997**, *9*, 67–74. [CrossRef]
65. Benson, V.; Ietswaart, M.; Milner, D. Eye Movements and Verbal Report in a Single Case of Visual Neglect. *PLoS ONE* **2012**, *7*, 11. [CrossRef]
66. Blangero, A.; Khan, A.Z.; Salemme, R.; Deubel, H.; Schneider, W.X.; Rode, G.; Vighetto, A.; Rossetti, Y.; Pisella, L. Pre-saccadic perceptual facilitation can occur without covert orienting of attention. *Cortex* **2010**, *46*, 1132–1137. [CrossRef] [PubMed]
67. Sato, T.R.; Schall, J.D. Effects of stimulus-response compatibility on neural selection in frontal eye field. *Neuron* **2003**, *38*, 637–648. [CrossRef]
68. Thompson, K.G.; Biscoe, K.L.; Sato, T.R. Neuronal basis of covert spatial attention in the frontal eye field. *J. Neurosci.* **2005**, *25*, 9479–9487. [CrossRef]
69. Thompson, K.G.; Bichot, N.P.; Schall, J.D. Dissociation of visual discrimination from saccade programming in macaque frontal eye field. *J. Neurophysiol.* **1997**, *77*, 1046–1050. [CrossRef]
70. Tehovnik, E.J. Electrical stimulation of neural tissue to evoke behavioral responses. *J. Neurosci. Methods* **1996**, *65*, 1–17. [CrossRef]
71. Juan, C.H.; Muggleton, N.G.; Tzeng, O.J.; Hung, D.L.; Cowey, A.; Walsh, V. Segregation of visual selection and saccades in human frontal eye fields. *Cereb. Cortex* **2008**, *18*, 2410–2415. [CrossRef] [PubMed]
72. Deubel, H.; Wolf, W.; Hauske, G. The evaluation of the oculomotor error signal. *Adv. Psychol.* **1984**, *22*, 55–62.
73. Weaver, M.D.; van Zoest, W.; Hickey, C. A temporal dependency account of attentional inhibition in oculomotor control. *NeuroImage* **2016**, *147*, 880–894. [CrossRef]
74. Remington, R.W. Attention and saccadic eye movements. *J. Exp. Psychol.* **1980**, *6*, 726–744. [CrossRef]
75. Stelmach, L.B.; Campsall, J.M.; Herdman, C.M. Attentional and ocular movements. *J. Exp. Psychol.* **1997**, *23*, 823–844. [CrossRef]
76. Born, S.; Mottet, I.; Kerzel, D. Presaccadic perceptual facilitation effects depend on saccade execution: Evidence from the stop-signal paradigm. *J. Vis.* **2014**, *14*, 1–10. [CrossRef]
77. Findlay, J.M. Global visual processing for saccadic eye movements. *Vis. Res.* **1982**, *22*, 1033–1045. [CrossRef]
78. Coren, S.; Hoenig, P. Effect of Non-Target Stimuli Upon Length of Voluntary Saccades. *Percept. Mot. Skills* **1972**, *34*, 499–508. [CrossRef] [PubMed]
79. Van der Stigchel, S.; de Vries, J.P. There is no attentional global effect: Attentional shifts are independent of the saccade endpoint. *J. Vis.* **2015**, *15*, 12. [CrossRef]
80. Wollenberg, L.; Deubel, H.; Szinte, M. Visual attention is not deployed at the endpoint of averaging saccades. *PLoS Biol.* **2018**, *16*, e2006548. [CrossRef]
81. Van der Stigchel, S.; de Vries, J.P. Commentary: Visual attention is not deployed at the endpoint of averaging saccades. *Front. Psychol.* **2018**, *9*. [CrossRef]
82. Bedard, P.; Song, J.H. Attention modulates generalization of visuomotor adaptation. *J. Vis.* **2013**, *13*, 12. [CrossRef] [PubMed]
83. Hunt, A.R.; Kingstone, A. Inhibition of return: Dissociating attentional and oculomotor components. *J. Exp. Psychol. Hum. Percept. Perform.* **2003**, *29*, 1068–1074. [CrossRef]
84. Belopolsky, A.V.; Theeuwes, J. When are attention and saccade preparation dissociated? *Psychol. Sci.* **2009**, *20*, 1340–1347. [CrossRef]
85. Smith, D.T.; Casteau, S. The effect of offset cues on saccade programming and covert attention. *Q. J. Exp. Psychol.* **2019**, *72*, 481–490. [CrossRef]
86. Smith, D.T.; Rorden, C.; Jackson, S.R. Exogenous orienting of attention depends upon the ability to execute eye movements. *Curr. Biol.* **2004**, *14*, 792–795. [CrossRef] [PubMed]
87. Gabay, S.; Henik, A.; Gradstein, L. Ocular motor ability and covert attention in patients with Duane Retraction Syndrome. *Neuropsychologia* **2010**, *48*, 3102–3109. [CrossRef] [PubMed]
88. Steele, J.C.; Richardson, J.C.; Olszewski, J. Progressive Supranuclear Palsy. A Heterogeneous Degeneration Involving the Brain Stem, Basal Ganglia and Cerebellum with Vertical Gaze and Pseudobulbar Palsy, Nuchal Dystonia and Dementia. *Arch. Neurol.* **1964**, *10*, 333–359. [CrossRef]
89. Posner, M.I.; Cohen, Y.; Rafal, R.D. Neural systems control of spatial orienting. *Philos. Trans. R. Soc. Lond. B* **1982**, *298*, 187–198. [CrossRef]
90. Rafal, R.D.; Posner, M.I.; Friedman, J.H.; Inhoff, A.W.; Bernstein, E. Orienting of Visual-Attention in Progressive Supranuclear Palsy. *Brain* **1988**, *111*, 267–280. [CrossRef] [PubMed]

91. Smith, D.T.; Archibald, N. Visual Search in Progressive Supranuclear Palsy. *Curr. Top. Behav. Neurosci.* **2018**. [CrossRef]

92. Smith, D.T.; Schenk, T.; Rorden, C. Saccade preparation is required for exogenous attention but not endogenous attention or IOR. *J. Exp. Psychol Hum. Percept. Perform.* **2012**, *38*, 1438–1447. [CrossRef] [PubMed]

93. Smith, D.T.; Ball, K.; Ellison, A. Covert visual search within and beyond the effective oculomotor range. *Vis. Res.* **2014**, *95*, 11–17. [CrossRef]

94. Smith, D.T.; Ball, K.; Ellison, A.; Schenk, T. Deficits of reflexive attention induced by abduction of the eye. *Neuropsychologia* **2010**, *48*, 1269–1276. [CrossRef]

95. Ball, K.; Pearson, D.G.; Smith, D.T. Oculomotor involvement in spatial working memory is task-specific. *Cognition* **2013**, *129*, 439–446. [CrossRef]

96. Pearson, D.G.; Ball, K.; Smith, D.T. Oculomotor preparation as a rehearsal mechanism in spatial working memory. *Cognition* **2014**, *132*, 416–428. [CrossRef] [PubMed]

97. Boon, P.J.; Theeuwes, J.; Belopolsky, A.V. Eye abduction reduces but does not eliminate competition in the oculomotor system. *J. Vis.* **2017**, *17*, 15. [CrossRef] [PubMed]

98. Morgan, E.J.; Ball, K.; Smith, D.T. The role of the oculomotor system in covert social attention. *Atten. Percept. Psychophys.* **2014**, *76*, 1265–1270. [CrossRef]

99. Kuhn, G.; Tatler, B.W.; Cole, G.G. You look where I look! Effect of gaze cues on overt and covert attention in misdirection. *Vis. Cogn.* **2009**, *17*, 925–944. [CrossRef]

100. Ricciardelli, P.; Bricolo, E.; Aglioti, S.M.; Chelazzi, L. My eyes want to look where your eyes are looking: Exploring the tendency to imitate another individual's gaze. *Neuroreport* **2002**, *13*, 2259–2264. [CrossRef]

101. Michalczyk, L.; Paszulewicz, J.; Bielas, J.; Wolski, P. Is saccade preparation required for inhibition of return (IOR)? *Neurosci. Lett.* **2018**, *665*, 13–17. [CrossRef] [PubMed]

102. MacLean, G.H.; Klein, R.M.; Hilchey, M.D. Does oculomotor readiness mediate exogenous capture of visual attention? *J. Exp. Psychol. Hum. Percept. Perform.* **2015**, *41*, 1260–1270. [CrossRef]

103. Balslev, D.; Newman, W.; Knox, P.C. Extraocular Muscle Afferent Signals Modulate Visual AttentionEye Proprioception and Visual Attention. *Investig. Ophthalmol. Vis. Sci.* **2012**, *53*, 7004–7009. [CrossRef]

104. Casteau, S.; Smith, D.T. Covert attention beyond the range of eye-movements: Evidence for a dissociation between exogenous and endogenous orienting. *Cortex* **2018**. [CrossRef]

105. Paap, K.R.; Ebenholtz, M. Perceptual consequences of potentiation in the extraocular muscles: An alternative explanation for adaptation to wedge prisms. *J. Exp. Psychol. Hum. Percept. Perform.* **1976**, *2*, 457–468. [CrossRef]

106. Gilligan, T.M.; Cristino, F.; Bultitude, J.H.; Rafal, R.D. The effect of prism adaptation on state estimates of eye position in the orbit. *Cortex* **2019**, *115*, 246–263. [CrossRef]

107. Casteau, S.; Smith, D.T. Is pre-attentive search restricted to the range of eye-movements? *Under Review.* **2018**.

108. Godijn, R.; Theeuwes, J. Programming of endogenous and exogenous saccades: Evidence for a competitive integration model. *J. Exp. Psychol.* **2002**, *28*, 1039–1054. [CrossRef]

109. Van der Stigchel, S.; Meeter, M.; Theeuwes, J. Eye movement trajectories and what they tell us. *Neurosci. Biobehav. Rev.* **2006**, *30*, 666–679. [CrossRef] [PubMed]

110. Van der Stigchel, S. Recent advances in the study of saccade trajectory deviations. *Vis. Res.* **2010**, *50*, 1619–1627. [CrossRef] [PubMed]

111. McSorley, E.; Haggard, P.; Walker, R. Time course of oculomotor inhibition revealed by saccade trajectory modulation. *J. Neurophysiol.* **2006**, *96*, 1420–1424. [CrossRef]

112. Desimone, R. Visual attention mediated by biased competition in extrastriate visual cortex. *Philos. Trans. R. Soc. Lond. B* **1998**, *353*, 1245–1255. [CrossRef] [PubMed]

113. Bisley, J.W.; Goldberg, M.E. Attention, intention, and priority in the parietal lobe. *Annu. Rev. Neurosci.* **2010**, *33*, 1–21. [CrossRef]

114. Paré, M.; Dorris, M.C. The role of posterior parietal cortex in the regulation of saccadic eye movements. In *The Oxford Handbook of Eye Movements*; Liversedge, S.P., Gilchrist, I., Everling, S., Eds.; Oxford University Press: New York, NY, USA, 2011; pp. 257–278.

115. Bisley, J.W.; Mirpour, K.; Arcizet, F.; Ong, W.S. The role of the lateral intraparietal area in orienting attention and its implications for visual search. *Eur. J. Neurosci.* **2011**, *33*, 1982–1990. [CrossRef] [PubMed]

116. Li, F.F.; VanRullen, R.; Koch, C.; Perona, P. Rapid natural scene categorization in the near absence of attention. *Proc. Nat. Acad. Sci. USA* **2002**, *99*, 9596–9601. [CrossRef]

117. Dunne, S.; Ellison, A.; Smith, D.T. Rewards modulate saccade latency but not exogenous spatial attention. *Front. Psychol.* **2015**, *6*, 1080. [CrossRef] [PubMed]
118. McCoy, B.; Theeuwes, J. Overt and covert attention to location-based reward. *Vis. Res.* **2018**, *142*, 27–39. [CrossRef]
119. McFadden, S.A.; Khan, A.; Wallman, J. Gain adaptation of exogenous shifts of visual attention. *Vis. Res.* **2002**, *42*, 2709–2726. [CrossRef]
120. Schneider, W.X. VAM: A neuro-cognitive model for visual attention control of segmentation, object recognition, and space-based motor action. *Vis. Sel. Atten.* **1995**, *2*, 331–376. [CrossRef]

Review

What Neuroscientific Studies Tell Us about Inhibition of Return

Jason Satel [1,*], Nicholas R. Wilson [1] and Raymond M. Klein [2]

[1] Division of Psychology, School of Medicine, College of Health and Medicine, University of Tasmania, Launceston, Tasmania 7250, Australia; nr.wilson@utas.edu.au
[2] Department of Psychology and Neuroscience, Faculty of Science, Dalhousie University, Halifax, NS B3H 4R2, Canada; ray.klein@dal.ca
* Correspondence: jsatel@gmail.com

Received: 1 April 2019; Accepted: 1 October 2019; Published: 29 October 2019

Abstract: An inhibitory aftermath of orienting, inhibition of return (IOR), has intrigued scholars since its discovery about 40 years ago. Since then, the phenomenon has been subjected to a wide range of neuroscientific methods and the results of these are reviewed in this paper. These include direct manipulations of brain structures (which occur naturally in brain damage and disease or experimentally as in TMS and lesion studies) and measurements of brain activity (in humans using EEG and fMRI and in animals using single unit recording). A variety of less direct methods (e.g., computational modeling, developmental studies, etc.) have also been used. The findings from this wide range of methods support the critical role of subcortical and cortical oculomotor pathways in the generation and nature of IOR.

Keywords: inhibition of return; oculomotor system; orienting

---

1. Introduction

An inhibitory aftermath of orienting was discovered in the early 1980s by Posner and Cohen [1] and subsequently named "inhibition of return" (IOR) by Posner et al. [2]. In these early papers, Posner and colleagues proposed a novelty-seeking function for IOR and in the next decade or two the phenomenon was subject to intense investigation. Eventually, a variety of exciting behavioral findings provided converging evidence for this proposal. Beginning with Posner et al. [2], a wide range of neuroscientific tools have been utilized to explore the neural basis of IOR, and in some cases, to resolve questions that arose primarily from behavioral studies. Our goal in this paper is to describe what we have learned about IOR from this neuroscientific literature. To provide the context for our coverage of this neuroscientific literature, we will begin with a brief overview of IOR's behavioral manifestations including its cause and effects, its spatial and temporal properties, its role as a foraging facilitator (see [3], for a review that despite its age remains quite contemporary), and of disputes in the IOR literature that neuroscientific evidence might help to resolve.

2. Behavioral Manifestations

In a review such as this one, it is prudent to begin by noting that there is no widespread agreement about the nature of IOR [4]. The source of some disagreements (as anticipated by Klein [3]) may be in overextension(s) of the term. Most notably, a cue-induced reaction time delay in the processing of a subsequent target can be due to sensory adaptation when it is presented along the same pathway traversed by the cue [5], particularly when the interval between cue and target onset is ~500 ms or less. When such an "inhibitory" effect (also called "onset detection cost") is conflated with IOR there is bound to be confusion. Here, we will focus on the longer-lasting inhibitory aftereffect of orienting, which we believe comes in two forms (first clearly distinguished in [6]) depending on whether the reflexive oculomotor system is in a suppressed or activated state when it is generated. As described in their account of the history of IOR (Hilchey et al. [5]; see also [7]), this duality has engendered confusion since the phenomenon was named by Posner et al. [2].

2.1. Causes and Consequences

Possible causes of IOR include prior orienting of attention, peripheral stimulation, and activation of the oculomotor system. Posner and Cohen [1] rejected the attentional cause because they demonstrated that prior endogenous allocation of attention did not generate IOR, a finding that has been subsequently confirmed (e.g., Rafal et al. [8]). They endorsed sensory stimulation as the cause because they believed that IOR was generated when peripheral cues were balanced around fixation. However, with better control conditions, this finding was not supported [9] and Posner et al. [2] further demonstrated that IOR could be generated by an endogenously directed eye movement in the absence of a unique peripheral onset. Using visual, auditory, and tactile stimuli and a target-target design, Spence et al. [10] found that as long as a target was presented in the same spatial position as the previous one (regardless of whether the modality repeated or switched to a different modality), there was a reaction time delay compared to when the spatial location changed. This finding, that IOR occurs cross-modally between all pairings of vision, audition, and touch, suggests that IOR is not restricted to cues and targets being delivered along the same sensory pathway. Early studies [2,8] suggested that the cause of IOR was activation of the oculomotor system and even though endogenous preparation of an eye movement does not cause IOR [11], this is still the most likely cause.

Three broad loci have been proposed for the effects IOR might have on subsequent behavior at, or near the inhibited location—it could degrade or delay sensory processing, delay or discourage spatial responses, or delay or discourage the orienting of attention. Of course, these possible loci are not mutually exclusive and a direct inhibitory effect at one level of processing could result in indirect inhibitory effects at other levels. Particularly, whether IOR affects input or output levels of processing may depend on the activation state of the reflexive oculomotor system [12]. According to this view, when this system is suppressed, inputs or attention are affected; when it is not suppressed, IOR manifests as an output level response bias.

2.2. Spatio-Temporal Properties

The time course of IOR is usually measured in a paradigm wherein spatially uninformative peripheral cues precede the target by varying intervals. In this paradigm, as demonstrated in Samuel and Kat's [13] review, IOR can last for at least three seconds. The apparent onset of IOR depends on several variables, including the complexity of the task, whether the target is accompanied by distractors, and the importance of shifting attention away from the cue. As noted by many authors: "It is possible that inhibition is present but will not be seen in performance because its effects are obscured by other processes (such as facilitation or a response-repetition advantage) operating at the same time as the hypothesized inhibition" [3] (p. 141). Therefore, one explanation for the variability in the apparent onset IOR depends on the time required for attention to disengage from the location of the spatially uninformative cue where facilitation due to attentional capture might obscure any inhibitory effects of the cue. Moreover, processes that have a negative effect on reaction time may masquerade as IOR, leading to incorrect inferences about how early IOR begins when these processes operate as an immediate consequence of the cue (e.g., sensory adaptation [5] and onset detection cost [14]).

In terms of the spatial extent of IOR, when generated by a stimulus at a particular location in the periphery (i.e., the cue), many studies have reported a gradient of decreasing inhibition centered on this location (e.g., [15]; for a recent review see [16]). Interestingly, when IOR is generated by an array of a small number of cues, the location that seems to be most inhibited is the geometric midpoint of the cues (likened to a "center of gravity"), even if no stimulus was presented there [9,17]. Using a 50 ms interval between such an array of cues and a target calling for a saccade, Christie et al. [18]

discovered an inverse effect. At this very short cue-target interval, rather than being inhibited, eye movements were facilitated toward the center of gravity of the array. This finding provides evidence for the proposal that prior activation of the oculomotor system is the cause of the subsequent IOR.

As first demonstrated by Maylor and Hockey [19], when the eyes move between the presentation of a cue and a target, the target suffers from greater inhibition at the spatiotopic location of the cue than at its retinotopic location. More recently, it has been demonstrated by Pertzov et al. [20] and Hilchey et al. [21] that spatiotopic coding was present as early after the saccade as they tested, and Yan et al. [22] demonstrated that spatiotopic recoding happens just before the eye movement. Furthermore, when a cued object moves between the cue and target, IOR can be seen at the new location of the cued object rather than, or in addition to the originally cued location [23]. From these findings, it has been inferred that IOR is coded in a scene- or object-based representation. Supporting this inference are the findings in both visual search (for a review see [24]) and cue-target paradigms [25] that IOR is eliminated if the scene is removed when the target is presented.

2.3. IOR as Foraging Facilitator

In Posner's seminal studies [1,2], the function of IOR was proposed to be novelty seeking, and in [2] such novelty seeking was hypothesized to increase the efficiency of visual search by encouraging orienting toward new items. This proposal was supported by later studies that looked directly for IOR within visual search paradigms. Using a probe-following-search task, Klein [26] found inhibition at the locations attention had presumably visited during a search to determine that the items (distractors) were not the target. Klein and MacInnes [27] replicated this finding in a more ecologically valid, "camouflaged" search paradigm, employing scenes from "Where's Waldo". One major finding of such studies is that inhibition can be observed at multiple previously fixated locations. Based on these findings, Klein elaborated on Posner and Cohen's [1] proposal, theorizing that IOR acts as an inhibitory tagging system that marks multiple previously attended locations and proposed that such a system could function as a foraging facilitator [26,27]. For a review of IOR findings in visual search tasks, see [24].

3. Neuroscience

Due to the temporal characteristics of IOR, neuroscientific research has been largely reliant on the high temporal resolution of ERP techniques to investigate the neural substrate of IOR. However, other methods have also been employed, including studies with patients who have subcortical brain injuries, developmental studies on infants, fMRI, TMS, single-cell recordings of rhesus monkeys, and computational modelling. Here, we review research in these areas.

3.1. Patient Studies

Early findings [2] from patients with subcortical damage due to progressive supranuclear palsy and cortical lesions involving the parietal or frontal lobes suggested that subcortical, but not cortical, systems were involved in the manifestation of IOR. The involvement of one subcortical structure that is at the nexus of control of the oculomotor system, the superior colliculus (SC) was later confirmed to be critical for the generation of IOR in patients with localized damage to this structure [28,29]. As described next, cortical involvement was later demonstrated to also be important for the coding of IOR in environmental and object (or scene) coordinates. In retrospect, perhaps this is not surprising, once IOR had been found to be represented in these coordinates, because the SC controls eye movements in oculocentric (or retinotopic) coordinates.

Tipper and colleagues [30] employed the moving-objects paradigm in two split brain patients to explore cortical contributions to object-based IOR. They demonstrated that an intact corpus callosum is necessary to transfer object-based inhibitory tags from one hemisphere to the other. In a study of patients with visual form agnosia, Smith et al. [31] demonstrated that although object-based facilitation effects were impaired, object-based IOR remained intact.

Sapir et al. [32] employed a version of the Maylor and Hockey [20] paradigm, in which a saccade intervenes between the cue and target to explore environmental coding of IOR in patients with lesions to the right intraparietal sulcus. A cue was presented above or below fixation and participants either remained fixated or made a leftward or rightward saccade in response to a central arrow. Normal participants and patients showed similar magnitudes of IOR in the stationary condition. When eye movements were made, the target could appear at either the retinal or environmental location of the cue. Normal participants showed IOR at both locations, but patients with right parietal damage only exhibited inhibition at the retinal location of the cue. This finding suggests that circuitry in the parietal lobes is responsible for preserving, in environmental coordinates, the inhibitory tags laid down by the SC.

An interesting response modality dissociation was reported by Bourgeois et al. [33]. Patients with damage to the right hemisphere were tested using a target-target paradigm in which IOR was expected when the current target was presented in the same location as the previous target. One condition required manual responses when a target was detected; the other condition required saccadic responses to targets. In a group suffering from left neglect associated with right parietal damage and/or disconnection of parietal from frontal regions, IOR was observed in the good (right) visual field when saccadic responses were required. In striking contrast, when manual responses were made, instead of suffering from IOR, repeated targets in the good visual field benefited from facilitation. It was suggested that with manual responses, IOR depends on an intact cortical circuit (fronto-parietal attentional network) in the right hemisphere whereas with saccadic responses, IOR might be mediated by subcortical circuits (retinotectal visual pathway).

Smith et al. [34] tested a neurologically normal individual (AI) who was unable to make eye movements due to congenital opthalmoplegia (oculo-muscular atrophy). Whereas AI could orient her attention endogenously in response to informative central cues, her attention was not captured exogenously by uninformative peripheral cues. Nevertheless, such cues did generate significant IOR. Converging evidence for this dissociation (IOR without exogenous cueing) was obtained by Smith et al. [35] using eye abduction in normal participants. With this manipulation, wherein cues and targets can be presented at locations that the eyes cannot reach, normal IOR and endogenous orienting were observed, but attention was not exogenously captured by peripheral cues. Although we are aware of one study with conflicting results [36], the discrepancy may be rooted in the difficulty of firm conclusions from the eye-abduction manipulation [37].

3.2. Developmental Studies

Although most research on IOR in normal individuals has been conducted with college age participants, there are data on the early development of IOR and on how it might change with aging. Research on the early development of IOR, as summarized by Klein [38] is illustrated in Figure 1, along with Mark Johnson's [39] analysis of the relative rates of development of neural circuitry that controls orienting in adults. Among the important developmental findings is the observation of IOR in newborns less than four days old [40,41]. Because subcortical pathways are operational in infants whereas the cortex is still developing, this finding supports the conclusion from patient and behavioral studies that the SC is critical for the generation of IOR. It is noteworthy that in these newborn studies the effect of IOR was generated and measured with oculomotor behavior. Thus, in the context of the 2-forms of IOR described above, this form, almost certainly, would have been the output form.

Surprisingly, the IOR that is seen in newborns is absent in 1–2 month olds. According to Johnson et al. [42], inhibitory projections through the basal ganglia and substantia nigra that regulate SC activity become functional at around one month of age. It is thought that the development of this pathway, which encourages obligatory attention and response repetition would work against IOR. The subsequent development of frontal systems that control the inhibitory projections to the SC from the basal ganglia/substantia nigra is thought to mediate the reappearance of IOR at around 3–4 months of age. In later childhood, adulthood, and in studies of aging, IOR may not be seen or its appearance may be delayed when, for a variety of reasons, after capture of attention by the spatially uninformative peripheral cue, attention lingers or fails to disengage from this location [38].

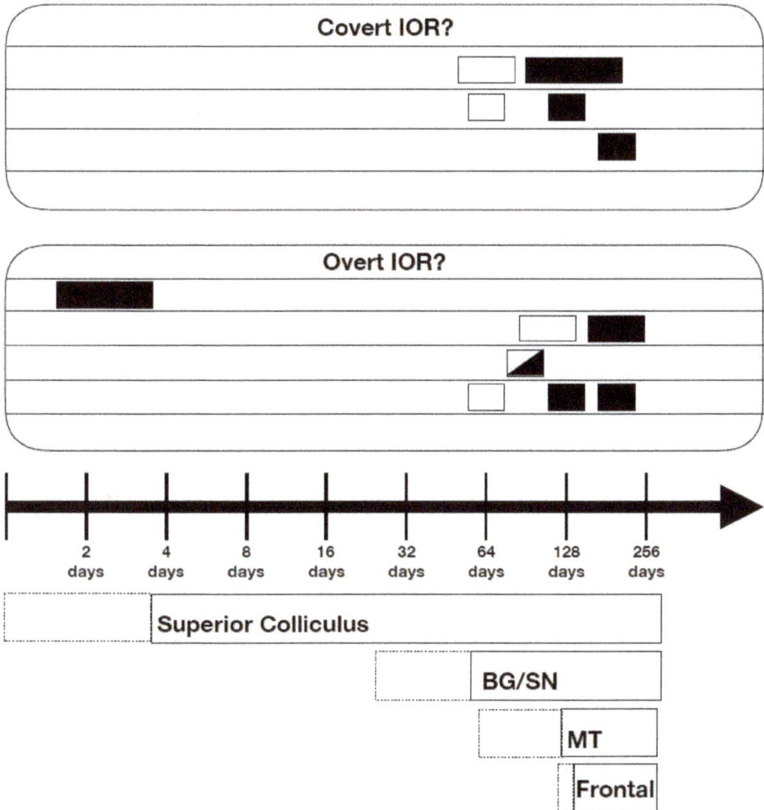

Figure 1. The relative time-course of maturation of different systems involved in orienting, as discussed by Johnson [39], is shown by the rectangles below the timeline (dotted lines simply reflect inter-individual variability in system maturation). These include the superior colliculus, the basal ganglia (BG), the substantia nigra (SN), the MT, and the frontal lobes. All the experiments illustrated here measured IOR using eye movements. Studies finding IOR (black rectangles) and failing to find IOR (open rectangles) in infants of different ages with covert (no eye movement to the cue) and overt (eye movements to the cue) orienting as the causal event are shown above the timeline. The split rectangle of reflects the fact that IOR was obtained when target eccentricities were 10 deg, but not when they were 30 deg. This figure has been redrawn from [38].

3.3. Human Brain Imaging

IOR has been extensively investigated with human brain imaging, particularly with electro-encephalography (EEG) using the event-related potential (ERP) technique. The vast majority of these studies have used a traditional Posner cueing task with peripheral stimuli requiring manual responses to the targets. When this literature was reviewed by Martin-Arevalo et al. [43], it was concluded that there was not one single ERP component that could serve as a "marker" for IOR. Although reduction of the early sensory P1 component was often seen in the literature reviewed by Martin-Arevalo et al. [43], P1 reductions are an unlikely reflection of IOR because, as was pointed out by Satel and colleagues [44,45], these modulations can occur without IOR and IOR can occur without P1 reductions.

Martin-Arevalo et al. [43] excluded a few studies from their review that required participants to make eye movements at some point during a trial. Indeed, in all included studies, participants were instructed not to make eye movements and researchers typically removed the data from any trials with obvious eye movements. One reason that there have been so few EEG studies investigating IOR with activated oculomotor systems is that eye movements can contaminate the ERPs due to the low signal-to-noise ratio of this technique [46]. However, as noted above, suppressing the reflexive oculomotor system in this way might lead to IOR affecting the input stages of processing rather than the output stages. These studies, which involved suppressed reflexive oculomotor systems have often found a reduction in the amplitude of the P1 ERP component for cued relative to uncued peripheral targets. Furthermore, this P1 effect has been shown to be correlated with behavioral measures of IOR [43,44].

However, there have also been a few ERP studies that have allowed eye movements in response to the cues, activating the oculomotor system on each trial. When cues and targets were both peripheral stimuli and eye movements were made to cues [44], although P1 effects were still observed, the P1 modulation was not correlated with behavioral IOR, unlike the case when the eye movement system was suppressed in this and in the earlier studies. Furthermore, this P1 cueing effect disappeared entirely when there was no repeated peripheral stimulation (i.e., eye movements were made in response to central arrow cues) [45], even though IOR was still observed behaviorally in response to the peripheral targets. Similarly, when the spatiotopic location was dissociated from the retinotopic location with an eye movement between cue and target, greater behavioral inhibition was observed at the spatiotopic location than at the retinotopic location, however cue-related P1 reductions were only observed in the retinotopic and not in the spatiotopic condition [46].

A later ERP component in the 220–300 ms post-cue period, the Nd, is also often modulated by cueing in IOR paradigms [43]. In addition to being observed when eye movements are forbidden and there is repeated peripheral stimulation, Nd modulations have also been observed when eye movements were made in response to central arrow cues [45] and at spatiotopic, but not retinotopic locations when an eye movement occurred between cue and target appearance [46]. However, these Nd effects are even more inconsistent than those related to P1. They are not always present and sometimes go in the 'wrong' direction, suggesting that although something is going on in this time range, the Nd component may not be the most appropriate marker for IOR [45].

Other studies, e.g., [47,48] have explored the possibility of IOR modulating the amplitude or latency of the N2pc component, which arises in a similar time range as the Nd component and is assumed to reflect a shift of attention. In the first such study, McDonald et al. [47] discovered that the N2pc component was reduced, but not delayed for targets presented at the cued location. Using a visual search paradigm, Pierce et al. [48] obtained converging evidence for this finding. As yet, we are not aware of any studies with eye movements that have investigated the association of N2pc modulations with IOR.

The majority of these ERP studies have focused on the brain's response to targets that might or might not have been suffering from IOR. It is important to point out that, in the cue-target paradigm, IOR is generated by the cue and measured by the target. The emphasis on the target in ERP experiments is designed to elucidate the nature of IOR's effect(s) on processing. Fewer studies have focused on the brain's response to the cues, an emphasis that, in principle, can tell us about the nature of IOR's cause(s). Using such an alternative approach to investigate neural activity during the cue-target period, Tian et al. [49] developed a theoretical model of IOR. This work attempted to identify the areas activated at different stages after cue onset using LORETA (low-resolution brain electromagnetic tomography) source localization algorithms. The main idea is that attending to cued stimuli stimulates neurons in early visual areas including the SC, which then sends signals to cortical areas such as the parietal and frontal eye fields, generating inhibitory tags that represent previously attended locations.

These tags then feed back down to the SC, inhibiting subsequent eye movements to the inhibited locations (i.e., IOR).

Another alternative approach to investigating sensory activity during the cue-target interval is to use the steady-state visual evoked potential (SSVEP) technique [50]. SSVEPs are periodic electrophysiological signals in the input pathway that are evoked by periodic stimulation and that share the same frequency as the stimulus. For visual stimuli, SSVEPs are observed over occipital cortex. A number of studies have demonstrated that these SSVEP signals are modulated by spatial attention (i.e., the signal is enhanced when attended), e.g., [51,52]. As far as we know, only one study [53] has employed this technique to explore the sensory consequences of an uninformative peripheral cue. This study found a biphasic pattern with enhanced signals from the cued location immediately after the appearance of the cue reversing to suppressed signals beginning about 200 ms later. Although the latter finding was assumed to be a reflection of IOR, this is challenged by the fact that IOR was observed behaviorally (in simple RT to targets) at 1200 ms post cue onset while the SSVEP suppression at the cued location was no longer present after 800 ms. We believe, therefore, that the SSVEP suppression observed at the cued location might have been a reflection of sensory adaptation. Regardless of one's interpretation, further research using this powerful methodology would be welcome.

In addition to using EEG, a few studies have also used the fMRI technique in an attempt to identify the neural circuits associated with IOR. Due to its low temporal resolution, fMRI is a poor method for exploring IOR. Moreover, as noted by Klein (2004, pp. 552 [54]): "it is difficult to generate a reasonable pair of conditions to generate a subtraction that might tap into the presence of IOR. This is because IOR may be generated by a cue whether or not a target is presented and regardless where it is presented and IOR is just as likely to follow orienting to a target as to follow orienting a cue [6]." Müller and Kleinschmidt [55], however, conducted a clever fMRI experiment, aimed at determining whether IOR might have a negative impact on processing in early visual pathways. They avoided sensory cue-target interactions on cued trials by presenting the target close to the cue but on the opposite side of the vertical meridian (i.e., opposite hemifield) and they compared the activity in occipital cortex for cued versus uncued targets. Strongly supporting an input form of IOR, they found that the responses in occipital areas stimulated by the target were reduced for targets suffering from IOR. In this study, although participants were not given trial-by-trial feedback on their oculomotor behavior in the magnet, they were trained with eye monitoring before the fMRI session and according to the authors, made eye movements on less than 1% of the trials during training. Therefore, it seems likely that their reflexive oculomotor systems were relatively suppressed during the fMRI recording sessions. According to Klein and Redden's view of the two forms of IOR [12], if an experiment like this were repeated with the IOR caused by a pro-saccade to the cue, suppression of visual cortex activity at the originally cued location would not be observed.

3.4. Manipulations Aimed at Exploring the Roles of Neural Structures and Pathways

Converging evidence for the special role of the superior colliculus in the generation of IOR can be found in two methods that have been used to differentially access this structure: nasal/temporal asymmetries and S-cone stimuli (for a review of the use of these methods in the study of visual attention, see [56]).

The retinotectal pathway sends visual inputs via the optic chiasm directly to the superior colliculus with more copious connections from the temporal hemifields (nasal hemiretinae) (for a review see [57]). Rafal et al. [8] hypothesized that if the SC played a special role in the generation of IOR, then stimulating a pathway with more copious connections to the SC ought to result in greater IOR. When they did this (Experiment 1, [8]) by presenting cues and targets monocularly they found, in accordance with their hypothesis, substantially more IOR when cues and targets were presented to the temporal hemifields than when they were presented to the nasal hemifields. Given our emphasis on the importance of the state of the reflexive oculomotor system for whether the input or output form of IOR might be generated, it is important to note that in this experiment eye position was not monitored and so it is likely that the output form was operating even though the response modality was manual.

An alternative strategy is to use S-cone stimuli that bypass the retinotectal pathway. Although such stimuli will not reach the SC directly, they will eventually get there via the geniculo-cortical pathway. Sumner et al. [58] used this strategy by presenting cues that were either luminant or S-cone followed 400–500 ms later by luminant targets. In one condition participants detected these targets using manual responses, and in another condition, they made saccades to the targets. With the typical luminant cues they found IOR regardless of the response modality. However, with S-cone cues, they only observed IOR when the eyes remained fixated for the duration of the task and target detection was signaled by a manual response. In other words, saccades were insensitive to the prior location of S-cone cues. This pattern of results suggests that the retinotectal pathway plays a special role in generating the output form of IOR when prosaccades are required. Moreover, if it is assumed that the reflexive oculomotor system was suppressed in their manual response condition, it would be reasonable to suggest that the input form of IOR was generated in the manual response condition.

Transcranial magnetic stimulation (TMS) is a powerful tool for inferring whether a targeted brain structure plays an important role in a particular behavior. A few studies have employed TMS in an effort to deduce which structures in the brain are integral to IOR. Three studies [59–61] administered TMS pulses over various brain structures on each trial during the time between the presentation of the cue and the target. Thus, if a certain structure was integral in generating IOR (through exposure to a cue), the TMS pulse to that area should nullify its effect and IOR would not be observed. Indeed, TMS pulses to the right frontal eye field [59], the temporoparietal junction (TPJ) [60], and the right intraparietal sulcus (IPS) [60] interfered with IOR. Importantly, using TMS in conjunction with a retinotopic/spatiotopic reference framing spatial cueing paradigm (as described in Section 2.2), van Koningsbruggen et al. [61] found an asymmetrical functionality for the right anterior intraparietal cortex. They showed that TMS pulses to this region diminished spatiotopic inhibition for targets both ipsilateral and contralateral to the pulse. Conversely, pulses to the left anterior intraparietal cortex had no effect on the usual IOR pattern. This finding provides strong converging evidence for the conclusion from patient work [33] described earlier, that the right parietal cortex is the neural substrate responsible for updating the locus of IOR to a spatiotopic representation in the presence of eye movements.

Seeking support for the response-modality dissociation reported in Section 2.2, Bourgeois et al. [62] administered 1200 TMS pulses over a 20-min period to create a temporary and reversible lesion of the right IPS or right TPJ. Similar to the findings from patients with neglect, for right-sided targets TMS-mediated disruption of either area decreased or eliminated manual, but not saccadic, IOR. By contrast, a later study by Bourgeois et al. [63], that used TMS to disrupt the left TPJ or left IPS showed no change in the IOR pattern for either manual or saccadic responses. Taken together, these results suggest both an asymmetrical control of visuospatial attention by the right parietal cortex and add converging evidence for the view that IOR may depend on different neural circuits depending on the activation state of the reflexive oculomotor system.

In an ingenious experiment, Gabay et al. [64] exposed the archer fish (which gets its name from the fact that when foraging for food it shoots down prey on low hanging vegetation by spitting water) to a Posner cueing paradigm. Fish are an interesting species for drawing inferences about neural structures required for generating IOR because they have such an underdeveloped cortex. Cues and targets were presented on a monitor mounted over the tank in which the fish were swimming and the latency of accurate spitting was measured. When the fish successfully shot a stream of water at the target on the screen, some food was dropped into the tank. Demonstrating IOR, when the interval between the cue and target was greater than one second, the spitting was slower for targets at the cued than at the uncued location. This finding supports the observation of IOR in newborn infants (see section above demonstrating that the generation of IOR does not require a fully developed cortex). In a subsequent study [65], archer fish were exposed to an endogenous version of the Posner cueing task in which the color of a central stimulus indicated the likely side of the upcoming target. The archer fish showed early facilitation, which the authors attributed to learning rather than volitional control. Interestingly, IOR was observed at later intervals. The authors concluded that when orienting is generated subcortically (as would be the case in this primarily sub-cortical species), IOR is observed even if the cue had been presented centrally.

3.5. Monkey Neurophysiology

A number of monkey single cell recording studies have recorded from oculomotor areas such as the superior colliculus (SC), while the animals performed spatial cueing tasks. Dorris et al. [66] demonstrated that at a CTOA of 200 ms, behavior was inhibited on cued trials as compared to uncued trials (as with humans), and that the activity of neurons in the SC was attenuated at the cued location (i.e., the target-related activity of neurons was lower when they had been previously been stimulated by a cue at the same location). Furthermore, when the same neurons were stimulated electrically (through the recording electrode), rather than by a visual stimulus to induce a saccade, facilitation rather than inhibition was observed, suggesting that the SC was not directly inhibited [65]. In later work [67], inhibition was observed behaviorally in monkeys at later CTOAs (100, 200, 500, and 1200 ms) while recordings were collected from both visual and visuomotor neurons in the SC. As in the previous study, target-related activity was reduced for cued neurons at 100 and 200 ms CTOAs, however, at the longer CTOAs this input attenuation was eliminated. These results suggest that the reduced responses of previously cued neurons in the SC at relatively short CTOAs are a reflection of sensory adaptation in the pathway projecting to the SC, whereas behavioral IOR observed at longer CTOAs reflects delays in pathways outside the SC.

Further neurophysiological data derived from recordings in monkeys provide additional converging evidence that, although the SC is crucial to the generation of IOR, higher cortical areas contribute importantly to output-based, oculomotor IOR. Mirpour and Bisley [68] recorded in the lateral intraparietal cortex (LIP) of monkeys while they performed a visual foraging task that allowed measurement of neural responses when new, or previously visited distractors entered the neuron's receptive field. Providing a neural correlate for the suggestion that IOR might function as a foraging facilitator [2,26,27], it was found that responses were reduced for previously fixated as compared to new distractors. More recently, when recording from the FEFs during such a search task, Mirpour et al. [69] identified neurons that maintain increased activity throughout trials once the location they represent had been fixated. The authors proposed that these neurons keep track of all fixated stimuli, later sending these signals to priority maps in parietal cortex. Such priority maps in parietal cortex, driven by FEF signals, are a likely locus for the inhibitory tags leading to the output form of IOR.

3.6. Computational Modeling

As noted by Klein [54]: "What is most needed to advance our cognitive-neuroscientific understanding are some comprehensive and computationally explicit theories of the inhibitory aftermath of orienting" (p. 556). In real-world applications of visual search, such as robot navigation, inhibitory algorithms must be implemented in order to avoid perseverance on highly salient stimuli. However, such computations are normally implemented by simply reducing the salience of previously attended stimuli to zero for a few seconds [70], which is clearly not how the primate brain accomplishes the task (see [71] for a recent review of such salience models).

Neurobiologically plausible computational implementations of IOR have tended to use dynamic neural field models simulating the activity of neurons in the SC, based on data obtained from monkey neurophysiology and human behavior [72]. This work has shown great success in reproducing behavioral data in humans as well as monkeys and has played an important role in making predictions for further empirical work. Early simulations attempted to determine the extent to which sensory adaptation and emergent properties of saccade dynamics could account for the behavioral effects of IOR [73,74]. Although a great deal of data could be reproduced with such implementations, it was determined that IOR at CTOAs greater than around 1000 ms could not be explained or accurately reproduced with such input-based mechanisms. More recent implementations of this model have incorporated a later inhibitory mechanism (i.e., IOR), presumably via pathways from cortical areas such as the frontal eye fields and/or posterior parietal cortex [75].

In a complementary approach, diffusion modelling considers the accumulation of evidence toward some decision threshold. Here, the delayed responses for targets suffering from IOR might be explained by a variety of model parameters, e.g., see [76,77] including sensory-level effects (e.g., slower rate of accumulation) or a later decision-level effect (higher evidentiary threshold). Although diffusion modeling can be done without consideration of the neural circuits that mediate the behavior of interest, such models can be fruitfully linked and applied to specific neural circuits [78].

4. Conclusions

The neuroscientific research described here points to the critical role of the oculomotor system in the generation of output-based IOR that facilitates novelty seeking. IOR arising when the reflexive oculomotor system is not suppressed, is probably generated by projections from the SC to cortical areas (FEF, PPC [or LIP]) but not implemented in the SC. It is represented in spatiotopic coordinates, seems to arise only after about 600 ms post-cue and is likely represented in cortical areas affecting spatial responses regardless of the output modality (manual or oculomotor). When the reflexive oculomotor system is actively suppressed, however, the input-based form of IOR is generated, affecting early sensory pathways in retinotopic coordinates rather than response outputs. Early sensory adaptation also occurs along input pathways but only affects behavior for up to around 600 ms post-cue, and only when there is repeated peripheral stimulation. Further studies of the inhibitory aftereffects of orienting should be careful to disentangle these multiple inhibitory cueing effects.

Funding: This research was funded by a NSERC Discovery grant awarded to RMK.

Conflicts of Interest: The authors declare no conflict of interest.

References

1. Posner, M.I.; Cohen, Y. Components of Visual Orienting. *Attent. Perform. X Control Lang. Processes* **1984**, *32*, 531–556.
2. Posner, M.I.; Rafal, R.D.; Choate, L.; Vaughan, J. Inhibition of return: Neural basis and function. *Cogn. Neuropsychol.* **1985**, *2*, 211–228. [CrossRef]
3. Klein, R.M. Inhibition of return. *Trends Cogn. Sci.* **2000**, *4*, 138–147. [CrossRef]
4. Dukewich, K.; Klein, R.M. Inhibition of return: A phenomenon in search of a definition and a theoretical framework. *Attent. Percept. Psychophys.* **2015**, *77*, 1647–1658. [CrossRef] [PubMed]
5. Hilchey, M.D.; Klein, R.M.; Satel, J. Returning to "Inhibition of Return" by dissociating long-term oculomotor IOR from short-term sensory adaptation and other nonoculomotor "inhibitory" cueing effects. *J. Exp. Psychol. Hum. Percept. Perform.* **2014**, *40*, 1606–1616. [CrossRef]
6. Taylor, T.; Klein, R.M. Visual and motor effects in inhibition of return. *J. Exp. Psychol. Hum. Percept. Perform.* **2000**, *6*, 1639–1655. [CrossRef]
7. Berlucchi, G. Inhibition of return: A phenomenon in search of a mechanism and a better name. *Cogn Neuropsychol.* **2006**, *23*, 1065–1074. [CrossRef]
8. Rafal, R.D.; Calabresi, P.A.; Brennan, C.W.; Sciolto, T.K. Saccade preparation inhibits reorienting to recently attended locations. *J. Exp. Psychol. Hum. Percept. Perform.* **1989**, *15*, 673–685. [CrossRef]
9. Klein, R.M.; Christie, J.; Morris, E. Vector averaging of inhibition of return. *Psychon. Bull.Rev.* **2005**, *12*, 295–300. [CrossRef]
10. Spence, C.; Lloyd, D.; McGlone, F.; Nicholls, M.E.R.; Driver, J. Inhibition of return is supramodal: A demonstration between all possible pairings of vision, touch, and audition. *Exp. Brain Res.* **2000**, *134*, 42–48. [CrossRef]
11. Chica, A.B.; Klein, R.M.; Rafal, R.D.; Hopfinger, J.B. Endogenous saccade preparation does not produce Inhibition of Return: Failure to replicate Rafal, Calabresi, Brennan, & Sciolto (1989). *J. Exp. Psychol. Hum. Percept. Perform.* **2010**, *36*, 1193–1206. [PubMed]

12. Klein, R.M.; Redden, R.S. Two "Inhibitions of Return" Bias Orienting Differently. In *Spatial Biases in Perception and Cognition*; Hubbard, T., Ed.; Cambridge University Press: Cambridge, UK, 2018; pp. 295–306.

13. Samuel, A.G.; Kat, D. Inhibition of return: A graphical meta-analysis of its time course and an empirical test of its temporal and spatial properties. *Psychon. Bull. Rev.* **2003**, *10*, 897–906. [CrossRef] [PubMed]

14. Lupiáñez, J. Inhibition of return. In *Attention and Time*; Nobre, A.C., Coull, J.T., Eds.; Oxford University Press: Oxford, UK, 2010; pp. 17–34.

15. Bennett, P.J.; Pratt, J. The spatial distribution of inhibition of return. *Psychol. Sci.* **2001**, *12*, 76–80. [CrossRef] [PubMed]

16. Wang, B.; Yan, C.; Klein, R.M.; Wang, Z. Inhibition of return revisited: Localized inhibition on top of a pervasive bias. *Psychon. Bull. Rev.* **2018**, *25*, 1861–1867. [CrossRef] [PubMed]

17. Christie, J.J.; Hilchey, M.D.; Klein, R.M. Inhibition of return is at the midpoint of simultaneous cues. *Attent. Percept. Psychophys.* **2013**, *75*, 1610–1618. [CrossRef] [PubMed]

18. Christie, J.J.; Hilchey, M.D.; Mishra, R.; Klein, R.M. Eye movements are primed toward the centre of multiple stimuli even when the interstimulus distances are too large to generate saccade averaging. *Exp. Brain Res.* **2015**, *233*, 1541–1549. [CrossRef] [PubMed]

19. Maylor, E.A.; Hockey, R. Inhibitory component of externally controlled covert orienting in visual space. *J. Exp. Psychol. Hum. Percept. Perform.* **1985**, *11*, 777–787. [CrossRef]

20. Pertzov, Y.; Zohary, E.; Avidan, G. Rapid formation of spatiotopic representations as revealed by inhibition of return. *J. Neurosci.* **2010**, *30*, 8882–8887. [CrossRef]

21. Hilchey, M.D.; Klein, R.M.; Satel, J.; Wang, Z. Oculomotor inhibition of return: How soon is it "recoded" into spatiotopic coordinates? *Attent. Percept. Psychophy.* **2012**, *74*, 1145–1153. [CrossRef]

22. Yan, C.; He, T.; Klein, R.M.; Wang, Z. Predictive remapping gives rise to environmental inhibition of return. *Psychon. Bull. Rev.* **2016**, *23*, 1860–1866. [CrossRef]

23. Tipper, S.P.; Driver, J.; Weaver, B. Object-centred inhibition of return of visual attention. *Q. J. Exp. Psychol. A* **1991**, *43*, 289–298. [CrossRef] [PubMed]

24. Wang, Z.; Klein, R.M. Searching for inhibition of return in visual search: A review. *Vis. Res.* **2010**, *50*, 220–228. [CrossRef] [PubMed]

25. Redden, R.S.; Klages, J.; Klein, R.M. The effect of scene removal on inhibition of return in a cue target task. *Attent. Percept. Psychophys.* **2017**, *79*, 78–84. [CrossRef] [PubMed]

26. Klein, R.M. Inhibitory tagging system facilitates visual search. *Nature* **1988**, *334*, 430. [CrossRef]

27. Klein, R.M.; MacInnes, W.J. Inhibition of return is a foraging facilitator in visual search. *Psychol. Sci.* **1999**, *10*, 346–352. [CrossRef]

28. Briand, K.A.; Szapiel, S.V.; Sereno, A.B. Disruption of reflexive visual orienting in an individual with a collicular lesion. *J. Clin. Exp. Neuropsychol.* **2003**, *28*, 145–166.

29. Sapir, A.; Soroker, N.; Berger, A.; Henik, A. Inhibition of return in spatial attention: Direct evidence for collicular generation. *Nat. Neurosci.* **1999**, *2*, 1053–1054. [CrossRef]

30. Tipper, S.P.; Rafal, R.; Reuter-Lorenz, P.A.; Starrveldt, Y.; Ro, T.; Egly, R. Object-based facilitation and inhibition from visual orienting in the human split brain. *J. Exp. Psychol. Hum. Percept. Perform.* **1997**, *23*, 1522–1532. [CrossRef]

31. Smith, D.T.; Ball, K.; Swalwell, R.; Schenk, T. Object-based attentional facilitation and inhibition are neuropsychologically dissociated. *Neuropsychologia* **2016**, *80*, 9–16. [CrossRef]

32. Sapir, A.; Hayes, A.; Henik, A.; Danziger, S.; Rafal, R. Parietal lobe lesions disrupt saccadic remapping of inhibitory location tagging. *J. Cogn. Neurosci.* **2004**, *16*, 503–509. [CrossRef]

33. Bourgeois, A.; Chica, A.B.; Migliaccio, R.; de Schotten, M.T.; Bartolomeo, P. Cortical control of inhibition of return: Evidence from patients with inferior parietal damage and visual neglect. *Neuropsychologia* **2012**, *50*, 800–809. [CrossRef] [PubMed]

34. Smith, D.T.; Rorden, C.; Jackson, S.R. Exogenous orienting of attention depends upon 805 the ability to execute eye movements. *Curr. Biol.* **2004**, *14*, 792–795. [CrossRef] [PubMed]

35. Smith, D.T.; Schenk, T.; Rorden, C. Saccade preparation is required for exogenous attention but not endogenous attention or IOR. *J. Exp. Psychol. Hum. Percept. Perform.* **2012**, *38*, 1438. [CrossRef] [PubMed]

36. Michalczyk, Ł.; Paszulewicz, J.; Bielas, J.; Wolski, P. Is saccade preparation required for inhibition of return (IOR)? *Neurosci. Lett.* **2018**, *665*, 13–17. [CrossRef]

37. Casteau, S.; Smith, D.T. Associations and Dissociations between Oculomotor Readiness and Covert attention. *Vision* **2019**, *3*, 17. [CrossRef]

38. Klein, R.M. On the role of endogenous orienting in the inhibitory aftermath of exogenous orienting. In *Developing Individuality in the Human Brain: A Tribute to Michael I. Posner*; Mayr, U., Awh, E., Keele, S., Eds.; APA Books: Washington, DC, USA, 2005; pp. 45–64.

39. Johnson, M.H. Cortical maturation and the development of visual attention in early infancy. *J. Cogn. Neurosci.* **1990**, *2*, 81–95. [CrossRef]

40. Valenza, E.L.; Simion, F.L.; Umilta, C.L. Inhibition of return in newborn infants. *Infant Behav. Dev.* **1994**, *17*, 293–302. [CrossRef]

41. Simion, F.; Valenza, E.; Umilta, C.; Dalla, B. Inhibition of return in newborns is temporo-nasal asymmetrical. *Infant Behav. Dev.* **1995**, *8*, 189–194. [CrossRef]

42. Johnson, M.H.; Posner, M.I.; Rothbart, M.K. Facilitation of saccades toward a covertly attended location in early infancy. *Psychol. Sci.* **1994**, *5*, 90–93. [CrossRef]

43. Martín-Arévalo, E.; Chica, A.B.; Lupiáñez, J. No single electrophysiological marker for facilitation and inhibition of return: A review. *Behav. Brain Res.* **2016**, *300*, 1–10. [CrossRef]

44. Satel, J.; Hilchey, M.D.; Wang, Z.; Story, R.; Klein, R.M. The effects of ignored versus foveated cues upon inhibition of return: An event-related potential study. *Attent. Percept. Psychophys.* **2013**, *75*, 29–40. [CrossRef] [PubMed]

45. Satel, J.; Hilchey, M.D.; Wang, Z.; Reiss, C.S.; Klein, R.M. In search of a reliable electrophysiological marker of oculomotor inhibition of return. *Psychophysiology* **2014**, *51*, 1037–1045. [CrossRef] [PubMed]

46. Satel, J.; Wang, Z.; Hilchey, M.D.; Klein, R.M. Examining the dissociation of retinotopic and spatiotopic inhibition of return with event-related potentials. *Neurosci. Lett.* **2012**, *524*, 40–44. [CrossRef]

47. McDonald, J.J.; Hickey, C.; Green, J.J.; Whitman, J.C. Inhibition of return in the covert deployment of attention: Evidence from human electrophysiology. *J. Cogn. Neurosci* **2009**, *21*, 725–733. [CrossRef] [PubMed]

48. Pierce, A.M.; Crouse, M.D.; Green, J.J. Evidence for an attentional component of inhibition of return in visual search. *Psychophysiology* **2017**, *54*, 1676–1685. [CrossRef] [PubMed]

49. Tian, Y.; Klein, R.M.; Satel, J.; Xu, P.; Yao, D. Electrophysiological explorations of the cause and effect of inhibition of return in a cue-target paradigm: A spatio-temporal theory. *Brain Topogr.* **2011**, *24*, 164–182. [CrossRef] [PubMed]

50. Regan, D. Chromatic adaptation and steady-state evoked potentials. *Vis. Res.* **1968**, *8*, 149–158. [CrossRef]

51. Morgan, S.T.; Hansen, J.C.; Hillyard, S.A.; Posner, M. Selective attention to stimulus location modulates the steady-state visual evoked potential. *Neurobiology* **1996**, *93*, 4770–4774. [CrossRef]

52. Müller, M.M.; Picton, T.W.; Valdes-Sosa, P.; Riera, J.; Teder-Sälejärvi, W.A.; Hillyard, S.A. Effects of spatial selective attention on the steady-state visual evoked potential in the 20–28 Hz range. *Cogn. Brain Res.* **1998**, *6*, 249–261. [CrossRef]

53. Li, A.-S.; Zhang, G.-L.; Miao, C.-G.; Wang, S.; Zhang, M.; Zhang, Y. The time course of inhibition of return: Evidence from steady-state visual evoked potentials. *Front. Psychol.* **2017**, *8*, 1–10. [CrossRef]

54. Klein, R.M. Orienting and inhibition of return. In *The Cognitive Neurosciences*, 3rd ed.; Gazzaniga, M.S., Ed.; MIT Press: Cambridge, MA, USA, 2004; pp. 545–560.

55. Müller, N.G.; Kleinschmidt, A. Temporal dynamics of the attentional spotlight: Neuronal correlates of attentional capture and inhibition of return in early visual cortex. *J. Cogn. Neurosci.* **2007**, *19*, 587–593. [CrossRef] [PubMed]

56. Mizzi, R.; Michael, G.A. Exploring visual attention functions of the human extrageniculate pathways through behavioral cues. *Psychol. Rev.* **2016**, *123*, 740–757. [CrossRef] [PubMed]

57. Jóhannesson, Ó.I.; Tagu, J.; Kristjánsson, Á. Asymmetries of the visual system and their influence on visual performance and oculomotor dynamics. *Eur. J. Neurosci.* **2018**, *48*, 3426–3445. [CrossRef]

58. Sumner, P.; Nachev, P.; Vora, N.; Husain, M.; Kennard, C. Distinct cortical and collicular mechanisms of inhibition of return revealed with S cone stimuli. *Curr. Biol.* **2004**, *14*, 2259–2263. [CrossRef]

59. Ro, T.; Farnè, A.; Chang, E. Inhibition of return and the human frontal eye fields. *Exp. Brain Res.* **2003**, *150*, 290–296. [CrossRef] [PubMed]

60. Chica, A.B.; Bartolomeo, P.; Valero-Cabré, A. Dorsal and ventral parietal contributions to spatial orienting in the human brain. *J. Neurosci.* **2011**, *31*, 8143–8149. [CrossRef] [PubMed]

61. Van Koningsbruggen, M.G.; Gabay, S.; Sapir, A.; Henik, A.; Rafal, R.D. Hemispheric asymmetry in the remapping and maintenance of visual saliency maps: A TMS study. *J. Cogn. Neurosci.* **2010**, *22*, 1730–1738. [CrossRef] [PubMed]

62. Bourgeois, A.; Chica, A.B.; Valero-Cabré, A.; Bartolomeo, P. Cortical control of inhibition of return: Causal evidence for task-dependent modulations by dorsal and ventral parietal regions. *Cortex* **2013**, *49*, 2229–2238. [CrossRef]

63. Bourgeois, A.; Chica, A.B.; Valero-Cabre, A.; Bartolomeo, P. Cortical Control of Inhibition of Return: Exploring the Causal Contributions of the Left Parietal Cortex. *Cortex* **2013**, *49*, 2927–2934. [CrossRef]

64. Gabay, S.; Leibovich, T.; Ben-Simon, A.; Henik, A.; Segev, R. Inhibition of return in the archer fish. *Nat. Commun.* **2013**, *4*, 1657. [CrossRef]

65. Saban, W.; Sekely, L.; Klein, R.M.; Gabay, S. Endogenous orienting in the archer fish. *Proc. Nat. Acad. Sci.* **2017**, *114*, 7577–7581. [CrossRef] [PubMed]

66. Dorris, M.C.; Klein, R.M.; Everling, S.; Munoz, D.P. Contribution of the primate superior colliculus to inhibition of return. *J. Cogn. Neurosci.* **2002**, *14*, 1256–1263. [CrossRef] [PubMed]

67. Fecteau, J.H.; Munoz, D.P. Correlates of capture of attention and inhibition of return across stages of visual processing. *J. Cogn. Neurosci.* **2005**, *17*, 1714–1727. [CrossRef] [PubMed]

68. Mirpour, K.; Bisley, J.W. Anticipatory remapping of attentional priority across the entire visual field. *J. Neurosci.* **2012**, *32*, 16449–16457. [CrossRef] [PubMed]

69. Mirpour, K.; Bolandnazar, Z.; Bisley, J.W. Neurons in FEF keep track of items that have been previously fixated in free viewing visual search. *J. Neurosci.* **2019**, *39*, 2114–2124. [CrossRef] [PubMed]

70. Itti, L.; Koch, C. A saliency-based search mechanism for overt and covert shifts of visual attention. *Vis. Res.* **2000**, *40*, 1489–1506. [CrossRef]

71. Krasovskaya, S.; MacInnes, J. Salience models: A computational cognitive neuroscience review. *Vision* **2019**, In press.

72. Trappenberg, T.P.; Dorris, M.C.; Munoz, D.P.; Klein, R.M. A model of saccade initiation based on the competitive integration of exogenous and endogenous signals in the superior colliculus. *J. Cogn. Neurosci.* **2001**, *13*, 256–271. [CrossRef]

73. Satel, J.; Wang, Z.; Klein, R.M.; Trappenberg, T.P. Modeling inhibition of return (IOR) as short-term depression of early sensory input to the superior colliculus. *Vis. Res.* **2011**, *51*, 987–996. [CrossRef]

74. Wang, Z.; Satel, J.; Trappenberg, T.P.; Klein, R.M. Behavioral aftereffects of a saccade explored in a dynamic neural field model of the superior colliculus. *J. Eye Mov. Res.* **2011**, *4*, 1–16.

75. Lim, A.; Eng, V.; Janssen, S.M.J.; Satel, J. Sensory adaptation and inhibition of return: Dissociating multiple inhibitory cueing effects. *Exp. Brain Res.* **2018**, *236*, 1369–1382. [CrossRef] [PubMed]

76. Ludwig, C.J.; Farrell, S.; Ellis, L.A.; Gilchrist, I.D. The mechanism underlying inhibition of saccadic return. *Cogn. Psychol.* **2009**, *59*, 180–202. [CrossRef] [PubMed]

77. MacInnes, W.J. Multiple diffusion models to compare saccadic and manual responses for inhibition of return. *Neural Comput.* **2017**, *29*, 804–824. [CrossRef] [PubMed]

78. Smith, P.L.; Ratcliff, R. Psychology and neurobiology of simple decisions. *Trends Neurosci.* **2004**, *27*, 161–168. [CrossRef] [PubMed]

Article

Saccadic Suppression of Displacement Does Not Reflect a Saccade-Specific Bias to Assume Stability

Sabine Born

Faculté de Psychologie et des Sciences de l'Education, Université de Genève, 1205 Geneva, Switzerland; sabine.born@unige.ch; Tel.: +41-22-379-9122

Received: 28 February 2019; Accepted: 17 September 2019; Published: 24 September 2019

Abstract: Across saccades, small displacements of a visual target are harder to detect and their directions more difficult to discriminate than during steady fixation. Prominent theories of this effect, known as saccadic suppression of displacement, propose that it is due to a bias to assume object stability across saccades. Recent studies comparing the saccadic effect to masking effects suggest that suppression of displacement is not saccade-specific. Further evidence for this account is presented from two experiments where participants judged the size of displacements on a continuous scale in saccade and mask conditions, with and without blanking. Saccades and masks both reduced the proportion of correctly perceived displacements and increased the proportion of missed displacements. Blanking improved performance in both conditions by reducing the proportion of missed displacements. Thus, if suppression of displacement reflects a bias for stability, it is not a saccade-specific bias, but a more general stability assumption revealed under conditions of impoverished vision. Specifically, I discuss the potentially decisive role of motion or other transient signals for displacement perception. Without transients or motion, the quality of relative position signals is poor, and saccadic and mask-induced suppression of displacement reflects performance when the decision has to be made on these signals alone. Blanking may improve those position signals by providing a transient onset or a longer time to encode the pre-saccadic target position.

Keywords: saccades; masking; displacement perception; spatial stability; motion perception

1. Introduction

1.1. Saccadic Suppression of Displacement and Visual Stability across Eye Movements

In our everyday experience, we have the impression that no important change in our environment escapes us. Contrary to this impression, it has been demonstrated that even large changes of the visual scene can go unnoticed if the change is not accompanied by a visible and localizable transient that draws attention towards it [1]. For instance, the change might be camouflaged when it occurs simultaneously with a saccadic eye movement. In particular, small displacements ('jumps') of a visual object that are easily seen during fixation may be missed when they happen during a saccade, even when the object was the target of the eye movement [2–7]. This effect is known as saccadic suppression of displacement. It is thought to be intrinsically linked to the problem of visual stability across eye movements. There are several facets to this problem [8]. Most relevant to saccadic suppression of displacement is the question how we can keep track of an object's location across saccades, despite the fact that each eye movement we make results in drastic shifts of the visual image projected onto our retina. It has long been assumed that the visual system integrates extraretinal information about the size of the saccade (e.g., efference copy or proprioceptive eye position signals) with the sensory information about the object's position, see [9] for a historical perspective. However, if we have such a mechanism, why are small displacements nonetheless more difficult to detect or discriminate across saccades than during fixation?

1.2. The Blanking Effect

Probably the simplest explanation is that the addition of the extraretinal signal adds further noise compared to a situation without concurrent eye movement. In a simple signal/noise ratio model, this alone could explain the increase in displacement detection thresholds [5,10]. This simple model has been challenged, however, by the observation that displacement perception is strongly improved when the target is removed during the saccade, reappearing in its displaced position only after the saccade has landed [6,11–16]. This so-called blanking effect described by Deubel, Schneider, and Bridgeman in 1996 [6] has been a seminal finding in saccadic suppression of displacement. In the current contribution, I will therefore focus on studies that have been published in or after 1996, see [8] and related commentary for discussion of earlier work. The blanking effect suggests that location information across saccades is more precise than initially presumed. There is now evidence that the available information may even include the trial-by-trial saccadic motor error, that is, the difference between the planned and the actual saccade size [17,18]. Thus, theories of saccadic suppression of displacement do not only have to explain why displacement perception is worse during saccades than during fixation. Maybe more importantly, they also have to explain the counter-intuitive blanking effect; that is, why performance is better when the target is absent when the eyes land.

1.3. Saccadic Suppression of Displacement as the Result of a Bias for Stability

One prominent line of argument today posits that saccadic suppression of displacement reflects a perceptual decision bias to assume stability across saccades [6,7,11–13]. Based on an object's pre-saccadic location and extraretinal signals of the size of the saccade, a prediction of the object's post-saccadic location may well be formed. In the case of the saccade target, the prediction does not even need to incorporate extraretinal signals, as the target should always be located close to the fovea after the eyes have landed [6,7,19]. The prediction could then be compared to the object's incoming visual signal after the saccade. However, it is assumed that this information is often not used, as long as the target was found near the predicted location after the saccade. Unless the discrepancy between prediction and incoming signal is very large (as is the case for larger jumps), the visual system per default assumes the null hypothesis that no displacement has occurred [6,7,11–13]. Alternatively, Niemeier and colleagues [16,20] have proposed a prior for stability that biases the perceived size of displacements. The specific shape of the prior leads to underestimations of jump size over a range of smaller displacements. For the smallest displacements, the prior leads to almost complete superposition of the pre- and post-saccadic signals ('contraction'). Consequently, the smallest jumps remain undetected.

The idea of a perceptual null hypothesis has already been expressed earlier, for instance by MacKay [21] as cited in [8]. The blanking effect, however, has helped the idea of a decision bias or prior for stability to achieve widespread acceptance. To explain the blanking effect, it has been proposed that the stability bias can be relaxed, when there is evidence suggesting a break in object continuity or object correspondence across the saccade: for instance, if the target is not present upon saccade landing, as in the blanking effect [6,11–16]. In other words, if the visual system has already one clear indication that the object has undergone a change, signals that suggest a concurrent change in position are more easily trusted or receive more weight in the estimation of the displacement.

1.4. Is the Stability Bias Saccade-Specific?

Although the proposed bias may not only be effective across eye movements, it is quite remarkable that suppression of displacement is almost exclusively discussed in the context of saccades. The specifications 'saccadic', 'across saccades' or similar terms are found in the titles of almost all studies [6,7,11–16,22–26]. Moreover, the specific challenge of spatial updating across saccades, that is, having to assess the displacement across a large shift of the entire retinal image brought on by the eye movement, is mentioned in most introductory passages [7,11–16,19,20,22–26]. Due to this emphasis on the saccadic context, one may be tempted to assume that the proposed mechanisms are saccade-specific, and closely related to the saccade-specific challenge of spatial updating. Taking the idea of a null hypothesis of stability as an example, one may be tempted to assume that the saccadic context raises the decision threshold for displacement perception: A difference in an object's position signal may be interpreted as a displacement during fixation, but a same-sized difference between the predicted location of the object and its actual post-saccadic signal may not be sufficient to reject the perceptual null hypothesis in the context of a saccade. However, many saccadic suppression of displacement studies did not compare performance to any kind of non-saccadic control condition [11–15,20,22–26]. It remains thus unclear whether similar mechanisms may operate during fixation and whether estimations across eye movements are more prone to the suggested stability bias or not.

1.5. Saccadic Suppression of Displacement as the Result of Saccadic Suppression of Vision

It is important to note that estimations across saccades are not only difficult because of the large shift of the retinal image. It is well established that saccades have detrimental effects on the visual input stream, e.g., [27–29]. These effects have been termed saccadic suppression of vision or visibility and have been largely examined separate from saccadic suppression of displacement, as they have been linked more closely to another facet of visual stability across saccades (namely why we do not perceive motion blur, that is, the entire scene sweeping across the retina, when we make saccades). Nevertheless, vision is of course also impoverished during situations of saccadic suppression of displacement. Thus, saccadic suppression of displacement may primarily reflect poor vision during saccades. To assess how much of saccadic suppression of displacement can be explained by poor vision and how much may be specific to the challenge of spatial updating, a fixation control condition with similarly impoverished vision of the target jump may be helpful.

Indeed, recent experiments have reported difficulties in discriminating small jumps of a peripheral target in a situation without saccades, when the jump was obscured by a full-screen pattern mask [30,31]. The effects were of similar magnitude as the saccadic effect. Further, prolonging the preview duration of the target before the displacement led to similar improvements with masks and saccades [31] and an improvement with blanking was also found in both conditions [30]. When it comes to masking, the general take on observed performance decrements is not that they primarily reflect a bias, but that they reflect the reduced visibility of the target [32]. Similarly, rather than highlighting a stability bias, saccadic suppression of displacement may primarily reflect the saccade's masking effect on the visual input stream [30]. Specifically, the masking of transient onset or motion signals (usually perceived when the target jumps) may play a crucial role. In this framework, saccadic suppression of displacement reflects poor sensitivity due to missing clues, not a bias to assume stability. That is, position signals alone are coarse to begin with, even without saccades, and especially when the time available for their encoding is short [31]. Consequently, when having to rely on position signals alone in saccade or mask conditions, displacement perception is poor [33]. In turn, the improvement with blanking may not reflect a relaxed bias, but could be due to better visual evidence for a displacement, for instance, by providing a clear onset transient when the target reappears [6,25,31], or by providing more time for the pre-jump position to be encoded [31].

1.6. Do Saccades and Masks Affect Displacement Perception in Similar Ways?

In accordance with many saccadic suppression of displacement studies, displacement direction discrimination tasks were used in previous masking experiments [30,31]. Participants had to decide on each trial, whether the target had jumped forward (i.e., further out into the periphery) or backward (i.e., towards fixation). Responses are hard to interpret in these tasks: If participants gave a false response, did they truly see a jump opposite to the actual displacement, or did they just guess because they did not see a displacement at all? Could it be that a higher decision threshold is applied during saccades, whereas poor performance in masking conditions rather reflect more noise in the signal compared to the control condition without mask? In the current experiments, I therefore asked participants to estimate the size of the displacement by indicating the target's starting position through a mouse click. Similar continuous measures have already been used to measure suppression of displacement, but only in saccade conditions [16,20,26]. Continuous report techniques may provide more information about the processes underlying the responses. For instance, in the literature on visual working memory, continuous report techniques have successfully provided tools to quantify the probability and fidelity of memory representations [34,35]. Similarly, a continuous report technique may be better suited to assess the probability and fidelity of perceiving a displacement in control, saccade and mask conditions.

2. Experiment 1: Displacement Estimates in Control, Saccade and Masking Conditions

Experiment 1 compares response distributions across control, saccade and mask conditions. The procedure in Experiment 1 is illustrated in Figure 1A. Details on the methods can be found after the General Discussion. Demo movies of the control (Video S1) and mask (Video S2) conditions can be found in Supplementary Online Material.

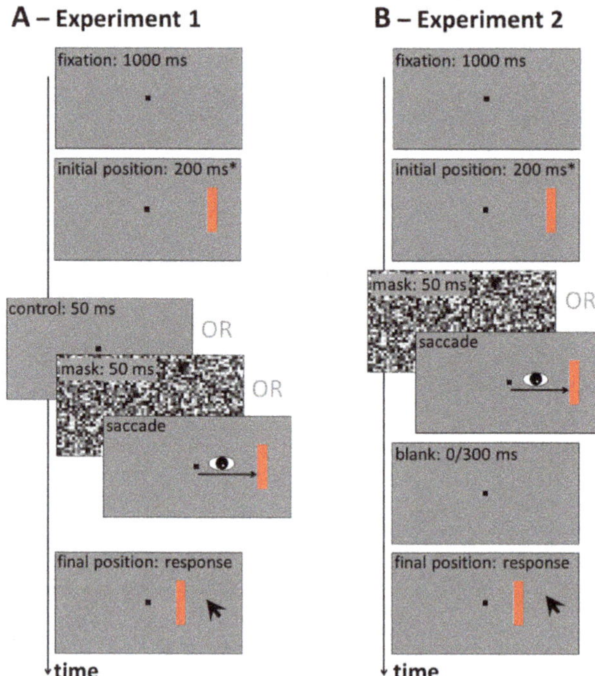

Figure 1. Sequence of events in Experiment 1 (**A**) and Experiment 2 (**B**). Participants fixated the central black square throughout a trial, or made a saccadic eye movement towards the red target bar. The target was displaced either after 200 ms, * or as soon as the saccade was detected. At the end of each trial, participants indicated the perceived displacement size by clicking on the target's initial (before displacement) position on the screen.

2.1. Direction Discrimination: Suppression of Displacement (Recoded Data)

First, to check whether changing the response mode from direction discrimination to mouse pointing gives results that are comparable to previous suppression of displacement experiments, data were first recoded into binary responses, that is, into reported 'forward' and 'backward' displacements. Figure 2 illustrates the percentage of trials recoded as a reported 'forward' displacement for the two jump directions ('backward' or 'forward' as indicated by negative or positive numbers, respectively) and four jump sizes (0.5, 1, 2, 3 deg) in the three conditions (control, mask, saccade). The slope of the data pattern can be taken as an estimate of performance: The steeper the curve, the better participants' discrimination of jump direction. To summarize performance, logistic functions were fit to the data of each individual observer in the three conditions, using a maximum-likelihood method in Matlab (see https://osf.io/atr6y/). Average slopes of these functions are illustrated in the bar graph of Figure 2. A one-way ANOVA on these slope values revealed a significant main effect, $F(2,54) = 30.56$, $p < 0.001$, partial $\eta^2 = 0.531$, and subsequent paired-samples t-tests and Bayesian t-tests (default Bayesian t-tests as offered in JASP [36,37]) confirmed significantly shallower slopes in the mask and saccade compared to the control condition, $ts(27) > 5.52$, $ps < 0.001$, $BF_{10} > 2762.75$. Thus, by recoding the data, the classic suppression of displacement effect emerged: We see impaired performance in direction discrimination in the saccade, but also the masking condition, compared to control. The slopes in the mask and saccade condition also showed a significant difference, $t(27) = 2.67$, $p = 0.013$, $BF_{10} = 3.78$, indicating a slightly shallower slope in the mask compared to the saccade conditions. However, this difference was much smaller compared to the large differences to the control condition.

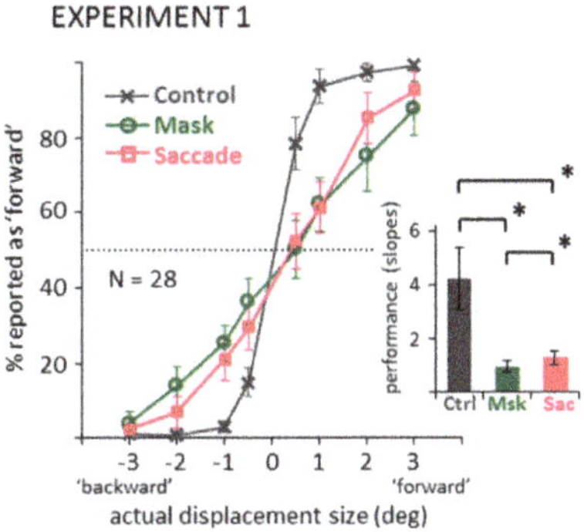

Figure 2. Data from the control, mask and saccade conditions of Experiment 1, recoded into binary displacement direction responses. Line graph shows actual displacements (negative: 'backward' vs. positive: 'forward') against the percentage of reported 'forward' responses (dotted line: chance level). Bar graph shows the average slopes of the logistic functions fitted to each participant's individual data. All error bars: between-subjects 95% confidence intervals. Brackets marked with * illustrate significant differences in post-hoc paired-samples t-tests ($p < 0.05$).

2.2. Distributions of Displacement Estimates

Next, I visualized the distributions of the continuous mouse click responses. Figure A1 in Appendix A shows the distributions separate for each displacement size. The essential aspect, however, can also be appreciated in Figure 3 which shows the response distributions separated only by backward (Figure 3A) and forward (Figure 3B) displacements. For the density functions, a kernel smoothing technique based on a Gaussian kernel function (ksdensity in Matlab) was applied to the combined data of all twenty-eight participants (kernel bandwidth: 0.15 deg). The shaded areas around each line represent bootstrapped 95% confidence intervals (based on 1000 samples with replacement). Distributions in the control condition (grey lines) are unimodal and responses around zero (i.e., misses) or false responses (i.e., on the "wrong" side of the vertical line marking zero displacement) are rare. In contrast, the curves in the saccade (magenta) and mask (green) conditions are clearly bimodal. There is always one peak around zero, subtending approximately ±0.5 deg, with flanks falling off symmetrically on both sides. The remaining part of the distributions fall under the one from the control condition. The two peaks are separated by troughs, which suggests that the peak around zero represents the proportion of trials on which participants completely missed the displacement, whereas the second peak represents the proportion of trials on which a displacement was perceived correctly as a backward or forward displacement. (Figure A1 in Appendix A shows that the location of the non-zero peak roughly follows the actual displacement size.) The graphs suggest that suppression of displacement (Figure 2) for the saccade and mask conditions may to a large part be explained by higher proportions of missed displacements compared to the control condition, as evident in the marked peaks around zero. (Figure A1 in Appendix A also shows that the size of the peaks around zero scale with displacement size: smaller displacements are more often missed. However, missed displacements did not only occur for the smallest displacements.)

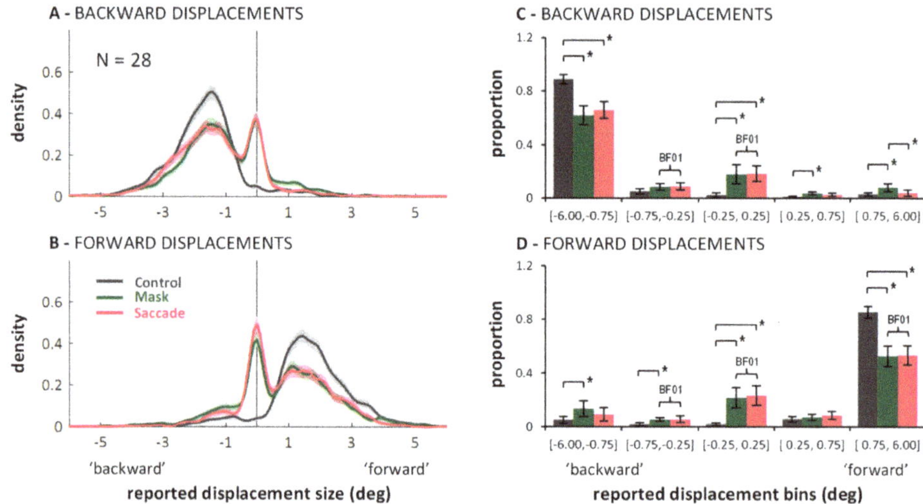

Figure 3. Distribution of responses in the three conditions (color-coded) of Experiment 1: Kernel density functions showing data combined across participants (panels (**A,B**); shadings represent bootstrapped 95% confidence intervals), and proportion of responses falling into the specified bins showing data averaged across participants (panels (**C,D**); error bars represent between-subjects 95% confidence intervals). Brackets marked with * illustrate significant differences in post-hoc within-subjects paired t-tests, $p < 0.0017$. Curly braces marked with BF01 illustrate Bayes factors in favor of H0, $BF_{01} > 3.00$. Data are shown separately for backward (upper row) and forward (lower row) displacements.

To see whether the observed pattern is robust across participants, I split each participant's responses in the three conditions into five bins, defined post-hoc on the basis of the kernel density function: The three central bins have the same size: $[-0.75, -0.25]$, $[-0.25, 0.25]$, and $[0.25, 0.75]$, representing missed displacements and the two adjacent bins where responses are ambiguous; the two outer bins are larger $[-6.00, -0.75]$ and $[0.75, 6.00]$, representing the trials on which participants clearly indicated a displacement, either in the correct direction, or in the wrong direction. The data (illustrated in Figure 3C,D) was entered into a 2 (displacement direction: backward or forward) × 3 (condition: control, saccade, mask) × 5 (bins) repeated measures ANOVA. Importantly, the three-way interaction was significant, $F(8,216) = 58.08$, $p < 0.001$, partial $\eta^2 = 0.683$. To examine how the different conditions changed the distribution of responses across bins, post-hoc paired t-tests were conducted, with p-level corrected to $0.05/30 = 0.0017$. Significant differences on this level are highlighted in Figure 3C,D with brackets and an asterisk. These reveal that in saccade and mask conditions, the proportion of correctly reported displacements (i.e., >0.75 deg and in the right direction) diminished compared to control, whereas the proportion of missed displacements in the range $[-0.25, 0.25]$ increased significantly. While these comparisons cannot attest that distributions were bimodal for all participants, importantly, they do affirm that the peak around zero as found in the kernel density functions does not result from only a small subset of participants. No differences across conditions were found in the bin in-between, corresponding to the trough in the kernel density function (i.e., [−0.75, −0.25] for backward, and [0.25, 0.75] for forward displacements). Additionally, masking, but not saccades, led to higher proportions of displacement reports in the wrong direction. Figure 3C,D also highlights which comparisons resulted in Bayes factors in favor of H0 larger than $BF_{01} > 3.00$ (again: default Bayesian *t*-tests as offered in JASP [36,37]). Those include the comparison between saccade and mask conditions for the central [−0.25, 0.25] bin for both forward and backward displacements.

To summarize, saccades and masks modulated the response distributions in quite similar ways. Both primarily seemed to increase the proportion of missed displacements. When taken in isolation, the bimodality in the distributions in the saccade condition can be easily explained by accounts that emphasize a stability bias across saccades [6,7,11–13,16]. A similar distribution was observed in the masking condition, however. Thus, the bias does not seem to be contingent on the saccadic situation.

3. Experiment 2: Estimates of Displacement Size in Blanking Conditions with Saccades or Mask

If saccades result in a larger proportion of missed trials, how does a blank introduced during a saccade produce a performance increase, as has been reported in many previous studies [6,11–15,30]? One prominent explanation for the blanking effect in saccades posits that the absence of the target upon saccade landing signals a break in object continuity e.g., [6,11,13,16]. As the visual system has already one clear indication that the object has undergone a change, the strong bias towards spatial stability is relaxed. Consequently, the proportion of reported misses should decrease. Given that the mask condition produced a similar pattern of responses in Experiment 1, the same reduction of misses may also be observed with masking. Further, one may wonder whether blanking has only positive effects on displacement perception. It has been argued that a bias towards stability across saccades could optimize perception by correctly discarding small mismatches between prediction and actual post-saccadic position. In particular, small deviations between the extraretinal signals and the actual saccade should result in the perception of small illusory jumps of the visual scene with almost every saccade we make, as discussed e.g., in [8,12,19]. If the bias is relaxed, then this additional noise should show through more often. In other words, the relaxed decision threshold should also lead to a slightly higher proportion of trials in which participants report the wrong jump direction. Indeed, it has been found that blanking conditions increase the false alarm rate in displacement detection tasks [24]. To examine these issues, Experiment 2 compares blank and no-blank conditions in saccade, but also mask conditions (see Figure 1B). A demo movie of the 300 ms blank with mask condition (Video S3) can be found in Supplementary Online Material.

3.1. Direction Discrimination: Suppression of Displacement (Recoded Data)

Figure 4 illustrates the results when recoding the data into binary responses (backward/forward displacement direction discrimination). In saccade and mask conditions, curves are steeper, that is, discrimination performance increased when introducing a 300 ms blank after the saccade or mask. The slope difference between the 0 and 300 ms blank conditions was larger in the saccade condition. Figure 4A suggests that in the masking condition, facilitation by blanking only occurred for forward, but not for backward displacements. A two-way ANOVA on the corresponding slopes (see bar graphs in Figure 4) confirmed significant main effects of blanking, $F(1,21) = 75.22$, $p < 0.001$, partial $\eta^2 = 0.782$, and condition, $F(1,21) = 20.65$, $p < 0.001$, partial $\eta^2 = 0.496$, and a significant interaction, $F(1,21) = 29.62$, $p < 0.001$, partial $\eta^2 = 0.585$. Although smaller in the masking condition, subsequent paired-samples *t*-tests confirmed a blanking effect, that is, a significantly steeper slope in the blanking condition in both the saccade as well as the mask conditions, $ts(21) > 4.37$, $ps < 0.001$, $BF_{10} > 114.20$. Also, Figure 4 shows data from an additional Experiment 3 in which a fixation condition without mask was tested with a 300 ms blank across the displacement (see Appendix B for the full data set; Video S4 for a demo). Independent-samples *t*-tests comparing this 300 ms blank during control condition (light blue data) to the 300 ms blank during masking condition revealed a significant result, $t(48) = 2.10$, $p = 0.041$, pointing to slightly better performance without masking, although the Bayes factor suggests that evidence in favor of a difference was very weak, $BF_{10} = 1.65$. In contrast, the slopes of the psychometric functions were significantly steeper, that is, performance was better, in the 300 ms blank during saccade than the control condition from the additional experiment, $t(48) = 3.80$, $p < 0.001$, $BF_{10} = 66.78$.

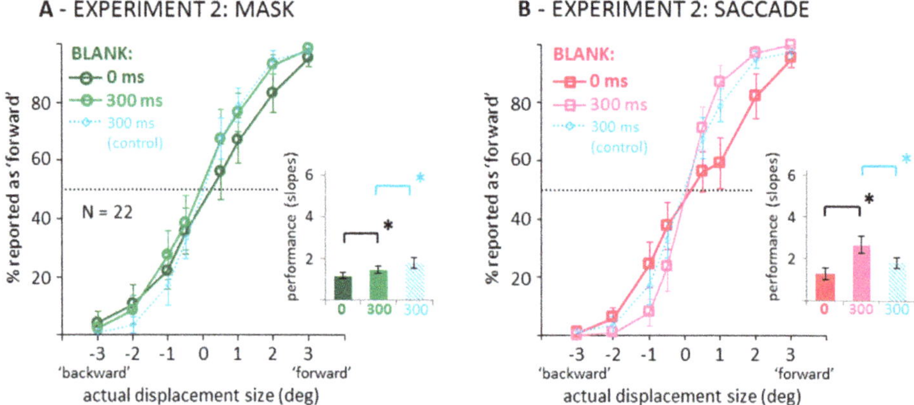

Figure 4. Data from the 0 and 300 ms blank conditions in the mask (**A**) and saccade (**B**) conditions of Experiment 2, recoded into binary displacement direction responses. Conventions as in Figure 2. Thin blue lines and striped blue bars represent data from an additional experiment (see Appendix B) with a 300 ms blank during fixation condition without mask.

3.2. Distributions of Displacement Estimates

Can the poor performance in saccade and mask conditions again be explained by a high proportion of missed displacements? Is the improvement through blanking due to a reduction in missed displacements? And does the benefit of blanking come at the expense of more false direction responses in saccade or mask conditions? Figure 5 (left column) shows the distributions of jump size estimates in each condition with data from all twenty-two participants combined.

In general, the curves show again two peaks: One around zero perceived displacement, representing the proportion of missed displacements, and a second peak, representing the trials on which a displacement was perceived in the correct direction. Comparing data from the no blank (saturated colors) to the blanking conditions (lighter colors), the improvements in direction discrimination performance indeed went along with an altered balance between those two peaks: A lower proportion of missed trials and a higher proportion of perceived displacements. Entering the corresponding proportions (see Figure 5, right column) into a 2 (displacement direction: backward or forward) × 2 (condition: mask or saccade) × 2 (blank: 0 vs. 300 ms) × 5 (bins) repeated-measures ANOVA, a significant four-way interaction emerged: $F(4,84) = 8.62$, $p < 0.001$, partial $\eta^2 = 0.291$. Conducting separate ANOVAs for the mask and saccade conditions still resulted in significant three-way interactions in both cases: $Fs(4,84) > 9.82$, $ps < 0.001$, partial $\eta^2 > 0.319$. Thus, although the modulations in the saccade condition may be larger, there were also significant modulations in the distributions of trials across the bins through blanking in the masking condition. Results of post-hoc paired samples t-tests are illustrated by the brackets in Figure 5 (p-level corrected: $0.05/20 = 0.0025$). In the saccade condition, the change in balance (more correctly estimated displacements, less missed displacements) resulted in significant differences for both jump directions (see Figure 5G,H; black brackets). Further, the bin in-between correctly judged and missed displacement contained a similar proportion of trials in blank and no blank conditions as indicated by Bayes factors in favor of H0, $BF_{01} > 3.00$ (black curly braces). In the masking condition, significantly more responses in the bin containing the larger jump estimates, paired with significantly fewer responses around zero with blanking were only evident for forward displacements (see Figure 5F; black brackets). Due to small variations in the distributions, Bayes factors in favor of H0, $BF_{01} > 3.00$ (black curly braces), were revealed only for two bins not immediately relevant for the current questions. False direction responses remained overall rare. (They occurred occasionally for small displacements in the masking conditions; see: distributions for ±0.5 deg displacements in Figure A2 in Appendix A: they extend to values up to 2–3 deg to the other side, although the proportions of those false direction estimates do not seem substantially higher with blanking). In the saccade condition, however, distributions drop sharply on the 'wrong' side and blanking did not evoke more false direction responses.

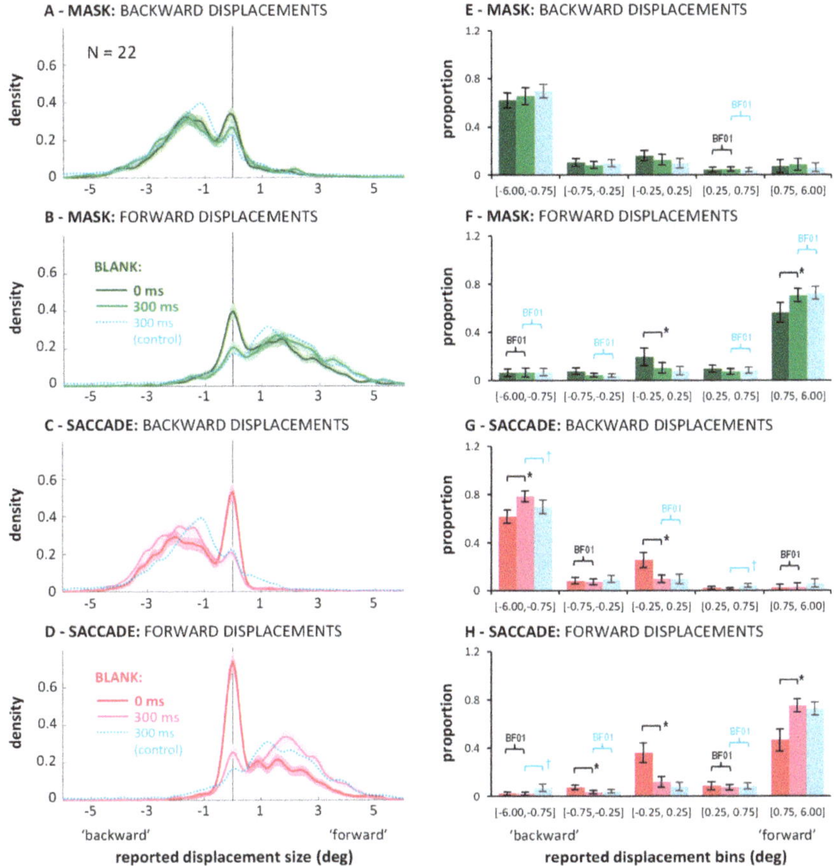

Figure 5. Distribution of responses in the saccade (magenta) and mask (green) conditions of Experiment 2, with (light colors) or without (saturated colors) a 300 ms blank: Kernel density functions (panels (**A–D**), and proportion of responses falling into the specified bins (panels (**E–H**). Conventions as in Figure 3. Thin blue lines and striped blue bars represent data from a 300 ms blank during fixation condition without mask (Appendix B). Brackets marked with * illustrate significant differences in post-hoc within-subjects paired *t*-tests ($p < 0.0025$). Brackets marked with † illustrate significant differences in post-hoc between-subjects *t*-tests ($p < 0.05$). Curly braces marked with BF01 illustrate Bayes factors in favor of H0, $BF_{01} > 3.00$.

The faint blue lines in Figure 5 and the striped blue bars represent data from the 300 ms blank during fixation condition from the additional experiment without mask (Appendix B). In the masking condition, distributions are similar to the blank distributions with masking. When entering the proportions of the two 300 ms blank conditions into a 2 (displacement direction: backward or forward) × 5 (bins) × 2 (experiment: with or without masking) mixed-measures ANOVA, neither the main effect nor any interaction with the between-subject factor experiment reached significance, $Fs < 1.65$, $ps > 0.205$. Thus, one may say that for forward displacements, blanking after the mask improved performance roughly up to the level without masking. This is further highlighted by the fact that most bins show Bayes factors in favor of H0, $BF_{01} > 3.00$, for forward displacements (see blue curly braces in Figure 5F; the central bin is at $BF_{01} = 2.57$). In contrast, comparing the 300 ms blank during fixation condition to the blank with saccade condition, the corresponding ANOVA did reveal a significant three-way interaction, $F(4,192) = 5.74$, $p < 0.001$, partial $\eta^2 = 0.107$. The small effect size and the post-hoc t-tests indicated that differences were subtle: None of the t-tests comparing proportions across experiments reached significance for any of the bins when correcting for multiple comparisons (p-level corrected: $0.05/20 = 0.0025$). Figure 5G,H (see blue brackets illustrating significant tests at $p < 0.05$) suggest the three-way interaction may be partly explained by less false direction responses in the saccade compared to the blanking during fixation condition (see also Figure A2). Bayes factors in favor of H0, $BF_{01} > 3.00$ (blue curly braces), occurred for the central bin for backward displacements and the two adjacent, ambiguous bins for forward displacements.

In sum, blanking during saccades improved jump estimates and this improvement did not come at the expense of false direction responses driven by noise: If the displacement was not missed (which happened less often), its direction was picked up correctly. In fact, blanking after saccades led to even better performance than blanking during fixation. This effect has already been reported by Deubel and colleagues [6], but has received little attention in the literature, but see [16]. The authors argued that the saccade brings the target close to the fovea where position signals are more accurate. The advantage seems to be greater than the error in the extraretinal signal. In fixation conditions, however, the comparison between two peripheral signals may introduce more spatial error. To my knowledge, the current data is the first to replicate the finding. Note, however, that the lack of replication for this saccade advantage up to date has a very simple reason: most (if not all) subsequent saccadic suppression of displacement studies examining the blanking effect did not include a fixation control condition [11–15,23–25,38].

Improvements with blanking in the masking conditions were smaller and restricted to forward displacements. Performance never exceeded the level from a 300 ms blank condition without mask (Appendix B), which is not surprising if one assumes that performance in the fixation condition reflects the precision and accuracy limits of peripheral vision. Also, on a small proportion of trials, the direction of small displacements was misjudged with masking, just as in fixation conditions without mask. Thus, it seems that just as Deubel and colleagues [6] proposed, comparing two peripheral signals is more error-prone than the comparison of a peripheral and a close-to-foveal signal, combined with extraretinal information about the saccade.

4. General Discussion

When participants are asked to judge the 'jump' direction of a visual target, performance is poor when the displacement occurs simultaneously with a saccadic eye movement [2–7], an effect known as saccadic suppression of displacement. Prominent theories of saccadic suppression of displacement postulate that the poor performance is to a large part due to a default assumption or bias for stability across saccades [6,7,11–13,16]. However, not surprisingly, poor performance in displacement perception can also be observed without saccades, for instance, when the jump is obscured by a large pattern mask [30,31]. In the current experiments, I additionally examined continuous judgments of the displacement's size and direction. Results suggest that saccade- as well as mask-induced suppression of displacement is primarily due to a larger proportion of trials on which participants report to have entirely missed the displacement (compared to a fixation control condition; Experiment 1). When introducing a 300 ms blank period straddling the jump, displacement detection was improved and the proportion of missed displacements reduced in the context of saccades and masks (Experiment 2). Complementing previous masking experiments [30,31], the current findings caution against the assumption of a saccade-specific stability bias.

4.1. A General Bias Revealed under Conditions of Impoverished Vision

The distributions of responses in the current experiments (and in particular the bimodality) support the idea of a stability bias or prior. Most simply put, small jumps are often missed across saccades and mask, but almost never in the control condition. In the saccadic context, this fits well with the idea that small discrepancies between prediction and incoming post-saccadic signal are discarded, in favor of a null hypothesis of stability [6,7,11–13]. It may also be in line with the more gradual influence of the bias and 'contraction' of jump estimates observed by Niemeier and colleagues [16,20]. It is important to note, however, that the masking conditions showed similar response distributions. This finding suggests that the bias for stability is not saccade-specific.

Estimating the size of a displacement across saccades is challenged by two things: On the one hand, the visual system must integrate extraretinal signals or at least establish object correspondence across a large shift of the retinal image. This challenge is saccade-specific. On the other hand, visual perception is suppressed during saccades, as has been demonstrated in many previous studies, e.g., [27–29]. This challenge is not specific to situations of self-motion. Studies of saccadic suppression of displacement rarely made attempts to estimate the influence of those two challenges on performance, but see, e.g., [16,25,30,31]. Moreover, given that the context in which suppression of displacement is discussed is usually restricted to eye movements [6,7,11–16,19,20,22–26], one could have been tempted to conclude that the proposed mechanisms are saccade-specific. For instance, one may have assumed that the visual system adjusts the decision threshold for displacement perception across saccades to compensate for imprecisions introduced through the spatial updating mechanism. In other words, more evidence would be required across saccades for the perceptual null hypothesis to be rejected. Such a saccade-contingent adjustment provides indeed an intuitive account during eye movements. However, it is a less obvious explanation in the masking case when no spatial updating occurs. Alternatively, the current masking results suggest that it is not the decision threshold that is raised through saccades and masks, but that the visual evidence for a displacement is poorer. Thus, with smaller displacements, there is a higher probability that the amount of evidence remains below a general decision threshold, applied by the visual system across many situations and not specific to saccades.

In the control condition of the current experiments, participants almost never missed displacements. It seems as if evidence for a displacement in the control condition is abundant, almost never below the threshold. (In line with this idea, small displacements were frequently overestimated: see Figure A1 in Appendix A.) How can saccades or masks so drastically impoverish this evidence? Note that the target itself is highly visible, see [30] for a visibility control experiment. Then again, not the target, but its displacement needs to be judged. It has been argued that saccades and masks both reduce the offset and onset transients usually perceived when the target jumps [6,25,30]. These transient signals alone might already facilitate the judgment of the displacement. Moreover, without strong transients, the impression of apparent motion that accompanies the jump should also be greatly reduced [6,25,30,33]. Perceived motion is a powerful cue, as it also conveys direction (which is all that is needed in a direction-discrimination task). Further, numerous illusions attest the influence of motion or apparent motion on position estimates [39–41]. In the control conditions, those signals are clearly seen and can reinforce the perception of a displacement. When degraded through saccades, masks or other circumstances, a crucial movement cue is lacking and thus displacements may be missed or underestimated. Indeed, Gysen and colleagues [22] have found that the detection of trans-saccadic jumps is better for a moving than a stationary object, demonstrating that violations along a smooth, predictable motion trajectory can be detected more easily than simple jumps. It is important to note that many authors have acknowledged that displacement estimates during fixation and saccades could rely on different sources of information: Whereas motion information is suppressed during saccades, apparent motion, as well as position information, is available during fixation [6,16,25,33,42]. For instance, saccadic suppression of displacement in the model by Niemeier and colleagues [16] is not due to a saccade-specific setting for the prior, but occurs due to a combination of suppressed motion signals (i.e., sole reliance on position information) and the Bayesian prior for jumps: The poorer the sensory information (e.g., due to motion suppression) the greater the influence of the Bayesian prior. Nevertheless, although a role for the quality of sensory information has been acknowledged, the idea of a stability bias has received much more attention in the saccadic suppression of displacement literature.

4.2. Improvement with Blanking: Relaxed Bias or Better Evidence?

Crucial for the prominent idea of a (variable) stability bias has been the improvement with blanking across saccades ([6] as replicated in Experiment 2). This counterintuitive finding has been taken as evidence that extraretinal signals seem to be more precise than initially assumed, but apparently not used in the non-blank conditions because of the default assumption of stability [6,7,11–13]. In blanking conditions, however, the stability bias is relaxed because the visual system has evidence that object continuity has been broken. Thus, it is argued, a concurrent change in position becomes, a priori, more likely as well, see also [16]. While the current findings do not speak against this account, an alternative could be that the blank does not change the bias, but actually improves the quality of the position signals. The blank makes it possible to perceive the onset transient upon reappearance of the displaced target in saccade as well as mask conditions. Again, this has not been overlooked in stability bias accounts. Quite to the contrary, the visible transient is thought to be one important source of evidence for a break in object continuity responsible for relaxing the bias [6,11–13,16]. Alternatively, the transient signal could actually improve displacement sensitivity. One possibility is that the transient enhances the precision of position signals, for instance, by engaging the magnocellular pathway [25]. In line with this assumption, it has been shown that reducing the onset's strength by diminishing the target's contrast to the background reduces or even abolishes the positive effect of blanking during saccades [25]. Another possibility is that the blank interval puts the post-saccadic or post-mask stimulus outside a critical temporal window for accumulation or integration of position or motion information [30,31,42,43]. Zimmermann and colleagues [31] argue that precise encoding of the pre-jump target location takes some time. The blank interval may provide that time. Further, they demonstrated that saccade- as well as mask-induced suppression of displacement is reduced when the pre-saccadic target is shown for a longer period of time before it is displaced. Others hypothesized that saccades, as well as masks, may lead to temporary distortions in the incoming position signal of the post-saccadic or post-mask target [44]. If the re-appearance of the target after the saccade or mask is delayed, though, the processes responsible for the distortions have already ceased and thus do not hamper displacement perception anymore. Additionally, blanking may also put pre- and post-jump target outside a critical temporal interval for perceiving apparent motion. This could explain why blanking during fixation (without mask) is detrimental to displacement perception ([6]; see also the additional Experiment 3 in Appendix B). Further, blanking has been found to be detrimental to transsaccadic displacement perception with moving objects [23], confirming that blanking interferes with the integration of motion information.

Taken together, during fixation without any event or blank (control) we are most sensitive to displacements as the offset and onset transients of the target and apparent motion are clearly perceived (see Supplementary Video S1). Blanking during fixation probably reduces the motion signal somewhat (see Supplementary Video S4), but the strong offset and onset transients are preserved. Without blank, saccades and masks degrade the transients as well as the motion signal (see Supplementary Video S2). Blanking during saccades and masks makes at least the onset signal available and delays the post-jump target, thus improving performance in saccade and mask conditions (see Supplementary Video S3). To explain any of these modulations, assuming a relaxation in the stability bias is not necessary. Thus, the 'bias' may, in the end, reflect a general, fixed perceptual or decision threshold for displacements.

4.3. Saccade-Specific Effects?

One may argue that the current emphasis on the saccade context in the literature, although maybe slightly misleading, is not really problematic. After all, if saccadic suppression of displacement reveals a general stability bias that operates over a wide range of situations, it is, of course, legitimate to examine it exemplarily in the saccadic context. However, as mentioned before, perception across saccades does face the additional challenge of spatial updating. By not trying to disentangle effects due to spatial updating from the results of poor vision, we may actually miss the opportunity to measure truly saccade-specific effects.

In the current experiments, although masks had overall similar effects to saccades, only in the saccade condition did blanking improve performance up to a level that was actually better than in a blanking control condition (Experiment 2). Deubel and colleagues [6] had already observed that performance in saccadic blanking conditions can be better than in blanking conditions during fixation. They speculated that bringing the stimulus from the periphery into the fovea by means of the saccade improves the spatial representation of the target after displacement. Combined with information about the saccade amplitude, the judgment of the jump size could thus be more precise than when the target remains entirely in the periphery. This may also explain why blanking effects were weaker with masking, where stimuli remain in the periphery. Quite strikingly, there were only very few direction errors in the saccade conditions. This could suggest that some weak directional or motion signal survives across the saccade. Indeed, it has been shown that apparent motion, although weakened, can be perceived across saccades [45–47]. In the masking conditions, motion signals may be suppressed more thoroughly, which may also be the reason why masking produced more false direction responses.

Further, it has now been repeatedly observed that not only blanking, but also a change in target features (e.g., polarity, shape or orientation) across the saccade results in weaker saccadic suppression of displacement [11,13,14,30]. The principal idea is again that the feature change lowers the decision threshold for displacement detection or attenuates a bias for stability. Because a change in the target is noticed across the saccade, concurrent differences in position signals are trusted. In principle, establishing object continuity or correspondence is a problem that arises in many situations, not only across saccades, e.g., [48,49]. The question is whether the visual system is more strongly biased to affirm spatial stability when object correspondence is established in the saccadic context. Here, I have argued against such saccade-specific settings. However, we have previously observed that an orientation change, although beneficial across saccades, did not improve performance across a mask [30] (see also Supplementary Experiment 4 for another potential dissociation). Future experiments may be needed to clarify whether the feature change effect is due to a saccade-specific bias setting, or, for instance, a difference in masking efficiency between saccades and our random-luminance mask.

Finally, even though saccadic suppression of displacement as such may not reveal much about how our own movements are integrated into visual experiences, this, of course, is not to say that those processes are not in effect in the paradigm. Quite to the contrary, if updating did not take place, we would not be able to estimate the displacement with such high precision across saccades (see Experiment 2; [6,17,18]). Interestingly, some studies have combined the suppression of displacement paradigm with manipulations that should affect the signals conveying motor information about the saccadic eye movement more directly. Of note, no matter whether the motor feedback (mostly, the corollary discharge was targeted) was altered through saccadic adaptation [17,50], transcranial magnetic stimulation over the frontal eye fields [51], a selective lesion of the thalamus [18], or inactivation of the assumed feedback route through muscimol [52], the resulting data pattern was always better characterized by a uniform shift in the psychometric function. It is thus maybe not so much a question of the experimental paradigm, but of the measure to be taken as an index for the processes involved in spatial stability and spatial updating across eye movements.

5. Materials and Methods

Data available on OSF: https://osf.io/z4pcm/.

5.1. Experiment 1

5.1.1. Participants

Twenty-eight first-year psychology students (eight men) between 17 and 28 years of age completed Experiment 1 in a single one-hour session. They reported normal or corrected-to-normal vision and received course credit for their participation. For all experiments in this study, observers gave written informed consent prior to participating and the procedures followed the principles laid down in the Code of Ethics of the World Medical Association (Declaration of Helsinki) and were approved by the Ethics committee of the *Faculté de Psychologie et des Sciences de l'Education* of the University of Geneva (Project submitted: PZ00P1_161224, entitled "Space without motion—judging object locations and relative distance in the absence of movement"; date of acceptance: February 19, 2016).

5.1.2. Apparatus

Experiments were programmed in Matlab (The MathWorks Inc., Natick, MA, USA) using the Psychophysics and Eyelink Toolbox extensions [53,54]. Stimuli were displayed on a 21" CRT monitor (NEC MultiSync FE2111SB) running at 85 Hz with a resolution of 1280 × 1024 pixels. Eye movements were recorded using an EyeLink1000 desk-mounted eye tracker (SR Research Ltd., Ottawa, ON, Canada) at a sampling rate of 1000 Hz. Participants were seated in a dimly lit room. Viewing was binocular but only the right eye was monitored. The participant's head was stabilized by a chin and a forehead rest at 45 cm from the monitor.

5.1.3. Stimuli, Design and Procedure

The procedure in Experiment 1 is illustrated in Figure 1A. Stimuli were displayed on a gray background (16 cd/m^2). Each trial started with the presentation of a black fixation square (subtending 0.5 deg) at the screen center. After 1000 ms, the red vertical target bar (9 cd/m^2, 0.6 × 3 deg) appeared either left or right from fixation on the horizontal meridian. Target eccentricity from fixation was on average 10 deg, but a random jitter was added on each trial, such that eccentricity ranged between 8 and 12 deg. In the fixation blocks, the target remained at its initial position for 200 ms and then was either followed by a 50 ms blank screen (control condition) or a 50 ms mask (full-screen pattern mask, made up of gray squares of 0.5 deg side length and randomized luminance). Then, the target bar was displaced (it 'jumped') randomly either leftward or rightward (possible displacement sizes: 0.5, 1, 2 or 3 deg). Participants were required to maintain gaze on the central fixation square until the displacement had occurred. Then, they were free to move their eyes. In the saccade blocks, participants were instructed to make an eye movement towards the red target bar upon its appearance. It was displaced as soon as a saccade was detected. In both fixation and saccade blocks, the target bar in its final position was shown as a reference until participants gave a response: They were asked to report the direction and size of the target jump by indicating the displacement's starting point with respect to its endpoint through a mouse click on the respective location on the screen. A feedback message was displayed on the response screen if a fixation or saccade error was registered during the trial (see Section 5.3 below for criteria). In total, participants completed three blocks of 160 trials: They always started with a fixation block (including randomly interleaved control and masking trials), then ran one saccade block and finished with a second fixation block. Trials with the target presented to the left or right from fixation were collapsed and left/right displacements recoded into 'forward' (jump further out into the periphery), and 'backward' (jump back towards the initial fixation location at the center of the screen). For each combination of condition (control, saccade, mask), displacement size and direction, 20 data points were collected from each participant.

5.2. Experiment 2

Another twenty-two first-year psychology students (six men) between 18 and 45 years of age completed Experiment 2 in a single one-hour session. Apparatus, stimuli, design and procedure were similar to Experiment 1, but there was no control condition without saccades or masks. Instead, saccade and masking trials without blank were compared to conditions in which a 300 ms blank followed the mask or saccade (see Figure 1B). Participants completed four blocks of 144 trials: Two blocks in the saccade, and two blocks in the masking condition in random order. Within these blocks, blank and no blank trials were randomly interleaved across trials, resulting in 18 data points per combination of condition (mask vs. saccade), blanking (no blank vs. 300 ms blank) and displacement size and direction.

5.3. Saccade Detection and Trial Exclusion Criteria for Both Experiments

In saccade conditions, the displacement was initialized as soon as the horizontal gaze coordinate taken online deviated by more than 1.5 deg from an initial sample taken at target onset of each trial. After the experiment, fixation periods, saccade onsets, offsets and blinks were extracted offline from the data of all conditions using the Eyelink parser with a velocity criterion of $30°/s$ and an acceleration criterion of $8000°/s^2$ for saccades. Trials in fixation conditions (control or mask) were discarded if a saccade was detected or the horizontal gaze coordinate drifted more than 1.5 deg away from screen center in a time window from 100 ms before target onset until 100 ms after the jump (breaks of fixation: 10.9% of trials in Experiment 1, 11.1% of trials in Experiment 2), or when a blink was detected during the same time window (3.1% of trials in Experiment 1, 2.0% in Experiment 2). Taken together, 12.0% of trials were discarded as errors from the fixation blocks (control or mask) in Experiment 1, and 12.1% of trials were discarded from the fixation (mask) blocks in Experiment 2. In saccade conditions, trials were excluded if gaze at the time of saccade onset deviated more than 1.5 deg from the screen center (breaks of fixation: 5.7% in Experiment 1, 5.4% in Experiment 2), if a saccade was detected less than 100 ms after target onset (anticipations: 3.7% in Experiment 1, 5.6% in Experiment 2), if the saccade was made into the wrong direction (0.8% in Experiment 1, 1.0% in Experiment 2), if the saccade amplitude was smaller than 5 deg (2.2% in Experiment 1, 3.2% in Experiment 2), if no saccade was detected within 500 ms of target onset (11.5% in Experiment 1, 7.8% in Experiment 2), or when a blink occurred (3.3% in Experiment 1, 2.2% in Experiment 2). Taken together, 20.8% of trials were discarded from the saccade blocks in Experiment 1, and 19.6% of trials from the saccade blocks in Experiment 2. Saccades were executed with an average latency of 236 ms in Experiment 1, and 219 ms in Experiment 2. The displacement occurred on average 28 ms after saccade onset and 22 ms before saccade offset in Experiment 1. In Experiment 2, the displacement occurred on average 29 ms after saccade onset and 22 ms before saccade offset. Average saccade amplitude was 9.21 deg in Experiment 1 and 9.62 deg in Experiment 2.

Supplementary Materials: The following are available online at http://www.mdpi.com/2411-5150/3/4/49/s1, Video S1: Control, Video S2: Mask, Video S3: Blank, Mask, Video S4: Blank, Control, Supplementary File S5: Experiment 4.

Funding: This work was funded by the Swiss National Science Foundation PZ00P1_161224 granted to S. Born.

Acknowledgments: I would like to thank Damien Jordan for collecting part of the data.

Conflicts of Interest: The author declares no conflict of interest.

Appendix A. Distributions across Different Jump Sizes

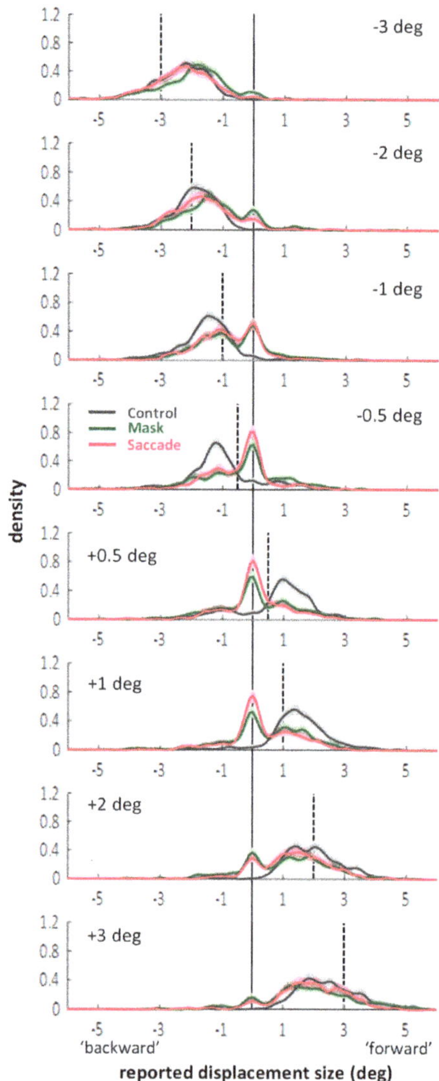

Figure A1. Distribution of responses (kernel density functions) in the three conditions of Experiment 1 (color-coded). Rows represent the different displacement sizes tested. Data are collapsed across all 28 participants. Shadings represent bootstrapped 95% confidence intervals. Vertical black continuous line: zero line. Vertical black dashed lines: actual displacement for each graph. Compared to actual displacements, the larger ±2 and ±3 deg displacements were underestimated in all conditions (distributions shifted towards zero compared to the dashed lines). The smaller ±0.5 and ±1 deg displacements, however, were overestimated in the control condition. Missed displacements, i.e., peaks around zero emerged for all displacements in the saccade and mask conditions and became more frequent with smaller displacements. All in all, distributions in mask and saccade conditions resembled each other, while distributions in the control conditions were markedly different for smaller displacements as virtually no misses occurred.

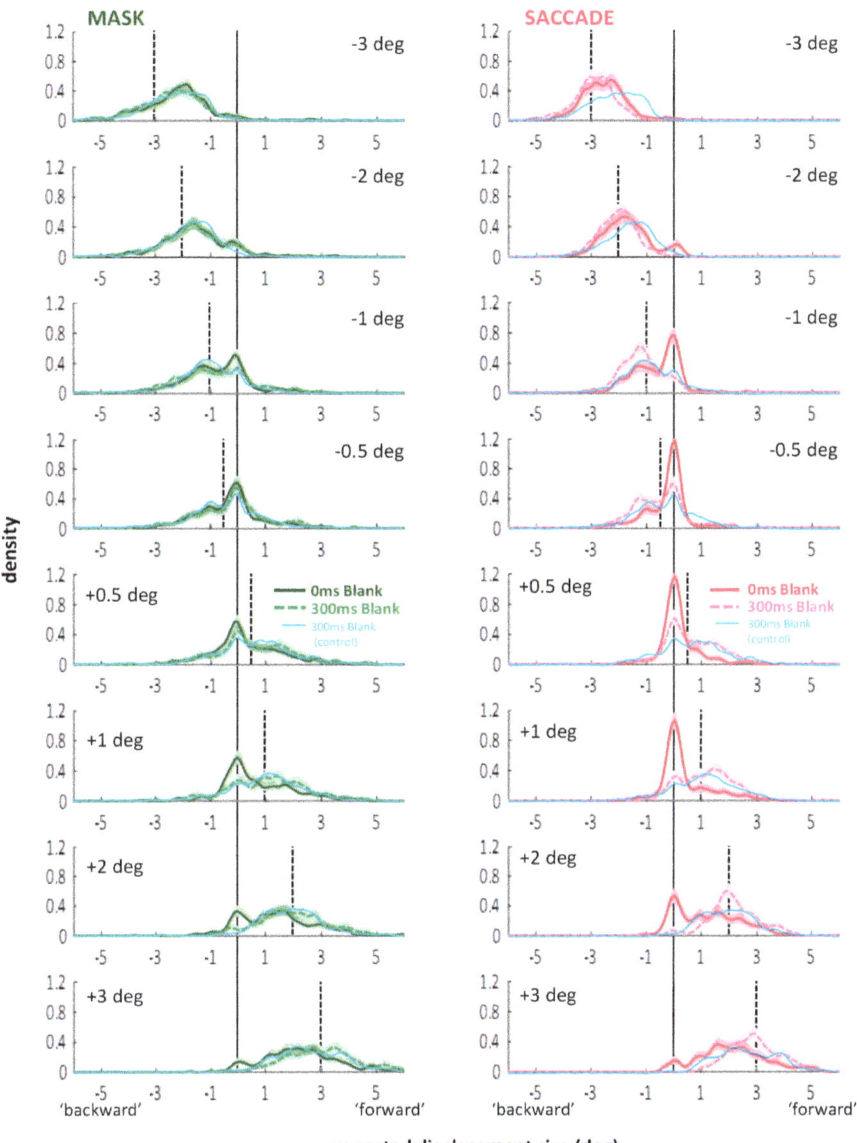

Figure A2. Distribution of responses (kernel density functions) in the mask (left) and saccade (right) conditions of Experiment 2. Continuous lines: no blank conditions, dashed lines: blank conditions. Thin blue lines represent data from the 300 ms blank condition during fixation from Experiment 3. Improvements with blanking (i.e., less misses) in the masking conditions were most visible in the +1 and +2 deg displacement conditions (with smaller modulations in the −1 and +0.5 deg conditions). Distributions overlap almost perfectly with the ones from the 300 ms blank conditions of Experiment 3. In the saccade conditions, less misses compared to the no blank condition occurred for all displacements (except for the −3 deg displacements where no misses occurred). Differences between the 300 ms blank conditions (Experiment 2 vs. Experiment 3) were subtle, but showed, for instance, in larger estimates in the saccade condition and an absence of false direction responses (see −0.5 deg: some displacements were judged to be around +1 deg in the blank during fixation condition of Experiment 3).

Appendix B. Experiment 3—Displacement Estimates in Blanking Conditions without Saccades or Mask (Control for Experiment 2)

Blanking effects have mostly been studied in saccade conditions as they produce a somewhat surprising increase in discrimination performance. Less studied has been the finding that in fixation conditions, performance decreases when a blank is introduced between the offset of the target at its initial position and its onset at the final position [6]. Demo movies of the 0 ms blank (Video S1) and the 300 ms blank (Video S4) conditions can be found in Supplementary Material online.

Appendix B.1. Methods

Another twenty-eight psychology students (sixteen women) between 18 and 45 years of age completed Experiment 3 in a single one-hour session. Apparatus, stimuli, design and procedure were similar to the control condition of Experiment 1. Instead of comparing control, mask and saccade conditions, I compared three different blanking intervals (0, 100, and 300 ms) between the target's offset at its initial position and its onset at the final position. Participants completed three blocks of 168 trials. The blanking conditions were randomly interleaved across trials in all three blocks. As in the fixation blocks of Experiment 1, participants were required to maintain gaze on the central fixation square during stimulus presentation and this was monitored by means of the eyetracker. For each combination of blanking condition (0, 100, 300 ms), displacement size and direction, 21 data points were collected from each participant.

Appendix B.2. Results and Discussion

Appendix B.2.1. Direction Discrimination: Suppression of Displacement (Recoded Data)

Again, data was first recoded into binary responses and the percentage of reported 'forward' jumps for each displacement direction and size was calculated. Figure A3 illustrates the results and shows average slopes of the logistic functions fitted to the data of each individual observer. Slopes were less steep with longer blanking intervals. A one-way ANOVA confirmed a significant main effect, $F(2,54) = 20.03$, $p < 0.001$, partial $\eta^2 = 0.426$ and subsequent paired-samples t-tests revealed significant differences between all conditions, $ts(27) > 3.19$, $ps < 0.004$, $BF_{10} > 11.08$. Thus, we see significantly worse performance with increasing blank interval in the fixation control condition. Numerically, the performance decrease was, however, much smaller compared to the saccade and mask conditions in Experiment 1, which is in line with previous results [6]. Figure A4 additionally shows the distributions of responses separate for each displacement size.

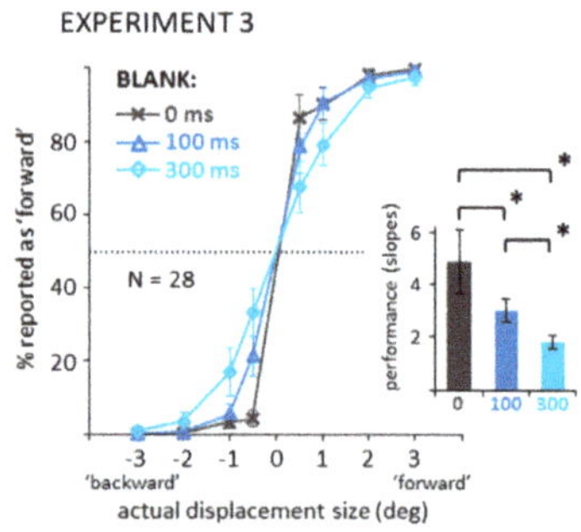

Figure A3. Data recoded into binary displacement direction responses. A: Data from the 0, 100, and 300 ms blank during fixation conditions of Experiment 3. Line graph shows actual displacements (negative: 'backward' vs. positive: 'forward') against the percentage of reported 'forward' responses (dotted line: chance level). Bar graph shows average slopes of the logistic functions fitted to each participant's individual data. All error bars: between-subjects 95% confidence intervals. Brackets marked with * illustrate significant differences in post-hoc paired-samples t-tests ($p < 0.05$).

Appendix B.2.2. Distributions of Displacement Estimates

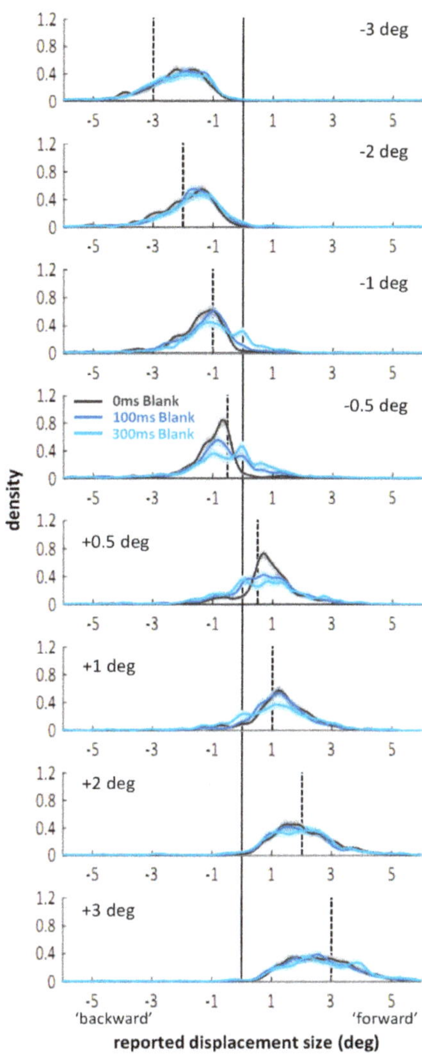

Figure A4. Distribution of responses (kernel density functions) in the three conditions of Experiment 3 (color-coded). Rows represent the different displacement sizes tested. Data are collapsed across all 28 participants. Shadings represent bootstrapped 95% confidence intervals. Vertical black continuous line: zero line. Vertical black dashed lines: actual displacement for each graph. Overall, distributions in the 0 ms blank control condition resembled the ones from Experiment 1 (except that overestimations of smaller displacements seemed somewhat smaller; as for Experiment 1, almost no misses occurred, though). In the 100 and 300 ms blank conditions, some of the smaller ±0.5 and ±1 deg displacements were missed. In the 300 ms blank conditions, this is visible in small peaks around zero, while the 100 ms blank conditions rather showed a spread of the distribution towards zero, compared to the no blank condition.

References

1. Simons, D.J.; Rensink, R.A. Change blindness: Past, present, and future. *Trends Cogn. Sci.* **2005**, *9*, 16–20. [CrossRef]
2. Wallach, H.; Lewis, C. The effect of abnormal displacement of the retinal image during eye movements. *Percept. Psychophys.* **1966**, *1*, 25–29. [CrossRef]
3. Mack, A. An investigation of the relationship between eye and retinal image movement in the perception of movement. *Percept. Psychophys.* **1970**, *8*, 291–298. [CrossRef]
4. Bridgeman, B.; Hendry, D.; Stark, L. Failure to detect displacement of the visual world during saccadic eye movements. *Vis. Res.* **1975**, *15*, 719–722. [CrossRef]
5. Li, W.X.; Matin, L. The influence of saccade length on the saccadic suppression of displacement detection. *Percept. Psychophys.* **1990**, *48*, 453–458. [CrossRef] [PubMed]
6. Deubel, H.; Schneider, W.X.; Bridgeman, B. Postsaccadic target blanking prevents saccadic suppression of image displacement. *Vis. Res.* **1996**, *36*, 985–996. [CrossRef]
7. McConkie, G.W.; Currie, C.B. Visual stability across saccades while viewing complex pictures. *J. Exp. Psychol. Hum. Percept. Perform.* **1996**, *22*, 563–581. [CrossRef] [PubMed]
8. Bridgeman, B.; Van der Heijden, A.H.C.; Velichkovsky, B.M. A theory of visual stability across saccadic eye movements. *Behav. Brain Sci.* **1994**, *17*, 247–292. [CrossRef]
9. Grüsser, O.J. Early concepts on efference copy and reafference. *Behav. Brain Sci.* **1994**, *17*, 262–265. [CrossRef]
10. Li, W.X.; Matin, L. Saccadic suppression of displacement: Influence of postsaccadic exposure duration and of saccadic stimulus elimination. *Vis. Res.* **1990**, *30*, 945–955. [CrossRef]
11. Tas, A.C.; Moore, C.M.; Hollingworth, A. An object-mediated updating account of insensitivity to transsaccadic change. *J. Vis.* **2012**, *12*, 18. [CrossRef] [PubMed]
12. Wexler, M.; Collins, T. Orthogonal steps relieve saccadic suppression. *J. Vis.* **2014**, *14*, 13. [CrossRef] [PubMed]
13. Demeyer, M.; De Graef, P.; Wagemans, J.; Verfaillie, K. Object form discontinuity facilitates displacement discrimination across saccades. *J. Vis.* **2010**, *10*, 17. [CrossRef] [PubMed]
14. Poth, C.H.; Herwig, A.; Schneider, W.X. Breaking object correspondence across saccadic eye movements deteriorates object recognition. *Front. Syst. Neurosci.* **2015**, *9*, 176. [CrossRef] [PubMed]
15. Irwin, D.E.; Robinson, M.M. How post-saccadic target blanking affects the detection of stimulus displacements across saccades. *Vis. Res.* **2018**, *142*, 11–19. [CrossRef] [PubMed]
16. Niemeier, M.; Crawford, J.D.; Tweed, D.B. Optimal transsaccadic integration explains distorted spatial perception. *Nature* **2003**, *422*, 76–80. [CrossRef]
17. Collins, T.; Rolfs, M.; Deubel, H.; Cavanagh, P. Post-saccadic location judgments reveal remapping of saccade targets to non-foveal locations. *J. Vis.* **2009**, *9*, 29. [CrossRef] [PubMed]
18. Ostendorf, F.; Liebermann, D.; Ploner, C.J. Human thalamus contributes to perceptual stability across eye movements. *Proc. Natl. Acad. Sci. USA* **2010**, *107*, 1229–1234. [CrossRef]
19. Currie, C.B.; McConkie, G.W.; Carlson-Radvansky, L.A.; Irwin, D.E. The role of the saccade target object in the perception of a visually stable world. *Percept. Psychophys.* **2000**, *62*, 673–683. [CrossRef]
20. Niemeier, M.; Crawford, J.D.; Tweed, D.B. Optimal inference explains dimension-specific contractions of spatial perception. *Exp. Brain Res.* **2007**, *179*, 313–323. [CrossRef]
21. MacKay, D.M. Visual stability and voluntary eye movements. In *Central Processing of Visual Information A: Integrative Functions and Comparative Data*; Jung, R., Ed.; Springer: Berlin/Heidelberg, Germany, 1973; pp. 307–331. [CrossRef]
22. Gysen, V.; De Graef, P.; Verfaillie, K. Detection of intrasaccadic displacements and depth rotations of moving objects. *Vis. Res.* **2002**, *42*, 379–391. [CrossRef]
23. Gysen, V.; Verfaillie, K.; De Graef, P. The effect of stimulus blanking on the detection of intrasaccadic displacements of translating objects. *Vis. Res.* **2002**, *42*, 2021–2030. [CrossRef]
24. Irwin, D.E.; Robinson, M.M. Detection of stimulus displacements across saccades is capacity-limited and biased in favor of the saccade target. *Front. Syst. Neurosci.* **2015**, *9*, 161. [CrossRef] [PubMed]
25. Matsumiya, K.; Sato, M.; Shioiri, S. Contrast dependence of saccadic blanking and landmark effects. *Vis. Res.* **2016**, *129*, 1–12. [CrossRef] [PubMed]

26. Atsma, J.; Maij, F.; Koppen, M.; Irwin, D.E.; Medendorp, W.P. Causal inference for spatial constancy across saccades. *PLoS Comput. Biol.* **2016**, *12*, e1004766. [CrossRef] [PubMed]

27. Breitmeyer, B.G.; Ganz, L. Implications of sustained and transient channels for theories of visual pattern masking, saccadic suppression, and information processing. *Psychol. Rev.* **1976**, *83*, 1–36. [CrossRef]

28. Burr, D.C.; Morrone, M.C.; Ross, J. Selective suppression of the magnocellular visual pathway during saccadic eye movements. *Nature* **1994**, *371*, 511–513. [CrossRef]

29. Matin, E. Saccadic suppression: A review and an analysis. *Psychol. Bull.* **1974**, *81*, 899–917. [CrossRef]

30. Zimmermann, E.; Born, S.; Fink, G.R.; Cavanagh, P. Masking produces compression of space and time in the absence of eye movements. *J. Neurophysiol.* **2014**, *112*, 3066–3076. [CrossRef]

31. Zimmermann, E.; Morrone, M.C.; Burr, D.C. Spatial position information accumulates steadily over time. *J. Neurosci.* **2013**, *33*, 18396–18401. [CrossRef]

32. Breitmeyer, B.G.; Öğmen, H. *Visual Masking: Time Slices through Conscious and Unconscious Vision*, 2nd ed.; Oxford University Press: Oxford, UK, 2006; ISBN 9780198530671.

33. Shioiri, S.; Cavanagh, P. Saccadic suppression of low-level motion. *Vis. Res.* **1989**, *29*, 915–928. [CrossRef]

34. Bays, P.M.; Catalao, R.F.; Husain, M. The precision of visual working memory is set by allocation of a shared resource. *J. Vis.* **2009**, *9*, 7–11. [CrossRef] [PubMed]

35. Zhang, W.; Luck, S.J. Discrete fixed-resolution representations in visual working memory. *Nature* **2008**, *453*, 233–235. [CrossRef] [PubMed]

36. Rouder, J.N.; Speckman, P.L.; Sun, D.; Morey, R.D.; Iverson, G. Bayesian *t* tests for accepting and rejecting the null hypothesis. *Psychon. Bull. Rev.* **2009**, *16*, 225–237. [CrossRef] [PubMed]

37. JASP (Version 0.8.2). Computer Software. Available online: https://jasp-stats.org/ (accessed on 11 September 2019).

38. Weiss, K.; Schneider, W.X.; Herwig, A. A "blanking effect" for surface features: Transsaccadic spatial-frequency discrimination is improved by postsaccadic blanking. *Atten. Percept. Psychophys.* **2015**, *77*, 1500–1506. [CrossRef] [PubMed]

39. Whitney, D. The influence of visual motion on perceived position. *Trends Cogn. Sci.* **2002**, *6*, 211–216. [CrossRef]

40. Li, H.H.; Shim, W.M.; Cavanagh, P. Backward position shift in apparent motion. *J. Vis.* **2014**, *14*, 16. [CrossRef] [PubMed]

41. Kerzel, D.; Gegenfurtner, K.R. Spatial distortions and processing latencies in the onset repulsion and frohlich effects. *Vis. Res.* **2004**, *44*, 577–590. [CrossRef] [PubMed]

42. Bergelt, J.; Hamker, F.H. Suppression of displacement detection in the presence and absence of eye movements: A neuro-computational perspective. *Biol. Cybern.* **2016**, *110*, 81–89. [CrossRef]

43. Cicchini, G.M.; Binda, P.; Burr, D.C.; Morrone, M.C. Transient spatiotopic integration across saccadic eye movements mediates visual stability. *J. Neurophysiol.* **2013**, *109*, 1117–1125. [CrossRef]

44. Ziesche, A.; Bergelt, J.; Deubel, H.; Hamker, F.H. Pre- and post-saccadic stimulus timing in saccadic suppression of displacement—A computational model. *Vis. Res.* **2017**, *138*, 1–11. [CrossRef] [PubMed]

45. Szinte, M.; Cavanagh, P. Spatiotopic apparent motion reveals local variations in space constancy. *J. Vis.* **2011**, *11*, 4. [CrossRef] [PubMed]

46. Fracasso, A.; Caramazza, A.; Melcher, D. Continuous perception of motion and shape across saccadic eye movements. *J. Vis.* **2010**, *10*, 14. [CrossRef] [PubMed]

47. Rock, I.; Ebenholtz, S. Stroboscopic movement based on change of phenomenal rather than retinal location. *Am. J. Psychol.* **1962**, *75*, 193–207. [CrossRef] [PubMed]

48. Anstis, S. The perception of apparent movement. *Philos. Trans. R. Soc. Lond. B Biol. Sci.* **1980**, *290*, 153–168. [CrossRef] [PubMed]

49. Enns, J.T.; Lleras, A.; Moore, C.M. Object updating: A force for perceptual continuity and scene stability in human vision. In *Space and Time in Perception and Action*; Nijhawan, R., Khurana, B., Eds.; Cambridge University Press: Cambridge, UK, 2010; pp. 503–520. [CrossRef]

50. Klingenhoefer, S.; Bremmer, F. Saccadic suppression of displacement in face of saccade adaptation. *Vis. Res.* **2011**, *51*, 881–889. [CrossRef]

51. Ostendorf, F.; Kilias, J.; Ploner, C.J. Theta-burst stimulation over human frontal cortex distorts perceptual stability across eye movements. *Cereb. Cortex* **2012**, *22*, 800–810. [CrossRef]

52. Cavanaugh, J.; Berman, R.A.; Joiner, W.M.; Wurtz, R.H. Saccadic corollary discharge underlies stable visual perception. *J. Neurosci.* **2016**, *36*, 31–42. [CrossRef]
53. Kleiner, M.; Brainard, D.; Pelli, D. What's new in psychtoolbox-3? *Perception* **2007**, *36*, 14.
54. Cornelissen, F.W.; Peters, E.M.; Palmer, J. The eyelink toolbox: Eye tracking with matlab and the psychophysics toolbox. *Behav. Res. Methods Instrum. Comput.* **2002**, *34*, 613–617. [CrossRef]

Review

What Can Eye Movements Tell Us about Subtle Cognitive Processing Differences in Autism?

Philippa L Howard [1], Li Zhang [2] and Valerie Benson [3,*]

[1]   Department of Psychology, Bournemouth University, Bournemouth BH12 5BB, UK;
      plhoward@bournemouth.ac.uk
[2]   Academy of Psychology and Behaviour, Tianjin Normal University, Tianjin 300074, China; zzli1992@126.com
[3]   School of Psychology, University of Central Lancashire, Preston PR1 2HE, UK
*   Correspondence: VBenson3@uclan.ac.uk; Tel.: +44-0-1772-892-021

Received: 27 March 2019; Accepted: 18 May 2019; Published: 24 May 2019

Abstract: Autism spectrum disorder (ASD) is neurodevelopmental condition principally characterised by impairments in social interaction and communication, and repetitive behaviours and interests. This article reviews the eye movement studies designed to investigate the underlying sampling or processing differences that might account for the principal characteristics of autism. Following a brief summary of a previous review chapter by one of the authors of the current paper, a detailed review of eye movement studies investigating various aspects of processing in autism over the last decade will be presented. The literature will be organised into sections covering different cognitive components, including language and social communication and interaction studies. The aim of the review will be to show how eye movement studies provide a very useful on-line processing measure, allowing us to account for observed differences in behavioural data (accuracy and reaction times). The subtle processing differences that eye movement data reveal in both language and social processing have the potential to impact in the everyday communication domain in autism.

Keywords: autism; eye movements; cognitive processing; social and everyday communication

---

1. What Can Eye Movements Tell Us about Subtle Cognitive Processing Differences in Autism?

Eye tracking is widely used to examine information processing [1] since it is well established that eye movement patterns provide detailed insight into on-going cognitive processing [2]. Autism spectrum disorder (ASD) is a heterogeneous developmental condition characterised by difficulties engaging in everyday social interaction/communication, restricted and repetitive patterns of behaviour, and sensory processing sensitivities [3]. It is widely accepted within the field that these behavioural symptoms are underpinned by information processing differences [4]. Therefore, eye tracking provides an opportunity to examine the nature of on-going cognitive processing in ASD, and to evaluate how any cognitive processing differences might underpin behavioural symptoms in this population.

In 2009 the research into autism that had utilised eye tracking was reviewed and at this point there were approximately 60 articles published in the field [5]. The chapter reviewed how eye-tracking had been used to explore low-level eye-movement characteristics, perception of complex stimuli, and processing of and attention to social information. The review concluded that basic oculomotor control such as smooth pursuit and saccadic programming appeared to be intact in ASD. However, subtle differences in attention allocation were thought to be present for tasks that required higher-level cognitive and social processing. For example, consistent observations reported enhanced local processing during visual search and atypical allocation of attention for social scenes. The heterogeneity of the disorder, changes across development and the effect of general ability and linguistic level were all shown to impact upon findings from studies reviewed in that chapter. For example, visual sampling or scanning in autism was shown to be affected by the complexity of stimuli social content, task complexity, and symptom profile, including age, symptom severity, the presence of language delay, and social competence. It was proposed that future research ought to take account of specific sub groups of the ASD population, and must also employ more naturalistic stimuli and settings, such as the presentation of dynamic information and investigation of processing in one to one social interactions.

Since the review was published, there has been a surge in experiments that have used eye tracking to study ASD. A search in Web of Science for "autism" AND "eye tracking" indicates more than 600 research papers have been published on this topic in the last decade. Many of the suggestions for future research, addressed in the previous review, have been taken on board in these new studies. However, many of the issues raised in relation to inconsistent findings reported in the previous review, are also apparent in the studies reported in the current review.

The current article reviews some of this more recent literature, with a focus upon what has been learnt about the nature of language and social processing in ASD over the past 10 years. The aim of the review will be to evaluate the contribution of the research to the understanding as to how cognitive processing differences could relate to behavioural symptoms in day-to-day communication in ASD.

Key findings will be presented at the end of each section, and these will outline observed processing differences in ASD for different paradigms or behavioural comparisons. A summary of how the findings from the social and language processing studies will follow each of these separate parts of the review, and each summary will attempt to evaluate how the eye movement patterns has advanced understanding of cognitive processing in ASD. Specifically, we will evaluate whether there are any consistent patterns of eye movements, that reveal subtle processing differences across the language and social processing domains that could account for the well documented characteristics of ASD in everyday communication.

2. Eye-Movement Studies and Language Processing in Autism

Language development and processing has been widely reported to be different for autistic individuals, relative to typically developing (TD) individuals. For example, autistic children may have a delayed onset of language production in comparison to TD peers and may demonstrate differences in pragmatic and higher-level language processing throughout adulthood [6,7]. Such differences have clear potential to contribute to social and communicative difficulties that are characteristic of ASD. Below, we summarise the eye tracking research that has examined the cognitive underpinnings of language processing differences in ASD, and in this section the summaries are presented according to the different paradigms adopted for language investigations in this field.

3. Referential Word Learning Paradigms

Individuals on the autism spectrum are often delayed in the development of oral language and until recently, this was considered a key component of autistic disorder [3]. Research has suggested that atypical social attention and pragmatic understanding in ASD may influence language acquisition. For example, the importance of interlocutor engagement and joint attention, to support the mapping of novel words to unfamiliar objects has been well documented [8,9]. Several researchers have used eye tracking to examine exactly how autistic children allocate attention during language learning contexts. These experiments typically involve a *learning phase* whereby children with ASD are exposed to novel (and familiar) words. During this phase, eye movements are monitored to compare how children with and without ASD attend to the available information. A *test phase* follows the learning phase to examine word learning which is assessed in different ways in different studies e.g., pointing or naming tasks. Word acquisition rates appear to be similar between ASD and TD children in referential learning contexts, but eye movement data has revealed subtle differences in the routes through which individuals with ASD attain this information.

Most predominantly, research has focused upon examining how gaze cues support word learning in young autistic children. Norbury et al. [10] asked children (aged 6–7 years) to view dynamic scenes of an interlocutor and three objects whilst concurrently hearing novel (and familiar) words. Interlocutor gaze was either biased (directed towards the referred object) or neutral (directed towards the camera). An example of the experimental procedure can be seen in Figure 1. Participants were tasked with clicking on the object that matched the spoken form.

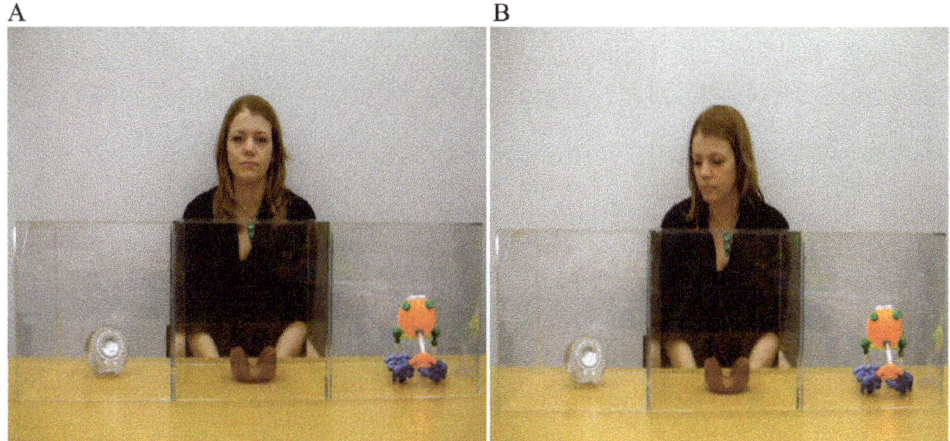

Figure 1. Example of the word learning paradigm used by Norbury et al [10]. Reproduced with permission from Elsevier. Image (**A**) is an example of the gaze neutral condition and image (**B**) is an example of the gaze bias condition.

Both TD and ASD participants used gaze direction to accurately identify target objects. This was demonstrated by faster and more accurate responses when gaze was biased, in comparison to when gaze was neutral. In addition, the interlocutor's face initially captured the attention of both ASD and TD children. However, eye movements revealed qualitative differences in the use and understanding of gaze direction. TD children fixated the face more than ASD children in the biased condition and TD children made more gaze-object contingent looks (fixations upon the target immediately following a fixation upon the interlocutor's face) in the gaze-biased condition, in comparison to the neutral condition. For ASD children, gaze-object contingent looks were not modulated by gaze condition. Furthermore, autistic children demonstrated higher initial recall of phonological information, but reduced recall of semantic features, relative to TD participants. Over time these recall rates remained relatively stable for ASD participants, but TD participants' performance increased for both phonological and semantic recall. The authors suggest that the autistic children's understanding of the social intention of gaze direction may be less well developed than TD peers, and that gaze may instead be used as an associative learning cue. This difference in the use of eye gaze as a social reference cue to support word learning has the potential to interfere with language learning and development. Although initial word learning rates may be similar in children with ASD, the routes to attaining this information may be qualitatively different, with sound potentially being prioritised over semantic information and social cues in ASD.

Since Norbury et al. [10] which was the first study to examine referential word learning using eye tracking, several studies have extended these findings using similar paradigms. Akechi et al. [11] used a word learning paradigm to examine attention during referential word learning in Japanese speaking children with ASD (aged 6–11 years) and Tenenbaum et al. [12] examined word learning in young children who were just starting to produce language (aged 2–5 years). In contrast to Norbury et al. [10], neither Akechi et al. [11] nor Tenenbaum et al. [12] reported group differences in fixations to the interlocutor's face or in saccades made from the face towards the target. Akechi et al. [11] used a schematic image of a face, which may have influenced the children's willingness to direct their attention to this stimulus, and Tenenbaum et al. [12] reported that attention allocation towards the eye and mouth region, whilst the interlocutor verbally labelled objects, was predictive of faster recognition during the test phase for autistic, but not TD participants (when adjusting for age). Akechi et al. [11] reported that both TD and ASD children fixated the target more than the distractor in a 'follow in' condition (when the schematic face followed the child's gaze to fixate the same object), but that only TD children fixated the target more than a distractor in a discrepant condition (when the schematic face was fixating a different object to the child). Fixations upon both the target and distractor were approximately equal for ASD participants in the discrepant condition. This suggested the autistic children may not have identified that the object the face was directing its gaze towards was 'special'.

Both Akechi et al. [11] and Tenenbaum et al. [13] extended these findings to examine whether word learning in ASD could be improved if alternative methods were used to direct attention towards a target object. Akechi et al. [11] made target items more salient via movement and this increased word learning for both ASD and TD groups, with differences in attention allocation between groups for the discrepant conditions disappearing. Tenenbaum et al. [13] demonstrated that word learning can be improved in both TD and ASD groups by explicitly increasing the likelihood of attention allocation to both the interlocutor's mouth and to the target object, by asking the interlocutor to hold the target item near to their face. These results indicate that alternative non-gaze methods to increase target saliency have the potential to support and improve referential word learning, for both autistic and TD children.

In contrast to the studies reported above, Lucas and Norbury [14] did not examine the use of gaze direction, but instead, investigated whether orthographic information supports word learning in children with ASD (aged 7–12 years). Participants heard words and concurrently viewed pictures of these words that either included or excluded the orthographic form (if excluded, the 'word' area of the screen remained blank). During the learning phase, gaze behaviour was very similar between TD and ASD groups with only subtle differences being detected. These subtle differences included increased duration (but not proportion) of fixations upon the target image for TD participants in the orthography present condition, in comparison to ASD participants. In addition, in the orthography absent condition ASD participants made a higher proportion of fixations upon the blank region where a word could be presented, even though there was no word present. This finding was interpreted to suggest that autistic children may have relied more heavily on orthographic form during word learning, and support for this reliance effect was mirrored in the off-line learning data which demonstrated that autistic participants had higher levels of facilitation, than TD participants, for orthography present conditions. In addition, superior phonological coding was found for children with ASD relative to TD children in the initial test phase, supporting Norbury et al.'s [10] work. However, in contrast to Norbury et al.'s [10] work, no differences in semantic learning or consolidation were detected. One factor that could explain the reason for the inconsistencies between these two studies is that, in the Lucas and Norbury [14] study learning was examined following a 24-h break, and this length of time is considerably shorter than the four week break used by Norbury et al. [10]. It is possible that either orthography facilitates consolidation for participants with ASD, or that 24 h is not sufficient to detect group differences in longer-term consolidation.

Key Findings

- Overall, referential word learning rates in children with ASD without additional language impairment appear to be comparable to TD controls;
- Children with ASD may rely more heavily on phonology and orthographic form during word-learning contexts;
- Eye tracking has revealed mixed findings in relation to the use of gaze information to support word learning in ASD. Some studies report subtle differences in the use of gaze cues, whereas others do not;
- Referential word learning may be improved for ASD and TD children if orthography is present and if attention is directed to target objects using extra cues in addition to gaze direction e.g., movement.

4. Intermodal Preferential Looking Paradigm

The intermodal preferential looking paradigm (IPL) has also been used widely to examine language processing in young children. Typically, in this paradigm children freely view two videos as they concurrently hear spoken language [15]. Children that comprehend the spoken language accurately, typically direct their gaze towards the video that reflects the meaning of the concurrently spoken language. The advantage of IPL is that the paradigm provides information as to the incremental processing of spoken language for very young children, without the requirement of additional task demands. However, since gaze direction and latencies are typically coded offline, this paradigm does not provide the level of temporal and spatial precision that can be gained from video-based eye tracking systems. Such precision might be important if there are subtle group processing differences that could account for the failure to develop certain language skills that could foster language communication abilities in ASD.

The IPL paradigm has demonstrated that the time-course of familiar word identification and the noun learning bias are intact in young children with ASD [16,17]. This indicates that early lexical acquisition and processing occur similarly in ASD and TD children. In contrast, Tek et al. [17] found no evidence of the shape learning bias for children with ASD (aged 2–3 years) across four time points over a year, relative to TD children. This bias refers to children's mapping of a novel word to an object's shape, as opposed to other features (e.g., colour) and provides insight into the development of semantic categorisation. This effect was reported to be related to the severity of ASD symptoms which suggests that the categorisation mechanisms adopted to learn new words may be different in autistic relative to TD peers, or that the onset with which autistic children adopt this mechanism is delayed. Note however that no difference in vocabulary size was detected between TD and ASD participants at any time point, indicating that an omission to use a shape bias does not appear to have influenced the speed of language acquisition in ASD. In addition, this finding should be interpreted with caution given that oral language skill was not measured and may have confounded these effects.

Beyond lexical processing, IPL has demonstrated evidence of intact grammatical processing for young children with ASD in the form of comparable subject-verb-object (SVO) structure comprehension [16,18], aspect morphology/tense processing [19], and syntactic bootstrapping [20] *Wh*-questions (e.g., where, who, when, what) are minimally used by children with ASD and provide grammatical and pragmatic challenges. These questions often deviate from SVO structure, involve an understanding that the *wh* word represents information absent from the sentence, and involve the speaker assuming the knowledge of another prior to producing a *wh*-question. IPL has been used to demonstrate that the onset of *wh*-question comprehension precedes production for both TD and ASD children, but that the onset of this stable comprehension is chronologically delayed for autistic children, in comparison to TD counterparts, at approximately 54 months compared to 28 months [21]. Note though that this effect occurs when ASD and TD children have similar levels of language, and, onset of stable comprehension at the individual level appears to be related to early linguistic and pragmatic competence [18]. Therefore, it would seem that the developmental delay in language production, coupled with pragmatic challenges, contribute to differences in language use which has the potential to feed forward and impact upon everyday communication difficulties in ASD.

Key Findings

- Lexical acquisition and processing appear to be similar between TD and ASD participants;
- Autistic children may not adopt a shape learning bias, or they may be delayed in the development of this bias, but this does not appear to impede language acquisition;
- Comprehension of basic syntactic form develops similarly in young children with and without ASD;
- Comprehension and production of *wh*-questions is chronologically delayed for autistic children relative to TD children and is likely a result of delayed language development and pragmatic challenges.

5. Visual World Paradigm

The visual world paradigm [22] requires participants to listen to spoken sentences (e.g., *The boy will eat the cake*) whilst concurrently viewing a visual scene that typically includes four objects (e.g., cake, boy, ball, bike). One of these objects will always match a word contained within the sentence, and that object can be predicted by prior linguistic content, such that participants make anticipatory eye movements towards this object before it has been spoken. Given the established relationship between eye movements and on-going linguistic interpretation [2], the speed with which sentences are processed can be readily observed. Studies employing this paradigm have demonstrated that individuals with ASD incrementally process verb information and make on-line predictions about the constraints of upcoming linguistic input. For example, similar proportions of anticipatory eye movements made towards a target item (e.g., hamster) upon hearing a biased verb (e.g., stroked) in comparison to a neutral verb (e.g., moved) have been reported for English speaking adolescents [23], young children [24], and Mandarin speaking children with ASD [25] in comparison to TD samples.

Hahn, Snedeker, and Rabagliati [26] used the visual world paradigm to examine the on-line processing of ambiguous words in ASD (e.g., star) when embedded in contexts that suggested that either the dominant (star in the sky) or the subordinate (movie star) meaning should be accessed. Participants with and without ASD both demonstrated evidence of using early sentence context to inhibit inappropriate ambiguous word meanings. Specifically, both groups showed evidence of reduced anticipatory eye movements towards an object that reflected the dominant word meaning within the first 500 ms of hearing a word, when the context was biased towards the subordinate meaning. This paradigm has also demonstrated that children with ASD use prosody to disambiguate syntactic ambiguities as efficiently as TD peers [27]. Moreover, Bavin et al. [24] found that autistic children were as effective as TD children in the use of context to detect and override an initial implausible sentence interpretation for sentences that contained ambiguous preposition phrases (e.g., The girl cut the cake with the candle). What should be evident from these studies is that when eye tracking tasks are used to examine incremental processing of language as speech unfolds, autistic children and adolescents do not appear to differ to TD comparison groups in the processing of context to predict, disambiguate, or update interpretations of incoming auditory information. This finding contradicts cognitive theories of ASD that suggest that global contextual processing may be atypical in ASD [28] and indicates that communication difficulties in ASD are unlikely to be related to autistic individuals failing to process linguistic context or compute on-line predictions about up-coming input. Note that these null group effects are reported when there is no requirement for a social response. Social demands may interfere with the efficiency of such processing in everyday communication.

Key Findings

- Visual world experiments demonstrate that children and adolescents with ASD use context and verb information to incrementally predict upcoming linguistic information;
- Visual world experiments demonstrate that children and adolescents with ASD disambiguate lexical and syntactic information at a similar time-course to TD comparison groups.

6. Listening Whilst Looking Paradigm

This paradigm is similar to the visual world paradigm in that it requires participants to look at images whilst hearing spoken language; however, in the listening whilst looking paradigm explicit auditory instructions are given to fixate one of the images. Bavin et al. [29] used this paradigm to examine how ASD symptoms severity influences the efficiency of lexical access. Children (aged 5–7 years) viewed four images including a target (e.g., boy), a phonological competitor (e.g., a box), and two unrelated distractor objects whilst concurrently hearing 'where is the *boy*?' Children with more prominent ASD symptom presentation were less likely and slower to fixate the target, compared to TD children and to ASD children with less prominent symptom presentation. Moreover, when autistic children did fixate the target, they shifted attention away from the target more quickly than TD children. These effects were not modulated by IQ and nor by oral language skill, and the authors suggest that this may indicate that the efficiency of lexical access is influenced by ASD symptom severity. In contrast, Venker, Eernisse, Saffran, and Weismer [30] who also adopted the listening-whilst-looking paradigm, found on-line accuracy and eye movement latencies for familiar words to be highly variable in children with ASD (aged 3–5 years old), following hearing questions such as *"Where's the __? Do you see it?"* but the findings were not associated with autistic symptoms. Instead, on-line accuracy was primarily associated with language competence. Note that the nature of this task involves participants comprehending and responding to *wh*-questions and, as described earlier, these types of questions (with high pragmatic and grammatical demands) are known to be particularly challenging for children with ASD [18,21]. The differences in task demands may explain why autistic symptomology was found to be predictive in Bavin et al. [29], but not in the Venker et al. [30] study, where prompts were not exclusively *wh*-questions.

This paradigm has also been extended, beyond investigating aspects of lexical processing, to examine incremental semantic and syntactic processing when listening to sentences that contain a noun modification (e.g., *"Look at the blue square with the dots"*) whilst simultaneously viewing four items on a screen, including the target object (e.g., a blue square with dots), a competitor object (e.g., a blue square) and two distractor objects [31]. The findings from that study show that both ASD and TD children fixated target and competitor objects more than distractor objects, upon hearing the noun phrase, and both groups fixated the target more than competitor when hearing the modifying information (e.g., with the dots). However, group differences revealed that the ASD group were slower to fixate the target, and that they had a lower proportion of looking time to the target overall, in comparison to TD participants. What the listening whilst looking experiments demonstrate is that whilst individuals with ASD can correctly comprehend auditory information and match this to the visual display, there are subtle differences in the speed of this processing in children with ASD, and, depending upon the pragmatic demands of a task, this efficiency may be related to symptom severity.

Key Findings

- The listening whilst looking paradigm demonstrates that the efficiency of lexical access may be mediated by symptom severity in ASD children;
- ASD children accurately process and comprehend noun modifiers; however, the time-course of this processing appears to be less efficient;
- The pragmatic demands of the listening whilst looking paradigm should be considered when interpreting findings.

7. Reading Paradigms

Reading skill is highly variable in ASD [32] and is determined by normative factors associated with reading (e.g., oral language, word decoding), in addition to ASD-specific higher-level language processing differences [33]. In general, research reports performance outcomes for reading comprehension tasks to be reduced in ASD in comparison to what would be predicted by age, IQ, or decoding skill [34,35]. However, the underpinning processing differences that contribute to comprehension differences in ASD remain unclear. In recent years, there has been an increase in the use of eye-tracking experiments designed to address this question. Typically, eye tracking tasks that examine reading, involve participants silently reading text on a monitor at their natural rate (typically one sentence or one small passage at a time). To detect word-level gaze behavior, reading paradigms are completed in very controlled environments, using head-stabilized tracking systems with high spatial and temporal accuracy, and involve attaining very precise calibrations typically within 0.25–0.50° accuracy. Note that the samples recruited in the studies reported below are predominantly adults that have received diagnoses of Asperger syndrome and, therefore, did not present with early language delay as children. As such the samples reported below may have fewer challenges associated with language than those diagnosed with autistic disorder (note that Asperger syndrome, autistic disorder, and pervasive developmental disorder were replaced in the most recent version of the DSM with a single diagnosis of ASD).

Studies that have examined the time course of low-level linguistic processing during reading have demonstrated comparable text processing between TD and ASD readers. For example, Howard, Liversedge, and Benson [36] demonstrated typical frequency effects (low-frequency words are fixated for longer than high-frequency words) for readers with ASD; Caruana and Brock [37] demonstrated typical predictability effects (fixations upon words that are predictable based upon previous sentence context are fixated for less time that unpredictable words) for TD individuals with high autistic traits; and Davidson and Weismer [38] demonstrated expected subordinate bias effects when processing ambiguous words for ASD participants (which were concurrently presented with auditory stimuli). In addition, Howard et al. [36] demonstrated that syntactic parsing preferences and the speed of recovery from syntactic misanalysis when reading sentences that contained ambiguous prepositional phrases was also comparable between TD and ASD adult readers. Thus, the time course of on-line lexical and syntactic processing during sentence reading appears to be comparable between TD and ASD adult readers.

There is also a body of work that has focused upon using eye-tracking methodology to examine higher-level aspects of reading. Given that these higher-level linguistic processing tasks are where differences in performance outcomes are predominantly reported [33], one would expect there to be differences in the eye movement measures between ASD and TD readers. However, many similarities between ASD and TD readers for these higher-level reading processes have been found when adopting this paradigm. For example, readers with and without ASD have been reported to demonstrate comparable irony processing [39], comparable counterfactual processing [40], comparable and even superior counterfactual emotion processing [41], comparable anomaly detection in real and fantasy worlds [42,43], and comparable co-referential processing [44].

Where differences are reported, these are subtle, and are almost exclusively reported for the reading of texts that require inferential processing. For example, Sansosti, Was, Rawson, and Remaklus [45] asked adolescents with and without ASD to read vignettes that evoked a causal inference. They reported more fixations, longer fixation durations, and more regressions back through the texts for the individuals with ASD in comparison to a TD comparison group. What these differences seem to indicate is that the processing of such text required more effort for ASD readers. However, Sansosti et al.'s [45] analyses were restricted to global eye movement measures which were not target related, but rather, the measures were calculated across the entire vignettes. This makes it more difficult to identify the source of such reading disruption, since we do not know when or where this occurred. However, it would seem likely that the disruption observed in the ASD group could be related to the inferential processing that the texts required.

In a later study, Micai, Joseph, Vulchanova, and Saldaña [46] examined inferential processing more directly by asking participants to read texts that required an inference to be formed, and then analysing eye movements for localised areas of these texts, areas that should have evoked the inference processing. Critically, TD readers demonstrated longer processing times upon portions of text where inferences were formed. The ASD participants in Micai et al.'s [46] experiment had longer gaze durations, in comparison to TD readers upon the critical words that informed the inference. In addition, ASD readers regressed back through the texts to words that supported and further informed this inference on a higher proportion of trials, in comparison to TD readers. Importantly, no differences in comprehension outcomes were detected. This suggests that ASD readers form inferences on-line during reading, but that inferential work specifically may require more effort for ASD readers relative to TD controls even when IQ and language skills are closely matched. Although the mechanistic explanation as to why inferential processing of this kind may be atypical in ASD remains unclear, two studies have provided some insight into why this atypicality may exist. Firstly, in a study where detection of implausibilities could only be successful if these were evaluated against situational world knowledge, it was reported that such detection is delayed in ASD relative to TD readers [43]. The implication from this finding suggests that the processing of world knowledge during reading, which is often necessary for inferences to be formed, may be less immediate in ASD. It is important to note, however, that more recent experiments that use tasks and stimuli which require world knowledge about fictional characters to be inferred, report an absence of group differences in this area [42]. Therefore, the suggestion that inferential processing differences are related to situational world knowledge processing, and are inherent in ASD, requires further investigation. The second study [47] reported that individuals with ASD were quicker to detect sentence-level anomalies (where the anomaly could be detected by reading the sentence in isolation) in comparison to paragraph level anomalies (where the anomaly could only be detected if the global context of the paragraph had been formed), compared to TD readers (see Figure 2 for an example of these stimuli).

```
Panel (a)

After three years of hard work on his degree in Oriental Studies, Scott finally graduated
from university. After his graduation, he received a job offer to work in Japan for a
year. He was really excited about this and he was planning to take the opportunity to
travel around Tokyo, one of the most vibrant cities in East Asia. However, Scott was
worried that his inability to speak Japanese/Chinese¹ would stop² him from communicating
with people. He really wanted to be able to make new friends out there³.

Panel (b)

There was a tourist flight travelling from Vienna to Barcelona. On the last leg of the
journey, it developed engine trouble. Over the Pyrenees, the pilot started to lose
control. The plane eventually crashed right on the border. Wreckage was equally strewn in
France and Spain. The authorities were trying to decide where to bury the dead/survivors¹
from the plane² crash. The families of the passengers were devastated about their losses³.
```

Figure 2. Example of the passage level (**a**) and sentence level (**b**) anomaly stimuli used within Au-Yeung et al.'s [47] experiment. Reproduced with permission from SAGE.

ASD readers were slower to begin to resolve passage-level anomalies, relative to sentence level anomalies, and relative to TD readers, as evidenced by regression path times (time from when a region is first fixated until a reader progresses to fixate information to the right of this region). This interaction is displayed in Figure 3 below and suggests that there may be broader differences in the time-course with which ASD readers integrate sentence meaning within discourse representations. Given that the integration of global information into the discourse model is often necessary for inferential processing, time course differences could potentially contribute to inferential and comprehension difficulties previously reported for autistic readers.

An unexpected yet consistent finding in studies that have used eye tracking to examine reading in adults with ASD is that autistic readers engage in a higher proportion of re-reading, in comparison to TD readers, e.g., [39,43,44,47]. It has been suggested that the re-reading in ASD could reflect a 'cautious' reading style [36]. A lack of any modulation of re-reading by text type (e.g., individual experiment conditions) and the finding that re-reading appears to be present even when ASD readers demonstrate expected first-pass reading effects in the eye movement record, suggests that the re-reading in ASD is unlikely to be a result of linguistic processing differences. However, the exact cause of this behaviour remains to be investigated. Since re-reading occurs for single sentences, short (three line) paragraphs and for longer texts, and since it reflects a propensity to 'go back' and re-read after having read through the text once in entirety, it is not yet clear whether this behaviour is necessary for full comprehension of what has been read, whether it reflects one arm of the diagnostic criteria for ASD (repetitive behaviours), or whether it reflects a strategy for coping with comprehension questions.

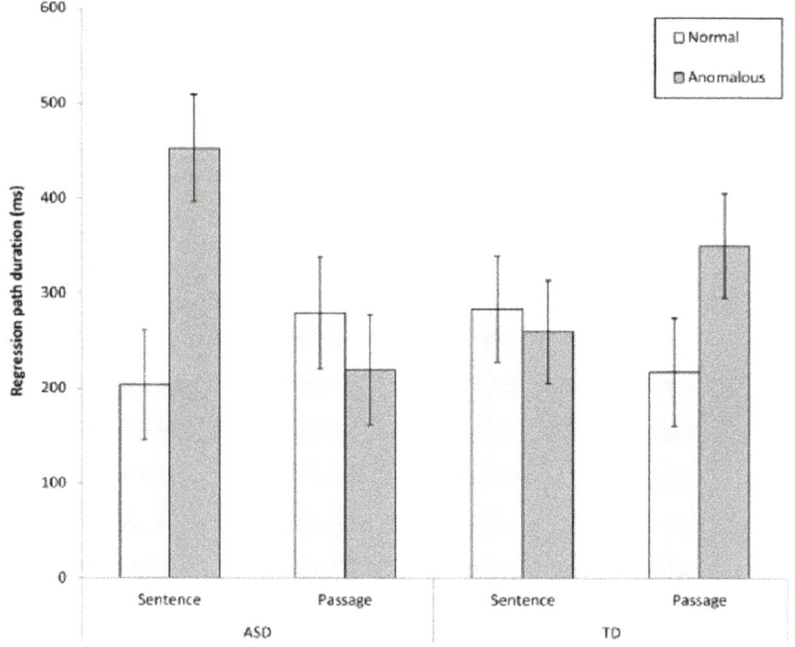

Figure 3. Group and anomaly type interaction for regression path durations reported by Au-Yeung et al. (2017) [47]. Reproduced with permission from SAGE.

Key Findings

- On-line lexical and syntactic processing is comparable during reading in ASD and TD adults;
- When adopting reading paradigms, TD and autistic adults do not differ in a range of higher-level language processes;
- The time-course of on-line inferential processing may require more effort for autistic adults;
- ASD readers tend to re-read texts more than TD readers.

8. Language-Processing Summary

Eye-tracking paradigms provide a valid way to measure incremental language comprehension in ASD with minimal task or response demands. Throughout all paradigms any apparent processing differences *predominantly* reflect quantitative delays, as opposed to qualitative deviances. Temporal processing differences, as revealed by eye movement measures, appear to be present when pragmatic and higher-level linguistic demands increase. Time-course differences in language comprehension at any level have clear potential to impede the fluidity and reciprocal nature of everyday communication, and such differences may be exacerbated or onset by increased pragmatic demands associated with conversational exchanges. Any such delays, even if very subtle, may result in autistic individuals comprehending language input later than TD comparison groups, particularly when rate of delivery cannot be controlled. The repeated reading in ASD may be needed for full text representation or comprehension, but in everyday communication there is no opportunity to 'go back' and resample what has been said. Thus, the eye movement studies investigating language processing over the past 10 years have provided insight into how subtle (temporal) processing differences might impact in everyday communication difficulties in ASD.

9. Eye-Movement Studies and Social Processing in Autism

Difficulties in social interaction and communication are a key characteristic of ASD and atypical social processing has been consistently reported in numerous studies [3,48,49]. Using eye-tracking technology, researchers have monitored and analysed the eye movements of infants, toddlers, children, adolescents and adults with autism, for a diverse range of social stimuli or social contexts, to examine the nature and the time course of any on-line processing differences between TD and ASD individuals. Using a variety of paradigms, numerous eye movement studies have provided ample evidence of how individuals with autism detect, attend to and show understanding of social information, and implicit social cues. The findings from these studies have revealed to some extent the relationship between visual attention to social information and ASD specific behaviour in everyday communication.

10. Joint Attention Paradigms

Joint attention is an important aspect of early social development and eye gaze can be used as a salient cue to help people understand the social world and to predict the actions of others. Several studies have investigated whether autistic individuals can engage in joint attention by monitoring eye movements, and results have shown that there are differences in joint attention and gaze-following behaviour in children on the autism spectrum. Gaze-following is a precursor to joint attention and has been found to be atypical in young children with ASD. As well as investigating eye gaze, some studies have also examined head following in ASD. For example, Vivanti, Trembath, and Dissanayake [50] investigated visual attention responses to head turns, and the findings showed that young children with autism (aged 46 months on average) responded less to turning heads compared to a TD group, with no significant increase in attention to the face and the target in head turning conditions relative to neutral conditions. Thorup et al. [51] showed that in a real interaction, compared to low-risk infants, infants (aged 10 months) who were at risk for autism tended to rely more on head turns, than on isolated eye gaze shifts, to follow another's gaze direction. One potential reason for this result may be that the cueing effect of eye gaze is less salient than a head-turning cue, such that autistic children may find it difficult to follow eye gaze in the presence of a turning head, or, that the development of gaze following is delayed in ASD children relative to TD children. There is also some evidence to suggest that engaging in joint attention can be explicitly improved in children on the autism spectrum. For example, Krstovska-Guerrero and Jones [52] reported that, following a 3–9 week training intervention in eye gaze behaviours, all toddlers with ASD mastered eye gaze following. Moreover, Navab, Gillespie-Lynch, Johnson, Sigman and Hutman [53] found correlations between responsive joint attention (RJA) to eye gaze as measured by eye-tracking and scores on the Early Communication Scales in infants, which provides evidence for the validity of eye tracking to assesses RJA, and also shows that joint attention is related to communication abilities in ASD.

When examined in older children, joint attention behaviours appear to have developed in ASD, and any group differences are more subtle than would be expected based upon the stark differences in early head and gaze following reported above. Swanson and Siller [54] used an adapted attentional cueing paradigm, whereby faces were presented at the central point of the screen with an object presented in peripheral vision. The eye gaze of the central face was shifted either to the direction of the object (congruent) or to the direction opposite from the object (incongruent). Participants without autism looked longer towards the object in the congruent condition, relative to the incongruent condition, as evidenced by increased early (first fixation duration, FFD) and later (total fixation time, TT) stages of processing. The ASD participants showed similar effects for TT, but not FFD. Similarly, Falck-Ytte, Thorup and Bölte [55] revealed a weaker initial processing bias for attended objects in young children with ASD compared to TD children and children with developmental delays. Using more naturalistic social stimuli, it has also been demonstrated that adolescents with ASD show intact global processing of eye gaze, but that the time-course with which gaze was followed is less immediate for adolescents with autism, relative to TD adolescents [56]. Furthermore, Riby et al. [57] found that, when given an explicit instruction, participants with autism showed no tendency to follow the cue and to increase fixations on the location of a gazed-at target. Together, these studies indicate a subtle difference in the initiation of gaze following behaviour in ASD.

The above studies examined the response to joint attention (RJA) cues in children with ASD. However, a recent study also examined the spontaneous initiation of joint attention (IJA) [58]. When toddlers were watching a person shift their gaze to one target, no group differences were found in response to the attentional allocation towards faces and objects (target and non-target), and to the transitions between the face and target. However, in IJA conditions, participants with ASD showed atypical fixation patterns. Toddlers with ASD looked longer to the face and made more transitions from the target/related object to the face but less transitions between the non-target/unrelated object and the face, compared to the TD group. Moreover, higher levels of atypical eye movement transitions in IJA conditions in autism were associated with more severe ASD symptoms. Based on the atypical viewing patterns in the IJA condition exclusively, it seems that JA differences in ASD may be related more predominantly to differences in initiation of JA, as opposed to the understanding of JA initiated by others.

A further early developing social behaviour related to JA is imitation, and, studies have revealed atypical imitation in ASD [59,60]. For example, Vivanti et al. [59] asked participants with autism to view short videos showing a goal-directed action being performed by an actor whose gaze was either directed towards the viewer or averted from the viewer. Results showed that for TD children, direct eye gaze from the person in the video increased participant fixation time to the person, and enhanced the performance of spontaneous social imitation. In contrast, for children with autism, there was no difference in social attention and social imitation accuracy between the direct and averted eye-gaze condition. This finding indicates that reduced imitation in ASD may be associated with reduced attention towards a person's communicative signals. Since imitation has been reported to be reduced in autism for direct eye gaze conditions, but not for averted eye gaze conditions [60], it could be inferred that the use of direct gaze may hinder imitation in ASD. The relationship of atypical visual processing for salient social cues (direct eye gaze or head turning) and poorer imitation performance in autism, as revealed in these two studies [59,60], support the hypothesis that subtle differences in social processing have the potential to impact in social interaction behaviours in everyday communication in autism, and that hypothesis is also supported by many other studies [61–63].

Key Findings

- Infants and young children with ASD, or at risk of ASD may rely more heavily upon head turns, as opposed to eye-gaze shifts as social cues;
- Children and adolescents with ASD engage in joint attention behaviours, but the initial onset of gaze following is delayed, relative to that observed for TD individuals;
- Similar to what is found for TD children, engaging in joint attention behaviours supports social development in ASD;
- Children with ASD show a different pattern of attention when initiating joint attention, in comparison to TD children;
- Direct gaze may not facilitate imitation in ASD and may even reduce spontaneous social imitation behaviour.

11. Free Viewing Paradigms for Faces and Social Scenes

Attentional biases to social information play a fundamental role in shaping typical development of social cognition and behaviour in humans. However, a number of studies which have utilized the free viewing paradigm report both typical and atypical social attention in ASD, relative to TD individuals. In a free-viewing paradigm, participants are presented with a visual stimulus which they look at without instruction, as their eye movements are monitored. Findings from this paradigm have reported that individuals with autism show decreased spontaneous attention to faces, and to people [55,57,62,64–74], and there is a reduced preference to look at the eyes or mouth regions when attending to faces [57,74–81], regardless of age of autistic individuals. Several studies have also revealed that autistic individuals show an attentional preference for less-salient social elements (e.g., bodies) and objects (e.g., backgrounds) in scenes [75,82]. This attenuated social attention in autism is mainly evidenced by two categories of eye movement measures, which are shorter viewing time (or the percentage of viewing time) and fewer fixations on social items. These two eye-movement measures are thought to reflect general processing of the social world, and the findings from the above studies seem to suggest that there is a reduction in general attention towards socially relevant information in ASD. However, and in contrast to those findings, other studies have reported evidence for preserved social orienting in autism. For example, in some studies individuals with autism have been reported to show the same probability as TD controls to execute their first saccade (rapid eye movement from one location to another) to social stimuli [83], and they have been reported in other studies to take the same time as TD viewers to initially direct and move their eyes to the social stimuli [84,85]. Furthermore, once the social stimulus has been fixated, the ASD group then shows intact social engagement (fixation duration), which is equivalent to counterpart TD controls [86–92]. Dicriscio et al., [93] adopted the anti-saccade paradigm [94], where the task is to look to the opposite direction of a peripherally presented target, to examine attentional control in autism for social (happy faces) and non-social stimuli (e.g., cars, shoes). This study found no differences in saccadic inhibition for social stimuli, as indicated by similar error rates (eye movements directed towards a social stimulus instead of away from it) in both groups, providing further evidence that social information can capture initial attention in ASD.

Several other studies report evidence for subtle differences in spontaneous social orientation in autism, with no atypicality reported for overall attentional processing of social cues [95,96]. For example, Freeth, Ropar, Mitchell, Chapman and Loher [95] found that although adolescents with autism took longer to first fixate on the person presented in social scenes, compared to age and IQ (full IQ and verbal IQ)-matched TD adolescents, the total fixation time measure indicated that both groups attended to social information similarly overall. Using the preferential looking paradigm, Guillon et al. [96] examined attentional biases to face-like objects paired with inverted face-like objects, which are usually not perceived as face-like. The results showed that young children with autism were less likely to direct their eyes initially to the upright face-like objects relative to an age matched TD group, but overall they spent a similar amount of time fixating on the upright face-like objects as the TD group. This result is consistent with the finding of Freeth et al. [95] for social scenes stimuli and does not suggest there is any overall general social viewing deficit in individuals with autism. The one subtle group difference that appears to exist from these studies is a reduced social prioritization in ASD participants, in relation to where their first saccade is directed.

Several studies have tried to provide detailed analyses of eye movement patterns for dynamic social information, by adopting analyses approaches which are different from the classic region of interest (ROI) analysis method (as reported for the above studies). These alternative analyses examine both the temporal and the spatial eye movement patterns simultaneously, across the whole duration and display of dynamic social interaction videos in ASD [97–99]. These studies consistently report that people with autism do not follow social events in dynamic interactions as efficiently as their TD peers. When attending to video clips showing several people conversing with each other, or when people in video clips speak to the audience in turn, participants within the TD group showed similar temporal–spatial gaze patterns, indicating that the TD participants tend to look at the same place in a specific moment in time. However, ASD participants did not show this viewing pattern. Further frame-to-frame analyses examined the potential causes of this atypicality and found that individuals with autism fixated less on people, accompanied by increased attention to non-social stimuli in the videos. Critically, participants with ASD tended to shift their attention from the speaking person ahead of the TD group [97,98]. This viewing pattern is observed in both ASD children and in ASD adults and demonstrates that ASD viewers are not processing the information in the same way as their counterpart TD peers. It is not known from these studies whether the ASD participants are similar to each other in the way they allocate attention during these tasks, but this would be worthy of investigation. Another finding has shown differences in how TD participants sample the information in these tasks. It appears that young TD children prefer to monitor the mouth rather than the eyes, while the TD adults showed a reversed pattern. This finding might potentially help to explain the lack of difference in fixation on eyes or mouth regions between both the TD and ASD participants. In Falck-Ytter et al.'s study [99], participants were presented with videos of semi-naturalistic social interactions between two young children. The results showed that, compared to the TD group, children with autism had decreased preference to look toward to the person who was likely to guide next interaction.

These findings indicate that children with ASD may be inattentive to social content when attending to dynamic social interactions. This result is consistent with the findings from studies showing inadequate attention to salient social stimuli in complex social scenes or dynamic videos [57,73]. The studies employing dynamic stimuli illustrate how eye movements are able to reveal subtle but potentially significant foundations of social-processing differences in autism, and, failures to follow or act upon available social cues in dynamic conversation may result in failures in reciprocity in every day communication in ASD.

Key Findings

- Autistic individuals may have decreased spontaneous attention to social information in free viewing paradigms;
- Autistic individuals may have a reduced preference to process eye and mouth regions of faces, relative to TD individuals;
- Autistic individuals show atypical attention to, and processing of, social information in dynamic interactions, relative to TD individuals.

12. Circumscribed Interest and Geometric Pattern Paradigms

An increased interest in circumscribed interests (CI, e.g., trains, electronics) or geometric patterns processing has been offered as an alternative account to illustrate social-processing characteristics in autism [100–106]. Results from these studies suggest that the presence of CI stimuli attract more attention in autism [100,101], resulting in decreased viewing time on social images [102] compared to when CI stimuli are present. Using paired dynamic social videos and geometric pattern displays, studies [103,104] found that toddlers (14 months) with autism spent significantly less time viewing social images compared to toddlers without autism. Furthermore, more than 69% of total time spent looking at geometric patterns could predict an autism diagnosis with 100% accuracy [103]. Moore et al. [105] investigated this issue with a larger group of toddlers and replicated the results from Pierce et al. [103], and this preference for geometric stimuli continues from infancy into childhood [106], with autistic children making fewer fixations on social interaction videos when these are presented with dynamic geometric stimuli simultaneously. From these studies it appears that the viewing pattern of social stimuli in autism depends on the type of competing non-social stimuli that are also present. The high level of attentional bias to either CI or geometric stimuli may result in reduced social attention in autism, and hence a greater likelihood of relevant social information going undetected.

However, this bias is not equivalent for all individuals with autism. Moore et al. [105] identified a geometric pattern-preference ASD subgroup and a social stimulus-preference ASD subgroup, and showed that, the increased attention to geometric patterns in autism is related with symptom severity, potentially indicating that increased attention to this kind of stimulus can be used as a prognostic tool. This result reveals the important influence of individual differences within the ASD group in relation to social processing. However, it is important to highlight that low-level stimulus features (e.g., spatiotemporal frequency, contrast, and luminance) are often not controlled across stimuli in the experiments reported above, which means that low-level differences in the visual characteristics of stimuli cannot be ruled out as likely contributors to some of the observed attentional differences.

Key Findings

- Differences observed between TD and ASD children in the visual processing of social information may not reflect atypical social processing per se, but instead increased attention towards alternative stimuli of interest;
- The attentional capture of CI interests and geometric patterns may be associated with ASD symptom severity.

13. Stimulus and Task Complexity

A further factor that has been shown to modulate allocation of attention to social information in ASD is the complexity of the social stimuli being viewed [69,76,106–109]. For example, Hanley et al. [76] found that when faces were presented in isolation, individuals diagnosed with Asperger syndrome spent a similar amount of time fixating faces compared to age and IQ matched TD individuals. However, in social scenes including more than one person, ASD participants showed reduced attention to eyes relative to the TD group. Further evidence comes from Chevallier et al. [69] who found that, compared to an age-matched TD group, ASD children reduced their attention to faces and increased their fixations on objects exclusively for complex dynamic social interaction situations. Using dynamic social stimuli, Chawarska et al. [107] investigated the modulation of social salience on the attentional allocation in toddlers with high risk for autism. The results showed that the high-risk group (who were later diagnosed with autism) attended to social stimuli typically in a goal-directed action condition (an actor was making a sandwich and made no direct eye contact with viewers) and in a moving toys condition (actor directing gaze towards moving mechanical toys) in which the salience of social behaviour was relatively low. In contrast, in conditions involving more salient social cues or interactions, like the dyadic bid (direct interaction from the actor in a video towards the participant) and joint attention conditions (the actor was directing her eye gaze towards the participants and then towards specific objects), ASD toddlers fixated less on the screen, and less on the face and mouth regions of the actor compared to the TD toddlers.

By developing a novel data-driven method of analysing eye movements for dynamic information, Wang et al. [108] investigated the time intervals as to when clinical and non-clinical toddlers looked at the same content in the social video context depicted in the Chawarska et al. [107] study. The results from this analysis showed that in the dyadic bid and in the sandwich-making condition, ASD toddlers had lower converging attention allocation to the same spatial location at specific moments in time, compared to the both a TD and a developmental delay (DD) toddler control group. However, in a moving toys condition, there were no group differences. No difference was found between the DD and TD group in any condition. The finding from this on-line analysis has been interpreted to suggest that ASD toddlers showed atypical gaze patterns in response to social bids. Furthermore, this atypical attention pattern in early life, which seems to have little to do with intelligence, was related to more severe social-affective symptoms, as assessed by the Autism Diagnostic Observation Schedule (ADOS). The eye movement patterns from the dynamic video studies indicate that deficits in social attention from early life might impede the detection of key social information, and thus might adversely impact upon the acquisition of social experience in later development. The effects of different types of dynamic social stimuli [107] coupled with the temporal and spatial analysis of the eye movements during the video presentation [108] have been invaluable in showing that atypical social attention in autism is observed for more complex social situations, and this atypicality in ASD appears to be absent for more simple social situations.

Further support for differences between TD and ASD groups in viewing complex scenes comes from Shi et al. [106] who compared visual preferences for simple and complex dynamic social stimuli in preschool-aged children with and without autism. Each social stimulus was presented with simultaneous presentation of a dynamic geometric pattern. The results suggested that ASD participants viewed social images less than the TD children in the complex social stimulus condition (which included several people playing games), whereas, for the simple social stimulus condition the group difference was absent. Using the same paradigm, Crawford, Moss, Oliver, Elliott, Anderson and McCleery [109] presented participants with social and non-social video stimuli in two conditions (moving towards or moving past the viewer) and found reduced preference to social over non-social videos in ASD only when stimuli were moving towards the viewers.

It is clear from the above studies that stimulus complexity can affect social processing in ASD, but what about task complexity? It is well established that task instructions can affect the allocation of attention to information in scenes [110], and studies that have directly examined the influence of task instructions on social attention in ASD have also found a failure in modulating eye gaze to pre-specified target areas (e.g., eye regions or eye-gazed targets) where they were guided to attend to these under an explicit instruction [57,111]. In a cross- modal study, Grossman, Erin, Teresa and William [112] investigated the modulation of task demand on preferential attention to the auditory-visual (AV) synchronization of speech in children with and without autism. Two videos of a speaker's mouth were presented concurrently, and these were either synchronized or not synchronized with speech audio. Participants were either asked to view the display freely (implicit task) or they were guided to look at the synchronized speech (explicit task). Grossman et al. [112] found that although both groups increased their fixation duration on synchronized speech in the explicit task condition compared to the implicit task condition, this tendency was reduced in ASD children and this group also looked significantly less to the AV-congruent image, and to the mouth region than their TD peers.

The results from the eye-movement studies reported so far are not entirely consistent and the eye-movement measures reported suggest that there are both similarities and differences in social processing in autism. There are, however, very subtle differences in visual processing of social stimuli in different social contexts [61,113,114]. For example, Benson et al. [61] revealed that, compared to TD adults, when deciding whether a social scene is weird or normal, high-functioning adults with autism fail to recognize the socially weird information during their initial fixation on that information (see Figures 4 and 5). This study showed that the ASD group needed more look backs and longer total fixation time to confirm detection of a socially weird target in the scenes. However, there were no significant differences in detection and processing of a physically weird target event or item, between the ASD and TD groups. This subtle processing difference in the immediacy with which social cues are detected, coupled with the repeated scanning of the target area in autism, may reflect a reduced speed in processing the social contexts depicted in the scenes. The impact of these two consistent findings has relevance for the everyday communication domain in ASD, since any delay in the detection of crucial information would affect the ability to follow what is happening, and in everyday communication, if something is 'missed' it is not possible to go back and recheck what has already been sampled.

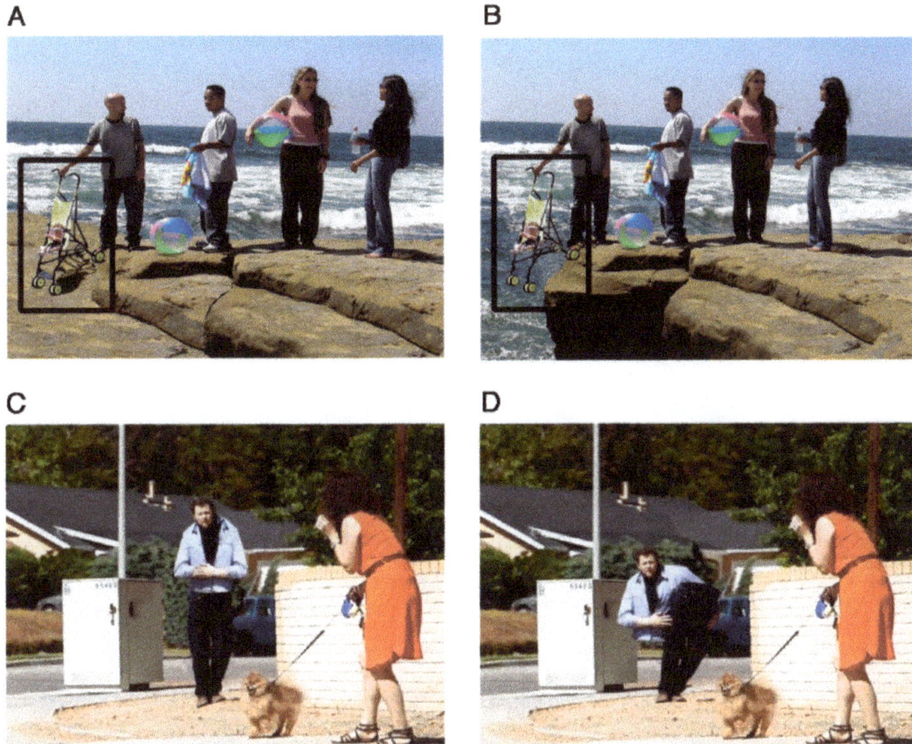

Figure 4. Example of social normal (**A**), social weird (**B**), physical normal (**C**), and physical weird (**D**) stimuli from Benson et al.'s experiment [61]. Reproduced with permission from John Wiley and Sons. The black rectangles represent the target region and were not visible during the experiment.

Limitations in the modulation of attention, according to task demands or stimulus complexity, could be an essential factor in shaping any social-processing differences in ASD, and, could potentially result in inappropriate attentional allocation to salient social cues, such as the eyes, or the eye direction for complex social scenes [72,81,115], which could result in failures to respond effectively or appropriately in everyday communication.

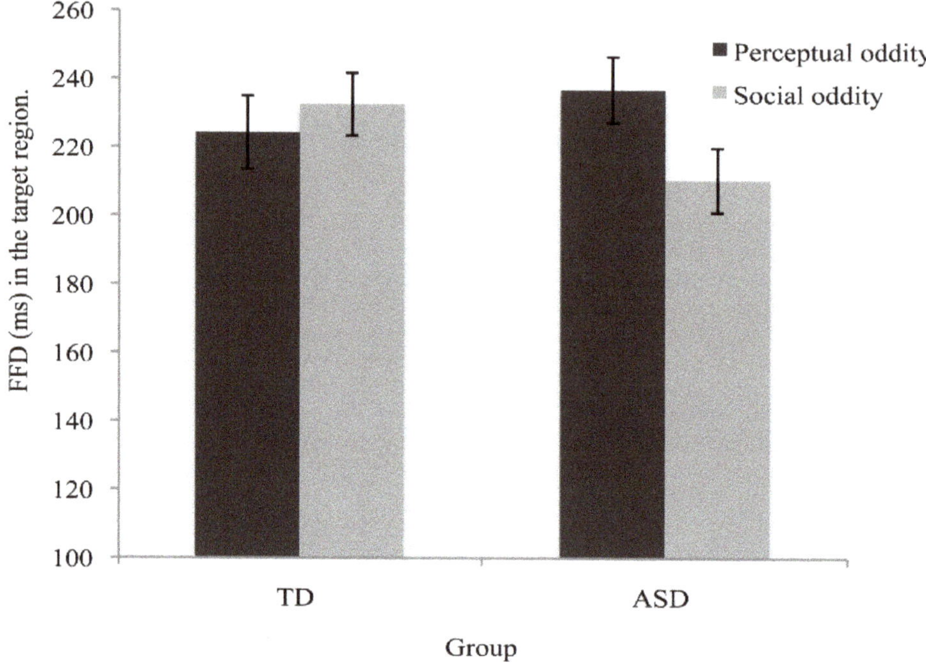

Figure 5. The interaction between oddity type and group for first fixation duration (FFD) detected in Benson et al.'s [61] experiment, indicating a temporal delay in the initial detection of social, but not physical/perceptual, oddities for autistic adults. Reproduced with permission from John Wiley and Sons.

Key Findings

- The allocation of attention to social information is comparable between TD and ASD participants when social stimuli are relatively simple, for example, include one person;
- The allocation of attention to social information is reduced in autistic children and adults when the complexity of social aspects of the scene increase e.g., include interaction;
- The exact determinants of what make a social scene simple or complex requires further examination and definition;
- In scenes that depict socially odd events autistic individuals fail to immediately detect these on initial fixation.

14. Face- and Emotion-Processing Paradigms

Atypical face processing is well documented in numerous studies in autism, and individuals with autism are found to show reduced accuracy in face identification and in emotion recognition [92,116–120]. This appears to be a face specific processing difference, which is absent for object recognition [64,117]. We reported, in earlier sections of this review, the differences in visual attention to faces between autistic and non-autistic individuals. These differences were observed most noticeably for naturalistic or dynamic social contexts. In this section we review the studies that have focused on exploring the visual coding characteristics of faces and face-processing cognition, with an aim to evaluate whether such differences might account for difficulties in social processing in autism.

Snow et al. [117] found that whilst a TD group made more fixations on faces compared to objects during a visual encoding phase, an ASD group did not show this pattern. Subsequently, in a recognition phase, the ASD group performed less accurately for a face-recognition test relative to TD participants.

This suggests that reduced fixations to faces in an encoding stage might have influenced performance in the recognition stage for autistic participants. Furthermore, Yi et al. [79,80] have shown atypical face scanning during a memory phase and report that ASD participants had a smaller frequency in scanning different core facial features relative to a TD group. The behavioural data from that study also showed the ASD group to have lower identification accuracy, compared to the TD group. Similarly, Liu, Li and Yi [118] used a variable analysis approach to investigate viewing patterns for faces in autism, and they found less fixation time on the right eye but longer fixation time on the region below the left eye (from the observer's view) in an ASD group compared to a TD group. It is not clear what this viewing pattern might indicate, and at this stage it is perhaps more important to accept that core facial features are fixated less by autistic participants. This reduced fixation for core facial features also fits with findings of reduced numbers of saccades made between core facial features in ASD, and Ellie, Romina, Jon and Dyer [119] have revealed that performance for face identification is positively associated with the number of saccades made between different facial features. Therefore, the widely found face identification differences in ASD appear to be related to atypical face scanning, and several studies seem to also point to a reduction in fixations on core facial features. Together this pattern of saccades and fixations for autistic individuals may indicate either a tendency to avoid eye gaze, or a reduced propensity to process faces holistically. However, what remains unclear is whether this is a face specific difference, or a consequence of general reduced attention allocation towards social information (such as faces), as reported in earlier sections of this review.

Recently, Evers, Van Belle, Steyaert, Noens and Wagemans [120] used a gaze-contingent display paradigm to examine whether children with ASD show a preference for holistic face processing. This paradigm allowed participants to view the whole face (full), or the remaining part of the face outside the fixated region (mask), or solely the fixated region (window) when participants were presented with two faces and were instructed to decide whether the faces were the same or different. The behavioural results suggested slowed and less efficient face processing in ASD, compared to the TD group. When comparing the group difference in three different presentation conditions, results showed younger ASD children (6–10 years old) had reduced performance in the full viewing and mask conditions, compared to the TD group, but not in the window condition. This pattern was not observed in the older ASD group (10–14 years old). The eye-movement data representing visual exploration during the task, as indicated by heat map analysis, showed that younger ASD children had a dense heat map, in which a smaller area was centred around the fixation peak compared to the TD group, but older ASD children did not differ to TD controls. Therefore, younger ASD children scanned faces more narrowly compared to the TD group, and this finding fits with the previous findings discussed in the preceding text. Noteworthy, a further two studies from Yi and colleagues [121,122] explored viewing patterns for a face identification task, with an own race face and another race face condition, in Chinese children with and without autism. The findings from both the behavioural and eye-movement measures suggest a typical processing of own race face information, regardless of the atypicality observed in general face identification in both studies. Specifically, the two studies found that although the ASD group's performance was reduced in the identification task compared to the TD group, the ASD group showed an expected advantage in identifying Chinese faces over Caucasian faces. Importantly, the eye movement analyses showed that all groups fixated for longer on the eyes but for shorter on the mouth in Caucasian faces compared to Chinese faces. These findings imply that any differences in face processing for faces from another race appear to be present in individuals with and without autism.

Further eye-movement studies designed to evaluate face processing differences in ASD have focused on multiple aspects of face information, including face identification, emotion recognition and the ability to infer mental states from faces. For example, Kirchner, Hatri, Heekeren and Dziobek [123] revealed that adults with autism spent less time fixating an emotional face presented in a naturalistic context and made more errors when they were asked to report the emotion of the face or to identify other characteristics from the faces (gender or age) compared to a TD control group. However, both groups showed longer fixation time on face and mouth regions in the emotion task relative to the identification task. Additionally, there was evidence of a positive correlation between fixation time on eyes and performance on an off-line face processing task (Reading the Mind in the Eyes Test, [124]) in the ASD group. This finding indicates that increased viewing of the eyes in a face is related to increased ability to infer the mental state of the face in the ASD group. Müller et al. [125] found the same relationship using dynamic social interaction stimuli and observed diminished pupil dilation only in ASD adults (pupil dilation can be used to infer an indication of interest or cognitive effort). The behavioural data from that study also showed the ASD group to have more difficulty to understand the mental states of the characters depicted in the interactions.

Some studies also support the view that longer fixation time on faces is related to emotion-processing skill in ASD [88]. For example, in an emotion-recognition task, Wieckowski and White [126] found a reduced ability to recognize disgust and sadness expressions in ASD, and this reduced ability was coupled with a finding of longer fixation times on the mouth in the ASD group compared to the TD group for these expressions. In further studies, autistic individuals have been shown to express emotion less appropriately in response to a person who was expressing emotions in a video, and they were also found to fixate less on the eyes of surprised faces. A comparison of fixation time between the correct and incorrect responses in both a recognition and expression condition revealed a more important role for the mouth than the eyes in modulating face-processing performance for specific expressions. A later study examined emotion recognition and face attention in ASD and schizophrenia across a broad range of contexts, where emotional faces were presented in isolation, digitally masked in emotional scenes, or were embedded in scenes with congruent or incongruent backgrounds [127]. Both clinical groups showed poorer recognition performance in all scene conditions, but this effect was absent for the isolated face condition. Compared to the masked scene condition, fixation time (%) to the face region was increased in unmasked scene conditions for all three groups. However, only the ASD group failed to show an increase of face viewing in the congruent scene condition relative to the incongruent scene condition, and this finding points to a subtle but specific atypicality of utilising face information in congruent contexts in ASD.

Neutral faces appear to have special significance in ASD. For example, Tottenham et al. [78] revealed atypical processing of the eyes in neutral faces, and this was observed in behavioural performance, eye-movement measures, and brain activity. They found that participants with ASD spent less time on the eyes when viewing neutral faces compared to the TD group. They also showed an increased response in Amygdala activity which was larger in the neutral face condition compared to an angry face condition in ASD, and especially in an eye-gaze manipulation condition, where participants were cued to look at the eyes directly. Behavioural data from the isolated off-line threat rating and expression coding tasks suggested that the ASD group made more errors to label the expression of neutral faces, showing a tendency to perceive them as negative stimuli. Increased error rates were also related with a higher likelihood to evaluate neutral faces as the threatening stimuli in the ASD group relative to the TD counterparts. More importantly, more threatening scores in neutral face ratings in ASD indicated shorter fixation time on eye regions. These findings from the behavioural, eye-movement and brain-activity measures consistently indicate that neutral faces are processed differently in the ASD group.

One explanation for this atypicality could be that neutral faces are perceived as more ambiguous in social information for the ASD group, and hence, shifting fixation away from the eyes of neutral faces may be adopted by autistic individuals as a compensatory strategy to alleviate any potential threatening feeling associated with neutral faces. However, a very recent study from Wang, Lu, Zhang, Fang, Zou and Yi [128] found evidence of eye avoidance exclusively for angry faces in ASD children compared to TD controls. There have also been other inconsistent reports for eye avoidance. For example, Moriuchi, Klin, and Jones [129] found evidence to support eye indifference but not avoidance, whereas Kliemann, Dziobek, Hatri, Steimke, and Heekeren [130] found evidence to support both avoidance and indifference to eyes in ASD. Despite these inconsistencies, the atypical visual processing of faces may be a significant indicator of emotional processing differences in ASD.

Key Findings

- Autistic individuals may show a reduced propensity to allocate attention to the eye region of faces, relative to TD individuals, which may reflect avoidance or indifference;
- Face processing may be less holistic in ASD with less allocation to core features and less saccades between these features, in comparison to TD individuals;
- Differences in the allocation of attention to facial features may be more pronounced for young children with ASD relative to older individuals;
- The use of face information to infer mental states results in longer processing times upon eye regions in autistic individuals relative to TD individuals;
- A reduced propensity to fixate the eyes has the potential to account for differences in emotion recognition in ASD.

15. Face-to-Face Interaction Paradigms

Some studies have further extended the findings of atypical social attention observed in static or dynamic stimuli, to actual face to face interactions, where individuals with autism are required to engage with an interlocutor [62,131,132]. Riby et al. [132] investigated the gaze patterns when participants with autism and individuals with Williams syndrome (WS) were engaged in a question and answer interaction. A video recorder was set up behind the experimenter to monitor the eye gaze behaviour of the participants. Similar to the TD group, both the ASD and the WS participants showed more gaze-aversion behaviours (GA, total viewing time spent on non-questioner areas) when the question difficulty increased, and, both groups showed reduced performance when they were asked to look directly at the questioner throughout the whole session relative to free viewing. In addition, the ASD group made more GA compared to the TD group when listening to the interlocutor [131], and they also showed a weaker tendency to maintain their attention to the questioner during the whole interaction compared to the TD group [132]. Hanley et al. [62] presented further evidence of reduced attention to an interlocutor's face, especially the eye regions, and this was coupled with increased attention to the non-face screen parts in an ASD group during a real live interaction in comparison to a TD group and to a group of participants with specific language impairment (SLI). A further subtle but significant finding showed that when an accidental event happened to a puppet in the interlocutor's hand, ASD participants took significantly longer to begin to monitor the face of the interlocutor compared to both TD and SLI participants (see Figure 6). These findings are indicative of an ASD-specific difference in the allocation of social attention, and this is consistent with previous studies [98,111,128,131]. Delays in looking to the interlocutor's face indicate that participants with ASD, unlike TD counterparts, did not immediately use the interlocutors face as a social reference marker to indicate detection of the 'accident'.

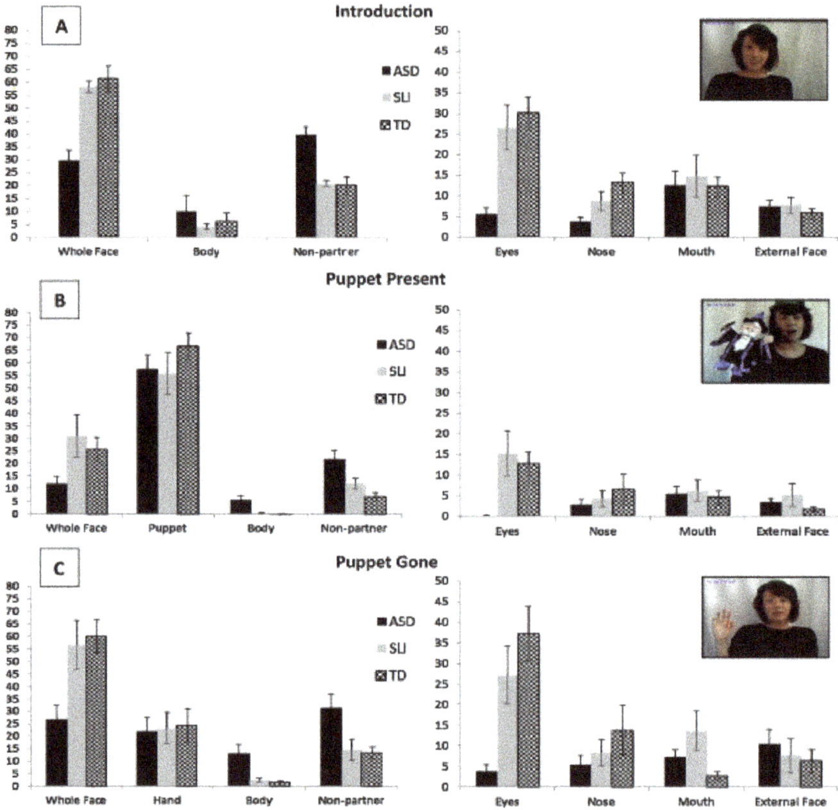

Figure 6. Percentage gaze to areas of interest for each section and for each group during a face-to-face interaction in Hanley et al.'s research [62]. Reproduced with permission from Elsevier.

Key Findings

- Research adopting real-time face to face interactions between autistic and non-autistic individuals corroborates previous findings that task complexity may mediate social attention differences in ASD, and, that a requirement to share gaze with another throughout a task may be detrimental to performance.

16. Individual Differences

Eye-movement studies have also contributed to identifying individual differences in the field of social processing in autism for infants, children and adults, and have led to accounts as to how these could explain the inconsistent findings reported in relation to social processing in ASD. In particular, studies have shown that autism symptom severity, gender and comorbidity with other disorders might be a potential factor contributing to the mixed reports of differences in social attention or processing in autism [65,81,125,133,134]. For example, Bird et al. [65] found that for participants diagnosed with ASD, reduced preference to look at faces over non-face regions, presented in dynamic social interaction videos, was related to greater autism symptom severity, while lower preference scores for eyes over mouths was associated with alexithymia (defined as problems in recognizing and describing emotions). This finding highlights how ASD symptom severity may modulate face-processing atypicalities and that research exploring emotional processing in autism should consider the role of alexithymia, in addition to ASD [135]. Social anxiety and ASD are highly co-morbid [136] and both are related to atypical social attention. Kleberg et al. [81] has shown that autistic traits are associated with longer eye movement latencies to fixate another's eyes, while social anxiety is associated with a greater tendency to avoid the eyes. Müller et al. [125] showed that there were two subgroups in their sample of ASD participants, with one group viewing less on the eyes and the other retaining similar fixation duration on the eyes as the TD group. These studies provide important evidence to promote consideration of the potential influence of autism symptom severity and other comorbidity on social processing in ASD.

Chawarska et al. [133] found that female infants at high risk for autism, between 6 and 12 months, showed increased attention to social scenes and to an interactive partner's face compared to male high-risk counterparts and to low-risk female infants. In addition, the increased social attention in the high-risk female infants was related prospectively with less severe social difficulties at 2 years of age. However, this characteristic in female autism, observed in the infant stage, may disappear as development progresses over time. For example, Ketelaars et al. [134] has reported that adult females with autism looked for significantly shorter durations on dynamic faces showing intense emotions, as well as on the inner facial features, compared to TD adult females. This study also found a positive correlation between the time to first fixate on faces and social impairments in females with ASD relative to males, which is consistent with the findings of increased social attention (e.g., to face, eyes) and appropriate social behaviours reported from studies that tested male-dominant ASD groups [66,68,70,71,73,89,99,107]. What the studies cited above suggest is that social communication in autism may develop differently in female and male infants, and that social competence is linked to early orientation and sustained attention for social stimuli. However, it is noteworthy to point out that regardless of gender; there consistently exists a relationship between symptom severity and social processing proficiency across a wide range of ages in ASD.

In terms of the relationship between looking at the mouths of dynamic faces and abilities of social communication, results have been mixed. Johnel et al. [82] found that, when watching dynamic films with a speaking person, young children with autism who showed longer viewing time on the mouth tended to have more severe symptoms of ASD, whereas, Ketelaars et al. [134] revealed a reverse pattern in female adults with autism, who showed that those with more severe symptoms of ASD fixated less on the mouth. Elsabbagh et al. [83] have suggested that the relationship between mouth viewing in 7 months and expressive language (EL) at 36 months was modulated by the complexity of social stimuli for all the toddler participants. Specifically, endogenous control of attention to the mouth in a complex condition is associated with superior EL, and increased exogenous attention, driven by the repetitive mouth moving, is related to poorer EL. However, Rice et al. [75] found that this relationship varied significantly based on intelligence quotient (IQ) score characteristics. ASD individuals with higher verbal IQ relative to non-verbal IQ looked longer on mouths and also showed fewer social impairments, while ASD individuals who had higher full-scale IQ scores and no differences between verbal and non-verbal IQ, showed a reverse pattern. What the findings from the individual difference studies indicate is that within the ASD groups, autism symptom severity is not the only factor to impact upon processing of social information. Furthermore, increasing attention to the social world may be adaptive in ASD to aid communication in social interactions.

Key Findings

- Social processing differences observed in autistic individuals may also be related to co-morbid conditions, such as social anxiety disorder and alexithymia;
- Based upon eye-tracking data, social processing differences may be more prominent for males with ASD in early life, relative to females;
- Symptom severity is associated with less adaptive social processing.

17. Social-Processing Summary

In summary, many studies employing eye tracking technology over the past 10 years have revealed subtle but significant differences in social attention in ASD. There is clear evidence for differences in the way autistic individuals allocate attention to social information and this may influence their response to such information. These differences predominantly appear to reflect either an *absence* of attention towards social information or *reduced propensity* or *delay* to sample this information. This atypicality in ASD could influence the seamless nature of interactions, or the detection of social cues in everyday exchanges. Ongoing events in everyday communication happen very rapidly, and, some social cues are implicit or ambiguous. Therefore, the subtle differences revealed by the eye-movement patterns in the studies reported in the social processing sections of this review, may be amplified in real social situations in ASD. This has clear potential to result in difficulties in understanding, following, keeping track of, and preparing and executing appropriate responses during ongoing communication.

18. Overall Conclusions

This review summarises the research that has used eye tracking over the past decade to examine cognitive processing differences in ASD, with a focus on social and language processing. The aim has been to identify how observed differences in processing could relate to behavioural symptoms that manifest in day to day communication.

In relation to the previous review into eye movements in ASD [5], issues that were raised from that review, such as the modulating effects of symptom severity, developmental stage, gender, co-morbidity, stimulus and social complexity, and, task demands on performance, are just as relevant for the studies presented in the current review, and will continue to be factors that need to be addressed in research into cognitive processing. A further factor when conducting such research relates to difficulties in recruitment of sufficient participants to carry out individual difference analyses. It is important that researchers work (possibly in a more collaborative way) to overcome such challenges to promote in-depth investigation into how these factors and diagnostic status influence on-line social and language processing, at an individual level, in the heterogenous autism population. Although there are many apparent inconsistencies in the research discussed in the current review, there are also some consistent eye-movement patterns, from both language and social processing studies that indicate how both temporal and spatial processing differences in autism might impact upon everyday communication.

From the language-processing literature, it would seem that there are subtle differences in how children with autism acquire language, and these differences may in part relate to how children with autism attend to and learn from the social environment. For example, in younger children some eye-movement studies have revealed that a reduced propensity to utilise an interlocutor's gaze to support the mapping of referents upon new words. For older children and adults, few differences in language processing are reported when paradigms have low pragmatic and social response demands (e.g., visual world, reading). In these instances, differences are predominantly reported for higher-level language processing such as inferential work and in those studies the eye-movement patterns reflect quantitative delays in the time-course of processing relative to comparison groups.

Future research would do well to focus upon identifying more specifically when and why differences in the time course of language processing occur in ASD and how this interacts with social and sensory processing. For example, it is important to consider the language demands of instructions for social tasks when considering these findings. Previously, social, language, and sensory processing have predominantly been examined independently, yet they are inextricably linked.

From the social-processing literature, one predominant finding is that the saliency of social information may not be prioritised in infants, toddlers, children, adolescents and adults with autism. For example, social information may not be initially fixated by autistic individuals, the time taken to fixate important social information is greater, social cues may not be followed in the same way, core facial features may not be fixated as frequently, and there may be a reduction in eye movements between core facial features. These differences in early attention to social information highlight the importance of analysing the time-course of attention allocation during experimental trials.

Since the previous review, much more research using dynamic stimuli to examine the influence of task and stimulus complexity has been conducted. For example, the nature of social information presented in the task influences processing efficiency (e.g., one person versus multiple people in dynamic videos, and physical versus social oddities in static scenes). However, there now needs to be a focus upon specifying more discretely how tasks and stimuli may increase in complexity and how such increases will influence on-line processing for both TD and autistic individuals. This may be achieved by systematically manipulating the information that is made available, by analysing whole scan paths for each trial rather than single regions of interest, and by adopting a range of different eye-movement measures that can inform as to whether any group differences reflect differences in early detection or processing, or differences in later processing.

The subtle differences in language and social processing reported in the current review have the power to influence the nature of communication and interaction in ASD. For example, a reduction in the use of eye contact and differences in the speed of detection of (often implicit) communication cues in both language and visual social stimuli have clear consequences for everyday interaction. Moreover, differences in the use of such information in children with ASD may impede the development and exposure to various social schemas and contexts, which may affect social development per se.

One fairly consistent eye-movement pattern in adults with autism clearly shows that there may be a reduction in the speed with which attention is initially oriented towards social information, and this finding is especially relevant when viewing complex social scenes, or when engaged in complex dynamic tasks. Furthermore, from the scene perception studies we know that when odd social events are depicted, there is a failure to detect this information upon first fixation in ASD. Similarly, in the language-processing research, any differences that are observed tend to reflect a delay in the detection of target words or phrases that have some higher-level linguistic manipulation e.g., some types of inferences and implausibilities. This finding highlights that processing differences for complex tasks and/or stimuli may extend across different processing domains in ASD and that the type of processing differences observed, when examined on a moment-by-moment level, may be similar in nature. A second important eye-movement pattern shows that adults with ASD repeatedly fixate information in scenes, and they repeatedly re-read text (after reading this through once in entirety). This pattern is absent in the TD population. It is not clear why this pattern occurs in ASD, but it may be necessary for a coherent representation of a scene or text passage to be developed. For many studies this detailed analysis of the eye movement data is not performed, which means that it is not known whether repeated sampling occurs as a matter of course in ASD, when there is unlimited time to complete the task at hand. Hence the emphasis here is to illustrate the importance of the quality of the temporal and spatial eye-movement data, and the very detailed and sophisticated analyses of the eye-movement data, if we are to improve our understanding of how differences in on-line cognitive processing might contribute to difficulties in communication in the real world for individuals with autism. If repeated sampling is informative and can support communication for autistic individuals, the lack of opportunity to engage in repeated scanning of information in everyday communication will impede interaction, since day-to-day communication is typically fast and dynamic.

In conclusion, eye-movement studies have been instrumental in increasing our understanding of the drivers that might underpin manifestations of communication differences in ASD. The findings from the language and social-processing fields align to a degree and highlight that some processing differences are similar in both of those domains. Consistent reports of a failure to immediately orient to, detect or comprehend significant social cues, coupled with a propensity to resample information in scenes and reading, means that information that may be important to ongoing processing in the real world, which may be 'missed', cannot be sampled again, and this will have important consequences for keeping track of events, and for preparing and executing appropriate responses in everyday communication.

Author Contributions: Conceptualization: V.B.; Literature review and writing of language processing section: P.H.; Literature review and writing of social processing section: Z.L.; Writing—Review and Editing: V.B. & P.H.

Funding: This research received no external funding.

Conflicts of Interest: The authors declare no conflict of interest.

References

1. Rayner, K. Eye movements and attention in reading, scene perception, and visual search. *Q. J. Exp. Psychol.* **2009**, *62*, 1457. [CrossRef]
2. Liversedge, S.P.; Findlay, J.M. Saccadic eye movements and cognition. *Trends Cogn. Sci.* **2000**, *4*, 6. [CrossRef]
3. American Psychiatric Association. *Diagnostic and Statistical Manual of Mental Disorders*, 5th ed.; American Psychiatric Publishing: Arlington, VA, USA, 2013.
4. Frith, U. Why we need cognitive explanations of autism. *Q. J. Exp. Psychol.* **2012**, *65*, 2073. [CrossRef] [PubMed]
5. Benson, V.; Fletcher-Watson, S. Eye movements in autism spectrum disorder. In *The Oxford Handbook of Eye Movements*; Liversedge, S.P., Gilchrist, I., Everling, S., Eds.; Oxford University Press: Hove, UK, 2009.
6. Tager-Flusberg, H. On the nature of linguistic functioning in early infantile autism. *J. Autism Dev. Disord.* **1981**, *11*, 45. [CrossRef] [PubMed]
7. Rapin, I.; Dunn, M. Update on the language disorders of individuals on the autism spectrum. *Brain Dev.* **2003**, *25*, 166. [CrossRef]
8. Baron-Cohen, S.; Baldwin, D.A.; Crowson, M. Do children with autism use the speaker's direction of gaze strategy to crack the code of language? *Child Dev.* **1997**, *68*, 48–57. [CrossRef]
9. Tomasello, M. Reference: Intending that others jointly attend. *Pragmat. Cogn.* **1998**, *6*, 229–243. [CrossRef]
10. Norbury, C.F.; Griffiths, H.; Nation, K. Sound before meaning: Word learning in autistic disorders. *Neuropsychologia* **2010**, *48*, 4012–4019. [CrossRef]
11. Akechi, H.; Senju, A.; Kikuchi, Y.; Tojo, Y.; Osanai, H.; Hasegawa, T. Do children with ASD use referential gaze to learn the name of an object? An eye-tracking study. *Res. Autism Spectr. Disord.* **2011**, *5*, 1230–1242. [CrossRef]
12. Tenenbaum, E.; Amso, D.; Abar, B.W.; Sheinkopf, S.J. Attention and word learning in autistic, language delayed and typically developing children. *Front. Psychol.* **2014**, *5*, 490. [CrossRef]
13. Tenenbaum, E.J.; Amso, D.; Righi, G.; Sheinkopf, S.J. Attempting to "increase intake from the input": Attention and word learning in children with autism. *J. Autism Dev. Disord.* **2017**, *47*, 1791–1805. [CrossRef] [PubMed]
14. Lucas, R.; Norbury, C.F. Orthography facilitates vocabulary learning for children with autism spectrum disorders (ASD). *Q. J. Exp. Psychol.* **2014**, *67*, 1317–1334. [CrossRef]
15. Naigles, L.R.; Tovar, A.T. Portable intermodal preferential looking (IPL): Investigating language comprehension in typically developing toddlers and young children with autism. *J. Vis. Exp.* **2012**, *70*, e4331. [CrossRef] [PubMed]
16. Swensen, L.D.; Kelley, E.; Fein, D.; Naigles, L.R. Processes of language acquisition in children with autism: Evidence from preferential looking. *Child Dev.* **2007**, *78*, 542–557. [CrossRef] [PubMed]
17. Tek, S.; Jaffery, G.; Fein, D.; Naigles, L.R. Do children with autism spectrum disorders show a shape bias in word learning? *Autism Res.* **2008**, *1*, 208–222. [CrossRef] [PubMed]
18. Jyotishi, M.; Fein, D.; Naigles, L. Investigating the grammatical and pragmatic origins of wh-questions in children with autism spectrum disorders. *Front. Psychol.* **2017**, *8*, 319. [CrossRef] [PubMed]
19. Tovar, A.T.; Fein, D.; Naigles, L.R. Grammatical aspect is a strength in the language comprehension of young children with autism spectrum disorder. *J. Speech. Lang. Hear. Res.* **2015**, *58*, 301–310. [CrossRef]
20. Naigles, L.R.; Kelty, E.; Jaffrey, R.; Fein, D. Abstractness and continuity in the syntactic development of young children with autism. *Autism Res.* **2011**, *4*, 422. [CrossRef] [PubMed]
21. Goodwin, A.; Fein, D.; Naigles, L.R. Comprehension of wh-questions precedes their production in typical development and autism spectrum disorders. *Autism Res.* **2012**, *5*, 109–123. [CrossRef]
22. Tanenhaus, M.K.; Spivey-Knowlton, M.J.; Eberhard, K.M.; Sedivy, J.C. Integration of visual and linguistic information in spoken language comprehension. *Science* **1995**, *268*, 1632–1634. [CrossRef] [PubMed]
23. Brock, J.; Norbury, C.; Einav, S.; Nation, K. Do individuals with autism process words in context? Evidence from language mediated eye-movements. *Cognition* **2008**, *108*, 896. [CrossRef] [PubMed]
24. Bavin, E.L.; Kidd, E.; Prendergast, L.A.; Baker, E.K. Young children with ASD use lexical and referential information during on-line sentence processing. *Front. Psychol.* **2016**, *7*, 171. [CrossRef] [PubMed]

25. Zhou, P.; Zhan, L.; Ma, H. Predictive Language Processing in Preschool Children with Autism Spectrum Disorder: An Eye-Tracking Study. *J. Psycholinguist. Res.* **2018**, 1–22. [CrossRef] [PubMed]
26. Hahn, N.; Snedeker, J.; Rabagliati, H. Rapid linguistic ambiguity resolution in young children with autism spectrum disorder: Eye tracking evidence for the limits of weak central coherence. *Autism Res.* **2012**, *8*, 717–726. [CrossRef] [PubMed]
27. Diehl, J.J.; Friedberg, C.; Paul, R.; Snedeker, J. The use of prosody during syntactic processing in children and adolescents with autism spectrum disorders. *Dev. Psychopathol.* **2015**, *27*, 867–884. [CrossRef]
28. Frith, U.; Happe, F. Autism—Beyond the theory of mind. *Cognition* **1994**, *50*, 115. [CrossRef]
29. Bavin, E.L.; Kidd, E.; Prendergast, L.; Baker, E.; Dissanayake, C.; Prior, M. Severity of autism is related to children's language processing. *Autism Res.* **2014**, *7*, 687–694. [CrossRef]
30. Venker, C.E.; Eernisse, E.R.; Saffran, J.R.; Weismer, S.E. Individual differences in the real-time comprehension of children with ASD. *Autism Res.* **2013**, *6*, 417–432. [CrossRef]
31. Bavin, E.L.; Prendergast, L.A.; Kidd, E.; Baker, E.; Dissanayake, C. Online processing of sentences containing noun modification in young children with high-functioning autism. *Int. J. Lang. Commun. Disord.* **2016**, *5*, 137–147. [CrossRef]
32. Nation, K.; Clarke, P.; Wright, B.; Williams, C. Patterns of reading ability in children with autism spectrum disorder. *J. Autism Dev. Disord.* **2006**, *36*, 911–919. [CrossRef]
33. McIntyre, N.S.; Solari, E.J.; Gonzales, J.E.; Solomon, M.; Lerro, L.E.; Novotny, S.; Mundy, P.C. The scope and nature of reading comprehension impairments in school-aged children with higher-functioning autism spectrum disorder. *J. Autism Dev. Disord.* **2017**, *47*, 2838–2860. [CrossRef] [PubMed]
34. Huemer, S.V.; Mann, V. A comprehensive profile of decoding and comprehension in autism spectrum disorders. *J. Autism Dev. Disord.* **2010**, *40*, 485–493. [CrossRef]
35. Brown, H.M.; Oram-Cardy, J.; Johnson, A. A meta-analysis of the reading comprehension skills of individuals on the autism spectrum. *J. Autism Dev. Disord.* **2013**, *43*, 932–955. [CrossRef] [PubMed]
36. Howard, P.L.; Liversedge, S.P.; Benson, V. Benchmark eye movement effects during natural reading in autism spectrum disorder. *J. Exp. Psychol. Learn. Mem. Cogn.* **2017**, *43*, 109–127. [CrossRef] [PubMed]
37. Caruana, N.; Brock, J. No association between autistic traits and contextual influences on eye-movements during reading. *PeerJ* **2014**, *2*, e466. [CrossRef] [PubMed]
38. Davidson, M.M.; Kaushanskaya, M.; Weismer, S.E. Reading Comprehension in Children With and Without ASD: The Role of Word Reading, Oral Language, and Working Memory. *J. Autism Dev. Disord.* **2018**, *48*, 3524–3541. [CrossRef]
39. Au-Yeung, S.K.; Kaakinen, J.K.; Liversedge, S.P.; Benson, V. Processing of Written Irony in Autism Spectrum Disorder: An Eye-Movement Study. *Autism Res.* **2015**, *8*, 749–760. [CrossRef]
40. Black, J.; Williams, H.J.; Ferguson, H. Imagining counterfactual worlds in autism spectrum disorder. *J. Exp. Psychol. Learn. Mem. Cogn.* **2018**, *44*, 1444. [CrossRef]
41. Black, J.; Barzy, M.; Williams, D.; Ferguson, H. Intact counterfactual emotion processing in autism spectrum disorder: Evidence from eye-tracking. *Autism Res.* **2018**, *12*, 422–444. [CrossRef]
42. Ferguson, H.J.; Black, J.; Williams, D. Distinguishing reality from fantasy in adults with autism spectrum disorder: Evidence from eye movements and reading. *J. Mem. Lang.* **2019**, *106*, 95. [CrossRef]
43. Howard, P.L.; Liversedge, S.P.; Benson, V. Investigating the use of world knowledge during on-line comprehension in adults with autism spectrum disorder. *J. Autism Dev. Disord.* **2017**, *47*, 2039–2053. [CrossRef] [PubMed]
44. Howard, P.L.; Liversedge, S.P.; Benson, V. Processing of co-reference in autism spectrum disorder. *Autism Res.* **2017**, *10*, 1968–1980. [CrossRef] [PubMed]
45. Sansosti, F.J.; Was, C.; Rawson, K.A.; Remaklus, B.L. Eye movements during processing of text requiring bridging inferences in adolescents with higher functioning autism spectrum disorders: A preliminary investigation. *Res. Autism Spectr. Disord.* **2013**, *7*, 1535–1542. [CrossRef]
46. Micai, M.; Joseph, H.; Vulchanova, M.; Saldaña, D. Strategies of readers with autism when responding to inferential questions: An eye-movement study. *Autism Res.* **2017**, *10*, 888–900. [CrossRef] [PubMed]
47. Au-Yeung, S.K.; Kaakinen, J.K.; Liversedge, S.P.; Benson, V. Would adults with autism be less likely to bury the survivors? An eye movement study of anomalous text reading. *Q. J. Exp. Psychol.* **2017**, 1–27. [CrossRef]
48. Chevallier, C.; Kohls, G.; Troiani, V.; Brodkin, E.S.; Schultz, R.T. The social motivation theory of autism. *Trends. Cogn. Sci.* **2012**, *16*, 231–239. [CrossRef] [PubMed]

49. Burnside, K.; Wright, K.; Poulin-Dubois, D. Social motivation and implicit theory of mind in children with autism spectrum disorder. *Autism Res.* **2017**, *10*, 1834–1844. [CrossRef]
50. Vivanti, G.; Trembath, D.; Dissanayake, C. Atypical monitoring and responsiveness to goal-directed gaze in autism spectrum disorder. *Exp. Brain Res.* **2014**, *232*, 695–701. [CrossRef] [PubMed]
51. Thorup, E.; Nyström, P.; Gredebäck, G.; Bölte, S.; Falck-Ytter, T.; EASE Team. Altered gaze following during live interaction in infants at risk for autism: An eye tracking study. *Mol. Autism* **2016**, *7*, 12. [CrossRef]
52. Krstovska-Guerrero, I.; Jones, E.A. Social-communication intervention for toddlers with autism spectrum disorder: Eye gaze in the context of requesting and joint attention. *J. Dev. Phys. Disabil.* **2016**, *28*, 289–316. [CrossRef]
53. Navab, A.; Gillespie-Lynch, K.; Johnson, S.P.; Sigman, M.; Hutman, T. Eye-tracking as a measure of responsiveness to joint attention in infants at risk for autism. *Infancy* **2012**, *17*, 416–431. [CrossRef]
54. Swanson, M.R.; Siller, M. Patterns of gaze behavior during an eye-tracking measure of joint attention in typically developing children and children with autism spectrum disorder. *Res. Autism Spectr. Disord.* **2013**, *7*, 1087–1096. [CrossRef]
55. Falck-Ytter, T.; Thorup, E.; Bölte, S. Brief report: Lack of processing bias for the objects other people attend to in 3-year-olds with autism. *J. Autism Dev. Disord.* **2015**, *45*, 1897–1904. [CrossRef]
56. Freeth, M.; Chapman, P.; Ropar, D.; Mitchell, P. Do gaze cues in complex scenes capture and direct the attention of high functioning adolescents with ASD? evidence from eye-tracking. *J. Autism Dev. Disord.* **2010**, *40*, 534–547. [CrossRef]
57. Riby, D.M.; Hancock, P.J.; Jones, N.; Hanley, M. Spontaneous and cued gaze-following in autism and williams syndrome. *J. Neurodev. Disord.* **2013**, *5*, 13. [CrossRef]
58. Billeci, L.; Narzisi, A.; Campatelli, G.; Crifaci, G.; Calderoni, S.; Gagliano, A.; Calzone, C.; Colombi, C.; Pioggia, G.; Muratori, F. Disentangling the initiation from the response in joint attention: An eye-tracking study in toddlers with autism spectrum disorders. *Transl. Psychiatry* **2016**, *6*, e808. [CrossRef] [PubMed]
59. Vivanti, G.; Dissanayake, C. Propensity to imitate in autism is not modulated by the model's gaze direction: An eye-tracking study. *Autism Res.* **2014**, *7*(3), 392–399. [CrossRef] [PubMed]
60. Vivanti, G.; Mccormick, C.; Young, G.S.; Abucayan, F.; Hatt, N.; Nadig, A.; Ozonoff, S.; Rogers, S.J. Intact and impaired mechanisms of action understanding in autism. *Dev. Psychol.* **2011**, *47*, 841–856. [CrossRef] [PubMed]
61. Benson, V.; Castelhano, M.S.; Howard, P.L.; Latif, N.; Rayner, K. Looking, seeing and believing in autism: Eye movements reveal how subtle cognitive processing differences impact in the social domain. *Autism Res.* **2016**, *9*. [CrossRef]
62. Hanley, M.; Riby, D.M.; Mccormack, T.; Carty, C.; Coyle, L.; Crozier, N.; McPhillips, M. Attention during social interaction in children with autism: Comparison to specific language impairment, typical development, and links to social cognition. *Res. Autism Spectr. Disord.* **2014**, *8*, 908–924. [CrossRef]
63. Flack-Ytter, T.; Carlström, C.; Johansson, M. Eye contact modulates cognitive processing differently in children with autism. *Child Dev.* **2015**, *86*, 37–47. [CrossRef] [PubMed]
64. Wilson, C.E.; Brock, J.; Palermo, R. Attention to social stimuli and facial identity recognition skills in autism spectrum disorder. *J. Intellect. Disabil. Res.* **2010**, *54*, 1104–1115. [CrossRef]
65. Bird, G.; Press, C.; Richardson, D.C. The role of alexithymia in reduced eye-fixation in autism spectrum conditions. *J. Autism Dev. Disord.* **2011**, *41*. [CrossRef] [PubMed]
66. Shic, F.; Bradshaw, J.; Klin, A.; Scassellati, B.; Chawarska, K. Limited activity monitoring in toddlers with autism spectrum disorder. *Brain Res.* **2011**, *1380*, 246–254. [CrossRef] [PubMed]
67. Chawarska, K.; Macari, S.; Shic, F. Decreased spontaneous attention to social scenes in 6-month-old infants later diagnosed with autism spectrum disorders. *Biol. Psychiatry* **2013**, *74*, 195–203. [CrossRef]
68. Guimard-Brunault, M.; Hernandez, N.; Roché, L.; Roux, S.; Barthélémy, C.; Martineau, J. Back to Basic: Do Children with Autism Spontaneously Look at Screen Displaying a Face or an Object? *Autism Res. Treat.* **2013**, 835247. [CrossRef] [PubMed]
69. Chevallier, C.; Parish-Morris, J.; Mcvey, A.; Rump, K.M.; Sasson, No.J.; Herrington, J.D.; Schultz, R.T. Measuring social attention and motivation in autism spectrum disorder using eye-tracking: Stimulus type matters. *Autism Res.* **2015**, *8*, 620–628. [CrossRef]
70. Rigby, S.N.; Stoesz, B.M.; Jakobson, L.S. Gaze patterns during scene processing in typical adults and adults with autism spectrum disorders. *Res. Autism Spectr. Disord.* **2016**, *25*, 24–36. [CrossRef]

71. Chawarska, K.; Ye, S.; Shic, F.; Chen, L. Multilevel differences in spontaneous social attention in toddlers with autism spectrum disorder. *Child Dev.* **2016**, *87*, 543–557. [CrossRef]
72. Vivanti, G.; Fanning, P.A.J.; Hocking, D.R.; Sievers, S.; Dissanayake, C. Social attention, joint attention and sustained attention in autism spectrum disorder and williams syndrome: Convergences and divergences. *J. Autism Dev. Disord.* **2017**, *47*, 1866–1877. [CrossRef]
73. Zantinge, G.; Van Rijn, S.; Stockmann, L.; Swaab, H. Psychophysiological responses to emotions of others in young children with autism spectrum disorders: Correlates of social functioning. *Autism Res.* **2017**, *10*, 1499–1509. [CrossRef] [PubMed]
74. Sumner, E.; Leonard, H.C.; Hill, E.L. Comparing attention to socially-relevant stimuli in autism spectrum disorder and developmental coordination disorder. *J. Abnorm. Child Psychol.* **2018**, *46*, 1717–1729. [CrossRef]
75. Rice, K.; Moriuchi, J.M.; Jones, W.; Klin, A. Parsing heterogeneity in autism spectrum disorders: Visual scanning of dynamic social scenes in school-aged children. *J. Am. Acad. Child Adolesc. Psychiatry* **2012**, *51*, 238–248. [CrossRef]
76. Hanley, M.; Mcphillips, M.; Mulhern, G.; Riby, D.M. Spontaneous attention to faces in asperger syndrome using ecologically valid static stimuli. *Autism* **2013**, *17*, 754–761. [CrossRef]
77. Irwin, J.R.; Brancazio, L. Seeing to hear? patterns of gaze to speaking faces in children with autism spectrum disorders. *Front. Psychol.* **2014**, *5*, 397. [CrossRef] [PubMed]
78. Tottenham, N.; Hertzig, M.E.; Gillespie-Lynch, K.; Gilhooly, T.; Millner, A.J.; Casey, B.J. Elevated amygdala response to faces and gaze aversion in autism spectrum disorder. *Soc. Cogn. Affect. Neurosci.* **2014**, *9*, 106–117. [CrossRef] [PubMed]
79. Yi, L.; Fan, Y.; Quinn, P.C.; Feng, C.; Huang, D.; Li, J.; Mao, G.; Lee, K. Abnormality in face scanning by children with autism spectrum disorder is limited to the eye region: Evidence from multi-method analyses of eye tracking data. *J. Vis.* **2013**, *13*, 1–13. [CrossRef]
80. Yi, L.; Feng, C.; Quinn, P.C.; Ding, H.; Li, J.; Liu, Y.; Lee, K. Do individuals with and without autism spectrum disorder scan faces differently? A new multi-method look at an existing controversy. *Autism Res.* **2014**, *7*, 72–83. [CrossRef]
81. Kleberg, J.L.; Högström, J.; Nord, M.; Bölte, S.; Serlachius, E.; Falck-Ytter, T. Autistic traits and symptoms of social anxiety are differentially related to attention to others' eyes in social anxiety disorder. *J. Autism Dev. Disord.* **2017**, *47*, 3814–3821. [CrossRef]
82. Johnels, J.A.; Gillberg, C.; Falckytter, T.; Miniscalco, C. Face viewing patterns in young children with autism spectrum disorders: Speaking up for a role of language comprehension. *J. Speech Lang. Hear. Res.* **2014**, *57*, 2246–2252. [CrossRef]
83. Elsabbagh, M.; Bedford, R.; Senju, A.; Charman, T.; Pickles, A.; Johnson, M.H. What you see is what you get: Contextual modulation of face scanning in typical and atypical development. *Soc. Cogn. Affect. Neurosci.* **2014**, *9*, 538–543. [CrossRef] [PubMed]
84. Fischer, J.; Koldewyn, K.; Jiang, Y.V.; Kanwisher, N. Unimpaired attentional disengagement and social orienting in children with autism. *Clin. Psychol. Sci.* **2014**, *2*, 214–223. [CrossRef]
85. Fischer, J.; Smith, H.; Martinezâ-Pedraza, F.; Carter, A.S.; Kanwisher, N.; Kaldy, Z. Unimpaired attentional disengagement in toddlers with autism spectrum disorder. *Dev. Sci.* **2016**, *19*, 1095–1103. [CrossRef]
86. Wilkinson, K.M.; Light, J. Preliminary study of gaze toward humans in photographs by individuals with autism, down syndrome, or other intellectual disabilities: Implications fr design of visual scene displays. *Augment. Altern. Commun.* **2014**, *30*, 130–146. [CrossRef]
87. Elsabbagh, M.; Gliga, T.; Pickles, A.; Hudry, K.; Charman, T.; Johnson, M.H. The development of face orienting mechanisms in infants at-risk for autism. *Behav. Brain Res.* **2013**, *251*, 147–154. [CrossRef]
88. Julia, P.M.; Coralie, C.; Natasha, T.; Janelle, L.; Juhi, P.; Schultz, R.T. Visual attention to dynamic faces and objects is linked to face processing skills: A combined study of children with autism and controls. *Front. Psychol.* **2013**, *4*, 1–7. [CrossRef]
89. Gillespie-Smith, K.; Doherty-Sneddon, G.; Hancock, P.J.B.; Riby, D.M. That looks familiar: Attention allocation to familiar and unfamiliar faces in children with autism spectrum disorder. *Cogn. Neuropsychiatry* **2014**, *19*, 554–569. [CrossRef]
90. Fujisawa, T.X.; Shiho, T.; Saito, D.N.; Hirotaka, K.; Akemi, T. Visual attention for social information and salivary oxytocin levels in preschool children with autism spectrum disorders: An eye-tracking study. *Front. Neurosci.* **2014**, *8*, 295. [CrossRef] [PubMed]

91. Finke, E.H.; Wilkinson, K.M.; Hickerson, B.D. Social referencing gaze behavior during a videogame task: Eye tracking evidence from children with and without ASD. *J. Autism Dev. Disord.* **2016**, *47*, 415–423. [CrossRef]

92. Mcpartland, J.C.; Webb, S.J.; Keehn, B.; Dawson, G. Patterns of visual attention to faces and objects in autism spectrum disorder. *J. Autism Dev. Disord.* **2011**, *41*, 148–157. [CrossRef]

93. Dicriscio, A.S.; Miller, S.J.; Hanna, E.K.; Kovac, M.; Turner-Brown, L.; Sasson, N.J.; Sapyta, J.; Troiani, V.; Dichter, G.S. Brief report: Cognitive control of social and nonsocial visual attention in autism. *J. Autism Dev. Disord.* **2016**, *46*, 2797–2805. [CrossRef]

94. Hallett, P.E. Primary and secondary saccades to goals defined by instructions. *Vis. Res.* **1978**, *18*, 1279–1296. [CrossRef]

95. Freeth, M.; Ropar, D.; Mitchell, P.; Chapman, P.; Loher, L. Brief Report: How Adolescents with ASD Process Social Information in Complex Scenes. Combining Evidence from Eye Movements and Verbal Descriptions. *J. Autism Dev. Disord.* **2011**, *41*, 364–371. [CrossRef] [PubMed]

96. Guillon, Q.; Rogé, B.; Afzali, M.H.; Baduel, S.; Kruck, J.; Hadjikhani, N. Intact perception but abnormal orientation towards face-like objects in young children with ASD. *Sci. Rep.* **2016**, *6*, 22119. [CrossRef] [PubMed]

97. Nakano, T.; Tanaka, K.; Endo, Y.; Yamane, Y.; Yamamoto, T.; Nakano, Y.; Ohta, H.; Kato, N.; Kitazawa, S. Atypical gaze patterns in children and adults with autism spectrum disorders dissociated from developmental changes in gaze behaviour. *Proc. R. Soc.* **2010**, *277*, 2935–2943. [CrossRef]

98. Hosozawa, M.; Tanaka, K.; Shimizu, T.; Nakano, T.; Kitazawa, S. How children with specific language impairment view social situations: An eye tracking study. *Pediatrics* **2012**, *129*, 1453–1460. [CrossRef]

99. Falck-Ytter, T.; Hofsten, C.V.; Gillberg, C.; Fernell, E. Visualization and Analysis of Eye Movement Data from Children with Typical and Atypical Development. *J. Autism Dev. Disord.* **2013**, *43*, 2249–2258. [CrossRef] [PubMed]

100. Sasson, N.J.; Elison, J.T.; Turner-Brown, L.M.; Dichter, G.S.; Bodfish, J.W. Brief report: Circumscribed attention in young children with autism. *J. Autism Dev. Disord.* **2011**, *41*, 242–247. [CrossRef]

101. Unruh, K.E.; Sasson, N.J.; Shafer, R.L.; Whitten, A.; Miller, S.J.; Turner-Brown, L.; Bodfish, J.W. Social orienting and attention is influenced by the presence of competing nonsocial information in adolescents with autism. *Front. Neurosci.* **2016**, *10*, 586. [CrossRef]

102. Sasson, N.J.; Touchstone, E.W. Visual attention to competing social and object images by preschool children with autism spectrum disorder. *J. Autism Dev. Disord.* **2014**, *44*, 584–592. [CrossRef] [PubMed]

103. Pierce, K.; Conant, D.; Hazin, R.; Stoner, R.; Desmond, J. Preference for geometric patterns early in life as a risk factor for autism. *Arch. Gen. Psychiatry* **2011**, *68*, 101–109. [CrossRef]

104. Fujioka, T.; Inohara, K.; Okamoto, Y.; Masuya, Y.; Ishitobi, M.; Saito, D.N.; Jung, M.; Arai, S.; Fujisawa, T.X.; Narita, K.; et al. Gazefinder as a clinical supplementary tool for discriminating between autism spectrum disorder and typical development in male adolescents and adults. *Mol. Autism* **2016**, *7*, 19. [CrossRef] [PubMed]

105. Moore, A.; Wozniak, M.; Yousef, A.; Barnes, C.C.; Cha, D.; Courchesne, E.; Pierce, K. The geometric preference subtype in ASD: Identifying a consistent, early-emerging phenomenon through eye tracking. *Mol. Autism* **2018**, *9*. [CrossRef] [PubMed]

106. Shi, L.; Zhou, Y.; Ou, J.; Gong, J.; Wang, S.; Cui, X.; Lyu, H.; Zhao, J.; Luo, X. Different visual preference patterns in response to simple and complex dynamic social stimuli in preschool-aged children with autism spectrum disorders. *PLoS ONE* **2015**, *10*, e0122280. [CrossRef]

107. Chawarska, K.; Macari, S.; Shic, F. Context modulates attention to social scenes in toddlers with autism. *J. Child Psychol. Psychiatry* **2012**, *53*, 903–913. [CrossRef]

108. Wang, Q.; Campbell, D.J.; Macari, S.L.; Chawarska, K.; Shic, F. Operationalizing atypical gaze in toddlers with autism spectrum disorders: A cohesion-based approach. *Mol. Autism* **2018**, *9*, 25. [CrossRef]

109. Crawford, H.; Moss, J.; Oliver, C.; Elliott, N.; Anderson, G.M.; McCleery, J.P. Visual preference for social stimuli in individuals with autism or neurodevelopmental disorders: An eye-tracking study. *Mol. Autism* **2016**, *7*, 24. [CrossRef] [PubMed]

110. Yarbus, A. *Eye Movements and Vision*; Plenum: New York, NY, USA, 1967.

111. Birmingham, E.; Cerf, M.; Adolphs, R. Comparing social attention in autism and amygdala lesions: Effects of stimulus and task condition. *Soc. Neurosci.* **2011**, *6*, 420–435. [CrossRef]

112. Grossman, R.B.; Steinhart, E.; Mitchell, T.; McIlvane, W. "Look who's talking!" Gaze Patterns for Implicit and Explicit Audio-Visual Speech Synchrony Detection in Children with High-Functioning Autism. *Autism Res.* **2015**, *8*, 307–316. [CrossRef]

113. Au-Yeung, S.K.; Benson, V.; Castelhano, M.; Rayner, K. Eye movement sequences during simple versus complex information processing of scenes in autism spectrum disorder. *Autism Res. Treat.* **2011**, 657383. [CrossRef]

114. Au-Yeung, S.K.; Kaakinen, J.K.; Benson, V. Cognitive perspective-taking during scene perception in autism spectrum disorder: Evidence from eye movements. *Autism Res.* **2014**, *7*, 84–93. [CrossRef] [PubMed]

115. Nele, D.; Ellen, D.; Petra, W.; Herbert, R. Social information processing in infants at risk for asd at 5 months of age: The influence of a familiar face and direct gaze on attention allocation. *Res. Autism Spectr. Disord.* **2015**, *17*, 95–105. [CrossRef]

116. Hedley, D.; Young, R.; Brewer, N. Using eye movements as an index of implicit face recognition in autism spectrum disorder. *Autism Res.* **2012**, *5*, 363–379. [CrossRef]

117. Snow, J.; Ingeholm, J.E.; Levy, I.F.; Caravella, R.A.; Case, L.K.; Wallace, G.L.; Martin, A. Impaired visual scanning and memory for faces in high-functioning autism spectrum disorders: It's not just the eyes. *J. Int. Neuropsychol. Soc.* **2011**, *17*, 1021–1029. [CrossRef]

118. Liu, W.; Li, M.; Yi, L. Identifying children with Autism Spectrum Disorder based on their face processing abnormality: A machine learning framework. *Autism Res.* **2016**, *9*, 888–898. [CrossRef] [PubMed]

119. Ellie, W.C.; Romina, P.; Jon, B.; Dyer, A.G. Visual scan paths and recognition of facial identity in autism spectrum disorder and typical development. *PLoS ONE* **2012**, *7*, e37681. [CrossRef]

120. Evers, K.; Van Belle, G.; Steyaert, J.; Noens, I.; Wagemans, J. Gaze-contingent display changes as new window on analytical and holistic face perception in children with autism spectrum disorder. *Child Dev.* **2018**, *89*, 430–445. [CrossRef]

121. Yi, L.; Quinn, P.C.; Fan, Y.; Huang, D.; Feng, C.; Li, J.; Lee, K. Children with Autism Spectrum Disorder scan own-race faces differently than other-race faces. *J. Exp. Child Psychol.* **2016**, *141*, 177–186. [CrossRef]

122. Yi, L.; Quinn, P.C.; Feng, C.; Li, J.; Ding, H.; Lee, K. Do individuals with autism spectrum disorder process own-and other-race faces differently? *Vis. Res.* **2015**, *107*, 124–132. [CrossRef] [PubMed]

123. Kirchner, J.C.; Hatri, A.; Heekeren, H.R.; Dziobek, I. Autistic symptomatology, face processing abilities, and eye fixation patterns. *J. Autism Dev. Disord.* **2011**, *41*, 158–167. [CrossRef]

124. Baron-Cohen, S.; Wheelwright, S.; Hill, J.; Raste, Y.; Plumb, I. The "Reading the Mind in the Eyes" test revised version: A study with normal adults, and adults with Asperger syndrome or high-functioning autism. *J. Child Psychol. Psychiat.* **2001**, *42*, 241–251. [CrossRef]

125. Müller, N.; Baumeister, S.; Dziobek, I.; BanASDhewski, T.; Poustka, L. Validation of the movie for the assessment of social cognition in adolescents with asd: Fixation duration and pupil dilation as predictors of performance. *J. Autism Dev. Disord.* **2016**, *46*, 2831–2844. [CrossRef]

126. Wieckowski, A.T.; White, S.W. Eye-gaze analysis of facial emotion recognition and expression in adolescents with asd. *J. Clin. Child Adolesc. Psychol.* **2017**, *46*, 110–124. [CrossRef] [PubMed]

127. Sasson, N.J.; Pinkham, A.E.; Weittenhiller, L.P.; Faso, D.J.; Simpson, C. Context effects on facial affect recognition in schizophrenia and autism: Behavioral and eye-tracking evidence. *Schizophr. Bull.* **2016**, *42*, 675–683. [CrossRef]

128. Wang, Q.; Lu, L.; Zhang, Q.; Fang, F.; Zou, X.; Yi, L. Eye avoidance in young children with Autism Spectrum Disorder is modulated by emotional facial expressions. *J. Abnorm. Psychol.* **2018**, *127*, 722–732. [CrossRef] [PubMed]

129. Moriuchi, J.M.; Klin, A.; Jones, W. Mechanisms of diminished attention to eyes in Autism. *Am. J. Psychiatry* **2017**, *174*, 26–35. [CrossRef] [PubMed]

130. Kliemann, D.; Dziobek, I.; Hatri, A.; Steimke, R.; Heekeren, H.R. Atypical reflexive gaze patterns on emotional faces in autism spectrum disorders. *J. Neurosci.* **2010**, *30*, 12281–12287. [CrossRef]

131. Doherty-Sneddon, G.; Riby, D.M.; Whittle, L. Gaze aversion as a cognitive load management strategy in autism spectrum disorder and williams syndrome. *J. Child Psychol. Psychiatry* **2012**, *53*, 420–430. [CrossRef]

132. Riby, D.M.; Doherty-Sneddon, G.; Whittle, L. Face-to-face interference in typical and atypical development. *Dev. Sci.* **2015**, *15*, 281–291. [CrossRef] [PubMed]

133. Chawarska, K.; Macari, S.; Powell, K.; Dinicola, L.; Shic, F. Enhanced social attention in female infant siblings at risk for autism. *J. Am. Acad. Child Adolesc. Psychiatry* **2016**, *55*, 188–195. [CrossRef]

134. Ketelaars, M.P.; In't Velt, A.; Mol, A.; Swaab, H.; Bodrij, F.; Van Rijn, S. Social attention and autism symptoms in high functioning women with autism spectrum disorders. *Res. Dev. Disabil.* **2017**, *64*, 78–86. [CrossRef] [PubMed]
135. Bal, E.; Harden, E.; Lamb, D.; Van Hecke, A.V.; Denver, J.W.; Porges, S.W. Emotion recognition in children with autism spectrum disorders: Relations to eye gaze and autonomic state. *J. Autism Dev. Disord.* **2010**, *40*, 358–370. [CrossRef] [PubMed]
136. Simonoff, E.; Pickles, A.; Charman, T.; Chandler, S.; Loucas, T.; Baird, G. Psychiatric disorders in children with autism spectrum disorders: Prevalence, comorbidity, and associated factors in a population derived sample. *J. Am. Acad. Child Adolesc. Psychiatry* **2008**, *47*, 921. [CrossRef] [PubMed]

 vision

Review

Eye Movements Actively Reinstate Spatiotemporal Mnemonic Content

Jordana S. Wynn [1,2], Kelly Shen [1] and Jennifer D. Ryan [1,2,3,*]

[1] Rotman Research Institute, Baycrest, 3560 Bathurst St., Toronto, ON M6A 2E1, Canada; jwynn@research.baycrest.org (J.S.W.); kshen@research.baycrest.org (K.S.)
[2] Department of Psychology, University of Toronto, 100 St George St., Toronto, ON M5S 3G3, Canada
[3] Department of Psychiatry, University of Toronto, 250 College St., Toronto, ON M5T 1R8, Canada
* Correspondence: jryan@research.baycrest.org

Received: 28 February 2019; Accepted: 10 May 2019; Published: 18 May 2019

Abstract: Eye movements support memory encoding by binding distinct elements of the visual world into coherent representations. However, the role of eye movements in memory retrieval is less clear. We propose that eye movements play a functional role in retrieval by reinstating the encoding context. By overtly shifting attention in a manner that broadly recapitulates the spatial locations and temporal order of encoded content, eye movements facilitate access to, and reactivation of, associated details. Such mnemonic gaze reinstatement may be obligatorily recruited when task demands exceed cognitive resources, as is often observed in older adults. We review research linking gaze reinstatement to retrieval, describe the neural integration between the oculomotor and memory systems, and discuss implications for models of oculomotor control, memory, and aging.

Keywords: eye tracking; eye movements; gaze; memory; retrieval; vision; aging

---

1. Eye Movements and Memory Encoding

The visual world is stunningly complex and taking it all in is no easy feat. Since our retina limits visual details mostly to the high-acuity fovea, we must move our eyes continuously to encode the world around us. Several times a second, visual items compete for our attention on the basis of exogenous and endogenous signals, with the winner determining which item will be selected for fixation and further processing. Models of overt visual attention (i.e., eye movements) capitalize on these features, using them to predict how real human observers will explore a given visual scene. Some notable selective attention models, such as the saliency map model [1], predict eye movements based solely on low-level (i.e., salient) visual features such as intensity, color, and edge orientation, and do so significantly better than chance. However, the power of purely bottom-up saliency-based models to predict naturalistic viewing is limited, with other work suggesting that endogenous features such as task instructions, e.g., "estimate the ages of the people in the painting" [2] (see also [3]), prior knowledge, e.g., an octopus does not belong in a barnyard scene [4], and viewing biases, e.g., the tendency to view faces and text [5] (see also [6]) can also be used to predict gaze allocation and to improve the performance of saliency-based models [6–9]. The combined influence of these cognitive factors on viewing can be summed into "meaning maps", an analogue to saliency maps generated by crowd sourcing ratings of "meaningfulness" (informativeness + recognizability) for each region of a scene [10]. When compared directly, meaning maps significantly outperform saliency maps in predicting eye movements during naturalistic scene viewing, suggesting that visual saliency alone is insufficient to model human gaze behavior.

Combined evidence from exogenous and endogenous viewing models suggests that real-world viewing behavior integrates both bottom-up and top-down signals to support selective attention. The results of this selection process have critical implications, not only for attention and perception, but also for memory. Since we cannot encode the entirety of the visual environment at once, fixations and saccades facilitate the alternating encoding and selection of relevant stimulus features. Where we look thus largely determines what we encode into memory and, as a result, what information is available for retrieval. Accordingly, behavioral findings show that recognition accuracy is significantly greater for images encoded under free viewing conditions compared to restricted viewing conditions [11,12]; see Figure 1. Furthermore, for images encoded under free viewing conditions, recognition accuracy is significantly correlated with the amount of visual sampling, i.e., mean number of fixations [13]. However, in cases of amnesia, in which a severe and lasting memory deficit arises due to damage to the hippocampus and its extended system (including the fornix, mammillary bodies and anterior thalami), the relationship between the amount of visual sampling and subsequent recognition is absent [13,14]. These case studies suggest that encoding-related eye movements support the accumulation and integration of visual information into a cohesive memory representation. Expanding on this work, recent neuroimaging findings with healthy young adults indicate that the amount of visual exploration during encoding predicts neural activity in the hippocampus, further suggesting that encoding-related eye movements are involved in the development of lasting memories [15]. Moreover, in nonhuman primates, saccades have been shown to modulate hippocampal neuron spiking and theta phase activity during visual exploration, and this modulation has been linked to memory formation [16,17]. In humans, evidence from intracranial recordings indicates that temporal coordination between saccades and alpha oscillatory activity in brain regions supporting scene perception and memory predicts successful memory encoding [18]. Thus, taken together, findings from behavioral, neuropsychological, neuroimaging, and electrophysiological research converge on a key role for eye movements in selecting and integrating visual information in the service of memory encoding. However, research has yet to reach consensus on whether and how eye movements support memory retrieval.

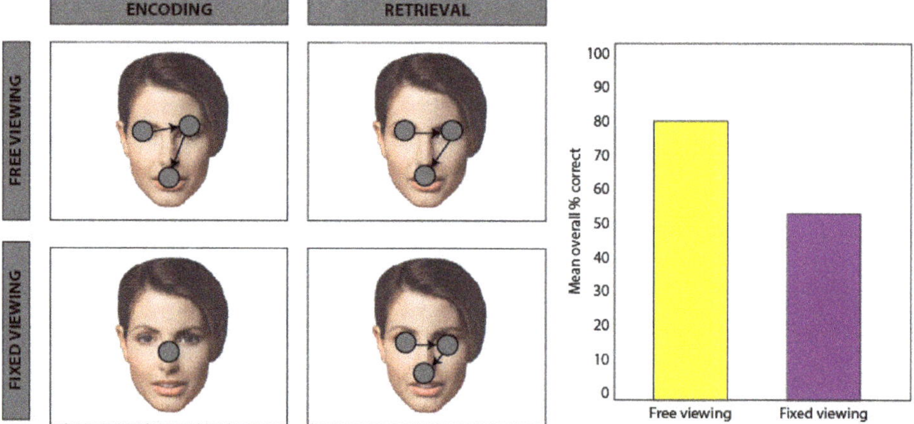

Figure 1. Schematic of encoding-related eye movement effects. Adapted from Henderson, William, and Falk, 2005 [12]. Participants viewed, and were subsequently tested on their memory for, a series of faces. During encoding (**left** column), participants were presented with images of faces. In the free viewing conditions (row 1, **left**), participants were able to move their eyes freely during learning, whereas in the fixed viewing condition (row 2, **left**), participants were required to maintain central fixation. During a recognition test (**right**), participants were presented with repeated and novel faces under free viewing conditions and were required to make an old/new recognition response. The mean percentage of correctly identified faces was significantly lower for faces encoded under the fixed viewing condition compared to faces encoded under the free viewing condition, suggesting that eye movements facilitate the binding of stimulus features at encoding for subsequent memory.

2. Eye Movements and Memory Retrieval

Although the role of eye movements in memory encoding has been well established, a relatively smaller literature suggests that eye movements may also reflect, and be guided by, the contents of memory at retrieval. For example, several studies have shown that humans make fewer fixations and view fewer regions during the examination of repeated images compared to novel images [19,20], and other studies have shown that eye movements during retrieval are disproportionately drawn to regions of a stimulus that reflect a previously learned association [21] or have changed from a prior viewing [20]. For example, a study by Hannula and Ranganath (2009) [21] found that following a scene cue, participants disproportionately directed viewing to a face that had been associated with that scene in a prior viewing compared to other previously viewed, but unpaired, faces. Moreover, this viewing effect was correlated with increased activity in the hippocampus, suggesting that eye movements reflect hippocampally-mediated memory regarding the relations among items. Notably, this eye movement-based expression of memory for relations, and associated activity in the hippocampus, was observed regardless of whether viewers had explicit recognition of the appropriate scene-face pairing. Thus, eye movements may reveal the contents of memory in an obligatory manner, such that it can occur outside of conscious awareness.

Whereas several studies have provided evidence that eye movements can reflect the contents of memory, other studies suggest that memory retrieval can *directly guide* further viewing behavior. For example, Bridge, Cohen and Voss (2017) [22] found that following the retrieval of one item in an array, participants strategically directed viewing to the other items in the display, presumably to allow for re-examination of previously studied items that were not as strongly represented in memory. This effect was associated with functional connectivity between the hippocampus and fronto-parietal control regions, supporting a possible mechanism by which memory retrieval might guide viewing behavior. Indeed, network analyses and computational modeling of anatomical connections between the hippocampus and oculomotor control regions (including frontal eye fields) indicate that there are several pathways by which memories might guide ongoing visual exploration [23,24]. However, whether eye movements can in turn influence memory retrieval remains unclear.

The characterization of eye movements as a *passive* reflection of mnemonic content has been generally agreed upon, and their guidance by memory in particular, though less extensively studied, has likewise received some support. However, the notion that eye movements *actively* facilitate memory retrieval has been a subject of contentious debate. In the sections that follow, we propose that eye movements play a functional role in retrieval by reinstating the spatial and temporal details of the encoding context, in alignment with past models of endogenous viewing guidance, current theories regarding memory function, and new evidence of memory–oculomotor network interactions.

Scanpath Theory

The notion that eye movements play a functional role in memory retrieval can be traced back to Hebb's musings on mental imagery: "If the image is a reinstatement of the perceptual process it should include the eye movements". Hebb went on to suggest that the oculomotor system, upon being presented with a "part-image", might serve as a link between perception and imagery by activating the next "part-image" [25,26], an idea that was subsequently formalized by Noton and Stark's (1971) [27,28] seminal scanpath theory. Based on the observation that participants repeatedly examining a simple line drawing produced similar patterns of eye movements with each viewing, scanpath theory proposed that image features are represented in memory along with the accompanying series of fixations and alternating saccades in a sensory-motor memory trace or *scanpath*. According to scanpath theory, recapitulation of the encoding scanpath during repeated viewing (or by extension, imagery) facilitates memory retrieval by comparing presented input with stored memory traces. Though subsequent research would depart from its strict predictions, scanpath theory's legacy lies in its groundbreaking proposal that eye movements play a functional role in retrieval, an idea that has been critical in shaping current models of memory and visual attention.

Although interpretations vary widely, scanpath theory makes two key predictions: (1) remembered stimuli should be accompanied by repetition of the scanpath enacted during encoding (see Figure 2, middle left), and (2) similarity between scanpaths during encoding and retrieval should predict memory performance (see Figure 2, bottom left). Confirming the first prediction, early studies using eye movement monitoring demonstrated that scanpaths are more similar for the same subject viewing the same image than for the same subject viewing different images or different participants viewing the same image [29–32]. Using string similarity metrics (see Figure 2 caption for description), these studies showed that eye movements during repeated stimulus presentations reinstate both the spatial locations and temporal order of eye movements enacted during novel viewing of those same stimuli. Moreover, repetition of the encoding scanpath during repeated viewing is greater than would be expected based on chance or visual saliency [33–35], or based on subject-invariant viewing biases, such as the tendency to fixate the center of an image [36,37], suggesting that eye movements reflect the content of memory above and beyond image- or subject-specific idiosyncrasies.

According to scanpath theory, retrieval-related eye movements not only reflect memory, but also play a critical role in supporting it. Consistent with this hypothesis, recent work has demonstrated that similarity between encoding and test fixations is greater for recognition hits than misses [38,39] and for "remember" responses (associated with a stronger recollection-based memory) relative to "know" responses (associated with a weaker, familiarity-based memory) [35]. These findings suggest that eye movement reinstatement is related to both objective memory and the subjective sense of memory strength. Several studies have shown that across a variety of tasks, the reinstatement of encoding-related eye movements supports memory retrieval [36,37,40–43], whereas maintaining fixation impairs it [12,42]. Yet, despite supporting evidence, and a revived interest and rapidly growing literature on eye-movement-based memory effects, scanpath theory has largely fallen out of favor in recent years.

According to scanpath theory, image recognition is achieved via "an alternating sequence of sensory and motor memory traces" matching perceived image features to remembered features, a process that is theoretically accompanied by complete reinstatement of the encoding scanpath [27,28]; see Figure 2 middle left. However, few studies have actually examined or found evidence for complete scanpath repetition [33,37,41,44]. In their initial experiments, Noton and Stark (1971) [27] noted that only initial fixations were reinstated during repeated stimulus viewings and only on 65% of trials, on average. Other studies have similarly shown that the temporal sequence of encoding fixations is not recapitulated in full during retrieval. Evidence of the reinstatement of previously sampled spatial regions is similarly varied, with some studies defining spatial similarity based on screen quadrants, e.g., [42,45–47] and others using more strictly defined grid patterns, e.g., [32,41] or experimenter-defined areas of interest, e.g., [30,45]. Despite wide variance in definitions and measures of scanpath similarity, multiple studies have found evidence for some amount of eye movement-based reinstatement during repeated stimulus presentations and retrieval. Yet, several critical questions remain unanswered. Primarily, how is the reinstatement of encoding-related gaze patterns related to underlying mnemonic content, and under what conditions does such reinstatement support memory retrieval?

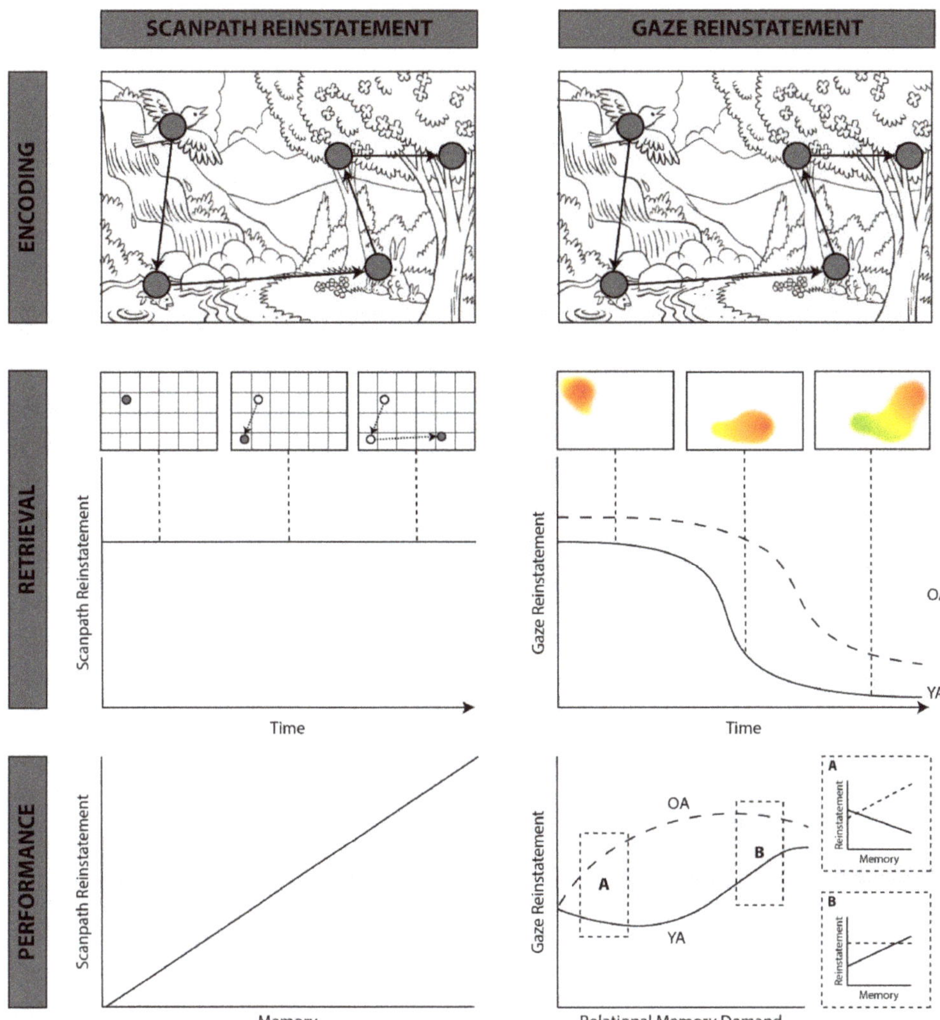

Figure 2. Schematic comparing the predictions of the standard scanpath model (**left**) and the proposed gaze reinstatement model (**right**). Scanpath model (**left**): Row 1: a simplified scanpath enacted during the encoding of a line drawing of a scene. The same encoding scanpath is used to illustrate the predictions of both the scanpath model and the gaze reinstatement model (row 2, **right**). Row 2: the predictions of the standard scanpath model regarding retrieval-related viewing. In the present example, retrieval consists of visualization while "looking at nothing". However, these predictions could similarly apply to repeated viewing of the stimulus. Early tests of scanpath theory used string similarity analyses to measure the similarity between encoding and retrieval fixation sequences [29,30,32]. These methods label fixations based on their location within predefined interest areas (often based on a grid, as shown here) and compute the number of transitions required to convert one scanpath into the other. Scanpath theory does not make any predictions regarding scanpath reinstatement over time or with memory decline [27,28]. Row 3: the predictions of the standard scanpath model regarding the relationship between reinstatement and mnemonic performance. The scanpath model predicts that

scanpath reinstatement will be positively correlated with mnemonic performance [27,28]. Gaze reinstatement model (**right**): Row 1: a simplified scanpath enacted during the encoding of a line drawing of a scene. This is the same scanpath that is used to make predictions regarding the scanpath model (top **left**). Row 2: the gaze reinstatement model proposes that retrieval-related viewing patterns broadly reinstate the temporal order and spatial locations of encoding-related fixations. In the present example, gaze reinstatement decreases across time. This would be expected in the case of image recognition, wherein reinstatement declines when sufficient visual information has been gathered, e.g., [27,28,35,47], or in the case of image visualization, when the most salient parts of the image have been reinstated, e.g., [43,48]. The duration of gaze reinstatement would be expected to change based on the nature of the retrieval task (e.g., visual search, [37]). The gaze reinstatement model additionally predicts that reinstatement will be greater and extended in time for older adults (OA), relative to younger adults (YA) [36,37]. Row 3: The gaze reinstatement model (**right**) predicts that the relationship between reinstatement and mnemonic performance is modulated by memory demands (i.e., memory for spatial, temporal, or object-object relations) and memory integrity (indexed here by age). When relational memory demands are low (A), older adults, and some low performing younger adults, use gaze reinstatement to support mnemonic performance [36]. As demands on relational memory increase (B), the relationship between reinstatement and mnemonic performance in older adults plateaus, whereas younger adults use gaze reinstatement to support performance [36,43]. Based on findings from the compensation literature [49], we predict that as relational memory demands overwhelm older adults, gaze reinstatement will not be sufficient to support performance and will thus decline, whereas in younger adults, the relationship between gaze reinstatement and mnemonic performance would plateau before eventually declining as well.

3. A New Theory of Functional Gaze Reinstatement

We propose that eye movements support online memory retrieval; that is, active retrieval that continuously updates as viewers move their eyes, by broadly reinstating the spatiotemporal encoding context based on the demands of the task and availability of cognitive resources; see Figure 2, right panel. Such reinstatement may include the spatial locations of salient encoded stimuli and/or the temporal order in which they were encoded and supports memory retrieval by reactivating additional features (e.g., semantic, perceptual) associated with sampled locations.

Whereas others have suggested that the reinstatement of any feature of the encoding event (including conceptual, linguistic, and visual features) is sufficient to reinstate the other features of that event (see [50]), we propose that gaze reinstatement specifically supports the reinstatement of the spatiotemporal context, which in turn supports the retrieval of associated event features. Indeed, spatial and temporal information have been proposed to play a foundational role in organizing and structuring percepts and memories [51,52], and the reinstatement of the encoding context has been proposed to facilitate the reinstatement of associated stimulus/event details [53,54]. Accordingly, we further propose that gaze reinstatement and the relationship between gaze reinstatement and memory retrieval are flexibly modulated by both the mnemonic demands of the task and the integrity of memory functions, such that gaze reinstatement is recruited to support memory when the mnemonic demands of the task exceed memory capacity.

Although eye movement reinstatement has traditionally been conceived as a fixation-by-fixation reactivation of associated stimulus features, widespread evidence of eye movement reinstatement across a variety of tasks and using different similarity measures suggests that the underlying mechanism is insensitive to small deviations in spatial or temporal scale. In other words, the mnemonic benefits conferred by reinstatement do not appear to rely on the reinstatement of the precise spatial locations or temporal order of encoding-related fixations, or even on complete recapitulation of the encoding scanpath. Therefore, we advocate for use of the term "gaze reinstatement" to broadly describe evidence of similarity (regardless of the measure employed) that is greater than would be expected by chance or based on subject- or image-specific characteristics. Regardless of the similarity measure used, gaze reinstatement, as defined here, reflects a specialized role for the reinstatement of subject- and image-specific gaze patterns in the retrieval of spatiotemporal contextual information. These gaze patterns need not be complete or exact reinstatements of the motor program enacted at encoding, but must contain some overlapping spatial and temporal information in order to facilitate memory retrieval (see Figure 2, middle right). In the following sections, we provide support for our proposal, drawing on evidence from eye movement monitoring, neuropsychology, neuroimaging, and emerging computational modeling to elucidate the relationship between eye movements, memory retrieval, and the neural systems involved in storing spatiotemporal relations.

4. How Does Gaze Reinstatement Support Memory Retrieval?

4.1. Spatial Reinstatement

By nature, scanpaths contain spatial and temporal information regarding the location and order of fixations. But the extent to which prior spatial and temporal information are embedded in the scanpath, and contribute to memory retrieval, remains unclear. Following the tradition of using eye movements to examine the contents of memory, e.g., [55], Ryan and colleagues (2000) [20] showed that removing or moving an item in a previously studied scene resulted in increased viewing to the now empty region that had previously contained that item, despite a lack of salient visual information at that location. Critically, this study provided one of the first demonstrations that viewers use their eyes to reinstate the memory for the exact position of a previously studied object by disproportionately viewing the region previously occupied by that object. Building on this research, several studies have since adopted a "looking at nothing" paradigm; for review, see [50], in which viewing patterns during perception are compared to viewing patterns during a period in which no stimulus is present; see Figure 3. In the absence of visual input, participants, often spontaneously, look at regions of a scene corresponding to locations described auditorily [42,46,56–60], or visually [32,40,41,61]. When cued to recall a previously presented stimulus on a blank screen, for example, participants spend a disproportionate amount of time looking in the screen quadrant in which the stimulus previously appeared relative to the other quadrants, despite them being equally devoid of visual information [41,42,46,56,57,61]. Moreover, preferential viewing of screen regions previously occupied by salient information has been correlated with an array of task performance measures, including imagery vividness [41], reaction time [37,42], memory accuracy [40,60], and change detection performance [36,43], further suggesting that the eye movement-based reinstatement of spatial contextual information supports mnemonic performance.

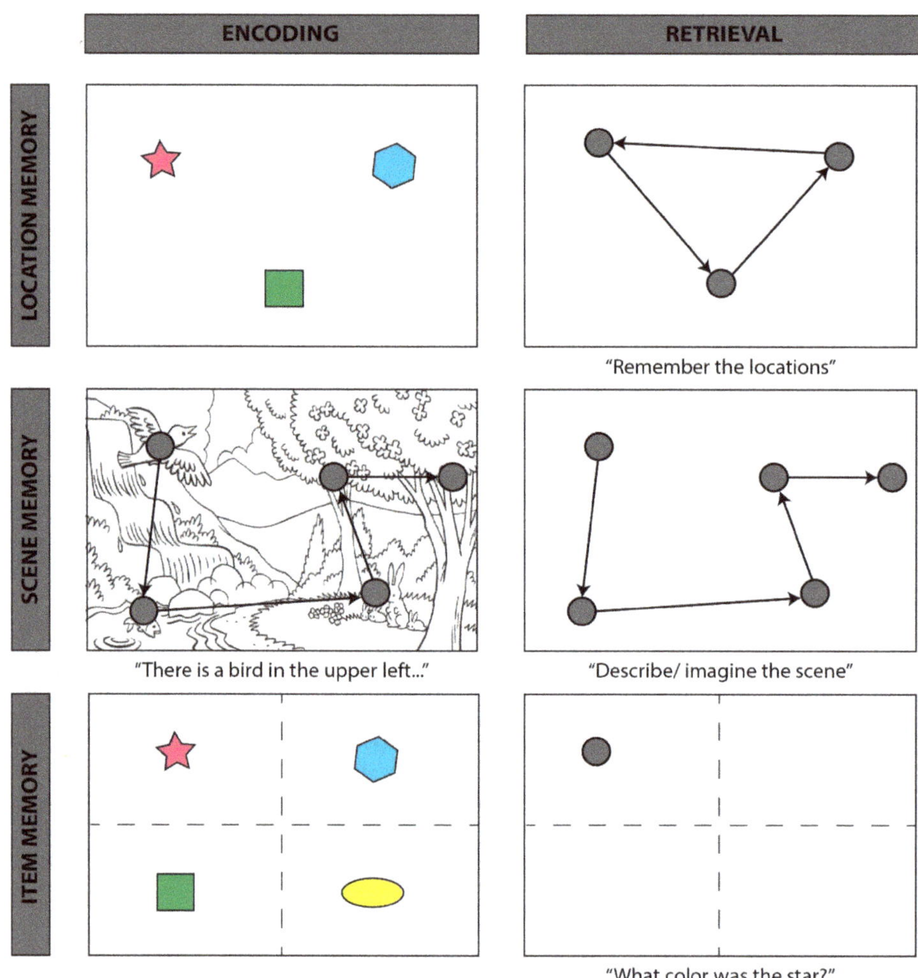

Figure 3. Schematic of "looking at nothing" behavior, whereby participants reinstate encoding-related eye movements during retrieval in the absence of visual input, across three task types. Row 1 depicts tasks in which participants are required to remember the relative locations of presented objects (**left**). During maintenance (whereby a representation is held in an active state in memory) or retrieval (**right**), participants' eye movements reinstate the locations and spatial relations among encoded objects, e.g., [36,43]. Row 2 depicts tasks in which participants are required to remember a complex scene that was presented either visually or auditorily (**left**). During retrieval (**right**), participants' eye movements return to regions that were inspected during encoding, e.g., [48,56]. Row 3 depicts tasks in which participants are required to answer questions or make judgments about previously presented items. During retrieval (**right**), participants look in the region of the scene that previously contained the target item, even when successful task performance does not require the retrieval of the previously observed spatial locations [42,45]. For within-item effects, see [40]; for words, see [46]; such effects persist even after a week-long delay [61].

4.2. Temporal Reinstatement

Like spatial reinstatement, research suggests that retrieval-related eye movements can reinstate the temporal order of the encoded scanpath, and that such reinstatement may be functional for memory retrieval. For example, using an old/new recognition task with yoked encoding (where the image is presented one part at a time) and free retrieval, Foulsham and Kingstone (2013) [48] found that participants were more likely to look at previously encoded image regions than non-encoded regions during retrieval, and that this effect followed a temporal trend, such that the region that was presented first at encoding was the region most likely to be viewed at retrieval. Given that image regions high in visual [62–64] or semantic [65] saliency are likely to be visited first during encoding, it is perhaps not surprising that these regions are also likely to be visited first during retrieval, as they facilitate the matching of present input with stored memory representations (see Figure 2, bottom right). Indeed, the preservation of temporal order in initial fixations has been widely reported in image recognition tasks [39,41,46,47,64]. Critically, however, the reinstatement of spatial locations and temporal order are often confounded. In order to tease these mechanisms apart, Rondina and colleagues (2017) devised an experiment in which studied items were manipulated in either their spatial locations or temporal order at test [52]. During the study phase, participants were presented with three unique items, presented one at a time in different locations on the screen. During the test phase, items could appear in the same locations, but a different order, or in the same temporal order, but different locations. Whereas changes in the spatial locations of test items did not affect memory for their temporal order, changes in the temporal order of test items did affect memory for their spatial locations. Moreover, changes in the temporal order of test items, but not changes in their spatial locations, lead to increased looking back (to regions previously occupied by presented test items) behavior. Extending previous work, these findings suggest that memory for temporal relations, preserved in eye movements, plays a critical organizing role in memory independent of memory for spatial relations.

Although the reinstatement of initial fixations is consistent with the time-course and cognitive demands of image recognition tasks, gaze reinstatement in more complex tasks such as visual search follows a different trajectory. Notably, targets in repeated search arrays are detected faster [66] and in fewer fixations [67] than targets in novel arrays, thus negating the possibility of complete scanpath repetition. Moreover, unlike recognition memory, visual search relies on the dual processes of image recognition and target detection, necessarily in that sequence. Accordingly, we hypothesized that repeating search arrays would result in speeded target detection and incomplete repetition of the encoding scanpath [37,68]. To evaluate the latter prediction, we measured the similarity between contiguous subsets of corresponding (same subject and image) novel and repeated viewing fixations at multiple time points across the search scanpath. In line with our hypothesis, gaze reinstatement during repeated search arrays was limited to the first few and last few search fixations, consistent with their respective roles in image recognition and target detection. Moreover, the reinstatement of initial fixations was positively predictive of repeated search performance whereas the reinstatement of fixations from the middle of the encoding scanpath was negatively predictive of repeated search performance. These findings suggest that recapitulation of the first few fixations in the scanpath provides enough information to facilitate the retrieval of the target location and subsequent target detection, while eliminating unnecessary (middle) fixations from the scanpath (i.e., those that do not contribute to image recognition or target detection). Taken together with previous work, these findings support an earlier suggestion made by Noton and Stark (1971) that the scanpath may prioritize "essential fixations at major points on the path" [27], such that only fixations that facilitate the current task goals are reinstated.

5. When Does Gaze Reinstatement Support Memory Retrieval?

Through recent theoretical and methodological advances, it is now clear that the reinstatement of encoding-related gaze patterns during memory maintenance and retrieval supports mnemonic performance by reinstating the spatiotemporal encoding context. When participants spontaneously reinstate encoding-related gaze patterns, memory performance benefits. On the other hand, when participants fail to reinstate encoding-related gaze patterns, either spontaneously or through restrictive gaze manipulations, memory suffers. Although ample evidence suggests that changes in gaze reinstatement can affect changes in memory performance, relatively fewer studies have investigated whether changes in memory can affect changes in gaze reinstatement. This question can be answered in two ways: (1) by looking at reinstatement in tasks in which memory demands are manipulated, for example by increasing memory load or delay, and (2) by looking at reinstatement in cases of memory impairment, such as that seen in healthy aging or disease. Although further research is needed to elucidate the conditions under which gaze reinstatement supports memory performance, current evidence suggests that gaze reinstatement is recruited to support performance when demands on memory exceed cognitive capacity, either by virtue of increased task difficulty or often-observed age-related memory decline.

5.1. Effects of Task Demands

In order to investigate how task-related changes affect visual sampling, and task performance, we examined gaze reinstatement across two experiments in which participants studied arrays of abstract objects and subsequently performed a change detection task on the same objects. Critically, the study and test phases were separated by a delay period of variable length, ranging from 650 ms–20 s across the two studies [36,43]. The relationship between gaze reinstatement and change detection in younger adults depended on, and increased with, the length of the delay, such that gaze reinstatement was negatively correlated with performance at short delays and positively correlated with performance at long delays. That is, when mnemonic demands were low (i.e., when objects only had to be held in memory over brief delays), only younger adults who performed poorly on the task recruited gaze reinstatement [36], but when mnemonic demands were high (i.e., when objects had to be held in memory over long delays, [43]), gaze reinstatement supported performance in younger adults by maintaining object locations in memory, (see Figure 2, bottom right). Indeed, varying task demands may explain several discrepancies in findings of gaze reinstatement. For example, shuffling the order in which encoded images are presented at test impairs memory when those images are complex stimuli [45], but has no effect when the images are simple grid patterns [48]. Together, these findings suggest that gaze reinstatement may be necessarily recruited when demands on memory exceed cognitive resources.

5.2. Effects of Memory Decline

Given that gaze reinstatement increases in response to increased task demands, we might expect similar effects in populations in whom memory function is already impaired, such as in older adults. Yet, research suggests that many endogenous viewing effects are impervious to age-related changes. For example, younger and older adults similarly benefit from intentional viewing instructions [69], active viewing (e.g., user-controlled) viewing versus passive viewing (e.g., following a moving window) [70], and free (versus constrained) viewing [11]. In line with findings of preserved top-down viewing guidance in older adults, several studies have now provided evidence for age-related increases in gaze reinstatement [36,37,71], even when performance is otherwise unaffected [36,37]. During repeated visual search, for example, older adults reinstate more initial encoding fixations relative to younger adults, although both groups show similar repetition-related increases in search efficiency [37]. Thus, whereas younger adults need only to reinstate a few encoding-related fixations in order to recognize the image and subsequently retrieve the target location from memory, older adults must refixate more of the presented image in order to gather sufficient information for comparison with internal mnemonic representations. Likewise, when older adults were tested on the previously described change detection task, they showed significantly greater delay-period gaze reinstatement than younger adults, despite similar performance [36]. This effect was strongest at short delays and decreased with increasing delay, while the opposite effect was observed for younger adults (i.e., gaze reinstatement was negatively correlated with performance at short delays, but a positive relationship was observed with increasing delay). Consistent with evidence of widespread memory dysfunction in age [72,73], these findings suggest that older adults might recruit gaze reinstatement to support mnemonic performance when task demands exceed cognitive resources. Moreover, drawing on related findings of age-related compensation, whereby older adults over-recruit cognitive and neural resources relative to younger adults to achieve similar levels of performance [49,74,75], these findings reflect part of a larger pattern of task and memory effects on gaze reinstatement. Like compensation, gaze reinstatement may follow an inverted u-shaped curve, increasing with increasing memory demands until a critical point and then decreasing. In other words, when a task is too easy or too difficult relative to memory capacity, gaze reinstatement may be either not necessary or not available to support performance; see Figure 2, bottom right. Finally, although the focus here in on aging, we expect similar, or perhaps more graded effects, in other memory-impaired populations such as those with dementia or amnesia.

5.3. What Are the Conditions under Which Gaze Reinstatement Supports Memory Retrieval?

Although the described evidence suggests that younger and older adults alike spontaneously reinstate encoding-related gaze patterns to support memory maintenance and retrieval, several studies have failed to observe these effects. Examined through the lens of the current proposal, these findings help further inform the conditions under which gaze reinstatement occurs and supports memory retrieval. For example, a study by Johansson, Holsanova, Dewhurst and Holmqvist (2012) [59] showed that participants who were forced to maintain fixation during encoding did not similarly maintain fixation at retrieval (oral recall) when they were permitted to freely move their eyes over a blank screen; see also [12], Figure 1. This finding was interpreted as evidence that, contrary to prior research, retrieval-related eye movements are not reinstatements of the specific eye movements enacted during encoding. Critically, eye movements made during retrieval did correspond to the spatial locations described by the participants during oral recall. When considered in light of the present proposal, these findings suggest that retrieval-related fixations may not reinstate encoding-related fixations when those fixations are uninformative, as in the case of fixed viewing, but, rather, reinstate the spatial context of the studied stimulus, which can be encoded overtly or covertly. Indeed, in a second experiment, participants who were forced to maintain fixation during retrieval showed significantly poorer (less detailed) recall than participants who were able to freely view the blank screen, further suggesting that although retrieval-related fixations may not precisely replicate the fixations that occur at encoding, reinstating the spatial context via gaze shifts plays an important role in memory retrieval.

In another study, Damiano and Walther (2019) [39] showed that although encoding-retrieval eye movement similarity was greater for remembered images compared to forgotten images, memory performance could only be predicted from gaze patterns at encoding, but not retrieval. Moreover, recognition hit rates were significantly reduced when viewing was restricted during encoding, whereas restricting viewing at retrieval had no effect. The authors interpreted these findings as evidence that eye movements during test play no role in memory retrieval. Critically, however, test fixations were predictive of memory when they were taken from the first 600 ms of the test phase. This finding is consistent with previous research suggesting that the reinstatement of initial fixations supports image recognition [27,28,35,37,48], and is in line with our proposal that the relationship between gaze reinstatement and memory may change across time depending on the demands of the task. Finally, although fixed viewing at test did not significantly affect the hit rate, it did significantly increase the rate of false alarms, further suggesting that gaze reinstatement is involved in the comparison of present input with stored memory representations.

Consistent with the present proposal, findings from the described studies suggest that gaze reinstatement, though not necessary for memory retrieval, can support retrieval when task demands exceed cognitive resources by reinstating the spatiotemporal encoding context in line with current goals. Extending this work, other research suggests that by binding and/or comparing spatial and temporal relations among stimulus elements, gaze reinstatement specifically supports performance on tasks that rely on memory for relations. If a task can be accomplished using non-relational mnemonic processes, e.g., semantic memory, [56] and object memory [42,43], gaze reinstatement may not be necessary or useful for performance. For example, using a simple change detection task, Olsen and colleagues (2014) [43] showed that the similarity between eye movements during the study and retention of a set of abstract visual objects correlated significantly with memory for relative, but not absolute object locations. Extending those findings to retrieval, Johansson and Johansson (2013) [42] demonstrated that constraining gaze during a memory test prolonged response times to questions regarding the relative location, but not orientation of studied objects, suggesting that retaining and reinstating the spatial index of encoding is important for recall of inter-object, but not intra-object spatial information. Taken together, these studies suggest that the relations among objects in, and with, the larger spatial context are reinstated via eye movements and may serve as a scaffold for further retrieval of detail-rich memories; for review, see [51].

5.4. What Are the Neural Correlates of Gaze Reinstatement?

The notion that spatiotemporal gaze reinstatement supports memory retrieval aligns with the purported role of the hippocampus, a region of the brain that is critical for memory and whose functions are disrupted in amnesia and in aging. It has been suggested that the primary role of the hippocampus is in the obligatory binding of relationships among elements of an encoding event (including spatial and temporal elements), which are stored independently within cortex, in order to form new memories [76–78]. Additionally, the hippocampus plays a role in the comparison of incoming relational information to relations already stored in memory in order to form integrated memory traces of complex spatial and temporal episodes [76,79–85]. Support for the role of the hippocampus in relational memory is well documented in neuropsychological studies, in which amnesic cases show impairments specific to relational memory [20,76,86], electrophysiological studies, in which cells within the hippocampus respond to space [87] and time [88], and neuroimaging studies, in which the hippocampus shows increasing activity in response to increasing task demands [89–92]. Thus, the hippocampus plays a critical role in the binding and comparison of information across space and time.

Although extensive research has documented the binding and comparison functions of the hippocampus during memory-guided behavior, research using eye movement monitoring suggests that these functions extend to active viewing behavior. As discussed earlier, the hippocampus supports the formation of lasting memories via encoding-related visual exploration [13,15–18], and the retrieval of those memories by means of memory-guided overt visual attention [22]. Extending these findings to gaze reinstatement, Ryals, Wang, Polnaszek, and Voss (2015) [38] looked at the similarity of eye movements ("exploration overlap") during study and test for novel and configurally similar computer-generated scenes. Although participants were unable to reliably discriminate between new and similar scenes at test, exploration overlap was significantly greater for similar scenes than novel scenes, and for similar scenes correctly identified as similar than those incorrectly endorsed as new. Notably, trial-by-trial variability in exploration overlap was correlated with activity in the right hippocampus, suggesting that, similar to other eye-movement-based retrieval effects, gaze reinstatement might reflect hippocampal relational memory. Moreover, activity related to exploration overlap was observed in cortical regions including middle frontal gyrus and inferior parietal cortex. Thus, gaze reinstatement may support retrieval by reactivating the spatiotemporal context (via the hippocampus) and, in turn, associated stimulus features (via hippocampal–neocortical interactions), although further research using combined eye tracking and neuroimaging will be required to determine whether this is indeed the case.

In a second study investigating the neural correlates of gaze reinstatement, Bone et al. (2018) [93] had participants study a set of images in preparation for an imagery test in which they were instructed to visualize those same images (on a blank screen) and rate their subjective vividness. Consistent with findings from other "looking at nothing" studies [32,40,41,50,61], gaze patterns during imagery were significantly similar to gaze patterns during encoding, suggesting that even in the absence of visual input, participants reinstate the context of encoding via gaze shifts. Interestingly, gaze reinstatement was significantly correlated with whole-brain neural reinstatement (i.e., similarity between image-specific patterns of brain activity evoked during perception and imagery), which was in turn correlated with subjective vividness ratings and subsequent memory performance. These findings suggest that gaze reinstatement contributes to the construction of a mental image during visualization. Taken together with the findings from Ryals et al. (2015) [38], these results suggest that gaze reinstatement is supported by the same neural mechanisms that support mental imagery and relational memory. However, further research will be required to fully elucidate the neural mechanisms underlying gaze reinstatement across different tasks and populations.

6. Implications for Models of Oculomotor Control, Memory, and Aging

Despite much research suggesting that eye movements and memory, and the neural regions underlying them, are intimately related, few models of oculomotor control account for memory. Most popular theories of oculomotor control model the guidance of eye movements based on a "priority map" that combines stimulus-driven features such as luminance, color and contrast with relevant top-down information such as task rules and prior knowledge [1,94,95]. The selection of the location of the next gaze fixation proceeds in a winner-take-all fashion where the peak/largest representation is selected, e.g., [96]. The neural instantiation of the priority map involves neural activity across a network of oculomotor control areas such as the frontal eye fields (FEF), the lateral intraparietal area (LIP), and superior colliculus (SC), which represent the location of an upcoming eye movement [97–101]. Most models also include a form of retention, which discourages the eyes from moving back to recently visited locations. This retention process is often considered as an attentional disengagement, i.e., inhibition of return [94] or mediated by visual working memory [102] via a fronto-parietal network. Yet, even with this retention process, existing models lack the power to fully explain the described effects of gaze reinstatement. These effects include looking at empty regions that previously held a salient stimulus in "looking at nothing" paradigms, and the relationship between gaze reinstatement and subsequent (relational) memory performance, particularly at longer delays. Thus, a comprehensive model of oculomotor control should consider inclusion of regions critical for long-term memory, such as the hippocampus and broader medial temporal lobe.

Recent work has suggested that the hippocampus and larger medial temporal lobe network might modulate activity in regions involved in the computation of visual saliency and selection of saccade targets [103]. Indeed, network analysis of anatomical connections between the hippocampal memory network and oculomotor network in the macaque brain indicates that there are several polysynaptic pathways by which relational memories might guide ongoing visual exploration [23]. Computational modeling of potential functional pathways connecting these two networks further indicates that stimulation of hippocampal nodes identified by the network analysis leads to observable responses in oculomotor control regions including the frontal eye fields [24]. Taken together, these findings point to a potential anatomical pathway by which spatiotemporal relational memories retrieved by the hippocampus may guide gaze reinstatement, and by which gaze reinstatement may support further memory retrieval. Although current evidence supporting the proposed role for functional gaze reinstatement is primarily behavioral, future work using neuroimaging, computational, and analytical techniques may help us to further determine whether feedback from reinstated gaze patterns can act back on the hippocampus and other memory regions to support and strengthen the retrieval of contextual and event details.

The research presented here suggests that gaze reinstatement is not only a passive reflection of the contents of memory, but that it also actively facilitates further memory retrieval by reinstating the spatiotemporal encoding context. During encoding, eye movements serve to bind and encode spatial and temporal relations among objects and the context in which they are embedded. When that encoded representation is subsequently cued, gaze reinstatement facilitates the reactivation of further details by reinstating the spatiotemporal context that links them. Over the past several decades, models of overt visual attention have begun to incorporate top-down effects along with bottom-up effects to predict eye movements. However, the same cannot be said about models of memory. In fact, studies of memory retrieval rarely examine eye movements or the possible effects of eye movements on mnemonic processes and performance. Recently, however, some reviews have called for greater incorporation of eye tracking and eye movement analysis in memory research [104,105]. Extending these appeals, we suggest that future memory research not only control for measures such as the number and duration of fixations, but also consider gaze patterns and the similarity between them. As discussed previously, retrieval-related gaze reinstatement is significantly correlated with neural reinstatement [93], which is commonly used as a measure of memory. Thus, it is possible that reports of neural reinstatement may be partially explained by overlap in eye movements between encoding and retrieval. Understanding the relationship between gaze reinstatement and neural reinstatement and other mnemonic effects will be critical to advancing memory theories.

Evidence of gaze reinstatement in younger adults critically extends ideas regarding oculomotor control and memory. But, gaze reinstatement not only supports memory performance in younger adults. In fact, gaze reinstatement shows the largest memory effects in older adults, a population that typically shows declines in hippocampal integrity (e.g., volume, structural and functional connectivity) and related deficits in relational memory [73]. Yet, despite the potential for age-related memory improvement, research on gaze reinstatement in older adults is limited, with only a few studies investigating how functional gaze reinstatement changes across the adult lifespan. Given that age-related cognitive deficits are often accompanied by significant behavioral changes, identifying early markers of age-related cognitive decline and possible strategies for overcoming it are critical targets for cognitive neuroscience research. Gaze reinstatement has the potential to address both of these questions. Studies investigating gaze reinstatement in older adults have shown that older adults recruit gaze reinstatement to a greater extent than younger adults to support memory performance [36,37]. Gaze reinstatement may be particularly invoked by older adults to reinforce the spatiotemporal context, which older adults have difficulty establishing at encoding, and to reduce internal cognitive demands. This is consistent with a more general age-related shift from reliance on internal representations to reliance on the external world for environmental support [106]. Thus, given the memory-enhancing effects of gaze reinstatement, future research may help to determine whether healthy older adults, and older adults with dementia, can be taught, or otherwise biased towards gaze reinstatement in order to boost memory performance. Answering these questions will be critical for advancing theories of aging and memory, and for developing applied interventions for aging.

7. Conclusions

Just over fifty years ago, Hebb (1968) [25] and, shortly thereafter, Noton and Stark (1971) [27,28] suggested that gaze behavior is important not only for seeing the world around us, but also for imagining and remembering it. Decades later, technological and theoretical advances have now made clear that overt visual attention and memory are intimately related. Consistent with the predictions made by Hebb [25] and Noton and Stark [27,28], research has established that eye movements carry information, not only about visual attentional selection, but also about the contents of memory. Expanding on this link between visual exploration and memory, other work suggests that eye movements play an *active* role in memory encoding and retrieval, facilitating the mnemonic processes of relational binding and comparison by shifting attention within and between salient stimulus features, or the locations previously occupied by them. We propose that this gaze reinstatement is obligatorily recruited to support memory retrieval when task demands exceed mnemonic resources by actively reinstating the spatiotemporal context of encoded stimuli, which in turn facilitates access to and the retrieval of associated stimulus features via relational memory. Future work should continue to investigate the conditions under which gaze reinstatement supports memory, including task requirements (e.g., spatial versus temporal), mnemonic demands (e.g., short term memory maintenance versus long-term memory retrieval), goal states (e.g., recognition versus visualization), and individual differences in cognitive abilities (e.g., younger adults versus older adults). In addition, future research should explore the boundary limits of gaze reinstatement, or more specifically, how much or how little of the spatial or temporal context must be reinstated in order to facilitate memory retrieval under these different conditions. For example, tasks that are more spatial in nature, such as remembering the relative locations of items in a visual array, may rely more heavily on the reinstatement of spatial relations than temporal relations, whereas tasks that are more temporal in nature, such as remembering the relative order of the appearance of items in an array, may show the opposite effect. Moreover, we might expect that gaze would be more faithfully reinstated during image repetition, wherein the spatial index of the encoded stimulus is preserved, compared to visualization, whereas gaze reinstatement may play a more significant role in memory retrieval during visualization when lack of visual input increases reliance on internal cognitive processes. Ultimately, further research will be required to better understand potential nuances in the relationship between the quality and features of gaze reinstatement, and mnemonic performance, across different task conditions. Finally, combining eye movement monitoring with other methodologies including functional magnetic resonance imaging (fMRI), magnetoencephalography (MEG), and electroencephalography (EEG) will be critical to understanding the neural mechanisms underlying the beneficial effects of gaze reinstatement on memory. Finally, although many questions remain regarding the relationship between gaze reinstatement and memory retrieval, the discussed research serves as a foundation for advancing a comprehensive understanding of visual exploration and memory as fundamentally related processes.

Funding: This work was supported by funding from the Natural Sciences and Engineering Research Council of Canada; NSERC; 482639.

Conflicts of Interest: The authors declare no conflict of interest.

References

1. Itti, L.; Koch, C. A saliency-based search mechanism for overt and covert shifts of visual attention. *Vis. Res.* **2000**, *40*, 1489–1506. [CrossRef]
2. Yarbus, A.L. *Eye Movements and Vision*; Springer US: New York, NY, USA, 1967.
3. Castelhano, M.S.; Mack, M.L.; Henderson, J.M. Viewing task influences eye movement control during active scene perception. *J. Vis.* **2009**, *9*, 1–15. [CrossRef] [PubMed]
4. Loftus, G.R.; Mackworth, N.H. Cognitive determinants of fixat ion locati on during picture viewing. *J. Exp. Psychol. Hum. Percept. Perform.* **1978**, *4*, 565–572. [CrossRef] [PubMed]
5. Cerf, M.; Frady, E.P.; Koch, C. Faces and text attract gaze independent of the task: Experimental data and computer model. *J. Vis.* **2009**, *9*, 1–15. [CrossRef]
6. Tatler, B.W.; Vincent, B.T. The prominence of behavioural biases in eye guidance. *Vis. Cogn.* **2009**, *17*, 1029–1054. [CrossRef]
7. Einhäuser, W.; Perona, P. Objects predict fixations better than early saliency. *J. Vis.* **2008**, *8*, 1–26. [CrossRef]
8. Torralba, A.; Oliva, A.; Castelhano, M.S.; Henderson, J.M. Contextual guidance of eye movements and attention in real-world scenes: The role of global features on object search. *Psy. Rev.* **2006**, *113*, 766–786.
9. Henderson, J. Human gaze control during real-world scene perception. *Trends Cogn. Sci.* **2003**, *7*, 498–504. [CrossRef] [PubMed]
10. Henderson, J.M.; Hayes, T.R. Meaning guides attention in real-world scene images: Evidence from eye movements and meaning maps. *J. Vis.* **2018**, *18*, 10. [CrossRef] [PubMed]
11. Chan, J.P.K.; Kamino, D.; Binns, M.A.; Ryan, J.D. Can changes in eye movement scanning alter the age-related deficit in recognition memory? *Front. Psychol.* **2011**, *2*, 1–11. [CrossRef]
12. Henderson, J.M.; Williams, C.C.; Falk, R.J. Eye movements are functional during face learning. *Mem. Cognit.* **2005**, *33*, 98–106. [CrossRef] [PubMed]
13. Olsen, R.K.; Sebanayagam, V.; Lee, Y.; Moscovitch, M.; Grady, C.L.; Rosenbaum, R.S.; Ryan, J.D. The relationship between eye movements and subsequent recognition: Evidence from individual differences and amnesia. *Cortex* **2016**, *85*, 182–193. [CrossRef] [PubMed]

14. Olsen, R.K.; Lee, Y.; Kube, J.; Rosenbaum, R.S.; Grady, C.L.; Moscovitch, M.; Ryan, J.D. The Role of Relational Binding in Item Memory: Evidence from Face Recognition in a Case of Developmental Amnesia. *J. Neurosci.* **2015**, *35*, 5342–5350. [CrossRef]

15. Liu, Z.; Shen, K.; Olsen, R.K.; Ryan, J.D. Visual sampling predicts hippocampal activity. *J. Neurosci.* **2017**, *37*, 1–11. [CrossRef] [PubMed]

16. Killian, N.; Jutras, M.; Buffalo, E. A Map of Visual Space in the Primate Entorhinal Cortex. *Nature* **2012**, *491*, 761–764. [CrossRef] [PubMed]

17. Jutras, M.J.; Fries, P.; Buffalo, E.A. Oscillatory activity in the monkey hippocampus during visual exploration and memory formation. *Proc. Natl. Acad. Sci.* **2013**, *110*, 13144–13149. [CrossRef]

18. Staudigl, T.; Hartl, E.; Noachtar, S.; Doeller, C.F.; Jensen, O. Saccades are phase-locked to alpha oscillations in the occipital and medial temporal lobe during successful memory encoding. *PLOS Biol.* **2017**, *15*, e2003404. [CrossRef]

19. Althoff, R.; Cohen, N. Eye-movement based memory effect: A reprocessing effect in face perception. *J. Exp. Psychol. Learn. Mem. Cogn.* **1999**, *25*, 997–1010. [CrossRef]

20. Ryan, J.D.; Althoff, R.R.; Whitlow, S.; Cohen, N.J. Amnesia is a Deficit in Relational Memory. *Psychol. Sci.* **2000**, *11*, 454–461. [CrossRef]

21. Hannula, D.E.; Ranganath, C. The eyes have it: Hippocampal activity predicts expression of memory in eye movements. *Neuron* **2009**, *63*, 592–599. [CrossRef]

22. Bridge, D.J.; Cohen, N.J.; Voss, J.L. Distinct hippocampal versus frontoparietal-network contributions to retrieval and memory-guided exploration Donna. *J. Cogn. Neurosci.* **2017**, *26*, 194–198.

23. Shen, K.; Bezgin, G.; Selvam, R.; McIntosh, A.R.; Ryan, J.D. An Anatomical Interface between Memory and Oculomotor Systems. *J. Cogn. Neurosci.* **2016**, *28*, 1772–1783. [CrossRef] [PubMed]

24. Ryan, J.D.; Shen, K.; Kacollja, A.; Tian, H.; Griffiths, J.; McIntosh, R. The functional reach of the hippocampal memory system to the oculomotor system. *bioRxiv* **2018**. [CrossRef]

25. Hebb, D.O. Concerning imagery 1. *Psychol. Rev.* **1968**, *75*, 466–477. [CrossRef]

26. Sheehan, P.W.; Neisser, U. Some variables affecting the vividness of imagery in recall. *Br. J. Psychol.* **1969**, *60*, 71–80. [CrossRef]

27. Noton, D.; Stark, L. Scanpaths in saccadic eye movements while viewing and recognizing patterns. *Vis. Res.* **1971**, *11*, 929–942. [CrossRef]

28. Noton, D.; Stark, L. Scanpaths in eye movements during pattern perception. *Science* **1971**, *171*, 308–311. [CrossRef]

29. Blackmon, T.T.; Ho, Y.F.; Chernyak, D.A.; Azzariti, M.; Stark, L.W. Dynamic scanpaths: Eye movement analysis methods. In *Human Vision and Electronic Imaging IV: SPIE proceedings*; Rogowitz, B.E., Pappas, T.N., Eds.; SPIE Press: Bellingham, WA, USA, 1999; Volume 3644, pp. 511–519.

30. Choi, Y.S.; Mosley, A.D.; Stark, L.W. String editing analysis of human visual search. *Optom. Vis. Sci.* **1995**, *72*, 439–451. [CrossRef] [PubMed]

31. Hacisalihzade, S.S.; Stark, L.W.; Allen, J.S. Visual perception and sequences of eye movement fixations: A stochastic modeling approach. *IEEE Trans. Syst. Man. Cybern.* **1992**, *22*, 474–481. [CrossRef]

32. Brandt, S.; Stark, L. Spontaneous eye movements during visual imagery reflect the content of the visual scene. *J. Cogn. Neurosci.* **1997**, *9*, 27–38. [CrossRef] [PubMed]

33. Foulsham, T.; Underwood, G. What can saliency models predict about eye movements? Spatial and sequential aspects of fixations during encoding and recognition. *J. Vis.* **2008**, *8*, 1–17. [CrossRef]

34. Underwood, G.; Foulsham, T.; Humphrey, K. Saliency and scan patterns in the inspection of real-world scenes: Eye movements during encoding and recognition. *Vis. Cogn.* **2009**, *17*, 812–834. [CrossRef]

35. Holm, L.; Mäntylä, T. Memory for scenes: Refixations reflect retrieval. *Mem. Cogn.* **2007**, *35*, 1664–1674. [CrossRef]

36. Wynn, J.S.; Olsen, R.K.; Binns, M.A.; Buchsbaum, B.R.; Ryan, J.D. Fixation reinstatement supports visuospatial memory in older adults. *J. Exp. Psychol. Hum. Percept. Perform.* **2018**, *44*, 1119–1127. [CrossRef] [PubMed]

37. Wynn, J.S.; Bone, M.B.; Dragan, M.C.; Hoffman, K.L.; Buchsbaum, B.R.; Ryan, J.D. Selective scanpath repetition during memory-guided visual search. *Vis. Cogn.* **2016**, *24*, 15–37. [CrossRef] [PubMed]

38. Ryals, A.J.; Wang, J.X.; Polnaszek, K.L.; Voss, J.L. Hippocampal contribution to implicit configuration memory expressed via eye movements during scene exploration. *Hippocampus* **2015**, *25*, 1028–1041. [CrossRef] [PubMed]

39. Damiano, C.; Walther, D.B. Distinct roles of eye movements during memory encoding and retrieval. *Cognition* **2019**, *184*, 119–129. [CrossRef] [PubMed]
40. Laeng, B.; Bloem, I.M.; D'Ascenzo, S.; Tommasi, L. Scrutinizing visual images: The role of gaze in mental imagery and memory. *Cognition* **2014**, *131*, 263–283. [CrossRef]
41. Laeng, B.; Teodorescu, D.-S. Eye scanpaths during visual imagery reenact those of perception of the same visual scene. *Cogn. Sci.* **2002**, *26*, 207–231. [CrossRef]
42. Johansson, R.; Johansson, M. Look here, eye movements play a functional role in memory retrieval. *Psychol. Sci.* **2013**, *25*, 236–242. [CrossRef] [PubMed]
43. Olsen, R.K.; Chiew, M.; Buchsbaum, B.R.; Ryan, J.D. The relationship between delay period eye movements and visuospatial memory. *J. Vis.* **2014**, *14*, 8. [CrossRef] [PubMed]
44. Humphrey, K.; Underwood, G. Fixation sequences in imagery and in recognition during the processing of pictures of real-world scenes. *J. Eye Mov. Res.* **2008**, *2*, 1–15.
45. Bochynska, A.; Laeng, B. Tracking down the path of memory: Eye scanpaths facilitate retrieval of visuospatial information. *Cogn. Process.* **2015**, *16*, 159–163. [CrossRef] [PubMed]
46. Spivey, M.J.; Geng, J.J. Oculomotor mechanisms activated by imagery and memory: Eye movements to absent objects. *Psychol. Res.* **2001**, *65*, 235–241. [CrossRef]
47. Kumcu, A.; Thompson, R.L. Less imageable words lead to more looks to blank locations during memory retrieval. *Psychol. Res.* **2018**, *0*, 0. [CrossRef]
48. Foulsham, T.; Kingstone, A. Fixation-dependent memory for natural scenes: An experimental test of scanpath theory. *J. Exp. Psychol. Gen.* **2013**, *142*, 41–56. [CrossRef]
49. Reuter-lorenz, P.A.; Cappell, K.A. Neurocognitive Aging and the Compensation Hypothesis. *Curr. Dir. Psychol. Sci.* **2008**, *17*, 177–182. [CrossRef]
50. Ferreira, F.; Apel, J.; Henderson, J.M. Taking a new look at looking at nothing. *Trends Cogn. Sci.* **2008**, *12*, 405–410. [CrossRef]
51. Robin, J. Spatial scaffold effects in event memory and imagination. *Wiley Interdiscip. Rev. Cogn. Sci.* **2018**, *9*, e1462. [CrossRef]
52. Rondina, R.; Curtiss, K.; Meltzer, J.A.; Barense, M.D.; Ryan, J.D. The organisation of spatial and temporal relations in memory. *Memory* **2017**, *25*, 436–449. [CrossRef]
53. Tulving, E.; Thomson, D.M. Encoding specificity and retrieval processes in episodic memory. *Psychol. Rev.* **1973**, *80*, 352–373. [CrossRef]
54. Kent, C.; Lamberts, K. The encoding-retrieval relationship: Retrieval as mental simulation. *Trends Cogn. Sci.* **2008**, *12*, 92–98. [CrossRef] [PubMed]
55. Parker, R.E. Picture processing during recognition. *J. Exp. Psychol. Hum. Percept. Perform.* **1978**, *4*, 284–293. [CrossRef]
56. Richardson, D.C.; Spivey, M.J. Representation, space and Hollywood Squares: Looking at things that aren't there anymore. *Cognition* **2000**, *76*, 269–295. [CrossRef]
57. Altmann, G.T.M. Language-mediated eye movements in the absence of a visual world: The "blank screen paradigm". *Cognition* **2004**, *93*, 79–87. [CrossRef] [PubMed]
58. Johansson, R.; Holsanova, J.; Holmqvist, K. Pictures and spoken descriptions elicit similar eye movements during mental imagery, both in light and in complete darkness. *Cogn. Sci.* **2006**, *30*, 1053–1079. [CrossRef]
59. Johansson, R.; Holsanova, J.; Dewhurst, R.; Holmqvist, K. Eye movements during scene recollection have a functional role, but they are not reinstatements of those produced at encoding. *J. Exp. Psychol. Hum. Percept. Perform.* **2012**, *38*, 1289–1314. [CrossRef]
60. Scholz, A.; Mehlhorn, K.; Krems, J.F. Listen up, eye movements play a role in verbal memory retrieval. *Psychol. Res.* **2016**, *80*, 149–158. [CrossRef]
61. Martarelli, C.S.; Mast, F.W. Eye movements during long-term pictorial recall. *Psychol. Res.* **2013**, *77*, 303–309. [CrossRef]
62. Parkhurst, D.; Law, K.; Niebur, E. Modeling the role of salience in the allocation of overt visual attention. *Vis. Res.* **2002**, *42*, 107–123. [CrossRef]
63. Tatler, B.W.; Baddeley, R.J.; Gilchrist, I.D. Visual correlates of fixation selection: Effects of scale and time. *Vis. Res.* **2005**, *45*, 643–659. [CrossRef] [PubMed]
64. O'Connell, T.P.; Walther, D.B. Dissociation of salience-driven and content-driven spatial attention to scene category with predictive decoding of gaze patterns. *J. Vis.* **2015**, *15*, 1–13. [CrossRef]

65. Castelhano, M.S.; Henderson, J.M. Initial scene representations facilitate eye movement guidance in visual search. *J. Exp. Psychol. Hum. Percept. Perform.* **2007**, *33*, 753–763. [CrossRef]

66. Chun, M.M.; Jiang, Y. Contextual cueing: Implicit learning and memory of visual context guides spatial attention. *Cogn. Psychol.* **1998**, *36*, 28–71. [CrossRef]

67. Chau, V.L.; Murphy, E.F.; Rosenbaum, R.S.; Ryan, J.D.; Hoffman, K.L. A Flicker Change Detection Task Reveals Object-in-Scene Memory Across Species. *Front. Behav. Neurosci.* **2011**, *5*, 1–13. [CrossRef] [PubMed]

68. Myers, C.W.; Gray, W.D. Visual scan adaptation during repeated visual search. *J. Vis.* **2010**, *10*, 1–14. [CrossRef] [PubMed]

69. Shih, S.-I.; Meadmore, K.L.; Liversedge, S.P. Aging, eye movements, and object-location memory. *PLoS ONE* **2012**, *7*, 1–7. [CrossRef]

70. Brandstatt, K.L.; Voss, J.L. Age-related impairments in active learning and strategic visual exploration. *Front. Aging Neurosci.* **2014**, *6*, 19. [CrossRef]

71. Vieweg, P.; Riemer, M.; Berron, D.; Wolbers, T. Memory Image Completion: Establishing a task to behaviorally assess pattern completion in humans. *Hippocampus* **2018**, 340–351. [CrossRef]

72. Old, S.R.; Naveh-Benjamin, M. Differential effects of age on item and associative measures of memory: A meta-analysis. *Psychol. Aging* **2008**, *23*, 104–118. [CrossRef]

73. Grady, C.L.; Ryan, J.D. Age-Related Differences in the Human Hippocampus: Behavioral, Structural and Functional Measures. In *The Hippocampus from Cells to Systems*; Hannula, D.E., Duff, M.C., Eds.; Springer International Publishing: Cham, Switzerland, 2017; pp. 167–208. ISBN 978-3-319-50406-3.

74. Grady, C. The cognitive neuroscience of ageing. *Nat. Rev. Neurosci.* **2012**, *13*, 491–505. [CrossRef]

75. Stern, Y. Cognitive reserve. *Neuropsychologia* **2009**, *47*, 2015–2028. [CrossRef]

76. Cohen, N.J.; Eichenbaum, H. *Memory, Amnesia, and the Hippocampal System*; The MIT Press: Cambridge, MA, USA, 1993; ISBN 0-262-03203-1.

77. Eichenbaum, H.; Cohen, N.J. *From Conditioning to Conscious Recollection: Memory Systems of the Brain*; Oxford University Press: New York, NY, USA, 2001; ISBN 0-19-508590-6.

78. Moses, S.N.; Ryan, J.D. A comparison and evaluation of the predictions of relational and conjunctive accounts of hippocampal function. *Hippocampus* **2006**, *16*, 43–65. [CrossRef]

79. Eichenbaum, H.; Otto, T.; Cohen, N.J. Two functional components of the hippocampal memory system. *Behav. Brain Sci.* **2010**, *17*, 449–472. [CrossRef]

80. Ryan, J.D.; Cohen, N.J. The nature of change detection and online representations of scenes. *J. Exp. Psychol. Hum. Percept. Perform.* **2004**, *30*, 988–1015. [CrossRef] [PubMed]

81. Ryan, J.D.; Cohen, N.J. Evaluating the neuropsychological dissociation evidence for multiple memory systems. *Cogn. Affect. Behav. Neurosci.* **2003**, *3*, 168–185. [CrossRef]

82. Olsen, R.K.; Moses, S.N.; Riggs, L.; Ryan, J.D. The hippocampus supports multiple cognitive processes through relational binding and comparison. *Front. Hum. Neurosci.* **2011**, *6*, 146. [CrossRef] [PubMed]

83. Yassa, M.A.; Stark, C.E.L. Pattern separation in the hippocampus. *Trends Neurosci.* **2011**, *34*, 515–525. [CrossRef]

84. Hunsaker, M.R.; Kesner, R.P. The operation of pattern separation and pattern completion processes associated with different attributes or domains of memory. *Neurosci. Biobehav. Rev.* **2013**, *37*, 36–58. [CrossRef] [PubMed]

85. Liu, K.Y.; Gould, R.L.; Coulson, M.C.; Ward, E.V.; Howard, R.J. Tests of pattern separation and pattern completion in humans—A systematic review. *Hippocampus* **2016**, *26*, 2–31. [CrossRef]

86. Hannula, D.E.; Ryan, J.D.; Warren, D.E. Beyond Long-Term Declarative Memory: Evaluating Hippocampal Contributions to Unconscious Memory Expression, Perception, and Short-Term Retention. In *The Hippocampus from Cells to Systems*; Springer International Publishing: Cham, Switzerland, 2017; pp. 281–336.

87. Rolls, E.T.; Wirth, S. Spatial representations in the primate hippocampus, and their functions in memory and navigation. *Prog. Neurobiol.* **2018**, *171*, 90–113. [CrossRef]

88. Eichenbaum, H. Time cells in the hippocampus: A new dimension for mapping memories. *Nat. Rev. Neurosci.* **2014**, *15*, 732–744. [CrossRef]

89. Cohen, N.J.; Ryan, J.; Hunt, C.; Romine, L.; Wszalek, T.; Nash, C. Hippocampal system and declarative (relational) memory: Summarizing the data from functional neuroimaging studies. *Hippocampus* **1999**, *9*, 83–98. [CrossRef]

90. Staresina, B.P.; Davachi, L. Mind the Gap: Binding Experiences across Space and Time in the Human Hippocampus. *Neuron* **2009**, *63*, 267–276. [CrossRef]

91. Henke, K.; Buck, A.; Weber, B.; Wieser, H.G. Human Hippocampus Establishes Associations in Memory. *Hippocampus* **1997**, *256*, 249–256. [CrossRef]

92. Mayes, A.; Montaldi, D.; Migo, E. Associative memory and the medial temporal lobes. *Trends Cogn. Sci.* **2007**, *11*, 126–135. [CrossRef]

93. Bone, M.B.; St-Laurent, M.; Dang, C.; McQuiggan, D.A.; Ryan, J.D.; Buchsbaum, B.R. Eye Movement Reinstatement and Neural Reactivation During Mental Imagery. *Cereb. Cortex* **2018**, 1–15. [CrossRef]

94. Wolfe, J.M. Guided Search 2.0 A revised model of visual search. *Psychon. Bull. Rev.* **1994**, *1*, 202–238. [CrossRef]

95. Hamker, F.H. Modeling feature-based attention as an active top-down inference process. *Biosystems* **2006**, *86*, 91–99. [CrossRef]

96. Klein, R.M. Inhibition of return. *Trends Cogn. Sci.* **2000**, *4*, 138–147. [CrossRef]

97. Thompson, K.G.; Biscoe, K.L.; Sato, T.R. Neuronal Basis of Covert Spatial Attention in the Frontal Eye Field. *J. Neurophysiol.* **2005**, *25*, 9479–9487. [CrossRef]

98. Ipata, A.E.; Gee, A.L.; Goldberg, M.E.; Bisley, J.W. Activity in the Lateral Intraparietal Area Predicts the Goal and Latency of Saccades in a Free-Viewing Visual Search Task. *J. Neurosci.* **2006**, *34*, 880–886. [CrossRef] [PubMed]

99. Shen, K.; Paré, M. Neuronal activity in superior colliculus signals both stimulus identity and saccade goals during visual conjunction search. *J. Vis.* **2007**, *7*, 15. [CrossRef]

100. Fecteau, J.H.; Munoz, D.P. Salience, relevance, and firing: A priority map for target selection. *Trends Cogn. Sci.* **2006**, *10*, 382–390. [CrossRef] [PubMed]

101. Bisley, J.W.; Mirpour, K. The neural instantiation of a priority map. *Curr. Opin. Psychol.* **2019**, *29*, 108–112. [CrossRef] [PubMed]

102. Shen, K.; McIntosh, A.R.; Ryan, J.D. A working memory account of refixations in visual search. *J. Vis.* **2014**, *14*, 11. [CrossRef]

103. Meister, M.L.R.; Buffalo, E.A. Getting directions from the hippocampus: The neural connection between looking and memory. *Neurobiol. Learn. Mem.* **2016**, *134*, 135–144. [CrossRef]

104. Hannula, D.E.; Althoff, R.R.; Warren, D.E.; Riggs, L.; Cohen, N.J. Worth a glance: Using eye movements to investigate the cognitive neuroscience of memory. *Front. Hum. Neurosci.* **2010**, *4*, 1–16. [CrossRef]

105. Voss, J.L.; Bridge, D.J.; Cohen, N.J.; Walker, J.A. A Closer Look at the Hippocampus and Memory. *Trends Cogn. Sci.* **2017**, *21*, 577–588. [CrossRef]

106. Lindenberger, U.; Mayr, U. Cognitive aging: Is there a dark side to environmental support? *Trends Cogn. Sci.* **2014**, *18*, 7–15. [CrossRef] [PubMed]

Review

The Nature of Unconscious Attention to Subliminal Cues

Seema Prasad * and Ramesh Kumar Mishra

Center for Neural and Cognitive Sciences, Science Complex, University of Hyderabad, Hyderabad,
Telangana 500046, India
* Correspondence: gp.seema@gmail.com; Tel.: +91-088-8619-8575

Received: 28 February 2019; Accepted: 31 July 2019; Published: 1 August 2019

Abstract: Attentional selection in humans is mostly determined by what is important to them or by the saliency of the objects around them. How our visual and attentional system manage these various sources of attentional capture is one of the most intensely debated issues in cognitive psychology. Along with the traditional dichotomy of goal-driven and stimulus-driven theories, newer frameworks such as reward learning and selection history have been proposed as well to understand how a stimulus captures attention. However, surprisingly little is known about the different forms of attentional control by information that is not consciously accessible to us. In this article, we will review several studies that have examined attentional capture by subliminal cues. We will specifically focus on spatial cuing studies that have shown through response times and eye movements that subliminal cues can affect attentional selection. A majority of these studies have argued that attentional capture by subliminal cues is entirely automatic and stimulus-driven. We will evaluate their claims of automaticity and contrast them with a few other studies that have suggested that orienting to unconscious cues proceeds in a manner that is contingent with the top-down goals of the individual. Resolving this debate has consequences for understanding the depths and the limits of unconscious processing. It has implications for general theories of attentional selection as well. In this review, we aim to provide the current status of research in this domain and point out open questions and future directions.

Keywords: attention; unconscious; subliminal; top-down; bottom-up; contingent capture

1. Introduction

We are constantly bombarded by a variety of stimuli in our daily life. At any given time, only some of it is relevant; a lot of other information may not be useful to us in any way. For instance, when you are reading a book, you need to focus on the text and not get distracted by your flatmate speaking loudly on the phone. How successfully can we ignore irrelevant objects? The last 30 years or so has witnessed a heated debate on this topic with some studies showing that the stimulus properties solely determine where and how attention is allocated (your flatmate speaking on the phone will inevitably distract you) and others proposing that attentional selection is contingent on the top-down goals of the individual (you will not get distracted if your goal is to read the book). This "attention capture debate" has mostly been restricted to attentional selection of consciously perceived information (e.g., [1–4]). However, attention can also act on unconscious information [5]. Much like the debate with consciously perceived cues, the unconscious attention capture debate (although not as extensively studied) has also been riddled with opposing viewpoints with no solution in sight.

For example, in one of the first studies to comprehensively demonstrate top-down attentional control of unconscious cues, Ansorge and Neumann [6] conclude that "the results confirm the notion of contingent attentional capture and extend previous finding to show that contingent capture also holds for nonconsciously registered information (p. 774)". In complete contrast, Schoeberl, Fuchs, Theeuwes, and Ansorge [7] claim that "the current findings unequivocally provide evidence that subliminal attentional capture is completely stimulus-driven (p. 746)". There is no clarity on why and how such differing viewpoints have arisen on a single phenomenon. In this article, we will first review the key findings regarding the relative contributions of stimulus-driven and goal-directed factors in attentional selection of unconscious stimuli. We will then address the discrepancies among these findings and propose possible explanations that might help resolve the debate. Specifically, we examine the evidence for the answer to the following question: can we selectively attend to information that is relevant to us, even when we are not consciously aware of such information?

Before getting into the question of top-down attentional selection of unconscious information, it is first and foremost necessary to examine whether it is at all possible to attend to objects in the absence of conscious awareness. Answering this question requires us to examine the relationship between attention and awareness, which is one of the most widely debated topics in cognitive psychology. Throughout this article, we will use "consciousness" and "awareness" interchangeably to denote the accessibility or the report-ability of the contents of consciousness. Researchers have long been interested in whether the mechanisms involved in attention and awareness are completely independent of each other or whether there is any relationship between them [5,8–12]. Some have argued that attention and conscious awareness are inseparable and depend on each other [10]. That is, attention is necessary for conscious awareness and vice versa. Others have argued for a clear independence suggesting that it is possible to attend to objects without being conscious of them and to be conscious of objects without attending to them [5]. Attentional modulations of unconscious stimuli provide strong evidence for such a dissociation between attention and consciousness at least in one direction—that we can attend to objects in the absence of conscious awareness.

2. Attentional Control Over Unconscious Processing

It has long been assumed that the processing of unconscious information proceeds in an "automatic" manner, independent of the goals of the individual. Strategic processing and cognitive control is assumed to be exclusive to the stimuli that are consciously perceived. This alleged association between automaticity and lack of consciousness finds its origins in some of the classic theories of attention and control [13–16]. For instance, in a highly influential paper, Posner and Snyder [15] argued that a process can be termed automatic if it fulfils the following criteria: (1) it is initiated without any deliberate intention; (2) it cannot be interfered with by external processes; and (3) it proceeds in the absence of conscious awareness. Bargh [17,18] similarly showed that any process can be termed automatic if it includes the following "four horsemen of automaticity"—efficient, unintentional, uncontrollable and unconscious. Although there is no universal definition of automaticity (see [19] for a review) so far, lack of awareness has been a defining feature of automaticity according to most of the early theories. Thus, unconscious processing has been considered to be a proto-typical example of an automatic process and the influence of this view can be seen in many areas of cognitive psychology—even today.

It is only recently that this view has been challenged in favour of newer theories of automaticity which posit that automatic processing in general depends on the configuration of the cognitive system and is susceptible to currently-active task intentions. That is, even automatic processes critically depend on higher-level, top-down factors such as attention, task-intentions and goals. This has led to more refined ideas which allow for flexible top-down control over the processing of unconscious information (e.g., [20,21]). The processing of such unconscious information is flexible in the sense that a given attentional or goal-state is necessary for unconscious stimuli to trigger further processes. However, they are "automatic" in the sense that once initiated, they proceed without awareness and are no longer under any intentional control. Thus, the control settings have to be specified in advance in the case of unconscious information—often known as pro-active control. This is in contrast to top-down control on conscious information [22] which is susceptible to both pro-active (specified in advance) and reactive control (triggered dynamically in response to ongoing stimulus processing). Thus, it has been suggested that both unconscious and conscious information are susceptible to top-down control—but they differ in terms of the temporal onset of the control processes [23].

Unconscious processing can be studied in different ways. In this review, we will focus on the unconscious cuing of attention—where stimuli that are not consciously reportable influence subsequent orienting of attention. Such studies are also typically accompanied by an objective or subjective awareness test of the stimuli to verify that they were indeed unconscious. This dissociation between the processing of unconscious stimuli and its access to conscious awareness is the hallmark of establishing unconscious priming [24]. The existence of such unconscious influences on behaviour has been hotly debated in the past [25,26]. While the phenomenon is no longer outrightly rejected, the controversy has shifted to the depths and limits of unconscious processing. Specifically, researchers are now interested in examining which aspects of a stimulus can be processed unconsciously, what is the level of interaction possible in the absence of awareness and what control mechanisms can be initiated in the absence of conscious awareness.

The extent to which attention and other factors modulate unconscious processing has been examined primarily using two different paradigms: the masked priming paradigm [27,28] and variants of the spatial orienting paradigm [29]. The masked priming paradigm (also referred to as "subliminal priming paradigm" when the primes are shown to be objectively/subjectively unconscious) has been extensively used to probe the influence of an unconscious stimulus on response behaviour [30–32]. Typically, masked primes are presented for a very short duration followed by target stimuli. The targets (for example, the numbers 1 or 2) are arbitrarily assigned to responses (say A and L). Crucially, there can be a match or a mismatch between the responses triggered by the target and the prime. When the delay between the prime and target presentations is short, it is commonly found that responses are faster when both the prime and the target are associated with the same response: prime '1' followed by target '1'. Such a paradigm—where a masked prime triggers a motor response—is typically known as the visuomotor priming paradigm. Semantic priming has also been demonstrated with subliminal primes where it is shown that the meaning of an invisible stimulus can also be accessed unconsciously. In a classic study, Marcel ([33] Experiment 4) presented a lexical decision task wherein two words were presented one after the other. The first word was pattern masked, energy masked or unmasked and the participants were not required to give any response. The second word was always unmasked and the participants had to judge whether it was a word or not. Marcel observed that participants were faster responding to the second word (BUTTER) if it was semantically associated with the first word (BREAD)—despite being unaware of the masked works. This demonstrated that semantic analysis of words can proceed without awareness. More recently, Kiefer and Spitzer [34] elucidated the time course of such an activation and observed that unconscious semantic activation decays rapidly after stimulus onset, in contrast to conscious semantic activation.

2.1. Spatial Orienting Studies with Unconscious Cues

Another line of research that has demonstrated the influence of subliminal cues on response behaviour are the spatial orienting studies using the Posner cuing paradigm [29]. In a typical experiment, on each trial a peripheral cue is flashed briefly on the screen followed by a target which requires a response (manual or ocular). The influence of the cue is usually measured in terms of the cuing effect which is the difference in response times (RTs) on invalid (when cue and target locations do not match) and valid (when cue and target locations match) trials. If the delay between the cue and target is short (usually less than 100 ms), the cue facilitates responses to the target appearing at the cued location, resulting in faster RTs on valid trials. At longer delays, the pattern is reversed and is referred to as inhibition-of-return (IOR; [35]).

McCormick [36] was the first study to use a Posner cuing task with exogenous cues below the threshold of conscious awareness. The aim of McCormick [36] was to examine the role of awareness in exogenous capture of attention. The argument was that if it can be shown that exogenous orienting can happen even without awareness, then it proves that it is indeed automatic. The theoretical antecedents to this line of thinking once again goes back to traditional theories of attention and control (e.g., [15]) which made a case for the automatic and unconscious nature of exogenous control of attention, contrasting it with deliberate, voluntary, endogenous attention. In his study, McCormick [36] presented cues either subliminally or supraliminally. Participants were told that the target would appear most often at the uncued location. Faster orienting to the target when the cue and target locations matched was seen for subliminal cues. For supraliminal cues, participants were faster when the target appeared at the expected location (that is, the uncued location) suggesting that participants were strategically able to reorient their attention to the expected target location based on the cue. In the absence of awareness, the attention capture at the cued location was more automatic.

Surprisingly, McCormick [36] failed to observe IOR at long stimulus-onset-asynchrony (SOA). Ivanoff and Klein [37] explained the absence of IOR by conducting a similar study, but with cue-report and no-report conditions. In the cue-report condition, participants were asked to report on whether they detected the presence of the unconscious cue on that trial. IOR was observed at long SOA (1005 ms) only in the no-report condition, but not in the cue-report condition. The inclusion of the cue-report task made the unconscious cues task-relevant and hence, part of the attentional set. Participants failed to disengage from the unconscious cues leading to the absence of IOR in the cue-report condition. This explained the lack of IOR in the McCormick [36] study as well because the participants in that study were additionally asked to report whether they had perceived the unconscious cue on each trial. Interestingly, Ivanoff and Klein [37] observed facilitation at short SOA (105 ms) only in the cue-report, but not in the no-report condition. The authors explained this finding by assuming that "early facilitation combined with early IOR leaving no net facilitation". This did not happen in the cue-report condition because attention never disengaged from the cue location (hence the absence of IOR). These sets of results showed that unconscious guidance of attention is susceptible to attentional control settings.

Similarly, Mulckhuyse, Talsma, and Theeuwes [38] also showed that cues presented subliminally can exogenously capture attention. A single grey circle presented for 16 ms served as the subliminal cue. The target (a black dot within a grey circle) could be presented either left or right of a central grey circle. The subliminal cue was presented exactly at the target location or at the opposite location. The task was to press the SPACE bar as soon as the target was detected. A short (0 ms) or long (1000 ms) SOA between the cue and the target was used. At short SOA, the subliminal cues facilitated responses to target on valid trials (when cue location matched the target location) as opposed to invalid trials. However, the effect was reversed at long SOA (IOR). The authors argued that this showed exogenous capture by subliminal onset cues. It was considered exogenous for several reasons: the cue was uninformative, it did not resemble the target and gave no information regarding the appropriate response for the target. Further, inhibition of return is considered to occur only when attention has initially shifted reflexively to a location (and not voluntarily). Thus, the authors argued that finding IOR at long SOA can be considered as evidence for exogenous capture of attention at short SOA by the subliminal cues (however, see [39]). This was the first study of unconscious spatial cuing used to show the classic facilitation-followed-by-inhibition effect for unconscious cues. Subsequently, several studies claimed to have shown exogenous attention capture by subliminal cues (see [40] for a review).

In contrast, Ansorge, Kiss and Eimer [41] claimed to show that subliminal stimuli can trigger goal-directed attention capture. The method used in this study was said to be based on the seminal study by Folk, Remington and Johnson [42] using the contingent-capture paradigm. In Folk et al. [42] participants searched either for a red coloured target among white distractors (colour singleton: a unique element in a display with homogenous items) or for a white cross in one of the locations (abrupt-onset: a sudden flashing of a stimulus for a brief duration). Crucially, before the target display, a cue was presented in any of the four possible target locations. The cue could be a colour singleton (one location surrounded by red dots and the others by white dots) or an onset cue (one locations surrounded by white dots). It was found that colour singleton cues captured attention when participants had to look for a colour target. Similarly, the onset cues captured attention when participants looked for an abrupt onset target. Thus, the current task goals of the participant determined what was selected for further processing. With these findings as the basis, in an ERP study, Ansorge et al. [41] presented a cue-display consisting of four circles for 17 ms. One of the colours matched the colour of the target to be searched for. The cue-display was followed by a target display on half of the trials, where the participants had to decide whether the coloured target was a diamond or a square. On the other half of the trials, no coloured target was presented. Participants were faster responding on trials when the cue location matched the target location. The key evidence came from the N2pc ERP component, which is a signature of attentional allocation to an item. Crucially, this component was observed for the target-coloured cues even when there was no target to be searched for. Based on this, the authors argued that the task-instructions ("look for x coloured target") lead to goal-oriented attentional capture by the target-coloured cues, irrespective of whether the task was to be subsequently performed on that trial or not. This was taken as evidence for the claim that the top-down attentional set a priori determines selection priority, even for unconscious stimuli. However, one major difference between this study and the contingent capture studies such as Folk et al. [42] needs to be noted. Ansorge et al. [41] only used target-matching colour cues. In contrast, Folk et al. [41] used both mismatching- and matching-colour cues and the the key argument for top-down control in their study comes from the fact that cue-validity effects were seen for colour-cue-colour-target—but not for abrupt-onset-cue-colour-target (and vice versa for abrupt-onset cues). Thus, the cue either matched/mismatched the task-set, which was not the case in Ansorge et al. [41]. To what extent this difference raises questions regarding the claim by Ansorge et al. [41] for top-down control will be discussed in later sections (Section 3.3).

2.2. Unconscious Effects on Oculomotor Selection

Can unconscious cues influence the oculomotor system? Recent evidences have demonstrated a dissociation between awareness and eye movements [43], where eye movement behaviour can be modulated by information that is not consciously accessible. One of the earliest demonstrations for this came from studies on blindsight patients with lesions in visual cortex who make saccadic eye movements to visual stimuli that they were not aware of (e.g., [44]). Superior colliculus (SC) has been implicated in meditating such visually guided behaviour in the absence of awareness through retinal projections that directly provide input to the SC bypassing the primary visual cortex [45]. Such a mechanism possibly exists because full perceptual analysis of any visual information takes time; whereas actions often need to be performed fast—sometimes even without the observer being fully perceptually aware. Eye movements provide a fast, orienting response and are continuously updated to reflect changes in the visual environment. Eye movements are also initiated at a faster time scale than manual response times, and are thus more susceptible to subtle modulations by invisible stimuli.

The influence of unconscious cues on eye movements was examined in a study by Van der Stigchel, Mulckhuyse, and Theeuwes [46]. Participants had to make a vertical saccade to a target (a diamond) placed either above or below the fixation in the presence of four placeholder circles on the screen. A subliminal cue was presented for 17 ms at one of the four locations just before the target screen. It was observed that subliminal cues in the opposite vertical field deviated the saccade trajectory to the target. This and similar other studies [47] were taken as evidence for the bottom-up capture of eye movements by subliminal cues. Weichselbaum, Fuchs, and Ansorge [48] argued that in [46] both the target and the cue were of the same contrast polarity (light grey cue/target against grey background) and thus the effects could be because of top-down capture. That is, since participants were expected to look for a light grey target, the light grey subliminal cues could have captured attention in a contingent manner because of this match. To test this hypothesis, Weichselbaum et al. [48] repeated the study with a specific manipulation: subliminal cues of two different contrast polarities were used. The contrast polarity of the cues either matched (e.g., white cues and white targets) or did not match (e.g., white cues and black targets) the contrast polarity of the targets. If subliminal cues capture attention in a top-down manner, then cues of mismatching contrast polarity should not affect eye movements. Instead, if subliminal capture is bottom up, then the contrast polarities should make no difference.

The results showed that saccade latencies to the target were slowed down by a subliminal distractor in the opposite field. Crucially, this effect was observed for both contrast polarities which the authors claim as evidence for the bottom-up nature of subliminal capture.

3. Possible Explanations for the Inconsistencies

All the studies—to the best of our knowledge—that have used variants of the Posner cuing paradigm with unconscious spatial cues and their relevant details have been tabulated in Table 1. As is evident from the table and the review of literature presented so far, the evidence for goal-oriented processing of unconscious stimuli has been mixed [40,49] within the spatial orienting paradigm. In contrast, evidence for top-down control of unconscious processing is more consistently found with the masked priming paradigm where top-down attention is mostly defined on the basis of current task-sets [20,21,50]). It is important to note that these two paradigms are designed to measure different cognitive processes. The masked priming paradigm measures the extent of unconscious influences on responses where the masked primes contain information relevant to the required response. In contrast, the spatial cuing paradigm with unconscious cues measures the orienting of attention to exogenous subliminal cues. These cues may or may not be informative regarding the location or other relevant features of the target. Below, we only review the findings obtained from the spatial cuing paradigm, attempt to fill the gap between them and reconcile the contradictory results. It is to be noted that we are only focusing on studies with unconscious spatial (peripheral) cues and will not be reviewing studies with masked central cues.

3.1. Top-Down Control over Unconscious Stimuli Is More Prominently Seen with Feature-Based Attention, Than with Spatial Attention

It is widely accepted that there are two major forms of attention: spatial and feature-based attention [51]. Spatial attention refers to the selection of particular locations in space whereas feature-based attention controls selection of particular attributes (colour/shape) of an object. These two forms of attention operate differently and are based on distinct neural mechanisms [52,53]. It is not clear which of these forms of attention is more susceptible to top-down control by goal-relevant information. Spatial attention is also known to be closely linked with the saccadic system. We constantly move our eyes—either based on where we want to look or because something in the visual field caught our eye due to its salience. This sensitivity to salient stimuli in the environment has been evolutionarily important to us in order to avoid sudden dangers in our environment. Studies that have examined the time-course of spatial vs. feature-based attention have found that the effects of feature-based attention is often delayed, whereas spatial-attention acts during the early stages of visual processing [54]. It is well known in the attention literature that slow, delayed responses reflect volitional, goal-driven processes [55] because it takes time to implement higher-order control. Similarly, based on single-neuron recordings, Hayden and Gallant [53] concluded that spatial attention is driven by both bottom-up and top-down factors whereas feature-based attention is mostly goal-driven (although this is not applicable to singleton or contrast detection which is mostly stimulus-driven, [56,57]). Thus, it might be argued that unconscious processing is perhaps more susceptible to top-down modulation when attentional deployment is feature-based. In support of this idea, Kanai, Tsuchiya, and Verstraten [58] found that top-down feature-based attention modulated the perception of invisible stimuli whereas spatial attention did not. This could be one of the reasons why masked priming studies that manipulate feature-based attention to the masked primes have predominantly found evidence for top-down control of unconscious priming, whereas the findings from spatial cuing paradigm is less consistent.

Few studies have been able to bridge this gap between masked priming and spatial cuing [6,59,60]. For instance, Scharlau and Ansorge [60] administered the temporal-order judgement task (ToJ) in which the participants were presented two stimuli (diamond and a square) with a short asynchrony and were asked to report which of the stimuli appeared first. It was found that a masked prime (a smaller replica of the target) facilitated the conscious perception of the stimulus at the primed location. This effect, known as perceptual latency priming (PLP), is considered to be attentional in nature and is said to arise from shifts of attention to the primed location ([61–63], however, see [64] for alternative non-attentional explanations of ToJ task). Crucially, Scharlau and Ansorge [61] observed that PLP was greater when the prime features matched the target features than when they matched the non-target features, suggesting that attention is captured more by unconscious information that match the participant's current task-goals. Thus, top-down attentional effects of unconscious stimuli have been seen not just on response times, but also on the latency of conscious perception.

In another classic study, Ansorge and Neumann [6] used masked primes within the peripheral cuing paradigm and comprehensively demonstrated the role of task-intentions on unconscious priming effects. The target display in this study was an array of two rectangular boxes arranged horizontally. Horizontal bars above and below one of the boxes marked the target. The prime display was a replica of the target display, but with smaller boxes. Similar to the target display, horizontal bars above/below one of the boxes marked the prime. The task was to search for a coloured target and respond based on its location. The unconscious primes were presented either left or right of the fixation cross. The targets could be presented at the primed location or in the opposite location. Priming effects were observed when the prime matched the current task-intentions. That is, a red-coloured prime (but not a black-coloured prime) resulted in priming effects when the task was to look for a red target. Thus, the task-goals are said to selectively enhance certain features of the masked primes depending on their relevancy to the task. This is in line with the findings from contingent-capture studies which demonstrate that deployment of spatial attention depends on feature-based attentional control settings [42]. As can be seen from Table 1, all the studies that have demonstrated goal-oriented processing of unconscious primes (in bold) have used a variant of the feature-based cuing paradigm (except [37], where the unconscious cues matched the top-down goals in the cue-report condition). In these studies, a particular target feature is cued (mostly, colour) and the participants have to perform a discrimination task on the target [41,59,65–67]. Thus, it is possible that top-down attentional control on unconscious priming is seen mostly when attentional deployment is feature-based—even within a spatial orienting paradigm.

Some researchers argue that the effects seen in feature-based attentional paradigms such as the contingent-capture paradigm can be attributed to inter-trial priming and not necessarily to top-down control ([68,69], however, see [70]). In Folk et al. [42], for instance, participants searched for a specific type of target for an entire block, which could have led to bottom-up inter-trial priming. That is, the target of the previous trial could have primed the target of the current trial. To test whether contingent capture is seen even when the trials (and target-types) are mixed, Belopolsky, Schreij and Theeuwes [69] replicated the Folk et al. study with a minor change—participants were cued before each trial to look for either an abrupt-onset target or a colour-target. Thus, the type of target could change from trial-to-trial and was not constant across a block of trials. The results showed no evidence for contingent-capture. This and other similar findings [71–73] have lead researchers to argue that so-called top-down effects seen in such feature-based attentional studies could originate from the establishment of attentional set as a result of previous (trial) experiences with the target and might have nothing to do with top-down control. A similar argument can be made for unconscious cuing studies showing contingent-capture [6,41]. Thus, it is possible that task-goals played little to no role in these studies and the observed effect could be attributed to repeated exposure to the same type of target leading to bottom-up activations associated with the target. To provide definitive evidence for establishment of top-down attentional sets, it is necessary to have mixed designs where the task-relevancy or task-goal changes dynamically from trial to trial. Such a design rules out contributions from inter-trial priming and can provide strong support for strategic, top-down attentional control over unconscious attention.

3.2. Abrupt-Onset Cues Automatically Capture Attention

Abrupt onset cues have been extensively used to study unconscious cuing. Most such studies have observed attention capture by cues irrespective of task-goals (e.g., [38,46] lending support to the claim that unconscious attention is primarily stimulus-driven. In fact, many have argued that supraliminal abrupt-onset cues are a special category of stimulus that capture attention automatically irrespective of task-goals [74]. In contrast, goal-driven theories have argued that abrupt-onset cues capture attention when the task is to respond to an abrupt-onset target [42,75]. In these cases, the abrupt-onset cues become part of the participants' attentional set thereby facilitating capture. Recently, Gaspelin, Ruthruff, and Lien [76] provided a parsimonious explanation of these various contradictory findings regarding the nature of attention capture by irrelevant supraliminal cues. They proposed that abrupt onset cues always capture attention, but the extent to which attention dwells on the cues and manifests as cuing effects depends on the difficulty of the task. When the search for the target is easy (e.g., to look for a red target among blue or green distractors), the abrupt onsets that do not match the top-down goals are rejected more easily and cuing effects are lesser. However, when the search is difficult (search for E among other letters), it is harder to decide whether the cue is really a distractor and thus the cue is slowly rejected resulting in increased cue effects. In short, the authors argued that evidence for bottom-up capture by abrupt onset cues is more commonly found in studies with a difficult search, because an easy search is an insensitive measure of attention capture.

It is to be noted that the attentional dwelling hypothesis is formulated based on the modified spatial cuing paradigm, where the task is to search for a target in the presence of distractors. Most studies with abrupt-onset unconscious cues have used detection or localisation tasks ([38,77–80] Experiments 1–3). The predictions of the attentional dwelling hypothesis can be tested on unconscious cues in a study where the task is to search for a target—which is either easy or difficult. For instance, in Fuchs and Ansorge ([77] Experiments 4 and 5), participants had to look for a colour-defined target (red among green and blue) following a subliminal abrupt-onset cue (in black colour). The authors observed no cuing effects indicating that the cues completely failed to capture attention because of the mismatch between the cue and target colours. This was taken as evidence that abrupt-onset cues do not necessarily always capture attention. However, based on the attentional dwelling hypothesis one could argue that because the study used an easy colour search, attentional dwelling effects at the cued location were masked resulting in much smaller, if not no cuing effects. The same study with a difficult letter search might have given rise to cuing effects, suggesting that attention-capture by the subliminal cues in this study was stimulus-driven and not goal-driven. Ansorge and Neumann [6] similarly used an easy colour search with abrupt-onset cues to argue for goal-driven modulation of the processing of unconscious cues.

Future studies are required to definitively assess whether the implications of the attentional dwelling hypothesis is also valid for unconscious attention capture. It is necessary to conduct studies in which the search difficulty is manipulated in a modified spatial cuing paradigm with abrupt-onset cues. Several studies have shown that colour cues are relatively less efficient in capturing attention in a bottom-up manner compared to abrupt-onset cues [81]. In line with this, most studies that have found evidence for goal-driven attention capture with unconscious cues have used colour singleton cues in a colour search task ([6,41,59,65–67] Experiment 5). It is possible that search difficulty plays a role in attention capture by subliminal onset cues, but not for colour cues. A single study comparing both in a paradigm designed to test the dwelling hypothesis will throw more light on how the type of subliminal cue plays a role in the orienting of unconscious attention.

3.3. Inclusion of Both Matching and Non-Matching Cues

What establishes the involvement of top-down control in attention capture? Conversely, when is attentional capture considered to be truly exogenous? For instance, attentional orienting in Posner cuing paradigm with abrupt peripheral cues has been considered the hallmark of exogenous, stimulus-driven attention capture. This is primarily because the abrupt cues are spatially uninformative of the target location. However, even though a sudden onset cue is spatially uninformative, it still matches the top-down goals of the participant when the task is to look for an onset singleton. For instance, the participants in Van der Stigchel et al. [46] were asked to make a saccade to a singleton target.

The subliminal cue was a grey circle, which was also a singleton. The participants can be said to adopt a singleton detection mode [82] where the presence of a singleton cue captures attention as it is part of the task-set to look for a singleton target. This criticism is applicable to many of the studies tabulated in Table 1 where both the cue and target were abrupt-onsets. Thus, the findings of Van der Stigchel et al. [46] and several subsequent studies ([48,59,78] Experiments 1–3), do not reflect purely stimulus-driven capture as they can be explained through top-down contingent attentional capture.

Schoeberl et al. [7] acknowledged this limitation of previous findings and addressed it in a study where they presented cues that were truly non-matching with the top-down goals and examined whether they facilitated the search for a colour target. The authors ensured that cues did not have any matching feature with the target by pairing a colour target with a singleton cue. If the contingent capture account is valid for subliminal cues, the authors argued that non-matching cues should not lead to cuing effects. If the bottom-up account is valid, then cuing effects should be observed even though the cues did not match the goals of the participant. The authors found robust cuing effects which led them to conclude that attention capture by subliminal cues is "completely stimulus-driven". The key difference between these studies and studies such as Ansorge et al. [41,65] is that Ansorge et al. [41,65] argued for the top-down contingent capture based on attention capture by matching cues, whereas the Schoeberl et al. [7] presented only non-matching cues. Is it possible that contingent capture is observed only when a matching cue condition is compared with non-matching cue condition? As can be seen from Table 1, all the studies so far that have found evidence for top-down contingent capture have compared matching and non-matching cues, except [41]—which only had a matching cue condition. However, even in the Ansorge et al. [41] study, the claim for top-down control may not be warranted because they only used target-matching colour cues. Not finding N2pc on mismatch trials would have been a strong evidence for contingent-capture. Since they did not have such a condition, it is not possible to draw firm conclusions on how effectively top-down control was exercised.

It is important to compare the effects of both types of cues to firmly establish their relative contributions to attention capture. For instance, Lamy et al. [66] conducted a study where a single subliminal colour cue either matched the target colour or the distractor colour. The task was to search for a target defined by its colour and report its orientation. They found that only target-coloured cues elicited cuing effects, indicating that attention capture is contingent on the participant's attention set. Including both matching and non-matching cues is important because attention is primarily a selective process. Selection occurs in the face of competition—when different stimuli compete for capacity-limited resources of the brain. The attentional system is forced to select only relevant or important information when there is competition from irrelevant information. For instance, attentional effects are found to emerge when targets are presented along with distractors—but not when they are presented alone [83–85]. Thus, it is possible that goal-relevant unconscious stimuli are more likely to be preferentially processed when there is competition for selection from goal-irrelevant stimuli. In the absence of this competition, there is no need for the participants to engage their control mechanisms. Barring few exceptions (see also [65,86], not many studies have included both matching and non-matching cues. Further, in all these studies, match/mismatch has been manipulated at the feature-level (colour/shape, etc.). It has been known for long that spatial attentional control settings can influence attention capture by supraliminal cues ([87] Experiment 2). No study so far—to the best of our knowledge—has examined this in the context of unconscious attention. It is necessary to conduct more such studies involving feature-based or spatial match/mismatch between the cue and the target to understand the role of attentional control settings on unconscious cuing.

Table 1. Studies examining the nature of attentional control over unconscious processing using variations of the spatial orienting paradigm.

Sl No.	Study	Cue	Target	Task
[36]	McCormick (1997)	Abrupt onset	Abrupt onset	Discrimination
[79]	Leuthold and Kopp (1998)	Abrupt onset	Abrupt onset	Localisation
[37]	Ivanoff and Klein (2003)	Abrupt onset	Abrupt onset	Go/No-go
[38]	Mulckhuyse, Talsma, and Theeuwes (2007)	Abrupt onset	Abrupt onset	Detection
[6]	Ansorge and Neumann (2005)	Abrupt onset (Colour: match/mismatch)	Colour-defined target	Search followed by localisation
[41]	Ansorge, Kiss and Eimer (2009)	Colour singleton (Colour: match)	Colour-defined target [1]	Search followed by discrimination
[65]	Ansorge, Horstmann, and Worschech (2010)	Colour singleton (Colour: match/mismatch)	Colour-defined target	Go/No-go followed by discrimination
[46]	Van der Stigchel, Mulckhuyse, and Theeuwes (2009)	Abrupt onset	Abrupt onset	Localisation (eye movement)
[59]	Held, Ansorge, and Mueller, 2010 (Experiments 1–4)	Feature-singleton	Feature-singleton	Localisation
[59]	Held, Ansorge, and Mueller, 2010 (Experiment 5)	Colour singleton (Colour: match/mismatch)	Colour- and shape defined target	Search followed by discrimination
[47]	Mulckhuyse and Theeuwes, 2010	Abrupt onset	Abrupt onset	Search followed by localisation (eye movement)
[39]	Fuchs and Ansorge, 2012a (Experiments 1–4)	Abrupt onset (contrast polarity: match/mismatch)	Abrupt onset	Detection
[77]	Fuchs and Ansorge, 2012b (Experiments 1–3)	Abrupt onset (contrast polarity: match/mismatch)	Abrupt onset	Detection
[77]	Fuchs and Ansorge, 2012b (Experiments 4 and 5) [2]	Abrupt onset (Colour: mismatch)	Colour-defined target (non-singleton in Expt 4 and singleton in Expt 5)	Search
[77]	Fuchs and Ansorge, 2012b (Experiments 6) [2]	Colour singleton (Colour: mismatch)	Colour-defined target	Search

Table 1. *Cont.*

Sl No.	Study	Cue	Target	Task
[78]	Fuchs, Theeuwes, and Ansorge, 2013(Experiments 1 and 2)	Abrupt onset (contrast polarity: match/mismatch)	Abrupt onset	Detection
[78]	Fuchs, Theeuwes, and Ansorge, 2013(Experiment 3)	Abrupt onset (contrast polarity: match/mismatch)	Abrupt onset	Go/No-go
[48]	Weichselbaum, Fuchs, and Ansorge, 2014	Abrupt onset (contrast polarity: match/mismatch)	Abrupt onset	Detection (eye movement)
[7]	Schoeberl, Fuchs, Theeuwes, and Ansorge (2015)	Abrupt onset	Colour-defined target	Search followed by discrimination
[66]	Lamy, Alon, Carmel, and Shalev (2015)	Colour singleton (colour: match/mismatch)	Colour-defined target	Search followed by discrimination
[80]	Schoeberl and Ansorge (2018)	Abrupt onset (temporally predictive/unpredictive)	Colour-defined target	Search followed by discrimination
[67]	Travis, Dux, and Mattingley (2018)	Colour singleton (colour: match/mismatch)	Colour-defined target	Search followed by discrimination
[86]	Schoeberl, Ditye and Ansorge (2018)	High or low frequency cue (frequency: match/mismatch)	Frequency-defined target	Search followed by discrimination

Note: the studies that found (or claimed to have found) evidence for top-down attentional capture are in bold. The text inside the brackets in the "Cue" column mention whether both matching and non-matching cues or only one of them was presented. The dimension along which the properties of the cue (colour or contrast polarity etc.) matched or mismatched the target properties is also mentioned. [1] Unless specified, "colour-defined target" refers to a non-singleton target. [2] No cuing effects observed.

3.4. Attention Capture Can Be Triggered by Both Stimulus and Goal-Driven Factors—However, Attentional Engagement Is Goal-Driven

Attentional selection first involves the shift of attention to an object/location ("attention capture") which is followed by enhanced processing of the information related to that object or location ("attentional engagement" [88,89]). More recent studies have shown that stimulus-driven and goals-driven factors might contribute differently to these two mechanisms. For instance, Zivony and Lamy [90] presented supraliminal abrupt-onset cues followed by a display of four letters. The task was to find the red-lettered target (e.g., "E") among distractors ("E" and "H" in pink/orange) and press a button accordingly. The target could appear either at the cued location or the uncued location, giving rise to valid and invalid trials. Crucially, distractor compatibility effects were calculated when the cue and the target appeared in different locations. The trials for which the cued distractor shared the identity of the target letter (e.g., E, but in pink or orange) were termed as compatible trials. The trials for which the cued distractor was different than the target letter (e.g., H in pink/orange) were termed incompatible trials. Thus, the compatibility effects were an index of the extent to which the identity of the distractor was processed at the cued location—reflecting the degree of attentional engagement triggered by the cue. The authors found attention capture (cuing effect) by both target-matching and non-matching cues; whereas attentional engagement (distractor compatibility effect) was observed only with matching cues. Thus, the authors argued that whereas the initial shift of attention to a location can proceed in a bottom-up manner, further engagement with the objects at that location occurs only if those objects or their features are relevant to the current goals.

It should be noted that this study was conducted with supraliminal cues. Most of the studies using the spatial cuing paradigm with unconscious cues have measured cuing effects, and thus only the extent of attention capture but not attentional engagement is known. The crucial distinction made by Zivony and Lamy [90] is that shifts of spatial attention during capture are "shallow" and can occur irrespective of top-down goals. However, further enhanced processing of the identity of the object at the cued location is costly and occurs only if it is necessary for the task at hand. If top-down control is possible only for attentional engagement, this could explain the bottom-up unconscious capture observed in studies like Schoeberl et al. [7]. The unconscious cues in Schoeberl et al. [7] could have captured attention because the study only measured the extent of the shift in spatial attention to the cue. It is possible that non-matching cues would not have had any influence if the participants were required to process/reject the identity of the cue. There have been other such studies that presented both matching and non-matching abrupt-onset cues [48,78] and found similar cuing effects for both types of cues. However, again, these studies only measured the extent of attention capture and not engagement. Thus, it might be necessary to revise the existing ideas of top-down attentional control by making clear distinctions between attention capture and engagement.

This hypothesis may seem to be in contrast with the findings of Ansorge et al. [65] who compared the effects of target-matching and non-matching cues on a task that required the participants to search for a colour-defined target and discriminate based on the shape. The subliminal cues could either be congruent (with response-relevant shape information) or incongruent (response-irrelevant shape information) with the target and could be presented at the target location (valid trials) or in a different location (invalid trials). The authors observed cuing effects only for the target-matching colour cues and not for the non-matching cues, suggesting that non-matching cues failed to capture attention. It would have been reasonable to expect cuing effects for both matching and non-matching cues based on the earlier argument regarding attention capture. Instead, no evidence of capture by non-matching cues were observed. However, this only suggests that both stimulus-driven and goal-driven factors are capable of influencing attention capture. The key argument based on the Zivony and Lamy [90] study is that attentional engagement is solely goal-driven. Interestingly, no response congruency effects were observed at all in Ansorge et al. [65] thus failing to observe any evidence for engagement. Further, the response congruency in Ansorge et al. [65] was manipulated on both valid and invalid trials because their aim was to see if response congruency effects can be seen irrespective of shifts of attention triggered by the cue. In contrast, the compatibility effects in Zivony and Lamy [90] were calculated only on invalid trials. A replication study of Zivony and Lamy [90] with unconscious cues will shed further light on whether the above-mentioned differences between attention capture and engagement are also applicable for unconscious guidance of attention.

4. Neural Mechanisms of Top-Down Attention to Unconscious Stimuli

Is there any neural evidence for top-down selection of unconscious stimuli? As mentioned before, Ansorge et al. [41] found a significant N2pc (which reflects spatial shifts in attention, [91]) for the relevant subliminal cues. Using the continuous flash suppression paradigm, Travis et al. [67] recently replicated the contingent capture findings of Ansorge et al. [41] to find the neural correlates of top-down suppression of unconscious cues. Using a similar feature-based cuing paradigm, the authors found that only subliminal cues that matched the target colour were successful in orienting the participants' attention to the cued location, thereby eliciting cuing effects. More importantly, authors found a significant N2pc for the target-colour cues and Pd (distractor positivity—a signature of distractor suppression, [92]) for the distractor coloured cues. The authors interpret the findings based on the signal suppression hypothesis [92,93] according to which top-down suppression mechanisms inhibit attention capture by distractor stimulus and enhance the processing of task-relevant stimuli. This position is a radical departure from earlier theories of visual selection, according to which attention is initially captured by salient distractors, and then redirected to task-relevant stimuli. Support for the signal-suppression hypothesis also comes from a recent study on monkeys, which used a combination of single neuron recordings and surface ERP measures [94]. Cosman et al. [94] found that neurons in the pre-frontal cortex and frontal eye field are responsible for both selection of relevant features and the suppression of irrelevant distractor items. Thus, the findings of Ansorge et al. [41] and Travis et al. [67] provides neural evidence of goal-oriented processing of relevant stimuli even in the absence of awareness.

Goal-directed processing of relevant stimuli or cognitive control has typically been associated with activations in the higher-order brain areas such as the pre-frontal cortex. Whereas such complex control of behaviour by the brain has been mostly restricted to the domain of consciousness [95], more recently, evidence has emerged that areas in the pre-frontal cortex can also be activated in response to unconscious stimuli [96,97]. Evidence has come from studies that have tracked the neural correlates of unconscious inhibition using tasks such as go no-go or stop-signal tasks, which tap into inhibitory control or conflict resolution [97–100]. Additionally, Ulrich, Adams, and Kiefer [101] showed dynamic functional connectivity between brain areas depending on the configuration of the task-set using fMRI. In a semantic categorisation task with subliminal primes, higher activation was found in regions responsible for semantic processing. This in turn led to enhanced processing of the semantic properties of the unconscious stimuli. This study provides neurobiological evidence for the attentional sensitisation model of unconscious cognition [21] and explains how the selective processing of unconscious stimuli based on their relevance to the current task goals is realised in the brain.

5. Conclusions

We have examined evidence for top-down attentional control of unconscious cuing based on studies from the spatial cuing literature. Our intention with this review was to bring out the inconsistencies in the unconscious cuing literature and offer possible explanations for them. It is not possible to offer a parsimonious explanation accounting for all the findings due to the paucity of the studies on this topic. We have examined which form of attention is more susceptible to top-down control (Section 3.1), if the studies claiming to have shown goal-driven (Section 3.2) or stimulus-driven (Section 3.3) attention capture have indeed done so, and the importance of distinguishing between attention capture and engagement (Section 3.4). Analysing the findings from all the studies covered in Table 1 suggests that evidence for top-down cuing effects with unconscious cues is more likely to be found when both matching and non-matching cues are presented on a task that typically involves some form of feature-search. However, most of such evidence comes from studies with colour singleton cues. All the studies with abrupt onset cues and abrupt-onset targets have found evidence for attention-capture which can be explained based on the contingent-capture account.

Our review has revealed that demonstration of top-down control over unconscious cuing depends on several crucial factors such as: the type of cue (abrupt onset vs. colour singletons), the relationship between the properties of the cue and the target features (whether they match or do not match) and finally, whether the task calls for attention capture or engagement. Teasing apart the individual contributions of each of these factors requires conducting experiments manipulating each of these factors with the same set of stimuli. Thus, it is necessary to conduct studies using a feature-based cuing paradigm involving both abrupt onset and colour singleton cues that match/ do not match the target features being searched for. Examining whether the differences in attention capture and attentional engagement can explain some of the inconsistent findings in the literature will be useful as well. While there are a handful of studies currently that have included one or more of these manipulations, it is necessary to conduct and replicate more studies to arrive at a comprehensive understanding. Additionally, especially in studies involving abrupt-onset cues, it is necessary to manipulate the search difficultly of the task and test the predictions of attentional dwelling hypothesis [76]. This will inform us whether unconscious abrupt-onset cues are indeed a special category of stimulus that capture attention automatically or if their effects can be explained based on the contingent capture account. There have also been other recent advances regarding the fate of irrelevant, distractor stimuli during visual processing. The signal suppression hypothesis [92], for instance, proposes that top-down signals (triggered by the setting up of the attentional set) proactively suppress the representation of irrelevant distractors, much before they capture attention. It is necessary to examine whether the signal-suppression hypothesis can also be extended to attentional selection of unconscious stimuli.

The evidence for top-down control over unconscious processing has far-reaching implications. Understanding the role of attention in the processing of unconscious stimuli assumes importance due to its potential in throwing more light on the relationship between attention and awareness. It is one of the most crucial pieces of evidence for the dissociation between attention and awareness as it illustrates that it is possible to attend to objects in the absence of conscious awareness. Demonstrations of top-down unconscious attention can also inform us regarding the nature of cognitive control and its dependence (or the lack of it) on conscious awareness. Importantly, it shows that unconscious processing is flexible and can adapt to the requirements of the individual.

Author Contributions: Writing—original draft preparation, S.P.; writing—review and editing, S.P. & R.K.M.; supervision, R.K.M.

Funding: The authors received no specific funding for this work.

Acknowledgments: The authors thank Vaishnavi Mohite and Aniruddha Ramgir for helpful discussions on some of the ideas discussed in the review.

Conflicts of Interest: The authors declare no conflict of interest.

References

1. Awh, E.; Belopolsky, A.V.; Theeuwes, J. Top-down versus bottom-up attentional control: A failed theoretical dichotomy. *Trends Cogn. Sci.* **2012**, *16*, 437–443. [CrossRef] [PubMed]
2. Lamy, D.; Leber, A.B.; Egeth, H.E. *Selective Attention. Handbook of Psychology*, 2nd ed.; John Wiley & Sons: Hoboken, NJ, USA, 2012; Volume 4.
3. Leber, A.B.; Egeth, H.E. It's under control: Top-down search strategies can override attentional capture. *Psychon. Bull. Rev.* **2006**, *13*, 132–138. [CrossRef] [PubMed]
4. Theeuwes, J. Top–down and bottom–up control of visual selection. *Acta Psychol.* **2010**, *135*, 77–99. [CrossRef] [PubMed]
5. Koch, C.; Tsuchiya, N. Attention and consciousness: Two distinct brain processes. *Trends Cogn. Sci.* **2007**, *11*, 16–22. [CrossRef] [PubMed]

6.   Ansorge, U.; Neumann, O. Intentions Determine the Effect of Invisible Metacontrast-Masked Primes: Evidence for Top-Down Contingencies in a Peripheral Cuing Task. *J. Exp. Psychol. Hum. Percept. Perform.* **2005**, *31*, 762–777. [CrossRef]

7.   Schoeberl, T.; Fuchs, I.; Theeuwes, J.; Ansorge, U. Stimulus-driven attentional capture by subliminal onset cues. *Atten. Percept. Psychophys.* **2015**, *77*, 737–748. [CrossRef] [PubMed]

8.   Chica, A.B.; Bartolomeo, P. Attentional Routes to Conscious Perception. *Front. Psychol.* **2012**, *3*, 1. [CrossRef]

9.   Chun, M.M.; Wolfe, J.M. Chapter nine visual attention. In *Blackwell Handbook of Sensation and Perception*; Wiley-Blackwell: Hoboken, NJ, USA, 2001.

10. Cohen, M.A.; Cavanagh, P.; Chun, M.M.; Nakayama, K. The attentional requirements of consciousness. *Trends Cogn. Sci.* **2012**, *16*, 411–417. [CrossRef]

11. De Brigard, F.; Prinz, J. Attention and consciousness. *Wiley Interdiscip. Rev. Cogn. Sci.* **2010**, *1*, 51–59. [CrossRef]

12. Dehaene, S.; Changeux, J.-P.; Naccache, L.; Sackur, J.; Sergent, C. Conscious, preconscious, and subliminal processing: A testable taxonomy. *Trends Cogn. Sci.* **2006**, *10*, 204–211. [CrossRef]

13. Neisser, U. *Cognitive Psychology (New York: Appleton-Century-Crofts)*; Yorkville University: Fredericton, NB, Canada, 1967.

14. Norman, D.A.; Shallice, T. Attention to action. In *Consciousness and Self-Regulation*; Springer: Boston, MA, USA, 1986; pp. 1–18.

15. Posner, M.I.; Snyder, C.R.R. Attention and cognitivecontrol. In *Information Processing and Cognition: The Loyola Symposium*; Solso, R.L., Ed.; Lawrence Erlbaum Associates: Hillsdale, NJ, USA, 1975; pp. 55–85.

16. Shiffrin, R.M.; Schneider, W. Controlled and automatic human information processing: II. Perceptual learning, automatic attending and a general theory. *Psychol. Rev.* **1977**, *84*, 127–190. [CrossRef]

17. Bargh, J.A. The Ecology of Automaticity: Toward Establishing the Conditions Needed to Produce Automatic Processing Effects. *Am. J. Psychol.* **1992**, *105*, 181–199. [CrossRef] [PubMed]

18. Bargh, J.A. The four horsemen of automaticity: Awareness, intention, efficiency, and control in social cognition. In *Handbook of Social Cognition*; Psychology Press: London, UK, 1994; Volume 1, pp. 1–40.

19. Moors, A.; De Houwer, J. Automaticity: A Theoretical and Conceptual Analysis. *Psychol. Bull.* **2006**, *132*, 297–326. [CrossRef] [PubMed]

20. Ansorge, U.; Kunde, W.; Kiefer, M. Unconscious vision and executive control: How unconscious processing and conscious action control interact. *Conscious. Cogn.* **2014**, *27*, 268–287. [CrossRef] [PubMed]

21. Kiefer, M. Executive control over unconscious cognition: Attentional sensitization of unconscious information processing. *Front. Hum. Neurosci.* **2012**, *6*, 61. [CrossRef] [PubMed]

22. Braver, T.S. The variable nature of cognitive control: A dual-mechanisms framework. *Trends Cogn. Sci.* **2012**, *16*, 106–113. [CrossRef] [PubMed]

23. Kiefer, M. Top-down modulation of unconscious 'automatic' processes: A gating framework. *Adv. Cogn. Psychol.* **2007**, *3*, 289–306. [CrossRef]

24. Kouider, S.; Faivre, N. Conscious and unconscious perception. In *The Blackwell Companion to Consciousness*; Wiley: Hoboken, NJ, USA, 2017; pp. 551–561.

25. Eriksen, C.W. Discrimination and learning without awareness: A methodological survey and evaluation. *Psychol. Rev.* **1960**, *67*, 279–300. [CrossRef]

26. Greenwald, A.G. New Look 3: Unconscious cognition reclaimed. *Am. Psychol.* **1992**, *47*, 766–779. [CrossRef] [PubMed]

27. Forster, K.I.; Davis, C. Repetition priming and frequency attenuation in lexical access. *J. Exp. Psychol. Learn. Mem. Cogn.* **1984**, *10*, 680–698. [CrossRef]

28. Greenwald, A.G.; Draine, S.C.; Abrams, R.L. Three Cognitive Markers of Unconscious Semantic Activation. *Science* **1996**, *273*, 1699–1702. [CrossRef] [PubMed]

29. Posner, M.I. Orienting of attention. *Q. J. Exp. Psychol.* **1980**, *32*, 3–25. [CrossRef] [PubMed]

30. Dehaene, S.; Naccache, L.; Le Clec'H, G.; Koechlin, E.; Mueller, M.; Dehaene-Lambertz, G.; Van De Moortele, P.-F.; Le Bihan, D. Imaging unconscious semantic priming. *Nature* **1998**, *395*, 597–600. [CrossRef] [PubMed]

31. Eimer, M.; Schlaghecken, F. Effects of masked stimuli on motor activation: Behavioral and electrophysiological evidence. *J. Exp. Psychol. Hum. Percept. Perform.* **1998**, *24*, 1737–1747. [CrossRef] [PubMed]

32. Neumann, O.; Klotz, W. Motor responses to nonreportable, masked stimuli: Where is the limit of direct parameter specification. In *Attention and Performance XV: Conscious and Nonconscious Information Processing*; MIT Press: Cambridge, MA, USA, 1994; pp. 123–150.

33. Marcel, A.J. Conscious and unconscious perception: Experiments on visual masking and word recognition. *Cogn. Psychol.* **1983**, *15*, 197–237. [CrossRef]

34. Kiefer, M.; Spitzer, M. Time course of conscious and unconscious semantic brain activation. *NeuroReport* **2000**, *11*, 2401–2407. [CrossRef] [PubMed]

35. Klein, R.M. Inhibition of return. *Trends Cogn. Sci.* **2000**, *4*, 138–147. [CrossRef]

36. McCormick, P.A. Orienting attention without awareness. *J. Exp. Psychol. Hum. Percept. Perform.* **1997**, *23*, 168–180. [CrossRef]

37. Ivanoff, J.; Klein, R.M. Orienting of attention without awareness is affected by measurement-induced attentional control settings. *J. Vis.* **2003**, *3*, 4. [CrossRef]

38. Mulckhuyse, M.; Talsma, D.; Theeuwes, J. Grabbing attention without knowing: Automatic capture of attention by subliminal spatial cues. *Vis. Cogn.* **2007**, *15*, 779–788. [CrossRef]

39. Fuchs, I.; Ansorge, U. Inhibition of return is no hallmark of exogenous capture by unconscious cues. *Front. Hum. Neurosci.* **2012**, *6*, 30. [CrossRef] [PubMed]

40. Mulckhuyse, M.; Theeuwes, J. Unconscious attentional orienting to exogenous cues: A review of the literature. *Acta Psychol.* **2010**, *134*, 299–309. [CrossRef] [PubMed]

41. Ansorge, U.; Kiss, M.; Eimer, M. Goal-driven attentional capture by invisible colors: Evidence from event-related potentials. *Psychon. Bull. Rev.* **2009**, *16*, 648–653. [CrossRef] [PubMed]

42. Folk, C.L.; Remington, R.W.; Johnston, J.C. Involuntary covert orienting is contingent on attentional control settings. *J. Exp. Psychol. Hum. Percept. Perform.* **1992**, *18*, 1030–1044. [CrossRef] [PubMed]

43. Spering, M.; Carrasco, M. Acting without seeing: Eye movements reveal visual processing without awareness. *Trends Neurosci.* **2015**, *38*, 247–258. [CrossRef] [PubMed]

44. Pöppel, E.; Held, R.; Frost, D. Residual Visual Function after Brain Wounds involving the Central Visual Pathways in Man. *Nature* **1973**, *243*, 295–296. [CrossRef] [PubMed]

45. Tamietto, M.; Cauda, F.; Corazzini, L.L.; Savazzi, S.; Marzi, C.A.; Goebel, R.; Weiskrantz, L.; De Gelder, B. Collicular Vision Guides Nonconscious Behavior. *J. Cogn. Neurosci.* **2010**, *22*, 888–902. [CrossRef]

46. Van Der Stigchel, S.; Mulckhuyse, M.; Theeuwes, J. Eye cannot see it: The interference of subliminal distractors on saccade metrics. *Vis. Res.* **2009**, *49*, 2104–2109. [CrossRef]

47. Mulckhuyse, M.; Theeuwes, J. Unconscious cueing effects in saccadic eye movements—Facilitation and inhibition in temporal and nasal hemifield. *Vis. Res.* **2010**, *50*, 606–613. [CrossRef]

48. Weichselbaum, H.; Fuchs, I.; Ansorge, U. Oculomotor capture by supraliminal and subliminal onset singletons: The role of contrast polarity. *Vis. Res.* **2014**, *100*, 1–7. [CrossRef]

49. Ansorge, U.; Horstmann, G.; Scharlau, I. Top-down contingent feature-specific orienting with and without awareness of the visual input. *Adv. Cogn. Psychol.* **2011**, *7*, 108–119. [CrossRef] [PubMed]

50. Kunde, W.; Reuss, H.; Kiesel, A. Consciousness and cognitive control. *Adv. Cogn. Psychol.* **2012**, *8*, 9–18. [CrossRef] [PubMed]

51. Carrasco, M. Visual attention: The past 25 years. *Vis. Res.* **2011**, *51*, 1484–1525. [CrossRef]

52. McAdams, C.J.; Maunsell, J.H.R. Attention to Both Space and Feature Modulates Neuronal Responses in Macaque Area V4. *J. Neurophysiol.* **2000**, *83*, 1751–1755. [CrossRef] [PubMed]

53. Hayden, B.Y.; Gallant, J.L. Time Course of Attention Reveals Different Mechanisms for Spatial and Feature-Based Attention in Area V4. *Neuron* **2005**, *47*, 637–643. [CrossRef] [PubMed]

54. Liu, T.; Stevens, S.T.; Carrasco, M. Comparing the time course and efficacy of spatial and feature-based attention. *Vis. Res.* **2007**, *47*, 108–113. [CrossRef] [PubMed]

55. Theeuwes, J. Visual Selection: Usually Fast and Automatic; Seldom Slow and Volitional. *J. Cogn.* **2018**, *1*, 21. [CrossRef]

56. Itti, L.; Koch, C.; Niebur, E. A model of saliency-based visual attention for rapid scene analysis. *IEEE Trans. Pattern Anal. Mach. Intell.* **1998**, *20*, 1254–1259. [CrossRef]

57. Nothdurft, H.-C. Salience from feature contrast: Additivity across dimensions. *Vis. Res.* **2000**, *40*, 1183–1201. [CrossRef]

58. Kanai, R.; Tsuchiya, N.; Verstraten, F.A. The Scope and Limits of Top-Down Attention in Unconscious Visual Processing. *Curr. Biol.* **2006**, *16*, 2332–2336. [CrossRef]

59. Held, B.; Ansorge, U.; Müller, H.J. Masked singleton effects. *Atten. Percept. Psychophys.* **2010**, *72*, 2069–2086. [CrossRef] [PubMed]

60. Scharlau, I.; Ansorge, U. Direct parameter specification of an attention shift: Evidence from perceptual latency priming. *Vis. Res.* **2003**, *43*, 1351–1363. [CrossRef]

61. Scharlau, I. Leading, but not trailing, primes influence temporal order perception: Further evidence for an attentional account of perceptual latency priming. *Percept. Psychophys.* **2002**, *64*, 1346–1360. [CrossRef] [PubMed]

62. Scharlau, I. Evidence against response bias in temporal order tasks with attention manipulation by masked primes. *Psychol. Res.* **2004**, *68*, 224–236. [CrossRef]

63. Scharlau, I. Perceptual latency priming: A measure of attentional facilitation. *Psychol. Res.* **2007**, *71*, 678–686. [CrossRef] [PubMed]

64. Jaśkowski, P. Selective attention and temporal-order judgment. *Perception* **1993**, *22*, 681–689. [CrossRef] [PubMed]

65. Ansorge, U.; Horstmann, G.; Worschech, F. Attentional capture by masked colour singletons. *Vis. Res.* **2010**, *50*, 2015–2027. [CrossRef]

66. Lamy, D.; Alon, L.; Carmel, T.; Shalev, N. The role of conscious perception in attentional capture and object-file updating. *Psychol. Sci.* **2015**, *26*, 48–57. [CrossRef]

67. Travis, S.L.; Dux, P.E.; Mattingley, J.B. Neural correlates of goal-directed enhancement and suppression of visual stimuli in the absence of conscious perception. *Atten. Percept. Psychophys.* **2018**, *81*, 1346–1364. [CrossRef]

68. Theeuwes, J. Feature-based attention: It is all bottom-up priming. *Philos. Trans. R. Soc. B Biol. Sci.* **2013**, *368*, 20130055. [CrossRef]

69. Belopolsky, A.V.; Schreij, D.; Theeuwes, J. What is top-down about contingent capture? *Atten. Percept. Psychophys.* **2010**, *72*, 326–341. [CrossRef] [PubMed]

70. Lamy, D.F.; Kristjansson, A. Is goal-directed attentional guidance just intertrial priming? A review. *J. Vis.* **2013**, *13*, 14. [CrossRef] [PubMed]

71. Kristjánsson, Á.; Wang, D.; Nakayama, K. The role of priming in conjunctive visual search. *Cognition* **2002**, *85*, 37–52. [CrossRef]

72. Lamy, D.; Bar-Anan, Y.; Egeth, H.E.; Carmel, T. Effects of top-down guidance and singleton-repetition priming on visual search. *Psychon. Bull. Rev.* **2006**, *13*, 287–293. [CrossRef] [PubMed]

73. Pinto, Y.; Olivers, C.N.L.; Theeuwes, J. Target uncertainty does not lead to more distraction by singletons: Intertrial priming does. *Percept. Psychophys.* **2005**, *67*, 1354–1361. [CrossRef] [PubMed]

74. Jonides, J.; Yantis, S. Uniqueness of abrupt visual onset in capturing attention. *Percept. Psychophys.* **1988**, *43*, 346–354. [CrossRef] [PubMed]

75. Lien, M.-C.; Ruthruff, E.; Johnston, J.C. Attentional capture with rapidly changing attentional control settings. *J. Exp. Psychol. Hum. Percept. Perform.* **2010**, *36*, 1–16. [CrossRef]

76. Gaspelin, N.; Ruthruff, E.; Lien, M.-C. The Problem of Latent Attentional Capture: Easy Visual Search Conceals Capture by Task-Irrelevant Abrupt Onsets. *J. Exp. Psychol. Hum. Percept. Perform.* **2016**, *42*, 1104–1120. [CrossRef]

77. Fuchs, I.; Ansorge, U. Unconscious Cueing via the Superior Colliculi: Evidence from Searching for Onset and Color Targets. *Brain Sci.* **2012**, *2*, 33–60. [CrossRef]

78. Fuchs, I.; Theeuwes, J.; Ansorge, U. Exogenous attentional capture by subliminal abrupt-onset cues: Evidence from contrast-polarity independent cueing effects. *J. Exp. Psychol. Hum. Percept. Perform.* **2013**, *39*, 974–988. [CrossRef]

79. Leuthold, H.; Kopp, B. Mechanisms of Priming by Masked Stimuli: Inferences from Event-Related Brain Potentials. *Psychol. Sci.* **1998**, *9*, 263–269. [CrossRef]

80. Schoeberl, T.; Ansorge, U. The impact of temporal contingencies between cue and target onset on spatial attentional capture by subliminal onset cues. *Psychol. Res.* **2018**, 1–10. [CrossRef] [PubMed]

81. Folk, C.L.; Remington, R.W. Unexpected Abrupt Onsets Can Override a Top-Down Set for Color. *J. Exp. Psychol. Hum. Percept. Perform.* **2015**, *41*, 1153–1165. [CrossRef] [PubMed]

82. Bacon, W.F.; Egeth, H.E. Overriding stimulus-driven attentional capture. *Percept. Psychophys.* **1994**, *55*, 485–496. [CrossRef] [PubMed]

83. Carrasco, M.; Tai, J.; Eckstein, M.; Cameron, E.L. Signal detection theory applied to three visual search tasks—Identification, yes/no detection and localization. *Spat. Vis.* **2004**, *17*, 295–325. [CrossRef]

84. Whiting, J.S.; Eckstein, M.P. Visual signal detection in structured backgrounds I.Effect of number of possible spatial locations and signal contrast. *J. Opt. Soc. Am. A* **1996**, *13*, 1777–1787.

85. Foley, J.M.; Schwarz, W. Spatial attention: Effect of position uncertainty and number of distractor patterns on the threshold-versus-contrast function for contrast discrimination. *J. Opt. Soc. Am. A* **1998**, *15*, 1036–1047. [CrossRef]

86. Schoeberl, T.; Ditye, T.; Ansorge, U. Same-location costs in peripheral cueing: The role of cue awareness and feature changes. *J. Exp. Psychol. Hum. Percept. Perform.* **2018**, *44*, 433–451. [CrossRef] [PubMed]

87. Yantis, S.; Jonides, J. Abrupt visual onsets and selective attention: Voluntary versus automatic allocation. *J. Exp. Psychol. Hum. Percept. Perform.* **1990**, *16*, 121–134. [CrossRef] [PubMed]

88. Folk, C.L.; Ester, E.F.; Troemel, K. How to keep attention from straying: Get engaged! *Psychon. Bull. Rev.* **2009**, *16*, 127–132. [CrossRef] [PubMed]

89. Nieuwenstein, M.R.; Chun, M.M.; Van Der Lubbe, R.H.J.; Hooge, I.T.C. Delayed Attentional Engagement in the Attentional Blink. *J. Exp. Psychol. Hum. Percept. Perform.* **2005**, *31*, 1463–1475. [CrossRef] [PubMed]

90. Zivony, A.; Lamy, D. Contingent Attentional Engagement: Stimulus- and Goal-Driven Capture Have Qualitatively Different Consequences. *Psychol. Sci.* **2018**, *29*, 1930–1941. [CrossRef]

91. Hickey, C.; McDonald, J.J.; Theeuwes, J. Electrophysiological Evidence of the Capture of Visual Attention. *J. Cogn. Neurosci.* **2006**, *18*, 604–613. [CrossRef] [PubMed]

92. Sawaki, R.; Luck, S.J. Capture versus Suppression of Attention by Salient Singletons: Electrophysiological Evidence for an Automatic Attend-to-Me Signal. *Atten. Percept. Psychophys.* **2010**, *72*, 1455–1470. [CrossRef] [PubMed]

93. Gaspelin, N.; Luck, S.J. The Role of Inhibition in Avoiding Distraction by Salient Stimuli. *Trends Cogn. Sci.* **2018**, *22*, 79–92. [CrossRef]

94. Cosman, J.D.; Lowe, K.A.; Woodman, G.F.; Schall, J.D. Prefrontal control of visual distraction. *Curr. Biol.* **2018**, *28*, 414–420. [CrossRef] [PubMed]

95. Sakai, K.; Passingham, R.E. Prefrontal interactions reflect future task operations. *Nat. Neurosci.* **2003**, *6*, 75–81. [CrossRef]

96. Lau, H.C.; Passingham, R.E. Unconscious Activation of the Cognitive Control System in the Human Prefrontal Cortex. *J. Neurosci.* **2007**, *27*, 5805–5811. [CrossRef]

97. Van Gaal, S.; Ridderinkhof, R.; Scholte, S.; Lamme, V. Unconscious Activation of the Prefrontal No-Go Network. *J. Neurosci.* **2010**, *30*, 4143–4150. [CrossRef]

98. Sumner, P.; Nachev, P.; Morris, P.; Peters, A.M.; Jackson, S.R.; Kennard, C.; Husain, M. Human Medial Frontal Cortex Mediates Unconscious Inhibition of Voluntary Action. *Neuron* **2007**, *54*, 697–711. [CrossRef]

99. Van Gaal, S.; Ridderinkhof, K.R.; Fahrenfort, J.; Scholte, H.S.; Lamme, V.A.F. Frontal Cortex Mediates Unconsciously Triggered Inhibitory Control. *J. Neurosci.* **2008**, *28*, 8053–8062. [CrossRef] [PubMed]

100. Van Gaal, S.; Ridderinkhof, K.R.; Wildenberg, W.P.M.V.D.; Lamme, V.A.F. Dissociating consciousness from inhibitory control: Evidence for unconsciously triggered response inhibition in the stop-signal task. *J. Exp. Psychol. Hum. Percept. Perform.* **2009**, *35*, 1129–1139. [CrossRef] [PubMed]

101. Ulrich, M.; Adams, S.C.; Kiefer, M. Flexible establishment of functional brain networks supports attentional modulation of unconscious cognition. *Hum. Brain Mapp.* **2014**, *35*, 5500–5516. [CrossRef] [PubMed]

Review

Salience Models: A Computational Cognitive Neuroscience Review

Sofia Krasovskaya [1,2,*] **and W. Joseph MacInnes** [1,2,*]

1 Vision Modelling Laboratory, Faculty of Social Science, National Research University Higher School of Economics, 101000 Moscow, Russia
2 School of Psychology, National Research University Higher School of Economics, 101000 Moscow, Russia
* Correspondence: svkrasovskaya@hse.ru (S.K.); jmacinnes@hse.ru (W.J.M.);
   Tel.: +7-495-771-32-32 (S.K.); +7-495-771-32-32 (W.J.M.)

Received: 18 March 2019; Accepted: 22 October 2019; Published: 25 October 2019

Abstract: The seminal model by Laurent Itti and Cristoph Koch demonstrated that we can compute the entire flow of visual processing from input to resulting fixations. Despite many replications and follow-ups, few have matched the impact of the original model—so what made this model so groundbreaking? We have selected five key contributions that distinguish the original salience model by Itti and Koch; namely, its contribution to our theoretical, neural, and computational understanding of visual processing, as well as the spatial and temporal predictions for fixation distributions. During the last 20 years, advances in the field have brought up various techniques and approaches to salience modelling, many of which tried to improve or add to the initial Itti and Koch model. One of the most recent trends has been to adopt the computational power of deep learning neural networks; however, this has also shifted their primary focus to spatial classification. We present a review of recent approaches to modelling salience, starting from direct variations of the Itti and Koch salience model to sophisticated deep-learning architectures, and discuss the models from the point of view of their contribution to computational cognitive neuroscience.

Keywords: salience; computational modelling; deep learning; Itti and Koch

---

1. Introduction

Roughly two decades ago, Laurent Itti and Cristoph Koch presented their model of bottom-up visual attention [1] based on the cognitive theoretical foundations of human visual attention. The model subsequently became a seminal work in the field, inspiring a multitude of researchers from various domains to propose their own models of visual attention. The Itti and Koch model was so influential because it combined several different aspects to reach one goal: to understand human visual processing by simulating the processing stages from scene level input to fixation selection. The model had a strong basis in cognitive theory and tried to replicate some neuronal mechanisms involved in visual attention by using many of the best approaches from computational vision available at the time. The result was a model that was to account for many of the spatial and the temporal aspects of human shifts of visual attention.

Subsequent models were numerous and improved the initial approach from the point of view of cognitive mechanisms, the biological accuracy, or simply augmented the bottom-up approach with additional semantic, object, or top-down information. Others were interested in using the general structure for application purposes such as computer vision and robotics. These critiques, improvements and augmentations of the original model have certainly added to what is now a rich literature in models of salience, and we do not intend to belittle their contribution. The fact that Itti and Koch are still being discussed, critiqued, and augmented 20 years after the original article is, in our opinion, a true sign of its importance in defining this field. We provide a review of the different approaches in modelling visual attention based on their contribution to computational cognitive neuroscience. We limit our review to models that were directly influenced by the Itti and Koch algorithm, and cover examples that model object attention, top-down influence, information theory, dual stream models, and conclude with recent advances in deep learning salience classifiers. We also include other methods of achieving the main goal: modelling image salience and the way it results in shifts of attention. The main questions this review tries to address are what contribution did these models make in our goal to explain human visual attention using computer simulations, and what directions are available for the next generation of models?

In this review, we approach the salience problem from a computational cognitive neuroscience perspective, which implies a combination of theoretical and methodological concepts from several domains. Such a perspective suggests that the perfect salience model should be based on a strong theoretical foundation, model the neurobiological processes underlying visual saliency, use explicit computational tools as a means of modelling these processes, and be generative by taking both spatial and temporal predictions of visual salience into account.

Studies of human visual salience have led to the creation of hundreds of computationally valid models, however, most of these models do not focus on all of the abovementioned components of salience simultaneously. We understand that with so many parameters to account for, like combining a broad cognitive theoretical scope and a focused precise neural approach, it is almost impossible to avoid trade-offs. Nevertheless, we will cover models that attempt to address at least several of the abovementioned aspects of visual salience simultaneously.

2. The Itti and Koch Model: Why Is It Seminal

One of the most significant works in the field of visual attention was the study of bottom-up attention by Christof Koch and Shimon Ullman [2] that attempted to explain the underlying biological mechanisms of visual attention. The result served as a foundation for the renowned salience model, the first implementation of which was realized in 1996 [3]. The model was later refined to the version now deemed classic [1,4,5]. A schematic of these general steps for salience computation is demonstrated in Figure 1.

According to the model, the basic features of a visual scene were organized into feature maps of colour, orientation and intensity based on eight scales of linear filtering. The feature maps were then normalized within each feature based on centre-surround differences [6] with the eight scales allowing for different receptive field sizes. Specifically, scales with greater downsampling (surround) could be contrasted with scales of lesser downsampling (centre). The result was 42 maps (six intensity, 12 colour, 24 orientation) containing local spatial contrasts. Neurons within each feature map competed for the most salient locations based on a mechanism of lateral inhibition. The term 'neuron' in the model is simply a mathematical function that is activated upon reaching a certain threshold. Neurons with higher activations are more likely to win against its surrounding competitors. Unlike neurons in a deep learning neural network (for details, see Section 5), the neurons in the Itti and Koch model were not organised hierarchically but were activation function representations distributed throughout the map with the sole purpose of burst activation from salient feature input signals. The feature maps were then combined into a conspicuity map for each of the three features and normalized according to that map's global maximum to prevent strong but common features in one modality from overriding weaker singletons in another feature modality. The conspicuity maps were finally united into a single 2D topographical map of general conspicuity or salience.

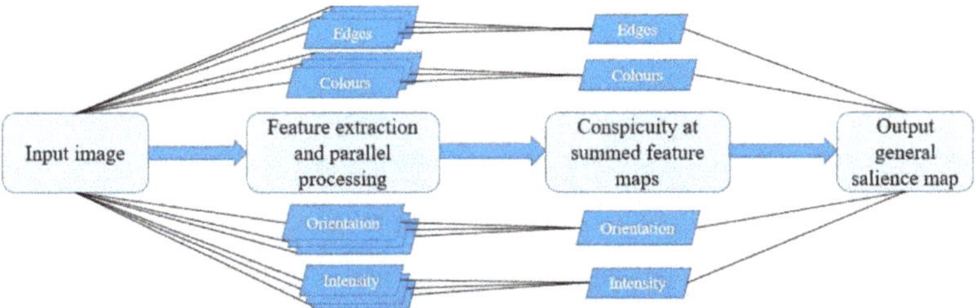

Figure 1. This figure is a demonstration of the salience computation algorithm present in many existing models meant to provide a general idea of the main steps required for bottom-up salience computation. In the current example, basic features like edges, colour, orientation and intensity are included, though the initial Itti and Koch model focused on three main channels: intensity, orientation and colour. Many implementations of the model take into account other features like edges, motion, junctions, shades, etc. Some implementations also include an additional temporal stage that computes fixation locations with a mechanism like inhibition of return (IOR) to prevent fixations at previously attended locations. The main steps present in the figure are as follows: (1) A model receives an input image; (2) features of the image are extracted and organized in a hierarchical manner, with separate feature channels consisting of several levels based on the scale of processing (usually features are downscaled). The features are extracted to individual feature maps using parallel processing, which means that separate feature channels are analysed simultaneously in parallel; (3) the feature maps extracted at every level are merged into one general feature map organized per feature, with the most conspicuous locations reflected on the maps; and (4) the resulting feature maps are combined into a single map of conspicuity, which is the output of the model in the form of a so-called salience map.

The merged salience map was then implemented as a two-dimensional array of Leaky Integrate-and-Fire (LIF) neurons (see reference [7] for a review of the LIF model) that represented the spatial topography of the input scene. A temporal winner-take-all algorithm was applied to the map in order to select a location of maximal salience as the first attended location [1,8]. This was an iterative process with subsequent fixations determined by a similar process. Previously attended high salient locations were discouraged from repeated fixations by an implementation of inhibition of return (IOR) [9–11]. Similar to the way biological centre-surround receptive fields work when distinguishing basic features of an object [6,12], the model distinguished salient object features from the background based on colour, orientation, and intensity in order to select a location for attention direction. The final layers combining LIF, winner-take-all (WTA), and IOR allowed the model to make predictions about temporal order and latency of fixations in addition to their spatial salience.

There are many reasons why this model has maintained relevance. It modelled the entire flow of visual information from visual input to fixation selection, it was easy to test against a variety of human experimental data, it has also garnered interest from applied vision researchers who wish to test more biologically plausible solutions for machine vision and robotics [13,14], and it had a strong theoretical basis that strived to implement our strongest theories of vision and attention. For example, the earliest processing stages were an implementation of the feature integration theory [15] with features processed pre-attentively and in parallel. Likewise, IOR was used as a search facilitator [10,16] to reduce the likelihood of repeated re-fixations of salient locations.

The model was based on biological mechanisms: centre-surround receptive fields sensitive to a specific set of features—pyramidal cells, projecting through the primary (V1), second (V2), dorsolateral (DL), and inferior temporal (IT) visual areas and serving as output pathways for the visual cortex [1], and leaky integrate-and-fire (LIF) neurons, which have strong global inhibition, where the entire population of neurons is inhibited in response to an activation of any single neuron in the population. Action potentials caused by such activation shift the focus of attention to the location where the signal is active. Finally, the model was generative and as such could simulate likely shifts of attention given any sample input. These shifts made spatial predictions that can be compared to real fixation locations during scene viewing but also temporal predictions in the order and latency of those attentional shifts. It is certainly possible for a model to contribute to our understanding of visual processing without addressing all of these areas, but models are particularly adept in overcoming the reductionism of controlled experiments and should continue to be evaluated in terms of how well they explicitly compute in terms of theory and biology in both the spatial and temporal domains.

3. What Is the Salience Problem?

It is impossible to process everything in our visual field simultaneously, so the notion of visual salience is used to describe how specific aspects of the visual field are selected for further processing. We define a salience map as a 2D accumulation of visuo-spatial information that includes, but is not limited to, bottom-up information from input image features. Since this definition includes the possibility of top-down influence, it could also be accurately referred to as a priority map [17], but we keep the term salience for consistancy with the original model. Like the many salience models suggest, this map provides a source for effective shifts of attention over time.

One candidate brain region responsible for mapping salience is the superior colliculus (SC), which is a layered structure in the dorsal part of the mammalian midbrain that directs behavioural responses toward specific locations in egocentric space. Activity in superficial layers of SC has been used to predict fixation locations in free scene viewing and visual search paradigms [18]. A salience map is represented as a two-dimensional array in the SC, where the field size increases with depth [19]. Another proposal for a neural map (for non-human primates) is LIP [20,21] (also reference [17], though they define their priority map as more strongly influenced by top-down attentional factors). This simple priority map may then work with FEF to tag previous locations [22] and SC as the result of the WTA [17].

To be effective, a salience map should be constructed from a combination of bottom-up and top-down information (the nature of this interaction is still a source of debate and beyond the scope of this review). The first area responsible for bottom-up visual attention after information is received on the retina is the lateral geniculate nucleus (LGN). Visual information from the LGN is sent toward V1, where it is then passed on to higher structures via two major pathways—the dorsal pathway, also known as the 'where' pathway responsible for identifying spatial locations and detecting movements, and the ventral pathway, known as the 'what' pathway, which deals with the identification of features and objects [23–25]. The dorsal pathway includes areas V1, V2, V3, MT, MST, and the posterior parietal cortex (PPC), from where projections reach the dorsolateral prefrontal cortex (dlPFC). The lateral intraparietal (LIP) area is also believed to be a part of this stream [26]. Areas comprising the ventral pathway include V1, V2, V4, the inferior temporal cortex (IT), and the ventrolateral prefrontal cortex (vlPFC) [27].

A top-down contribution includes information that is task-relevant and does not depend on simple visual salience of objects and locations [28]. Alfred Yarbus [29] demonstrated that fixation locations depended on the tasks that the participants carried out based on the directions given when studying natural scenes and the results of this study have been used to create a number of classifiers [20] and models [21,30] that predict the task carried out by observers based solely on their eye movement patterns. The evidence for task-based attention has served as basis for models of visual attention that take this factor into account [31,32]. Top-down attention is associated with a higher neuronal activation for behaviour-relevant stimuli and with neuronal inhibition for behaviour irrelevant stimuli [33]. As in bottom-up attention, such activations take place in areas of the dorsal and ventral visual streams, but if bottom-up processes mostly rely on feedforward connections and start in the lower areas, top-down attention is manifested through reciprocal feedback pathways that originate in the PPC and especially in PFC [34,35]. Several studies have also demonstrated that the fronto-parietal regions of the brain play a crucial role in attentional control [36–38]. Liu and colleagues [39] have highlighted the role of the intraparietal and precentral sulci in the choice of a fixation location, and there is evidence that the fronto-parietal network is active during visual search tasks [33,40–42]. Extensive research involving microstimulation [43], chemical inhibition [44], visual search tasks [45], and working memory and behavioral tasks [46] indicate a high level of involvement of the frontal eye fields (FEF) in top-down attention.

Such a strict division of processes into top-down or bottom-up is, of course, oversimplified [47,48], and both factors work together in order to produce a salience map that implements both top-down and bottom-up factors [1,49–52]. Wolfe and Horowitz [53] further highlighted five factors to support the idea that visual attention depends more than on just bottom-up and top-down approaches. The Itti and Koch model is primarily based on bottom-up feature processing, though it does contain a very limited option for top-down 'priority'. The model is able to assign weights to the different bottom-up features of the image, thus changing the degree of priority of these features (such as colour intensity or orientation, for instance) in the final conspicuity map. The model does not, however, model any other top-down components such as semantics, world knowledge, volitional control, or task-dependent behaviour, a limitation which is explicitly stated by the authors themselves [5].

In the Itti and Koch model, selection of attentional shifts are handled by the LIF and WTA layers. In humans and primates, the final control of attentional shifts is regulated by several brain regions that include structures in the occipital, parietal, temporal, and frontal cortices, such as the cerebral cortex, basal ganglia, thalamus, superior colliculus, brainstem reticular formation and cerebellum [54–57]. Deeper layers of the SC are responsible for the coding of eye movements [55]. In the Itti and Koch model, the LIF spiking layer is considered a valid implementation of the two-dimensional array of the SC [54], which consists of so-called peaks of excitation that react to projections from the frontal eye fields (FEF) and supplementary eye fields (SEF) [58]. The peaks of the array follow a 'winner-takes-all' (WTA) strategy, competing with each other in order to reach a certain threshold that would serve as a saccade initiator. This strategy is supported by lateral inhibition, which is responsible for the control of spatial attention by the process of inhibition of all neighbouring signals except for a single peak of neurons that would reach the threshold [1,15,59].

Consequently, the creation of an artificial model of human salience is a challenging goal, especially given the rich literature on the biological mechanisms of visual attention. An effective model of salience and visual processing, however, does not need to explicitly model the functional areas involved in bottom-up or top-down processing. In fact, a full simulation is likely beyond our current ability. However, the closer a model stays to biological realism, the closer we get to understanding the conection between neural function and resulting behaviour.

4. Computational Salience Models

The idea to model selective attention grew alongside the cognitive revolution in psychology [60], and grew with the sophistication of the inspiring theories and with the computational power to implement them. Many models emerged alongside and after Itti and Koch, and they provided extensions and alternate implementations of the process. Although the division of models into separate classes may not be categorical, we group them based on their most vivid characteristics. Thus, we highlight the following groups of models: models directly inspired by the Itti and Koch architecture, biologically driven models, top-down models, models based on the dorsal and ventral pathways, object models, and computationally driven models.

4.1. Direct Variations of the Itti and Koch Model

The initial Itti and Koch model served as a foundation for many subsequent models that may share the basic structure of the original, but deviate or expand from it in some important aspects. In this subsection, we focus on such models and list their similarities and discrepancies as compared to the Itti and Koch model.

One of the first variations was called Attentional Selection for Object Recognition [61], and like the name implies, it was based on the classical Itti and Koch model but with a hierarchical component for object recognition (HMAX) [62]. While the original model was able to focus on salient regions in space, HMAX added an explicit implementation of regions as objects. The HMAX module was hierarchically organised with initial layers matching the Itti and Koch model but then used feature combination and pooling functions to combine information as it moved up through the layers.

Keeping with the focus on neural plausibility, the mid layer of the HMAX module represented composite feature cells and functionally corresponded to area V4 in the visual cortex. The activity of units in this layer were subject to an attention modulation process responsible for approximating spatial location information and representing salient objects in the visual scene. The final layer was able to recognise specific objects with tolerance to changes in scale and position. This model also used IOR to prevent refixations, but instead of being based on spatial locations like in the Itti and Koch model, inhibition was object-based [63]. Although it shared many features with other salience models, HMAX was focused on the task of object recognition rather than spatial and temporal fixation simulation.

Since many early models focused on pixel level input, they were unable to compare salience between 2D transformations such as reflection, rotation and translation. The Selective Attention as a Front End (SAFE) model [64] accounted for these transformations by working horizontally across all feature channels for each feature scale, instead of vertically within each feature channel. For a better understanding, feature channels and scale levels may be regarded as a matrix, where feature channels are columns and scale levels are rows. In the vertical approach, features are analysed by separate columns, while the horizontal approach implies feature analysis by rows that include different features located on one row of the matrix. The model used transformed variants of the same image to calculate the error of fixations for the base image to fixations for the transformed image. Processing in the model was based on pyramids of intensity, colour opponency, and edges, and used difference of Gaussians to assess salience. Fixations were represented as a location and scale value inside the salience pyramid. Theoretically, the model supported Eriksen and James' [65] metaphor of visual attention as a zoom-lens by taking the scale of the image into account as opposed to the strict spotlight theory [66]. This resulted in an improved model for appearance-based object recognition systems that attended to the same features across different scales and geometric transformations.

As theories of attention and salience improved, so did the models that chose to implement them. Where feature integration theory allowed for preattentive processing based on basic features, Guided Search [67] proposed that attention could be guided by that preattentive information. Navalpakkam and Itti [68] created a model that used guided search to implement task-relevance, attention bias for low-level target features, target recognition based on basic features, and the creation of a task-relevant map of locations in the visual scene in addition to biased salience maps. The model had a working memory component responsible for memorizing task-relevant items and a long-term memory component used for hierarchical matching of items based on the given task. As a result, the model produced a topographic task-relevant map of locations and item relevance with regards to the task. The structure of the model used bottom-up salience [1,5] with separate feature maps in parallel fashion, but these maps underwent a process of top-down guidance via weight adjustment in order to separate task-relevant from task-irrelevant features.

The model did not incorporate a sufficient level of top-down control in order to make attentional shifts directed toward specific objects (a man's hand) or locations (bottom-right object) as this would require a spatial relation component. Moreover, it used fixed object signatures and was not able to adapt to distances and rotations of objects in the visual scene. However, it was a good example of the interaction between attention and working memory. There is enough evidence supporting the overlap between the two constructs during task-driven actions, which facilitates the choice of information relevant to the goal [69,70]. Moreover, there is evidence of visual working memory acting as an attentional guiding mechanism in visuospatial tasks [71] and of FEF neurons responding differently to stimuli that have already been attended to [22].

In addition to cognitive theories, some models chose to borrow aspects of mathematical theories. Graph-Based Visual Saliency (GBVS) [72] was a bottom-up model that was based on graph theory to create activation maps from a raw image. The idea is that, if maps are implemented as Markov chain graphs, then the equilibrium (stable) distribution of the chain can be interpreted as activation and salience values over the locations of the map. The model contained two major parts—activation and normalization. Activation maps were based on feature channels that represented a set of meaningful locations according to a specific criterion, like information from human fixations. The graph-based Markovian approach found regions of dissimilarity that were computed in parallel to calculate mass at a region and distribute it among neighbouring regions. Such parallel processing at different regions leads to a single equilibrium distribution of mass. The normalisation stage integrated mass concentration over all activation maps in order to identify a set of salient locations. The same Markovian approach took place, and regions with the highest activation attracted mass. The model was made as a part of the GVBS Toolbox (refer to Appendix A for information on toolbox implementations) and provided a good example of a computationally driven implementation of the initial model.

Overall, these direct variations of the original model demonstrate that theoretical and practical augmentations can be made to the classic salience model while maintaining key components of its essential structure. These models all share preattentive feature processing, location-based feature extraction, and a focus on bottom-up attention, yet make their own contribution to new functionality. For most of these models, feature and location selection coincide, but we know that feature-based attention can exist independent of spatial locations during visual search [73]. To be accurate, identification of location should happen at a later stage and with fewer locations based on the processed relevant features.

4.2. Biologically Driven Models

Most computational models will claim some degree of biological plausibility, and the differences are often in the level of granularity they choose for their model. Some advances in salience modelling, however, have made a point to focus on the neural underpinnings of some portion of the processing stream. In this subsection, we look at four biologically-inspired models that share a similar architecture with the original model and focus on attaining a high level of biological validity of visual salience processes.

Li [74] introduced a theory of salience based on the independent sensitivity of V1 cells to specific features, and this was implemented in a bottom-up salience model by Koene and Li [75]. The main idea was that the features were not summed across channels to identify the most salient location as in the Itti and Koch model, but salience was identified at the location to which a specific V1 cell activated the most to as compared to other V1 cells. Moreover, the model implemented conjunctive cells in V1 that were sensitive to combinations of motion and orientation (MO) and colour and orientation (CO) [76,77]. This involved V1 neurons with receptive fields sensitive to orientation (O), motion (M), colour (C), colour and orientation (CO), and motion and orientation (MO). These neurons were connected horizontally, and a mutual inhibition mechanism was applied to neurons that activated to uniform features. Receptive fields with the highest activation thus represented a salient location. The model resulted in lower reaction times to targets with combined features as compared to targets with independent features processed in a parallel manner. Overall, the model had a strong theoretical and neural foundation and matched the simpler feature processing in striate structures [78,79].

Salience models frequently use multiple maps for features and scales (a total of 42 in the original Salience model), but this adds much redundant information. Park and colleagues [80] focused on reducing redundancy of retinal cells by adding a new pre-processing stage simulating LGN [81]. Input from natural coloured scenes created four feature maps for edges, symmetry, and red-green (RG) and blue-yellow (BY) colour opponency. Redundancy was reduced by applying an independent component analysis on the feature maps. An unsupervised learning approach was then used to assess the optimal weights in each feature map in order to combine them into an optimal salience map. The proposed model provided an alternative approach with a strong neural basis to the classic bottom-up salience model, though it did not take into account important characteristics of vision, like IOR.

The flat, two-dimensional maps of most salience models are a useful simplification of retinal input, but they do not account for biological features such as retinal sampling and magnification in the visual cortex nor consider the covert visual angle included in visual processing. The model by Aboudib and colleagues [82] differed from the Itti and Koch model in replacing Gaussian subsampling and across-scale reductions of the input image with kernel-based Gabor filtering and receptive field emulations that mimicked biological cortical magnification. This filtering approach produced a total of 9 feature maps per image instead of the initial 42 in the Itti and Koch model and lead to a generation of 250 most salient locations chosen based on a WTA/IOR process. The model computed input layers such as cone receptor or retinal ganglion cell layers, going upward to areas LGN, V1, etc. The model also allowed feature layers to be stacked onto each other before WTA/IOR took place, thus providing a way to mimic the feed-forward stream in the visual system.

As processing progresses along the ventral pathway, neuron receptive fields get larger and more specific to complex patterns. Hamker [83] proposed that FEF, based on Duncan's theory on adaptive neural coding [84] might project complex templates backwards as a source of top-down attention. Specifically, the model implemented task-oriented feature-based attention in V4/TEO. It included top-down parameters into a bottom-up architecture in its feature processing stage, and contained 'match detection units' responsible for object detection in natural scenes. Processing used a population coding approach [85] to generate expectations based on prior knowledge, and thus, refresh the salience information of each feature based on top-down influences. The structure included several components: the early visual component for feature and conspicuity maps based on channels of colour opponency (two maps: RG, BY), intensities, orientations. and spatial resolutions; the V4 component, which coded the approximate salient stimuli locations; the TE component, which received pooled input from V4 in the form of feature conspicuity and location and resulted in several V4 features combining into one at TE; the FEF perceptual map, which integrated V4 and TE data across all five channels; and the FEF decision map, where several most salient locations competed for the highest saliency. A final IOR mechanism made sure that the model searched for objects in novel locations.

4.3. Top-Down Contributions

As mentioned earlier in Section 3, the Itti and Koch model did allow for basic top-down control through weighting of the different bottom-up features. Parameters could be set or learned that gave priority to colour or edges, for example, and these features would be weighted more strongly in the merged conspicuity map. The primary purpose of the model, however, was a demonstration of bottom-up salience, and it was left to future models to incorporate realistic aspects of top-down attention. To speak of a top-down contribution, however, we have to consider the attentional state of viewer and understand the intentions as defined by the viewing task [86]. Although task instructions are not the only manner of top-down influence, it is an established method of manipulating it experimentally. Although there have been many such contributions, we have chosen three models as exemplary.

Salience models are not restricted to visual search, but this specific task holds a prominent place in the literature. Jeremy Wolfe focused specifically on search instead of viewing without a specified task, and in doing so extended feature integration theory [15]. His Guided Search [87] model allowed searching for a target among a set of distractors and considered the bottleneck between visual input and object recognition. The model suggested that top-down information may be used to facilitate the search process by emphasizing bottom-up features of a target. The proposed architecture was divided into a preattentive stage, where a feature map distinguished targets from distractors using parallel processing, and a serial process stage, where each potential target was processed one by one using information from the first stage until the true target was found. Wolfe's model was based on several theories of visual attention [15,88,89]. There have been several revisions of the model [50,67,90], with Guided search 4.0 perhaps undergoing the largest changes, as it represented visual guidance as an independent control mechanism, whereas in earlier versions, the preattentive stage played the role of a guidance mechanism. The mechanism proposed was a guidance activation map, which contained a weighted sum of bottom-up and top-down activations. Wolfe and Horowitz [59] further addressed the issue of identifying attributes that guide attention and provide convincing arguments for the existence of a guiding representation. Attentional guidance is also affected by previous experience and may manifest in inhibiting mechanisms, such as IOR [9–11], or in facilitation of the search for relevant stimuli [91,92].

The Feature Gate model [93] was created in an attempt to understand specific mechanisms of visual attention like parallel and serial search, inhibition of distractors, bottom-up feature selection, split attention, and feature contrast variations. The model used a hierarchical architecture consisting of a network of spatial maps containing attentional gates that controlled the signals sent to higher levels of processing. The decision to pass a signal was based on the usual bottom-up features but also on top-down characteristics that separated relevant features from non-relevant ones. Activations for each location were calculated separately for bottom-up and top-down information and then summed to form a general activation map of locations. Gates with the highest location activation values were allowed to pass the signal to the next level. The structure of the Feature Gate model was similar to the structural organization of the visual cortex, with the lowest map corresponding to area V1, containing small receptive fields sensitive to specific features, and the highest map similar to higher layers of the cortex, with large receptive fields sensitive to complex features. The model also implemented an inhibiting mechanism that prevented distractions, which helped differentiate relevant and non-relevant signals, which is similar to the biological mechanism of lateral inhibition [94].

Generative models, often based on Bayesian probability, have the added benefit of being able to simulate response distributions to new unseen exemplars. Rao and colleagues [95], for example, introduced two such probabilistic models of visual attention that included top-down influence on salience. The first model was based on reproducing of overt shifts of attention in visual search using salience maps of natural scenes. The most salient target location was defined using a probabilistic Boltzmann distribution algorithm [96,97] as a weighted average value of all scene locations with regards to the weight of the most salient location. Salience maps were generated through bottom-up and top-down parameters with reference to both spatial and target representations. The second model was based on predictive coding. It was a generative model of image representation used to study response properties in the visual cortex [98] and mimicked feedforward and feedback mechanisms of the ventral and dorsal pathways. This model demonstrated how visual attention is globally organised throughout various brain structures by predicting probable consequences of damage to different areas of the system. Both models shared a common probabilistic approach, but the second model added an extra focus on shifts of covert attention.

Maps that combine top-down and bottom-up information may more correctly be referred to as a priority map [17]. Though these models do not address all of the aspects of visual attention, each of them has made its own contribution to the field, with the Guided Search model focusing on strong cognitive hypotheses, the Feature Gate model imitating the overall structure of the visual cortex, and the generative nature of Rao's probabilistic models. Hence, these models may be used as a basis for future models to further investigate biological visual attention processes.

4.4. The 'What' and 'Where' Models

In 1983, Mishkin, Ungerleider, and Macko [99] proposed a theory of two separate cortical pathways existing for object processing and spatial location processing and this was later elaborated by Milner and Goodale [23] as the 'Two Streams hypothesis'. According to the theory, visual areas are hierarchically organized into two separate processing streams running ventrally and dorsally and are responsible for object (the 'what' stream) and spatial location (the 'where' stream) processing, respectively [24]. Here, we discuss two models of visual salience that take the separate processing streams theory into account.

Models of object recognition need to consider properties of perception such as object invariance. Rybak and colleagues [100] proposed just such a model that was invariant to scale changes and geometric transformations, such as rotations and translations. It modelled visual processing through three subsystems: a low level subsystem mimicking V1 neurons and responsible for edge detection; an intermediate level that used frames of reference to translate basic features into second order features insensitive to transformations; and a high level module that was based on the cooperation of the dorsal and ventral streams for sensory and motor memory accordingly. This final layer was further capable of operating in three task-modes: memorization, search and recognition. The model was tested and performed well on both complex scene and facial recognition tasks.

Likewise, the Neurodynamical Model [101] implemented both object-based and space-based streams [23,102] of visual attention with natural scenes as input. It was able to model increased receptive field sizes in higher layers in the visual cortex and also the top-down modulation (reduction) of receptive field size in complex scenes as compared to plain backgrounds [103]. The model comprised five modules, structured in a way that resembled the dorsal and ventral pathways and connected by feedforward and feedback projections. A short-term memory block was used to integrate top-down bias via pathways simulating the prefrontal cortex. The model was able to simulate fMRI signals using pooling of neurons in a specific area of the cortex with regards to spatial and temporal parameters and demonstrated dynamic interactions between the dorsal and ventral pathways.

While the independence of the dual streams may be over exaggerated [104], the underlying premise of the pathways remains [105]. The idea of parallel pathways for different aspects of visual processing should also have an appeal for modelling, where partitioning of functionality offers both ease of development and greater explanatory value. These models show the way forward for providing a benefit for including the neuronal mechanisms underlying the separate avenues for visual attention.

4.5. Object Models

An interesting approach to the modelling of visual attention is one based on object rather than feature salience. Object detection models in natural vision have their own rich literature [106–108] but we focus on two such models that combine objects with the Itti and Koch bottom-up architecture as one of the steps in the overall structure.

The 'fingers of instantiation' (FINST) theory by Pylyshyn [109,110] suggests that the visual system is capable of tracking a limited number of salient locations over time and across multiple fixations. The NAVIS system [111] comprised a bottom-up salience component, a top-down attentional control component, and a behavioral component divided into two stages. The first stage was similar to the Itti and Koch model, though it produced several salient locations based on the idea of FINST. The second stage analysed the locations chosen during the first stage and chose one focus for overt attention. The aim of the double-stage selection in the model was to combine theories of early and late selection into one system.

The model implemented feature maps for edges, oriented areas, and colours feeding into conspicuity maps of symmetry, eccentricity, and colour contrast, respectively. The derived maps were then combined into one bottom-up salience map, which was updated by top-down mechanisms marking locations for increased relevance. Dynamic neural fields [112] were implemented as an inhibiting mechanism, and this was based on two approaches for separate local and global inhibition. Inhibition of return in this model was applied globally to objects rather than salient points in space, which is consistent with Tipper's [113] hypothesis of inhibiting mechanisms in humans.

Sun and Fisher [114] implemented a model based on object-based visual attention but incorporated components of multiple cognitive theories and studies of vision, such as the Integrated Competition Theory [115], the salience model of bottom-up attention [1,5], theories of top-down and bottom-up attention interaction, object and feature based salience, and others. The model suggested that competition for visual attention takes place both within an object and between objects and consisted of two main components. The first component was responsible for object salience and grouping and followed a similar strategy to other pre-attentive salience maps. The second component represented a hierarchical approach to attentional shifts. The first stage followed a similar strategy to other pre-attentive salience maps, and the second stage included hierarchical selection from the winners of the first stage from coarse to fine scales. This approach provided an integration of attentional selection from spatial locations, features, and feature conjunctions. The process took place between and within groupings across scales, and groupings at the final level marked a salient location.

Object recognition is a key component in vision processing. For example, the space between objects may be a key factor in spatial attention, search and reading [116]. Top-down factors, such as visual context and relevant task, are also important for object recognition, as they further facilitate the process of finding objects in certain contexts [117]. Object-based models might also provide some understanding of the interaction between feedforward and feedback mechanisms in the visual system, since such an interaction is considered important for the identification of the location of an object (for an example of such a model, see reference [118]). The feedback network could identify the location of objects, whereas the feedforward mechanism is responsible for the detection of shapes present in the visual field. The initial Itti and Koch model focused mainly on the spatial location of salient features, with object identification being a consequence of salient location detection, whereas object-based models deliberately concentrated on object salience by combining bottom-up and top-down approaches.

4.6. Computationally-Driven Models

Salience models have been a popular approach in psychology and neuroscience but also in applied fields like engineering and computer science where machine vision has shared many approaches with vision science. Linear combinations, normalisation algorithms, and Gaussian functions were instruments that helped the original model imitate bottom-up processes that take place in the visual cortex. Since then, many models that used computationally valid tools to predict visual salience have been proposed, and this subsection is dedicated to examples that have contributed to both applied and fundamental research.

The Selective Tuning model [119] shared a number of features with the Itti and Koch approach and was based on a hierarchical pyramid that initialised weights from a feedforward pass based on image input, followed by multiple feedback passes to tune selection. The selective tuning mechanism involved inhibition that behaved differently for spatial selection and feature selection. Spatial selection pruned irrelevant pyramid connections while feature selection inhibited components responsible for the computation of irrelevant features. After the inhibitory processes, a WTA mechanism was activated at the top layer of the hierarchy to determine a globally salient element. The model was one of the early implementations of feedback mechanisms as the signal was then sent back to the lower layers through direct connections, allowing the globally salient unit to influence lower level units, recursively identifying locally salient units at every level of the hierarchy.

Information theory in computer science and mathematics tries to quantify the value of any information by its ability to resolve uncertainty. Applying this idea to vision, we can think of fixation allocation as an extension of the question of how best choose locations that will maximize information gain. One example of this approach was AIM [120] that derived a self-information criterion from a natural image. An independent component analysis (ICA) was applied to patches of 7×7 pixels that were randomly selected from each image during the independent feature extraction stage. The likelihood per patch was then estimated and joined into a general likelihood for the entire image. The joint likelihood underwent Shannon's self-information measure [121], after which a salience map was created for the input image. In general, the model represented an information maximization approach to a set of images. Although it performed better than the Itti and Koch model based on the area under the ROC curve score, it lacked both the cognitive foundations, like the feature integration theory, and the biological mechanisms like WTA and IOR. The model did provide a strong computational background, was based on the idea of sparse coding [122], and was an effective functional implementation directed at computer and robotic vision.

Information gain is also suitable for coding top-down information, as, for example, the Schill [123] model based on the Dempster-Shafer belief theory [124]. The model contained three basic levels—a neural network preprocessing stage that analysed information extracted from the scene, a top-down component based on the Dempster-Shafer belief theory for uncertain reasoning that applied to information from the preprocessing stage, and a control component that predicted fixations that would maximize information gain. This architecture would constantly update retrieved information from a scene to predict the most promising fixation locations. Systems based on information gain and entropy have an interesting theoretical implication in that IOR is no longer needed to prevent repeat fixations since the information gain at these locations is automatically reduced.

A creative approach to visual attention modelling was based on Bayes' Theory. Torralba and colleagues [125] suggested using a Bayesian architecture, where attention was driven by global scene context. The authors proposed a contextual guidance model of attention based on two parallel pathways: one pathway for local feature salience, another for global feature processing that took scene characteristics into account. The salience pathway encoded each location independently, resulting in a bottom-up salience map. The global pathway extracted global features from the entire image filtered through six orientations and four scales. The result of the filtering was represented by a vector of 64 principal components. A 'scene prior' mechanism was then added based on the given task, and contextual modulation was applied to the bottom-up salience map, which resulted in a final context-sensitive, task-driven salience map. The global pathway made sure that most probable target locations were activated and that task-irrelevant salient locations were eliminated. The proposed model made accurate predictions of human visual behaviours based on a given task. It combined several theoretical approaches to visual attention, such as bottom-up salience and the effect of context on visual search.

Bayesian approaches are well suited for implementing context and task as the probability weights can shift based on evidence or prior knowledge. The SUN model [32], for example, regarded salience as differing based on the task being performed. In free-viewing, salience was based solely on bottom-up features, but with a specific task like target search, salience was based on the available knowledge about the visual features of the target class. The computation of the target probability was calculated per pixel and included bottom-up parameters, as well as visual appearance and location of a stimulus. Prior knowledge for the model was obtained using a set of images from the LabelMe dataset [126] which include prelabelled information on object location and identity. Testing of the model showed that it was able to predict human-like fixations, and even made mistakes similar to those made by people.

A simple but novel approach to salience modelling that addresses the role of foreground and background segmentation in visual attention [127] was proposed by Zhang and Sclaroff [128]. They used the Boolean map theory of visual attention [129] as a basis for their Boolean Map–based Salience model (BMS). Given an input image, BMS produced a set of Boolean maps based on binary figure-ground segregation according to randomly selected feature channels. The Boolean maps were then used to compute attention maps using Gestalt, figure-ground segregation principles to discover connected regions of the image. Normalised attention maps were then linearly combined into a single full-resolution salience map ready for object detection tasks or for the prediction of fixation locations. Despite its simple organisation, the BMS model maintained high levels of performance on several benchmark eyetracking datasets (see Appendix B).

The line between a strictly computational model and a cognitively plausible model is very much open to interpretation. One particular standout that has left a mark on approaches was a model proposed by Judd and colleagues [130]. It used a support vector machine (SVM) classifier method to learn how to estimate salience directly from human data. In the model, salient points on an image were defined by three layers of low-level features, which were then used to train the model. They also included 'mid-level' features such as automatic horizon line detectors and high-level features like face, people, and car detectors. The result was a model that was attracted to the same locations that attracted human fixations.

As seen from just several examples of computationally based models, there exist multiple approaches to mimicking visual salience with the help of computational tools. While computer vision and computational models of vision may have different goals, they have shared many algorithmic approaches over the years. Even computational approaches that are less biologically plausible, like ideal Bayesian observers [131], can be used to highlight the potential efficiencies of human search. Applied machine vision approaches do have an understandable focus on salience prediction through bottom-up salience since much top-down information would require some degree of common-sense knowledge.

5. Deep Learning Classifiers

In recent years, there has been a movement toward complex deep hierarchical structures for predicting salience in natural images [132,133]. This is because deep learning algorithms have been shown to be extremely accurate tools for modelling high levels of abstraction [134], such as vision and speech processes. They can also model human visual attention by drastically narrowing the gap in accuracy between model and human spatial predictions [135]. A neural network represents a set of elements called neurons, which are most commonly structured in the form of layers. Artificial neurons in neural networks are computationally similar to biological neurons, and early attempts to create neural networks were inspired by the human brain [136]. A biological network contains a collection of neurons that are electrically or chemically excited nerve cells and stores, processes, and passes signals to other neurons using electrical and chemical impulses. Artificial neurons may be described as a mathematical model of biological neurons, and these represent the basic element in an artificial neural network. Each neuron may receive several inputs, which are summed up and weighted to provide an output upon reaching a certain threshold. The input neurons are activated by a sensory signal that triggers evidence accumulation and pass weights to the next neural layers. The main advantage of a neural network is that it is possible to engage many neurons in parallel in order to solve different problems.

Deep learning is a type of machine learning algorithm that uses a non-linear function for parallel information processing [137]. It uses multi-layered (hence 'deep') neural networks and management of the neural weights to solve complex tasks or to replicate natural phenomena. Each layer of a deep network represents different levels of abstraction and is used to extract certain features of the data, such as images or audio. Deep learning architectures have proven to be useful in different spheres, such as biology, chemistry, finances, business, physics, neuroscience, and many others [138]. In vision research and spatial salience prediction, deep neural networks have become the leaders among other approaches as seen in the MIT saliency benchmark [139].

The class of deep learning networks most frequently used in visual processing is convolutional neural networks (CNN). CNNs have a multi-layered structure, but instead of learning all the weights between the nodes of neighbouring layers, they learn only the values of convolutional kernels, or filters, which are applied to the entire layer during the convolution operation. CNNs also have pooling layers and activation layers in the form of a rectified linear unit (ReLU) in addition to convolutional layers. Convolutional layers are responsible for the detection of local feature conjunctions, whereas pooling layers combine similar features into global features. The activation unit (ReLU) may be compared to a biological action potential that determines whether a neuron will fire or not. After a series of multiple sets of convolution, pooling and activation take place, the output progresses to a final, fully connected layer. The overall hierarchical organization of a CNN is said to be similar to the hierarchical organization of the visual cortex [140].

The applications of CNNs in salience detection are extensive. Some of the examples of these applications are Deep Gaze 2 [141], which presents two models for fixation predictions and object recognition; EML-NET [142] for salience feature detection using a modular approach; and DeepFix [143], which is sensitive to semantic information at different scales while using large receptive fields for global context analysis. The scope of existing models in the field is so broad that it is impossible to cover all of them in this review, but their accuracy in predicting areas of spatial salience is without question [144–149].

A recent trend in deep learning is the use of Generative Adversarial Networks (GANs), originally proposed by Ian Goodfellow and colleagues [150]. Unlike CNNs, GANs are generative, meaning that they are able to learn distributions of complex data and generate sets of new data that resembles the learned set. The term adversarial refers to the competitive process that takes place inside GANS—these networks contain two components, a generative and an adverse model, that are responsible for the generation and discrimination of data, respectively. SalGAN [151] used this technique to produce salience maps of an input image. GANs can also use conditional switches to generate maps for different images (SalGan) or different tasks [152] as an effective but simplistic mechanism for top-down attention. Although these networks are generative, to date, they have only been used to generate spatial saliency predictions and not temporal distributions of individual fixations like the more cognitive algorithms.

Recurrent neural networks (RNNs) have mostly been used for the processing of sequential data, like speech, but they have also been adapted for salience. RNNs contain units that are sensitive to temporal changes [153], namely looped connections that allow the information to be stored some period of time by being copied and passed to the next node. Their dynamic nature has allowed RNNs to serve as the basis for salience detection architectures in videos, such as the spatiotemporal grounding of evidence based on top-down factors [154] or the flow-guided approach to object detection [155].

Static scene salience detection have also used RNNs. For example, the DSRCNN model [156] uses recurrent network to enhance contextual information learning. Moreover, RNNs combined with CNNs have shown improved performance at salience detection based on different scales [157], as simple CNNs have a fixed receptive field size. The behaviour of RNNs has also been linked to biological lateral inhibition, as shown by the deep spatial contextual long-term recurrent convolutional network (DSCLRCN) [158] for fixation prediction in natural scenes.

RNNs can hold information for a short period of time, but result in the 'vanishing gradient' problem, which means that memory of a feature or input decays relatively quickly. This has led to the development of another artificial structure called long short-term memory (LSTM) networks, which were intended as explicit memory for recurrent architectures [159]. LSTM networks are a popular approach to solving long-term dependency problems, such as large-scale speech recognition [160,161] or object and salience detection in video sequences [162]. LSTMs have also proven to be useful in combination with other architectures in fixation prediction [158,163,164] and represent another possible implementation to include interactions between salience and working memory.

The leaps in accuracy for predicting areas of spatial salience have largely come from the deep learning approach of treating the image to salience map pipeline as a classification problem, which is a traditional strength of neural networks. However, to what degree can these classifiers be considered models in the computational cognitive neuroscience sense? The hierarchical organization of the visual cortex has certainly served as inspiration for deep learning models of visual attention, such as a deep belief network model for areas V1 and V2 [165]. Parallel processing is also an innate feature of deep learning neural networks, which allow them to process input in a way similar to the visual system. Another approach compares activation of artificial and biological networks at later layers of visual processing, such as area IT, to see how well deep learning architectures are able to match. For example, models as recent as Densenet-169 [166], created in 2016, have shown a trend for increased matching of human IT features in one of its layers [167]. Recently, further optimizations in classification accuracy for the ImageNet project [168] have produced models that have scored worse on brain benchmarks (Figure 2). Take, for example, popular models such as Inception [169] and PNASNet [170]. Despite their high performance on image recognition tasks, their predictivity of neural activity and human behavior falls short of anatomically simpler models such as CORnet [171].

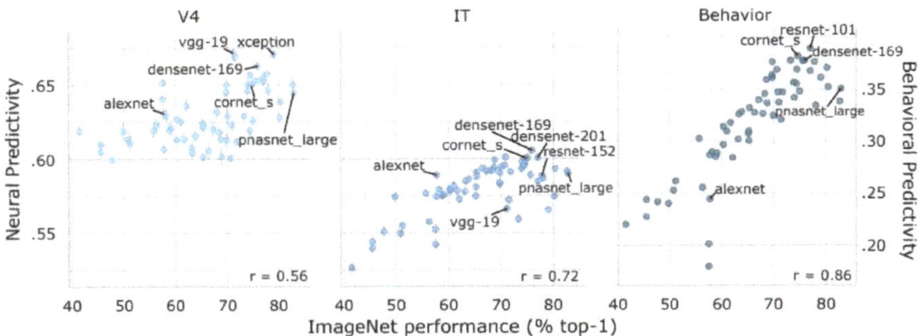

Figure 2. The performance of popular Deep Learning models based on their ability to match human V4 and inferior temporal (IT) regions, as well as human behavior performance. Adapted from "Brain-Score: which artificial neural network for object recognition is most brain-like?" by Schrimpf et al., 2018 [167], *BioRxiv*, 407007. Copyright 2018 by author. Reprinted with permission.

6. Metrics and Evaluation

Despite the variety of approaches to salience modelling, there is still a lack of a standardized evaluation. Different metrics have been suggested to assess model performance, and the most frequently used sources is the MIT Saliency Benchmark [139]. It provides an open and free source for model assessment and benchmark data based on eight different evaluation metrics. The benchmark website includes most recent attempts at salience performance but also interesting performance baselines like the original Itti and Koch [1,5], central bias alone, and 'infinite humans' as a gold standard.

Most methods of assessment use location-based algorithms that measure the error between predicted and human fixations. For example, a simple correlation coefficient, or Pearson's linear coefficient, measures the linear correlation or dependence between two salience maps. If the two maps are not correlated, the coefficient would be 0. One characteristic instrument of these metrics is the area under the Receiver Operating Characteristic curve (AUC ROC), which estimates the trade-off between true positive and false positive values at different discrimination thresholds by verifying that true positives are labelled before negative values [132,172]. One of the better AUC metrics and most commonly used for the evaluation task is the AUC-Judd [133,173]. The AUC-Judd interprets fixations as a classification task, where a pixel of the map may be either salient or not by applying a threshold over the intensity value of the saliency map [174]. Each salient pixel matching human fixations on the map is considered a true positive value, whereas salient pixels over non-fixation areas are classified as false positive values. The final AUC score is then calculated and plotted as a trade-off between true and false positive values. The highest possible score may be 1, whereas a 0.5 score is considered random. Another AUC approach is the AUC-Borji [175] metric that treats the salience map as a binary classifier. Positive and negative samples are differentiated through various thresholds. True positive values are then considered as values that exceed threshold at fixation locations, while false positive values are the values that exceed threshold at random locations. This approach differs from the AUC-Judd approach in that the threshold is set based on a fixed step size, and the false positive rate is calculated based on approximation of the Judd calculation. The shuffled AUC approach is also based on Ali Borji's approach [175] but is additionally sensitive to centre bias. This metric shows whether a model has used centre bias for its predictions and imposes a penalty for this.

The Kullback-Leibler (KL) divergence [176] is a measure based on distribution. It evaluates the difference between the distribution of two different saliency maps, and measures information lost during the estimation of a fixation map with regards to the salience map. The lower the KL divergence value, the better is the approximation to ground truth in a salience map. However, the KL measure is extremely sensitive to zero values, which leads to a large penalty on the overall score if the predictions are deemed insufficient and sporadic.

Other options have been suggested such as the Earth Mover Distance (EMD) and the Normalized scanpath Salience (NSS), but these evaluation metrics have been adapted mostly for computer vision and based solely on spatial performance, whereas shifts of visual attention happen over time. Many of the classic salience models based off of Itti and Koch do use temporal layers to make predictions of fixation latencies, but even these show poor performance against human latency distributions [177]. Additionally, the existing metrics focus on the level of similarity between ground truth based on human data and the model data, but they do not take into account the cognitive basis underlying visual salience.

Although most models focus on the accuracy of classification, we can learn much about the differences between biological and artificial networks by finding examples where the predictions differ. Nguyen and colleagues [178] demonstrate this by generating images that DLNNs are unable to recognize but that a person would easily be able to. This demonstrates that an important factor in making human-like models is in the types of mistakes that humans and DLNNs tend to make. One of such models has already been mentioned in the current review [32], and it predicts human fixations at a decent level, while making similar mistakes to humans.

7. Conclusions

Roughly 20 years of salience model research has produced many excellent approaches that have improved our understanding of the visual processing stream. Various models have been created after the seminal Itti and Koch architecture, each bringing their own contributions, such as prediction accuracy, modelling additional tasks, the addition of object features and the inclusion of additional theoretical frameworks. Not every model had the same focus on all aspects of the five contributions we highlighted, i.e., either on the theoretical, neural, spatial, temporal, or computational aspects, but the overall progress in the field has been to push the boundaries in all of these directions.

One trend we would like to highlight, however, has been a recent shift away from visual attention modelling and toward fixation classification. The trending deep learning approaches are a very powerful computational tool; however, their current focus is not to explain visual processes but to reach state-of-the-art performance on spatial predictions. Due to this, important characteristics of visual attention, such as its temporal component and error production, have been widely ignored. The spectrum of existing powerful neural architectures, such as convolutional and recurrent neural networks, should provide a means of considering the temporal factors inherent to visual processes. We propose that an area of opportunity in the field is the exploration evaluation metrics that would account for all the above-mentioned aspects. Most current metrics are aimed at purely classification accuracy and focus on their ability to predict spatial locations, matching, and even surpassing human performance. While this trend has shown a great deal of success, the field of computational cognitive neuroscience should continue its goal of making biologically accurate and computationally powerful improvements to the Itti and Koch model for a deeper understanding of human visual attention processes. While we recognize the importance of purely computational approaches to visual attention directed at robotic vision, we suppose that the cognitive neuroscience community should shift its focus to a biologically and theoretically plausible approach at modelling visual attention. Instead of testing the limits of deep learning architectures, deep learning should be adopted as a means of testing cognitive theories and incorporating neural foundations of visual attention into the models. In particular, we suggest that testing our models against temporal distributions of attentional shifts, typical errors that humans make in visual processing and matching neural signatures of specific tasks will only improve the field going forward.

Author Contributions: The following individual author contributions were made: writing—original draft preparation, S.K.; writing—review and editing, S.K. and W.J.M.; supervision, W.J.M.

Funding: This work is supported in part by the HSE academic fund program for the scientific research lab "Vision Modelling Lab".

Vision **2019**, *3*, 56

Acknowledgments: We would like to thank James DiCarlo, Christopher Shay, and Martin Schrimpf for their kind advice and permission to use materials. We are also grateful to our reviewers and editors for their extensive help and advice during revisions.

Conflicts of Interest: The authors declare no conflict of interest.

Appendix A

The GVBS Toolbox [179] is available for open download at http://www.vision.caltech.edu/~{}harel/share/gbvs.php

There have also been computer toolboxes created based on the Itti and Koch architecture. The practical implementations of the initial model of bottom-up visual salience may be found in the Neuromorphic Visual Toolbox (NVT) [180] available at http://ilab.usc.edu/toolkit/ and the Saliency Toolbox [181] avalable at http://www.saliencytoolbox.net/.

Appendix B

The benchmark datasets mentioned in the current article are the MIT Dataset [130], the Toronto Dataset [182], the Kootstra Dataset [183], the Cerf Dataset (FIFA dataset) [184], the ImgSal Dataset [185], and the CAT2000 dataset [186]

References

1. Itti, L.; Koch, C. Computational modelling of visual attention. *Nat. Rev. Neurosci.* **2001**, *2*, 194–203. [CrossRef] [PubMed]
2. Koch, C.; Ullman, S. Shifts in selective visual attention: Towards the underlying neural circuitry. *Hum. Neurobiol.* **1985**, *4*, 219–227. [PubMed]
3. Niebur, E.; Koch, C. Control of selective visual attention: Modeling the "where" pathway. In *Advances in Neural Information Processing Systems*; Touretzky, D.S., Mozer, M.C., Hasselmo, M.E., Eds.; MIT Press: Cambridge, MA, USA, 1996; Volume 8, pp. 802–808.
4. Itti, L.; Koch, C.; Niebur, E. A model of saliency-based visual attention for rapid scene analysis. *IEEE Trans. Pattern Anal. Mach. Intell.* **1998**, *20*, 1254–1259. [CrossRef]
5. Itti, L.; Koch, C. A saliency-based search mechanism for overt and covert shifts of visual attention. *Vis. Res.* **2000**, *40*, 1489–1506. [CrossRef]
6. Dacey, D.; Packer, O.S.; Diller, L.; Brainard, D.; Peterson, B.; Lee, B. Center surround receptive field structure of cone bipolar cells in primate retina. *Vis. Res.* **2000**, *40*, 1801–1811. [CrossRef]
7. Burkitt, A.N. A Review of the Integrate-and-fire Neuron Model: I. Homogeneous Synaptic Input. *Biol. Cybern.* **2006**, *95*, 1–19. [CrossRef]
8. Lee, D.K.; Itti, L.; Koch, C.; Braun, J. Attention activates winner-take-all competition among visual filters. *Nat. Neurosci.* **1999**, *2*, 375–381. [CrossRef]
9. Posner, M.I.; Rafal, R.D.; Choate, L.S.; Vaughan, J. Inhibition of return: Neural basis and function. *Cogn. Neuropsychol.* **1985**, *2*, 211–228. [CrossRef]
10. Klein, R.M.; MacInnes, W.J. Inhibition of Return is a Foraging Facilitator in Visual Search. *Psychol. Sci.* **1999**, *10*, 346–352. [CrossRef]
11. Klein, R.M. Inhibition of return. *Trends Cogn. Sci.* **2000**, *4*, 138–147. [CrossRef]
12. Kuffler, S.W. Discharge Patterns And Functional Organization Of Mammalian Retina. *J. Neurophysiol.* **1953**, *16*, 37–68. [CrossRef] [PubMed]
13. Li, Y.; Zhou, Y.; Yan, J.; Niu, Z.; Yang, J. Visual Saliency Based on Conditional Entropy. *Lect. Notes Comput. Sci.* **2010**, 246–257. [CrossRef]
14. Marques, O.; Mayron, L.M.; Borba, G.B.; Gamba, H.R. Using visual attention to extract regions of interest in the context of image retrieval. In Proceedings of the 44th Annual Southeast Regional Conference on-ACM-SE 44, Melbourne, FL, USA, 10–12 March 2006. [CrossRef]
15. Treisman, A.M.; Gelade, G. A feature-integration theory of attention. *Cogn. Psychol.* **1980**, *12*, 97–136. [CrossRef]
16. Klein, R. Inhibitory tagging system facilitates visual search. *Nature* **1988**, *334*, 430–431. [CrossRef]

17. Bisley, J.W.; Mirpour, K. The neural instantiation of a priority map. *Curr. Opin. Psychol.* **2019**, *29*, 108–112. [CrossRef]

18. Adeli, H.; Vitu, F.; Zelinsky, G.J. A Model of the Superior Colliculus Predicts Fixation Locations during Scene Viewing and Visual Search. *J. Neurosci.* **2016**, *37*, 1453–1467. [CrossRef]

19. Sparks, D.L.; Hartwich-Young, R. The deep layers of the superior colliculus. *Rev. Oculomot. Res.* **1989**, *3*, 213–255.

20. Henderson, J.M.; Shinkareva, S.V.; Wang, J.; Luke, S.G.; Olejarczyk, J. Predicting Cognitive State from Eye Movements. *PLoS ONE* **2013**, *8*, e64937. [CrossRef]

21. Haji-Abolhassani, A.; Clark, J.J. An inverse Yarbus process: Predicting observers' task from eye movement patterns. *Vis. Res.* **2014**, *103*, 127–142. [CrossRef]

22. Mirpour, K.; Bolandnazar, Z.; Bisley, J.W. Neurons in FEF keep track of items that have been previously fixated in free viewing visual search. *J. Neurosci.* **2019**, *39*, 2114–2124. [CrossRef]

23. Goodale, M.A.; Milner, A.D. Separate visual pathways for perception and action. *Trends Neurosci.* **1992**, *15*, 20–25. [CrossRef]

24. Ungerleider, L.G.; Haxby, J.V. 'What' and 'where' in the human brain. *Curr. Opin. Neurobiol.* **1994**, *4*, 157–165. [CrossRef]

25. Ungerleider, L.G.; Mishkin, M. Two cortical visual systems. In *Analysis of Visual Behavior*; Ingle, D.J., Goodale, M.A., Mansfield, R.J.W., Eds.; MIT Press: Cambridge, MA, USA, 1982; pp. 549–586.

26. Subramanian, J.; Colby, C.L. Shape selectivity and remapping in dorsal stream visual area LIP. *J. Neurophysiol.* **2014**, *111*, 613–627. [CrossRef] [PubMed]

27. Kravitz, D.J.; Saleem, K.S.; Baker, C.I.; Ungerleider, L.G.; Mishkin, M. The ventral visual pathway: An expanded neural framework for the processing of object quality. *Trends Cogn. Sci.* **2013**, *17*, 26–49. [CrossRef] [PubMed]

28. Corbetta, M.; Shulman, G.L. Control of goal-directed and stimulus-driven attention in the brain. *Nat. Rev. Neurosci.* **2002**, *3*, 201–215. [CrossRef] [PubMed]

29. Yarbus, A.L. *Eye Movements and Vision*; Plenum Press: New York, NY, USA, 1967.

30. MacInnes, W.J.; Hunt, A.R.; Clarke, A.; Dodd, M.D. A Generative Model of Cognitive State from Task and Eye Movements. *Cogn. Comput.* **2018**, *10*, 703–717. [CrossRef] [PubMed]

31. Borji, A.; Sihite, D.N.; Itti, L. Probabilistic learning of task-specific visual attention. In Proceedings of the IEEE Conference on Computer Vision and Pattern Recognition, Providence, RI, USA, 16–21 June 2012. [CrossRef]

32. Kanan, C.; Tong, M.H.; Zhang, L.; Cottrell, G.W. SUN: Top-down saliency using natural statistics. *Vis. Cogn.* **2009**, *17*, 979–1003. [CrossRef]

33. Donner, T.; Kettermann, A.; Diesch, E.; Ostendorf, F.; Villringer, A.; Brandt, S.A. Involvement of the human frontal eye field and multiple parietal areas in covert visual selection during conjunction search. *Eur. J. Neurosci.* **2000**, *12*, 3407–3414. [CrossRef]

34. Suzuki, M.; Gottlieb, J. Distinct neural mechanisms of distractor suppression in the frontal and parietal lobe. *Nat. Neurosci.* **2013**, *16*, 98–104. [CrossRef]

35. Buschman, T.J.; Miller, E.K. Top-Down Versus Bottom-Up Control of Attention in the Prefrontal and Posterior Parietal Cortices. *Science* **2007**, *315*, 1860–1862. [CrossRef]

36. Buschman, T.J.; Miller, E.K. Shifting the Spotlight of Attention: Evidence for Discrete Computations in Cognition. *Front. Hum. Neurosci.* **2010**, *4*, 194. [CrossRef] [PubMed]

37. Corbetta, M.; Miezin, F.; Shulman, G.; Petersen, S. A PET study of visuospatial attention. *J. Neurosci.* **1993**, *13*, 1202–1226. [CrossRef] [PubMed]

38. Corbetta, M.; Akbudak, E.; Conturo, T.E.; Snyder, A.Z.; Ollinger, J.M.; Linenweber, M.R.; Petersen, S.E.; Raichle, M.E.; Van Essen, D.C.; Drury, H.A.; et al. A common network of functional areas for attention and eye movements. *Neuron* **1998**, *21*, 761–773. [CrossRef]

39. Liu, T.; Slotnick, S.D.; Serences, J.T.; Yantis, S. Cortical mechanisms of feature-based attentional control. *Cereb. Cortex* **2003**, *13*, 1334–1343. [CrossRef]

40. Corbetta, M.; Shulman, G.L.; Miezin, F.M.; Petersen, S.E. Superior parietal cortex activation during spatial attention shifts and visual feature conjunction. *Science* **1995**, *270*, 802–805. [CrossRef]

41. Donner, T.H.; Kettermann, A.; Diesch, E.; Ostendorf, F.; Villringer, A.; Brandt, S.A. Visual feature and conjunction searches of equal difficulty engage only partially overlapping frontoparietal networks. *Neuroimage* **2002**, *15*, 16–25. [CrossRef]

42. Nobre, A.C.; Sebestyen, G.N.; Gitelman, D.R.; Frith, C.D.; Mesulam, M.M. Filtering of distractors during visual search studied by positron emission tomography. *Neuroimage* **2002**, *16*, 968–976. [CrossRef]
43. Moore, T.; Fallah, M. Microstimulation of the frontal eye field and its effects on covert spatial attention. *J. Neurophysiol.* **2004**, *91*, 152–162. [CrossRef]
44. Wardak, C.; Ibos, G.; Duhamel, J.R.; Olivier, E. Contribution of the monkey frontal eye field to covert visual attention. *J. Neurosci.* **2006**, *26*, 4228–4235. [CrossRef]
45. Zhou, H.; Desimone, R. Feature-based attention in the frontal eye field and area V4 during visual search. *Neuron* **2011**, *70*, 1205–1217. [CrossRef]
46. Miller, B.T.; D'Esposito, M. Searching for "the Top" in Top-Down Control. *Neuron* **2005**, *48*, 535–538. [CrossRef] [PubMed]
47. Egeth, H.E.; Leonard, C.J.; Leber, A.B. Why salience is not enough: Reflections on top-down selection in vision. *Acta Psychol.* **2010**, *135*, 130–132. [CrossRef] [PubMed]
48. Schall, J.D.; Cohen, J.Y. The neural basis of saccade target selection. In *The Oxford Handbook of Eye Movements*; Liversedge, S.P., Gilchrist, I.D., Everling, S., Eds.; Oxford University Press: Oxford, UK, 2012; pp. 357–374.
49. Rodriguez-Sanchez, A.J.; Simine, E.; Tsotsos, J.K. Attention and visual search. *Int. J. Neural Syst.* **2007**, *17*, 275–288. [CrossRef]
50. Wolfe, J.M.; Gancarz, G. Guided Search 3.0. In *Basic and Clinical Applications of Vision Science*; Springer: Dordrecht, The Netherlands, 1997; pp. 189–192.
51. Fecteau, J.; Munoz, D. Salience, relevance, and firing: A priority map for target selection. *Trends Cogn. Sci.* **2006**, *10*, 382–390. [CrossRef] [PubMed]
52. Desimone, R.; Duncan, J. Neural Mechanisms of Selective Visual Attention. *Annu. Rev. Neurosci.* **1995**, *18*, 193–222. [CrossRef]
53. Wolfe, J.M.; Horowitz, T.S. Five factors that guide attention in visual search. *Nat. Hum. Behav.* **2017**, *1*, 0058. [CrossRef]
54. Veale, R.; Hafed, Z.M.; Yoshida, M. How is visual salience computed in the brain? Insights from behaviour, neurobiology and modelling. *Philos. Trans. R. Soc. B Biol. Sci.* **2017**, *372*, 20160113. [CrossRef]
55. Schiller, P.H.; Stryker, M. Single-unit recording and stimulation in superior colliculus of the alert rhesus monkey. *J. Neurophysiol.* **1972**, *35*, 915–924. [CrossRef]
56. Schiller, P.H.; Chou, I.H. The effects of frontal eye field and dorsomedial frontal cortex lesions on visually guided eye movements. *Nat. Neurosci.* **1998**, *1*, 248. [CrossRef]
57. Frintrop, S.; Rome, E.; Christensen, H.I. Computational visual attention systems and their cognitive foundations. *ACM Trans. Appl. Percept.* **2010**, *7*, 1–39. [CrossRef]
58. Munoz, D.P.; Everling, S. Look away: The anti-saccade task and the voluntary control of eye movement. *Nat. Rev. Neurosci.* **2004**, *5*, 218–228. [CrossRef] [PubMed]
59. Wolfe, J.M.; Horowitz, T.S. What attributes guide the deployment of visual attention and how do they do it? *Nat. Rev. Neurosci.* **2004**, *5*, 495–501. [CrossRef] [PubMed]
60. Broadbent, D.E. *Perception and Communication*; Elsevier: Amsterdam, The Netherlands, 2013.
61. Walther, D.; Itti, L.; Riesenhuber, M.; Poggio, T.; Koch, C. Attentional selection for object recognition—A gentle way. In *International Workshop on Biologically Motivated Computer Vision*; Springer: Berlin/Heidelberg, Germany, 2002; pp. 472–479.
62. Riesenhuber, M.; Poggio, T. Neural mechanisms of object recognition. *Curr. Opin. Neurobiol.* **2002**, *12*, 162–168. [CrossRef]
63. Tipper, S.P.; Weaver, B.; Jerreat, L.M.; Burak, A.L. Object-based and environment-based inhibition of return of visual attention. *J. Exp. Psychol. Hum. Percept. Perform.* **1994**, *20*, 478. [CrossRef]
64. Draper, B.A.; Lionelle, A. Evaluation of selective attention under similarity transformations. *Comput. Vis. Image Underst.* **2005**, *100*, 152–171. [CrossRef]
65. Eriksen, C.W.; James, J.D.S. Visual attention within and around the field of focal attention: A zoom lens model. *Percept. Psychophys.* **1986**, *40*, 225–240. [CrossRef]
66. Posner, M.I. Orienting of attention. *Q. J. Exp. Psychol.* **1980**, *32*, 3–25. [CrossRef]
67. Wolfe, J.M. Guided Search 2.0 A revised model of visual search. *Psychon. Bull. Rev.* **1994**, *1*, 202–238. [CrossRef]
68. Navalpakkam, V.; Itti, L. Modeling the influence of task on attention. *Vis. Res.* **2005**, *45*, 205–231. [CrossRef]

69. Awh, E.; Vogel, E.K.; Oh, S.-H. Interactions between attention and working memory. *Neuroscience* **2006**, *139*, 201–208. [CrossRef]
70. Fougnie, D. The relationship between attention and working memory. In *New Research on Short-Term Memory*; Johansen, N.B., Ed.; Nova Science Publishers: New York, NY, USA, 2008; pp. 1–45.
71. Rosen, M.L.; Stern, C.E.; Somers, D.C. Long-term memory guidance of visuospatial attention in a change-detection paradigm. *Front. Psychol.* **2014**, *5*, 266. [CrossRef] [PubMed]
72. Harel, J.; Koch, C.; Perona, P. Graph-based visual saliency. In *Advances in Neural Information Processing Systems*; The MIT Press: Cambridge, MA, USA, 2007; pp. 545–552.
73. Hopf, J.M. Attention to Features Precedes Attention to Locations in Visual Search: Evidence from Electromagnetic Brain Responses in Humans. *J. Neurosci.* **2004**, *24*, 1822–1832. [CrossRef] [PubMed]
74. Li, Z. A saliency map in primary visual cortex. *Trends Cogn. Sci.* **2002**, *6*, 9–16. [CrossRef]
75. Koene, A.R.; Li, Z. Feature-specific interactions in salience from combined feature contrasts: Evidence for a bottom–up saliency map in V1. *J. Vis.* **2007**, *7*, 6. [CrossRef] [PubMed]
76. Horwitz, G.D.; Albright, T.D. Paucity of chromatic linear motion detectors in macaque V1. *J. Vis.* **2005**, *5*, 4. [CrossRef] [PubMed]
77. Ts'o, D.Y.; Gilbert, C.D. The organization of chromatic and spatial interactions in the primate striate cortex. *J. Neurosci.* **1988**, *8*, 1712–1727. [CrossRef]
78. Lennie, P.; Movshon, J.A. Coding of color and form in the geniculostriate visual pathway (invited review). *J. Opt. Soc. Am. A* **2005**, *22*, 2013. [CrossRef]
79. Garg, A.K.; Li, P.; Rashid, S.M.; Callaway, M.E. Color and orientation are jointly coded and spatially organized in primate primary visual cortex. *Science* **2019**, *364*, 1275–1279. [CrossRef]
80. Park, S.J.; Shin, J.K.; Lee, M. Biologically inspired saliency map model for bottom-up visual attention. In *International Workshop on Biologically Motivated Computer Vision*; Springer: Berlin/Heidelberg, Germany, 2002; pp. 418–426.
81. Swindale, N.V. Neurophysiology: Parallel channels and redundant mechanisms in visual cortex. *Nature* **1986**, *322*, 775–776. [CrossRef]
82. Aboudib, A.; Gripon, V.; Coppin, G. A biologically inspired framework for visual information processing and an application on modeling bottom-up visual attention. *Cogn. Comput.* **2016**, *8*, 1007–1026. [CrossRef]
83. Hamker, F.H. Modeling feature-based attention as an active top-down inference process. *BioSystems* **2006**, *86*, 91–99. [CrossRef] [PubMed]
84. Duncan, J. An adaptive coding model of neural function in prefrontal cortex. *Nat. Rev. Neurosci.* **2001**, *2*, 820–829. [CrossRef] [PubMed]
85. Wu, S.; Amari, S.I.; Nakahara, H. Population coding and decoding in a neural field: A computational study. *Neural Comput.* **2002**, *14*, 999–1026. [CrossRef] [PubMed]
86. Wilder, J.D.; Kowler, E.; Schnitzer, B.S.; Gersch, T.M.; Dosher, B.A. Attention during active visual tasks: Counting, pointing, or simply looking. *Vis. Res.* **2009**, *49*, 1017–1031. [CrossRef] [PubMed]
87. Wolfe, J.M.; Cave, K.R.; Franzel, S.L. Guided search: An alternative to the feature integration model for visual search. *J. Exp. Psychol. Hum. Percept. Perform.* **1989**, *15*, 419. [CrossRef]
88. Julesz, B. A brief outline of the texton theory of human vision. *Trends Neurosci.* **1984**, *7*, 41–45. [CrossRef]
89. Neisser, U. Visual search. *Sci. Am.* **1964**, *210*, 94–102. [CrossRef]
90. Wolfe, J.M. Guided search 4.0. In *Integrated Models of Cognitive Systems*; Gray, W.D., Ed.; Oxford University Press: New York, NY, USA, 2007; pp. 99–119.
91. Jiang, Y.V.; Swallow, K.M.; Rosenbaum, G.M. Guidance of spatial attention by incidental learning and endogenous cuing. Journal of experimental psychology. *Hum. Percept. Perform.* **2013**, *39*, 285–297. [CrossRef]
92. Soto, D.; Humphreys, G.W.; Rotshstein, P. Dissociating the neural mechanisms of memory based guidance of visual selection. *Proc. Natl. Acad. Sci. USA* **2007**, *104*, 17186–17191. [CrossRef]
93. Cave, K.R. The FeatureGate model of visual selection. *Psychol. Res.* **1999**, *62*, 182–194. [CrossRef]
94. Cohen, R.A. Lateral Inhibition. *Encycl. Clin. Neuropsychol.* **2011**, 1436–1437. [CrossRef]
95. Rao, R.P.; Ballard, D.H. Probabilistic models of attention based on iconic representations and predictive coding. In *Neurobiology of Attention*; Academic Press: New York, NY, USA, 2005; pp. 553–561.
96. Hinton, G.E.; Sejnowski, T.J. Learning and relearning in Boltzmann machines. *Parallel Distrib. Process. Explor. Microstruct. Cogn.* **1986**, *1*, 2.

97. Nowlan, S.J. Maximum likelihood competitive learning. In *Advances in Neural Information Processing Systems*; The MIT Press: Cambridge, MA, USA, 1990; pp. 574–582.

98. Rao, R.P.; Ballard, D.H. Predictive coding in the visual cortex: A functional interpretation of some extra-classical receptive-field effects. *Nat. Neurosci.* **1999**, *2*, 79. [CrossRef] [PubMed]

99. Mishkin, M.; Ungerleider, L.G.; Macko, K.A. Object vision and spatial vision: Two cortical pathways. *Trends Neurosci.* **1983**, *6*, 414–417. [CrossRef]

100. Rybak, I.A.; Gusakova, V.I.; Golovan, A.V.; Podladchikova, L.N.; Shevtsova, N.A. Attention-Guided Recognition Based on "What" and "Where": Representations: A Behavioral Model. In *Neurobiology of Attention*; Academic Press: New York, NY, USA, 2005; pp. 663–670.

101. Deco, G.; Rolls, E.T. A neurodynamical cortical model of visual attention and invariant object recognition. *Vis. Res.* **2004**, *44*, 621–642. [CrossRef]

102. Norman, J. Two visual systems and two theories of perception: An attempt to reconcile the constructivist and ecological approaches. *Behav. Brain Sci.* **2002**, *25*, 73–96. [CrossRef]

103. Rolls, E.T.; Aggelopoulos, N.C.; Zheng, F. The Receptive Fields of Inferior Temporal Cortex Neurons in Natural Scenes. *J. Neurosci.* **2003**, *23*, 339–348. [CrossRef]

104. Schenk, T.; McIntosh, R.D. Do we have independent visual streams for perception and action? *Cogn. Neurosci.* **2010**, *1*, 52–62. [CrossRef]

105. Milner, A.D.; Goodale, M.A. Two visual systems re-viewed. *Neuropsychologia* **2008**, *46*, 774–785. [CrossRef]

106. DiCarlo, J.J.; Zoccolan, D.; Rust, N.C. How Does the Brain Solve Visual Object Recognition? *Neuron* **2012**, *73*, 415–434. [CrossRef]

107. Stein, T.; Peelen, M.V. Object detection in natural scenes: Independent effects of spatial and category-based attention. *Atten. Percept. Psychophys.* **2017**, *79*, 738–752. [CrossRef] [PubMed]

108. Gauthier, I.; Tarr, M.J. Visual Object Recognition: Do We (Finally) Know More Now Than We Did? *Annu. Rev. Vis. Sci.* **2016**, *2*, 377–396. [CrossRef] [PubMed]

109. Pylyshyn, Z.W.; Storm, R.W. Tracking multiple independent targets: Evidence for a parallel tracking mechanism. *Spat. Vis.* **1988**, *3*, 179–197. [CrossRef] [PubMed]

110. Pylyshyn, Z.; Burkell, J.; Fisher, B.; Sears, C.; Schmidt, W.; Trick, L. Multiple parallel access in visual attention. *Can. J. Exp. Psychol./Rev. Can. Psychol. Exp.* **1994**, *48*, 260. [CrossRef]

111. Backer, G.; Mertsching, B.; Bollmann, M. Data-and model-driven gaze control for an active-vision system. *IEEE Trans. Pattern Anal. Mach. Intell.* **2001**, *23*, 1415–1429. [CrossRef]

112. Amari, S. Dynamics of pattern formation in lateral-inhibition type neural fields. *Biol. Cybern.* **1977**, *27*, 77–87. [CrossRef]

113. Tipper, S.P.; Driver, J.; Weaver, B. Object-centred inhibition of return of visual attention. *Q. J. Exp. Psychol.* **1991**, *43*, 289–298. [CrossRef]

114. Sun, Y.; Fisher, R. Object-based visual attention for computer vision. *Artif. Intell.* **2003**, *146*, 77–123. [CrossRef]

115. Duncan, J.; Humphreys, G.; Ward, R. Competitive brain activity in visual attention. *Curr. Opin. Neurobiol.* **1997**, *7*, 255–261. [CrossRef]

116. Pelli, D.G.; Tillman, K.A. The uncrowded window of object recognition. *Nat. Neurosci.* **2008**, *11*, 1129–1135. [CrossRef]

117. Oliva, A.; Torralba, A.; Castelhano, M.; Henderson, J. Top-down control of visual attention in object detection. In Proceedings of the 2003 International Conference on Image Processing (Cat. No.03CH37429), Barcelona, Spain, 14–17 September 2003.

118. Van der Voort van der Kleij, G.T.; van der Velde, F.; de Kamps, M. Learning Location Invariance for Object Recognition and Localization. *Lect. Notes Comput. Sci.* **2005**, 235–244. [CrossRef]

119. Tsotsos, J.K.; Culhane, S.M.; Wai, W.Y.K.; Lai, Y.; Davis, N.; Nuflo, F. Modeling visual attention via selective tuning. *Artif. Intell.* **1995**, *78*, 507–545. [CrossRef]

120. Bruce, N.; Tsotsos, J. Saliency based on information maximization. In *Advances in Neural Information Processing Systems*; The MIT Press: Cambridge, MA, USA, 2006; pp. 155–162.

121. Shannon, C.E. A mathematical theory of communication. *Bell Syst. Tech. J.* **1948**, *27*, 379–423. [CrossRef]

122. Olshausen, B.A.; Field, D.J. Emergence of simple-cell receptive field properties by learning a sparse code for natural images. *Nature* **1996**, *381*, 607–609. [CrossRef]

123. Schill, K. *A Model of Attention and Recognition by Information Maximization*; Neurobiology of Attention Academic Press: New York, NY, USA, 2005; pp. 671–676.

124. Shafer, G. Dempster-shafer theory. *Encycl. Artif. Intell.* **1992**, *1*, 330–331.

125. Torralba, A.; Oliva, A.; Castelhano, M.S.; Henderson, J.M. Contextual guidance of eye movements and attention in real-world scenes: The role of global features in object search. *Psychol. Rev.* **2006**, *113*, 766. [CrossRef]

126. Russell, B.C.; Torralba, A.; Murphy, K.P.; Freeman, W.T. LabelMe: A database and web-based tool for image annotation. *Int. J. Comput. Vis.* **2008**, *77*, 157–173. [CrossRef]

127. Rubin, N. Figure and ground in the brain. *Nat. Neurosci.* **2001**, *4*, 857–858. [CrossRef]

128. Zhang, J.; Sclaroff, S. Saliency detection: A boolean map approach. In Proceedings of the IEEE international conference on computer vision, Sydney, Australia, 1–8 December 2013; pp. 153–160.

129. Huang, L.; Pashler, H. A Boolean map theory of visual attention. *Psychol. Rev.* **2007**, *114*, 599. [CrossRef]

130. Judd, T.; Ehinger, K.; Durand, F.; Torralba, A. Learning to predict where humans look. In Proceedings of the 2009 IEEE 12th international conference on computer vision, Kyoto, Japan, 29 September–2 October 2009; pp. 2106–2113.

131. Najemnik, J.; Geisler, W.S. Optimal eye movement strategies in visual search. *Nature* **2005**, *434*, 387–391. [CrossRef]

132. Bylinskii, Z.; Judd, T.; Oliva, A.; Torralba, A.; Durand, F. What do different evaluation metrics tell us about saliency models? *IEEE Trans. Pattern Anal. Mach. Intell.* **2019**, *41*, 740–757. [CrossRef] [PubMed]

133. Judd, T.; Durand, F.; Torralba, A. *A Benchmark of Computational Models of Saliency to Predict Human Fixations*; Technical rep. MIT-CSAIL-TR-2012-001; Massachusetts Institute of Technology: Cambridge, MA, USA, 2012.

134. Bengio, Y. Learning deep architectures for AI. *Found. Trends®Mach. Learn.* **2009**, *2*, 1–127. [CrossRef]

135. Huang, X.; Shen, C.; Boix, X.; Zhao, Q. Salicon: Reducing the semantic gap in saliency prediction by adapting deep neural networks. In Proceedings of the IEEE International Conference on Computer Vision, Santiago, Chile, 7–13 December 2015; pp. 262–270.

136. McCulloch, W.S.; Pitts, W. A logical calculus of the ideas immanent in nervous activity. *Bull. Math. Biophys.* **1943**, *5*, 115–133. [CrossRef]

137. Deng, L.; Yu, D. Deep learning: Methods and applications. *Found. Trends®Signal Process.* **2014**, *7*, 197–387. [CrossRef]

138. Basheer, I.A.; Hajmeer, M. Artificial neural networks: Fundamentals, computing, design, and application. *J. Microbiol. Methods* **2000**, *43*, 3–31. [CrossRef]

139. Bylinskii, Z.; Judd, T.; Borji, A.; Itti, L.; Durand, F.; Oliva, A.; Torralba, A. Mit Saliency Benchmark. 2015. Available online: http://saliency.mit.edu/ (accessed on 24 October 2019).

140. Cadieu, C.F.; Hong, H.; Yamins, D.L.; Pinto, N.; Ardila, D.; Solomon, E.A.; Majaj, N.J.; DiCarlo, J.J. Deep neural networks rival the representation of primate IT cortex for core visual object recognition. *PLoS Comput. Biol.* **2014**, *10*, e1003963. [CrossRef]

141. Kummerer, M.; Wallis, T.S.; Gatys, L.A.; Bethge, M. Understanding low-and high-level contributions to fixation prediction. In Proceedings of the IEEE International Conference on Computer Vision, Venice, Italy, 22–29 October 2017; pp. 4789–4798.

142. Jia, S. Eml-net: An expandable multi-layer network for saliency prediction. *arXiv* **2018**, arXiv:1805.01047.

143. Kruthiventi, S.S.; Ayush, K.; Babu, R.V. Deepfix: A fully convolutional neural network for predicting human eye fixations. *IEEE Trans. Image Process.* **2017**, *26*, 4446–4456. [CrossRef]

144. Pan, J.; Sayrol, E.; Giro-i-Nieto, X.; McGuinness, K.; O'Connor, N.E. Shallow and deep convolutional networks for saliency prediction. In Proceedings of the IEEE Conference on Computer Vision and Pattern Recognition, Las Vegas, NV, USA, 26 June–1 July 2016; pp. 598–606.

145. Dodge, S.F.; Karam, L.J. Visual saliency prediction using a mixture of deep neural networks. *IEEE Trans. Image Process.* **2018**, *27*, 4080–4090. [CrossRef]

146. He, S.; Pugeault, N. Deep saliency: What is learnt by a deep network about saliency? *arXiv* **2018**, arXiv:1801.04261.

147. Zhang, J.; Zhang, T.; Dai, Y.; Harandi, M.; Hartley, R. Deep unsupervised saliency detection: A multiple noisy labeling perspective. In Proceedings of the IEEE Conference on Computer Vision and Pattern Recognition, Salt Lake City, UT, USA, 18–22 June 2018; pp. 9029–9038.

148. Li, G.; Yu, Y. Visual saliency based on multiscale deep features. In Proceedings of the IEEE Conference on Computer Vision and Pattern Recognition, Boston, MA, USA, 7–12 June 2015; pp. 5455–5463.

149. Li, G.; Yu, Y. Deep contrast learning for salient object detection. In Proceedings of the IEEE Conference on Computer Vision and Pattern Recognition, Las Vegas, NV, USA, 26 June–1 July 2016; pp. 478–487.

150. Goodfellow, I.; Pouget-Abadie, J.; Mirza, M.; Xu, B.; Warde-Farley, D.; Ozair, S.; Courville, A.; Bengio, Y. Generative adversarial nets. In *Advances in Neural Information Processing Systems*; The MIT Press: Cambridge, MA, USA, 2014; pp. 2672–2680.

151. Pan, J.; Sayrol, E.; Nieto, X.G.I.; Ferrer, C.C.; Torres, J.; McGuinness, K.; OConnor, N.E. SalGAN: Visual saliency prediction with adversarial networks. In Proceedings of the CVPR Scene Understanding Workshop (SUNw), Honolulu, HI, USA, 26 July 2017.

152. Fernando, T.; Denman, S.; Sridharan, S.; Fookes, C. Task specific visual saliency prediction with memory augmented conditional generative adversarial networks. In Proceedings of the 2018 IEEE Winter Conference on Applications of Computer Vision (WACV), Lake Tahoe, NV, USA, 12–15 March 2018; pp. 1539–1548.

153. Elman, J. Finding structure in time. *Cogn. Sci.* **1990**, *14*, 179–211. [CrossRef]

154. Adel Bargal, S.; Zunino, A.; Kim, D.; Zhang, J.; Murino, V.; Sclaroff, S. Excitation backprop for RNNs. In Proceedings of the IEEE Conference on Computer Vision and Pattern Recognition, Salt Lake City, UT, USA, 18–22 June 2018; pp. 1440–1449.

155. Li, G.; Xie, Y.; Wei, T.; Wang, K.; Lin, L. Flow guided recurrent neural encoder for video salient object detection. In Proceedings of the IEEE conference on computer vision and pattern recognition, Munich, Germany, 8–14 September 2018; pp. 3243–3252.

156. Tang, Y.; Wu, X.; Bu, W. Deeply-supervised recurrent convolutional neural network for saliency detection. In Proceedings of the 24th ACM international conference on Multimedia, Amsterdam, The Netherlands, 15–19 October 2016; ACM: New York, NY, USA, 2016; pp. 397–401.

157. Kuen, J.; Wang, Z.; Wang, G. Recurrent attentional networks for saliency detection. In Proceedings of the IEEE Conference on Computer Vision and Pattern Recognition, Las Vegas, NV, USA, 26 June–1 July 2016; pp. 3668–3677.

158. Liu, N.; Han, J. A deep spatial contextual long-term recurrent convolutional network for saliency detection. *IEEE Trans. Image Process.* **2018**, *27*, 3264–3274. [CrossRef] [PubMed]

159. Hochreiter, S.; Schmidhuber, J. Long short-term memory. *Neural Comput.* **1997**, *9*, 1735–1780. [CrossRef] [PubMed]

160. Li, X.; Wu, X. Constructing long short-term memory based deep recurrent neural networks for large vocabulary speech recognition. In Proceedings of the 2015 IEEE International Conference on Acoustics, Speech and Signal Processing (ICASSP), Brisbane, Australia, 19–24 April 2015; pp. 4520–4524.

161. Sak, H.; Senior, A.; Beaufays, F. Long short-term memory recurrent neural network architectures for large scale acoustic modeling. In Proceedings of the Fifteenth Annual Conference of the International Speech Communication Association, Singapore, 14–18 September 2014.

162. Liu, Y.; Zhang, S.; Xu, M.; He, X. Predicting salient face in multiple-face videos. In Proceedings of the IEEE Conference on Computer Vision and Pattern Recognition, Honolulu, HI, USA, 21–26 July 2017; pp. 4420–4428.

163. Cornia, M.; Baraldi, L.; Serra, G.; Cucchiara, R. Predicting human eye fixations via an lstm-based saliency attentive model. *IEEE Trans. Image Process.* **2018**, *27*, 5142–5154.

164. Wang, W.; Shen, J.; Dong, X.; Borji, A. Salient object detection driven by fixation prediction. In Proceedings of the IEEE Conference on Computer Vision and Pattern Recognition, Salt Lake City, UT, USA, 18–23 June 2018; pp. 1711–1720.

165. Lee, H.; Ekanadham, C.; Ng, A.Y. Sparse deep belief net model for visual area V2. In *Advances in Neural Information Processing Systems*; The MIT Press: Cambridge, MA, USA, 2008; pp. 873–880.

166. Huang, G.; Liu, Z.; Van Der Maaten, L.; Weinberger, K.Q. Densely connected convolutional networks. In Proceedings of the IEEE Conference On Computer Vision and Pattern Recognition, Honolulu, HI, USA, 21–26 July 2017; pp. 4700–4708. [CrossRef]

167. Schrimpf, M.; Kubilius, J.; Hong, H.; Majaj, N.J.; Rajalingham, R.; Issa, E.B.; Yamins, D.L. Brain-Score: Which artificial neural network for object recognition is most brain-like? *BioRxiv* **2018**, 407007. [CrossRef]

168. Krizhevsky, A.; Sutskever, I.; Hinton, G.E. ImageNet Classification with Deep Convolutional Neural Networks. In *Advances in Neural Information Processing Systems (NIPS)*; The MIT Press: Cambridge, MA, USA, 2012; ISSN 10495258. [CrossRef]

169. Szegedy, C.; Liu, W.; Jia, Y.; Sermanet, P.; Reed, S.; Anguelov, D.; Rabinovich, A. Going deeper with convolutions. In Proceedings of the IEEE Conference on Computer Vision and Pattern Recognition, Boston, MA, USA, 7–12 June 2015; pp. 1–9.

170. Liu, C.; Zoph, B.; Neumann, M.; Shlens, J.; Hua, W.; Li, L.-J.; Fei-Fei, L.; Yuille, A.; Huang, J.; Murphy, K. Progressive Neural Architecture Search. *arXiv* **2017**, arXiv:1712.00559v3.

171. Kubilius, J.; Schrimpf, M.; DiCarlo, J. CORnet: Modeling Core Object Recognition. *arXiv* **2018**, arXiv:1808.01405.

172. Ferri, C.; Hernández-Orallo, J.; Flach, P.A. A coherent interpretation of AUC as a measure of aggregated classification performance. In Proceedings of the 28th International Conference on Machine Learning (ICML-11), Washington, DC, USA, 28 June–2 July 2011; pp. 657–664.

173. Riche, N.; Duvinage, M.; Mancas, M.; Gosselin, B.; Dutoit, T. Saliency and human fixations: State-of-the-art and study of comparison metrics. In Proceedings of the IEEE International Conference On Computer Vision, Sydney, Australia, 1–8 December 2013; pp. 1153–1160.

174. Kachurka, V.; Madani, K.; Sabourin, C.; Golovko, V. From human eye fixation to human-like autonomous artificial vision. In Proceedings of the International Work-Conference on Artificial Neural Networks, Palma de Mallorca, Spain, 10–12 June 2015; Springer: Champaign, IL, USA, 2015; pp. 171–184.

175. Borji, A.; Sihite, D.N.; Itti, L. Quantitative analysis of human-model agreement in visual saliency modeling: A comparative study. *IEEE Trans. Image Process.* **2013**, *22*, 55–69. [CrossRef]

176. Kullback, S.; Leibler, R.A. On information and sufficiency. *Ann. Math. Stat.* **1951**, *22*, 79–86. [CrossRef]

177. Merzon, L.; Zhulikov, G.; Malevich, T.; Krasovskaya, S.; MacInnes, J.W. Temporal Limitations of the Standard Leaky Integrate and Fire Model. *High. Sch. Econ. Res. Pap. No. WP BRP* **2018**, *94*, 1–16. [CrossRef]

178. Nguyen, A.; Yosinski, J.; Clune, J. Deep neural networks are easily fooled: High confidence predictions for unrecognizable images. In Proceedings of the IEEE Conference on Computer Vision and Pattern Recognition, Boston, MA, USA, 7–12 June 2015; pp. 427–436.

179. Harel, J. A Saliency Implementation in MATLAB. Available online: :http://www.vision.caltech.edu/~{}harel/share/gbvs.php (accessed on 19 February 2019).

180. Itti, L. The iLab Neuromorphic Vision C++ Toolkit: Free tools for the next generation of vision algorithms. *Neuromorphic Eng.* **2004**, *1*, 10.

181. Walther, D.; Koch, C. Modeling attention to salient proto-objects. *Neural Netw.* **2006**, *19*, 1395–1407. [CrossRef] [PubMed]

182. Bruce, N.D.; Tsotsos, J.K. Saliency, attention, and visual search: An information theoretic approach. *J. Vis.* **2009**, *9*, 5. [CrossRef] [PubMed]

183. Kootstra, G.; Nederveen, A.; De Boer, B. Paying attention to symmetry. In *British Machine Vision Conference (BMVC2008)*; The British Machine Vision Association and Society for Pattern Recognition: Oxford, UK, 2008; pp. 1115–1125.

184. Cerf, M.; Harel, J.; Einhäuser, W.; Koch, C. Predicting human gaze using low-level saliency combined with face detection. In *Advances in Neural Information Processing Systems*; The MIT Press: Cambridge, MA, USA, 2008; pp. 241–248.

185. Li, J.; Levine, M.D.; An, X.; Xu, X.; He, H. Visual saliency based on scale-space analysis in the frequency domain. *IEEE Trans. Pattern Anal. Mach. Intell.* **2012**, *35*, 996–1010. [CrossRef] [PubMed]

186. Borji, A.; Itti, L. CAT2000: A Large Scale Fixation Dataset for Boosting Saliency Research. *arXiv* **2015**, arXiv:1505.03581.

Review

Seeing Beyond Salience and Guidance: The Role of Bias and Decision in Visual Search

Alasdair D. F. Clarke [1],*, Anna Nowakowska [2] and Amelia R. Hunt [2]

[1]   Department of Psychology, University of Essex, Colchester CO4 3SQ, UK
[2]   School of Psychology, University of Aberdeen, Aberdeen AB24 3FX, UK
*   Correspondence: a.clarke@essex.ac.uk; Tel.: +44-1206-873-941

Received: 28 February 2019; Accepted: 21 August 2019; Published: 11 September 2019

Abstract: Visual search is a popular tool for studying a range of questions about perception and attention, thanks to the ease with which the basic paradigm can be controlled and manipulated. While often thought of as a sub-field of vision science, search tasks are significantly more complex than most other perceptual tasks, with strategy and decision playing an essential, but neglected, role. In this review, we briefly describe some of the important theoretical advances about perception and attention that have been gained from studying visual search within the signal detection and guided search frameworks. Under most circumstances, search also involves executing a series of eye movements. We argue that understanding the contribution of biases, routines and strategies to visual search performance over multiple fixations will lead to new insights about these decision-related processes and how they interact with perception and attention. We also highlight the neglected potential for variability, both within and between searchers, to contribute to our understanding of visual search. The exciting challenge will be to account for variations in search performance caused by these numerous factors and their interactions. We conclude the review with some recommendations for ways future research can tackle these challenges to move the field forward.

Keywords: visual search; eye movements; attention; strategy; decision

---

Searching is a familiar, sometimes frustrating, part of daily life. We search our homes and offices for keys, glasses, wallets and bags. We look for street signs, cars in parking lots, items in the supermarket, children on a playground and friends in crowded restaurants. With all this searching, one might expect us to be experts, but searching is often effortful, time-consuming and error-prone. In the laboratory, even highly-simplified versions of search tasks illustrate why this is the case: unlike many other tasks used in vision and attention research, visual search is a complex task that involves visual processing in both central and peripheral vision, controlling our attention to filter out distracting visual features, coordination of vision and attention with eye (and sometimes head) movements and a dependent sequence of one or more decisions about when and where to move attention and the eyes to sample information. Finally, visual search involves high-level strategic decisions about how to search and for how long to search without finding anything before giving up. Seen in this light, it is not surprising that the performance of visual search is often slow and highly variable.

Despite its complexity, visual search is a realistic task that is easy to control and manipulate in experimental settings, and it is a popular tool for studying a range of questions about perception and attention. Less commonly, visual search can also be used to study strategy and decision-making. Previous studies of visual search can be loosely categorised by whether observers make a single fixation or if search unfolds over multiple fixations. Single-fixation search makes heavy use of psychophysical methods (i.e., [1,2]). Theories in this literature have established how representations of the stimulus interact with the limits of attention to inform decisions as to the presence or absence of the target. However, decision here is in the context of psychophysical signal detection theories. In our literature review below, we focus on multiple-fixation search (also sometimes referred to as "foraging", although this label is usually reserved for multiple-target search), and when we talk of decisions, we mean in the context of the wider field of cognitive psychology, that is, how we select between different courses of action. While Palmer et al. [1] (p. 1256), for example, discussed how their psychophysical theory of visual search could be extended to multiple-fixations, in terms of search strategy, they went no further than suggesting "The choice of the location of the next fixation is probably a function of the information in the periphery and the observer's strategy." In what follows, we hope to convince the reader that this second component, the observer's strategy, is highly idiosyncratic, accounts for a large proportion of variation in the speed and success of search and is heavily influenced by both oculomotor and decision biases.

In this review, we will first briefly describe some of the important theoretical advances about perception and attention that have been gained from studying visual search. Our understanding of how visual features and selective attention operate and interact during a single fixation of visual search is well developed, thanks to several decades of research. We also have an emerging understanding of the influence of higher level factors like reward and serial dependences on these within-fixation processes. However, search performance also depends on decisions about where to fixate, and over a series of these fixation decisions, the effect on performance accumulates rapidly. We will argue that a focus on understanding the contribution of biases, routines and strategies to visual search performance will help complete our understanding and lead to new insights, not only about how we search, but about how these decision-related processes interact with perception and attention more generally.

1. Perception and Attention in Visual Search: A Brief Summary

How perceptual processes and attention interact to help us define and find visual targets is the subject of a large body of research. Many excellent reviews have recently been published on this topic. For example, Volume 29 of *Current Opinion in Psychology* contains several papers that give up-to-date reviews and opinions on perception and attention in visual search [3–8]. We direct the reader to these papers for a more thorough and critical treatment of the themes covered in this section. In what follows, we briefly summarize three key themes of this literature as the background for the main topic of this review, which is the decision processes involved in search over multiple fixations.

1.1. Visual Salience

The influence of low-level visual information on search is typically taken as synonymous with the influence of visual salience. The first formal computational model for visual salience was developed over 20 years ago [9], and early work used visual search paradigms to explore the model's behaviour [10]. Since then, visual salience has developed into its own sub-field within vision science, focused less on explaining human performance in visual search tasks and more on accounting for, and predicting, the patterns of fixations that human observers make when viewing complex (usually photographic) stimuli. In many respects, this body of work is one of the success stories for psychology in the early 21st Century, with the current state-of-the-art models (based on machine learning and convolutional neural networks) approaching ceiling performance [11,12]. While these models offer impressive performance in predicting which regions of an image are likely to be visually inspected, the move to a computer vision approach focused on performance metrics has led to questions about whether these models are actually modelling visual salience or acting more as a weighted object detector [13,14].

Prior to these models, debate over the relative importance of low-level and mid-to-high-level information in guiding fixations and attention was common (e.g., [15,16]). However, the success of these models suggests that drawing distinctions between these levels of analysis may not be necessary. The processing of low and high level features in the brain is highly interconnected; we should question whether it is appropriate to model them as distinct sets [17].

1.2. Attentional Guidance and Control Settings

The literature on visual search over the past 30 years has been dominated by the guided search model [18], which is now in its fifth iteration [19]; the full version of the fourth iteration can be found here [20]. Many ideas in the guided search model evolved from Treisman's Feature Integration Theory of attention (FIT) [21], in which a small set of basic visual features, such as colour and orientation, are pre-attentive, that is, they are processed in the absence of attention. Conjoining these basic features, FIT proposes, is an operation that depends on attention. Targets that can be defined based on pre-attentive features will "pop out" and be found quickly, while a target defined by conjunctions of features will require directing spatial attention to potential targets until it is found, a much slower and more demanding process. The time to find a conjunction target increases in proportion to the number of items in the set. This per-item increase in reaction time with search set size, known as the search slope, is taken as an index of the attentional demands of finding the target. Guided search builds on FIT by modelling search as a process of elimination, whereby the target will have "guiding features", which can be used to limit attention to the subset of items in the scene that share this feature. For example, if your own car is red and you cannot remember where you parked it, you only need to check the other red cars in the parking lot. In guided search, a pre-attentive visual feature like colour can be identified and used by attention rapidly and in parallel across the visual field. Once you have narrowed down the set, a more labourious process of serial inspection begins. There is broad agreement about the general premise of the guided search model, although some of the details are the subject of lively debate.

The flip side of guidance is distraction. Some types of distractors can impede search even when they do not share obvious features with the target. For example, sudden onsets have consistently been shown to capture attention "reflexively", that is irrespective of how the target of search has been defined [22], and the presence of a sudden onset will slow reaction times to find a target. The existence of attentional capture has led to the suggestion that some biologically-relevant or visually-distinctive information might act as a kind of "circuit-breaker" on attentional guidance, attracting attention even when it is not relevant to the current task [23]. The list of visual features that produce truly reflexive capture is shorter than the list of features capable of guiding attention, and sudden onsets may be the only feature that stands uncontended. In most other circumstances, whether and which distractors capture attention can be argued to depend largely on what features are being used to guide attention [24,25] or how much attention is engaged in the task [26]. The setting or tuning of attention to detect particular features is commonly known as attentional control settings [27].

In both guidance and distraction, debate persists around many categories of stimuli that might be more or less easy to ignore or find, depending on their status in the visual system or presumed biological relevance [7,28,29]. These special features are often referred to as "pre-attentive", meaning they do not require attention to be detected and used to control behaviour [21]. They are also often referred to as "bottom-up", suggesting they control attention in a feed-forward (rather than re-entrant) manner (e.g., [30]). These interpretations suggest the list of pre-attentive or bottom-up features should be invariant, that is, while there may be some disagreement around how exactly to categorize a feature, applying the same criteria should lead to the same list across individuals and over time. A serious issue with this, however, is recent questions around how predictable and invariant the guiding properties of a visual feature really are (e.g., [31]). The ease with which certain features can be used to guide attention is determined in part by how frequently, or recently, we have used those features to find a target (e.g., [6]), implying these could vary over time. Objects that have been previously associated with rewards can also capture attention later, when they are no longer relevant [32]. Thus, the historical distinction between "bottom-up" and "top-down" control of attention has begun falling out of favour, for similar reasons as for the distinction between the relative contribution of "low-level" and "high-level" information in determining fixations. As we have come to understand the details of how attention is guided by information, these categories have become increasingly blurred.

1.3. Eye Movements in Visual Search

While the above body of work has been very useful in helping us uncover how our attentional systems deal with different types of information (features such as colour, orientation and shape), it has either overlooked eye movements in visual search or treated them as synonymous with attention. Hulleman and Olivers [33] argued that progress in the field has been hampered by this approach, which tends to treat individual search items as the unit of analysis and neglects the varying effect of peripheral vision on different types of features. They proposed replacing search items with fixations and a functional viewing field as the base unit of analysis in visual search experiments. One immediate advantage of such a shift in conceptual unit is that it allows theories of visual search to go beyond describing visual search through simple arrays of clearly-defined items to tasks such as searching X-ray scans, textures and photographs of natural scenes. An earlier example of this approach was the target acquisition model [34], a computational model based on a pyramid of filters that are used to generate an activation map that represents the similarity between points in the search image (after transforming to take the retina into account) and the target. The model's search strategy is to simply move the eyes to the centre of mass of the activation map. While this allows it to generate centre-of-gravity fixations that are often seen in human scan-paths (i.e., [35]), there is no notion of varying search strategies, and while the problem of speed/accuracy trade-offs with relation to absent target responses was discussed, this decision process was not implemented. It should also be noted that eye movements do not always help search (e.g., [36–38]). However, given the nature of our foveated visual system, accounting for eye movements is clearly an important part of any theory of visual search.

2. Biases and Strategy

Searchers typically have the freedom to move (or not) their eyes and bodies around during natural search. It is important to understand the contribution of all the factors described in the previous section—low and high-level visual information and how attention is guided and distracted by these properties—to the selection of fixations during search. However, there is another layer of search behaviour that is not captured by these factors. This is the layer that involves biases, decisions, and heuristics that influence how we position our eyes. This is the primary focus of the rest of our review. As we will demonstrate, fixation strategies have a large impact on our search speed and success and vary dramatically from one individual to another.

2.1. Optimality vs. Stochasticity

Given the amount of time we spend looking for things in our day-to-day lives, one might expect us to have evolved and/or developed efficient search strategies. Consistent with this possibility, Najemnik and Geisler developed an influential model of visual search based on an ideal search strategy, and the model matched the number of fixations it takes humans to find a target [39,40]. The model directs each fixation during search to the location that will provide the most information about the target's location, taking into account which locations have already been fixated and the difficulty of spotting the target at different eccentricities. As such, it finds a target in the smallest possible number of fixations. That humans can match this level of efficiency suggests that we may be approaching performance that is as good as it gets, given the limitations of our visual systems. This approach has been expanded upon by Hopper and Rothkopf [41] who argued that humans are capable of constructing multi-step plans for eye movements in visual search tasks that maximize information gain over more than just the next fixation.

There is an important caveat to note. Najemnik and Geisler [40] acknowledged that they did not consider their work a plausible model for human search mechanisms, but instead, the human visual system makes use of heuristics that produce some of the rational behaviour exhibited by their model. The chief reason for this is the extremely high computational load associated with calculating expected information gain for every possible fixation position ahead of each eye movement. For the artificial search scenes of $1/f$ noise used by Najemnik and Geisler [40] (see Figure 1), with uniform and easily-estimated levels of target visibility, the calculation of an optimal fixation strategy is tractable, but resource-intensive. Considering we move our eyes 2–3 times each second during search and estimating target visibility is not often so straightforward as this, it seems likely that we take some short cuts.

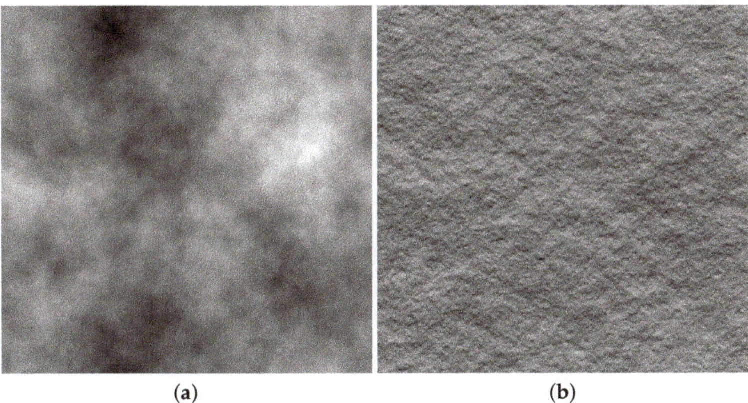

(a)   (b)

Figure 1. Example stimuli from [42]. (**a**) shows a two-dimensional $1/f$-noise stimulus, similar to those used by [40]. (**b**) shows the effect of treating this $1/f$ stimulus as a surface texture and rendering with illumination from above.

The idea of an "ideal" searcher also stands in contrast to our modelling work on visual search behaviour in $1/f$ textures [42] (see Figure 1) in which visual search is thought of as a random walk of saccades (see also [43]). After each saccade is made, there is a probability that the target will be detected, which is modulated by how far the target is from the current fixation location and the contrast of the texture. These detection probabilities are fitted to empirical psychophysical data, and importantly, while saccades are selected randomly, we sampled from the population of saccades participants made from the same region of the search array. Unlike the optimal model, this stochastic selection process has no information about the display or knowledge of what has already been fixated. Nonetheless, the stochastic search model, just like the optimal search model, takes a similar number of fixations to find the target as human searchers do. This suggests that the stochastic model also adequately describes human search, while being computationally far simpler than an ideal searcher.

2.2. Oculomotor Biases

It has been increasingly apparent that oculomotor biases have a large role to play in explaining the patterns of fixations made during scene viewing and visual search. At the very least, these should be considered a significant contribution to variance that needs to be taken into consideration when developing models of eye movement behaviour. However, a stronger statement is that these biases serve a functional role and can be thought of as evolved heuristics that allow observers to search their environment efficiently without needing to carry out the complex calculations necessary to implement the optimal strategy. The utility of these heuristics could explain why an optimal [40] and a stochastic [42] search model can describe human search behaviour equally well.

Perhaps the simplest demonstration of this point is the central bias [44,45]. Observers were tasked with searching photographs of everyday scenes for an embedded Gaussian luminance target. Recordings of eye movements made during search showed that observers preferentially fixate on the central region of the stimuli, even for photographs in which the interesting (in terms of salience and semantic content) information is located away from the centre. Furthermore, it has been demonstrated (e.g., [46]) that adding a central bias to salience models significantly improves their ability to predict which regions of an image will be fixated.

We also see biases in saccadic direction and amplitude. The distribution of directions shows a pronounced preference for saccades to the left and right (i.e., [47]). Human observers also exhibit a strong bias in the magnitude of the saccades they make, with distributions often having a lower mean and larger positive skew than what can be accounted for by a naive salience model [48]. A smaller, but robust, bias is the preference of observers to direct their attention initially to the left half of a scene [49–51]. Another behaviour that can be thought of as a heuristic and strategy is coarse-to-fine search [52]. When faced with a new search stimulus, observers start by making long saccades and short fixations. As the search progresses, saccadic amplitude decrease while the length of fixation duration increases. Interestingly, there was little influence of target salience on this behaviour, leading Over et al. [52] to suggest that an intrinsic coarse-to-fine heuristic for visual search is used, even when such a strategy is not optimal. Inhibition of return [53] and saccadic momentum (e.g., [54]) have been proposed to bias eye movements to new locations and avoid "wasting" eye movements by revisiting locations that have recently been fixated, though with different underlying mechanisms, time courses, and effects [55].

Following this line of thought, we have built on the stochastic search model discussed above [42] and developed a "blind" model of eye movements during scene viewing [56]. This model does not take any inputs and simply models the probability of making a saccade to (x_i, y_i) given a fixation at (x_{i-1}, y_{i-1}). These probabilities are estimated as a truncated multivariate Gaussian distribution, fitted over a range of datasets, and the model is intended to act as a strong baseline to which more sophisticated, image-processing models can be compared. Alternatively, it can be thought of as a computational model of some of the oculomotor biases outlined above.

2.3. Decision Biases

The division between eye movements guided by information gain on the one hand [40] and guided by heuristics and biases on the other [42] is reminiscent of a classic distinction in the animal cognition literature between actions and habits, where actions are responses guided by knowledge of the consequences and habits are repetitions of behaviours that have previously been reinforced [57]. We have been specifically considering visual search in this review and considering the process of selecting fixations as a way of measuring strategies and biases in how information is sampled during speeded search, as well as during visual inspection more generally. However, it is important to consider whether the principles and biases guiding fixation selection are specific to the visual and oculomotor systems, versus a reflection of principles and biases guiding human choices more generally. There are many reasons to expect the decision processes guiding eye movements to be unique to this system: these decisions are made with a very high frequency (2–3 times per second during visual search) and require very little effort to execute, relative to other decisions. Based on these properties, one might expect eye movements to be particularly "thoughtless", perhaps even meeting Dickinson's definition of a habit, and more likely to be guided by heuristics than by careful consideration of various options and their consequences.

A clever demonstration of "thoughtless" eye movement heuristics was devised by Morvan and Maloney [58]. In their task, there were two boxes in which a target (a small dot) could appear, presented at varying distances apart. A central box was always presented at the centre, but never contained the target. The participant moved their eyes to one of the boxes, and as soon the eye tracker had detected that one of the three boxes had been selected, the target was presented in one of the two flanking boxes. The authors were interested in which box people would select. To maximise accuracy, people should choose to fixate on the centre box when the boxes were close enough together that the target could still be discriminated in the flanking boxes. When the distance was too large for the target to be visible from the central box, participants should fixate on one or the other side box instead. Surprisingly, it was found that the choice of which box to fixate on did not vary systematically according to the distance between the boxes. The authors concluded that fixation decisions were guided by imperfect heuristics, even in this simplified context in which the best choice of the three fixation locations seemed intuitively obvious. However, is this decision failure specific to eye movements? To answer this question, we adapted their paradigm to study more deliberative, effortful decisions [59]. In one version of the task, we asked participants to choose a place to stand to throw a beanbag into one of two hoops, without knowing which of the two hoops was the target. The logic here was the same as for the detection task: to maximise accuracy, participants should choose to stand at the midpoint between the two hoops when they are close together and next to one hoop or the other when their expected odds of successfully hitting the hoops from the midpoint falls below 50%. Just as in the eye movement task, we found that participants did not modify their choice of standing position with the distance between the hoops. We also observed the same pattern of sub-optimal decisions in a memorization task, where participants were given two strings of digits and did not know which one they would be asked to report. When the digit strings were short, participants should try and memorize both, but when they are too long to remember both accurately, participants should focus on one. Again, we did not find a systematic variation in strategy with the length of the digit string. These tasks demonstrated that the failure to make fixation decisions that optimize accuracy is not unique to the eye movement task, but a more general problem with how we allocate limited resources when faced with multiple goals. This finding aligns with many other instances in the literature in which eye movements have been argued to be a valuable model system for understanding more complex decisions (e.g., [60–63]).

To explain this failure to optimize task performance in eye movements and other decisions, we suggest that under most natural circumstances, calculating the optimal strategy is not straightforward, as we have already seen from the high computational load associated with executing an optimal series of eye movements to find a Gabor patch in noise [40]. Analogously, in economics, as the number of investment options increases, the advantage of any particular investment strategy shrinks rapidly relative to a simple "$1/N$" strategy of dividing money equally over N options. In terms of risk-adjusted returns, the $1/N$ strategy was shown to be virtually indistinguishable from a set of 14 optimal asset allocation strategies for $N = 25$ [64]. For many decisions, from eye movements to investment strategies, there are more than just two goals to pursue, so a stochastic strategy will be computationally simpler, with outcomes that are indistinguishable from an optimal strategy. We can experimentally create situations where the optimal strategy is simple to calculate and has large benefits, but under most other circumstances, the calculations will be resource-intensive and the benefits far smaller and less predictable. If calculating the expected effects of the full set of potential actions comes at a greater cost under most natural circumstances, perhaps people rely on "habits" as a default to make most of their decisions and do not recognize and adapt their strategy when circumstances arise where better performance is possible.

Given that participants are clearly not recognizing and following optimal strategies in either making eye movements during visual search or in analogous decisions, an interesting question is, what do they do instead? The alternative based on the actions/habits distinction from the animal behaviour literature would be to rely instead on habits. However, "habits" implies rigid repetition of previously-reinforced responses. Defined this way, "habits" do not adequately describe the fixation choices observed in visual search, which are highly variable rather than fixed and rigid. We observed a similar striking range of variability in the choice tasks described above: given exactly the same decision dilemma, most participants exhibited a wide range of different choices from one trial to the next. While participants were consistent in terms of their average behaviour, which was largely insensitive to increasing difficulty, there was surprisingly wide variation around the mean within each individual. To account for this variability, we can again look to historical animal cognition research, where it has been suggested that variability may itself be a reasonable approach to solving problems under conditions of uncertainty [65]. Healthy rats tend to make variable choices when navigating mazes [66] and even seem to prefer mazes that are variable over ones that are fixed [67]. These researchers have suggested that making variable responses allows the animal to explore the "means-end-readiness" of the problem space, to understand the range of possible behaviours and their consequences under uncertain conditions. Bringing this idea back to visual search, within-condition variability is often swept aside as noise in the data, but it may contain important information about how we "solve" search problems, and this could be a useful model for understanding how we solve other problems as well.

2.4. Stopping Rules

One key aspect of effective visual search is deciding when the search should be terminated. For example, when searching for missing car keys, when should you stop searching one room and move on to the next? The decision when to stop, both in real life and in a laboratory search task, involves weighing the advantage of saving time (by not searching all possible locations) against missing the target (by not searching for long enough). The decision about when to stop looking can have important effects even in simple, single-fixation search tasks. For example, Dukewich and Klein [68] found shallower search slopes for making present/absent target decisions versus target localization and identification. Rather than being due to differences in the process of finding the target, as one might initially assume, this was likely due instead to the fact that larger sets of distractors cause participants to give up searching before they find the target, leading to reaction times that do not increase with set size as much as they would if people maintained the same level of accuracy across task difficulty. For identification and localization tasks, the reliable presence of a target led to stable error rates over set size. This demonstrates that even in very simple search tasks, it is important to understand the variance in performance that comes from when people decide to give up. The stopping decision will depend largely on people's knowledge of how likely it is that a target is present, in combination with how much they care about missing it. Circumstances also matter: target prevalence will bias people towards present or absent judgements (rare and extremely rare targets are much more likely to be missed than targets present in half of the trials), but when searching for tumours or weapons, which are rare, but life-or-death, this bias can be over-ridden [20,69]. For search over multiple fixations, there are two levels of stopping decisions. The first is the decision to leave the currently-fixated location and select a new one, which in many respects is an instance of single-fixation search. This level of decision was discussed in detail in a recent paper by Tatler et al. [70], who modelled the process as a threshold based on the relative benefit of the currently fixated target relative to a new fixation. The second level is the decision made after several fixations that it is time to stop looking for a given target altogether. In this discussion, we focus on the second of these.

Early attempts at formalizing stopping rules came from animal foraging models. Charnov [71], in his "marginal value theorem", postulated that when an animal searches for multiple targets, a single area is searched until the target acquisition rates (so-called "marginal rates") for this single area fall below the mean number of targets expected to be found in the entire search set. This main premise of the model constitutes its main failure because the forager is expected to know the distribution of the targets in the search array even before he/she searched through the array and because the observer does not differentiate between areas rich in targets and areas containing only a small number of targets. A later model [72] dropped the notions of average acquisition rates per area. Instead, the observer stopped searching an area if the time since the last target was acquired exceeded a certain point. Thus, observers were not expected to know the distribution of the targets across the search arrays prior to search: foragers spent more time searching areas rich in targets and dropped areas with few targets quickly.

Visual search and foraging differ in some fundamental ways: visual search laboratory tasks are normally presented on a computer screen, whereas foraging often involves immersing the participant within the search space. Foraging usually requires more energy, and often visual cues are not available for locating the target [73] (see also [74]). Lastly, foraging for food is more complex than visual search for a target singleton, as it often requires distinction between high- and low-quality patches of food and revisiting the areas that have previously been exhausted [75]. The guided search model addresses the question of when to decide when a single target is absent in a more traditional visual search paradigm, where a single target tends to be present in half the trials [76]. As with the general guided search model described above, the search array is evaluated pre-attentively, and items are selected according to their probability of being a target. Search is terminated when the remaining search items do not reach this probability threshold. This termination threshold is set by the observer to meet a specified speed-accuracy performance trade-off. The main advantage of this model over the earlier models is that the threshold is set flexibly: when error rates are too high, the threshold is lowered to allow longer search times, and when error rates are too low, the threshold is increased, thus increasing RT. Over the course of the experiment, observers acquire implicit knowledge about the duration of successful trials and are more likely to guess as a typical trial duration elapses and the target has not been found.

3. Search Strategy and Individual Differences

The majority of the visual search literature has concerned itself with understanding average performance, often expressed in terms of a set-size $\times$ reaction time slope. For many aspects of search, this is an entirely appropriate approach to hypothesis testing, but for other aspects, it is less appropriate. A recent paper by Hedge and colleagues [77] made the important point that "reliability" has two very distinct meanings in experimental psychology: on the one hand, it can mean how consistent a given effect size is across different samples, and on the other, it can mean how reliably the same sample produces the same results. The first definition of reliability is used in the context of interpreting averaged data and is important for ensuring a particular experimental manipulation is reproducible. The second definition of reliability depends on individual performance being stable over repeated measurements, which is important for correlational studies that attempt to understand variation in a particular effect size. Hedge and colleagues noted that not only are these two types of reliability very distinct conceptually, they are also to some extent mutually exclusive: low variation between individuals makes averages more reliable, but restricts the range for detecting correlations with other variables. Conversely, the large and stable differences between individuals that make correlations reliable only contribute noise to analyses of averaged data. In the current context, this is a particularly important distinction, because visual search involves a complicated set of sub-processes, some of which may be best studied using an averaging approach, and some of which need to be understood by examining individuals as the basic unit of analysis.

Data averaged across individuals not only hides important information, it can also sometimes lead to spurious conclusions. An example of this can be seen in our recent work on visual search strategies [78]. Participants searched an array of randomly-orientated line segments, as shown in Figure 2. The target was visible using peripheral vision on the homogeneous side of the array; eye movements to this side of the array provided no new information. We created these stimuli to distinguish between optimal (in which the searcher should fixate only on the heterogeneous half of the array) and stochastic strategies (in which both halves should be fixated equally often). If we use the average performance over our participants and compare to our hypotheses (Figure 3a), then it looks like the data support the stochastic search model. However, such a conclusion would be wrong, as when we examine how each individual approached this visual search task, we found large, and stable, differences, as shown in Figure 3b: about a third of our participants approached ideal performance, while another third did the opposite, leading to performance that was worse than if a purely random strategy had been followed (the remaining third was somewhere in-between).

Figure 2. Example stimulus from [78]. The target is the line segment that is perpendicular to the mean orientation of the distractors. In this example, it is nine items from the right, six down.

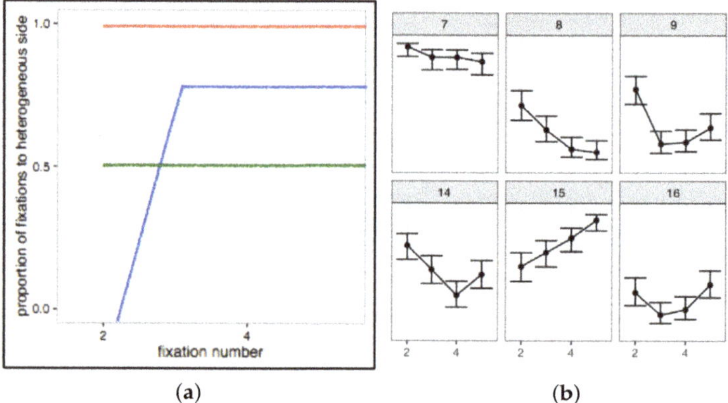

(a)   (b)

Figure 3. (**a**) The red and blue lines show how an optimal and close-to-optimal observer should search the split-half stimuli. The green line shows what we would expect from a stochastic searcher. (**b**) Data from six participants [78]. We can see that while Participant 7 approached the optimal strategy, and 15 could be considered close-to-optimal, other participants (14, 16) behaved in line with the predictions from the stochastic search strategy. Furthermore, Participants 8 and 9 appeared to be implementing a "counter-optimal" strategy.

A similarly striking range of individual differences in the search task was reported by Irons and Leber ([79], see also [80]) with their adaptive choice visual search task (ACVS) [79]. In their paradigm, participants had to search for a numeral 1–5 in a small red or blue square. On each trial, there was both a red and blue target, embedded within a collection of red, blue, green, and variable distractors. The variable coloured distractors started out red (in which case the optimal strategy was to search for a blue target) and, upon each successive trial, slowly turned from red to purple to blue. For trials in which there were more blue distractors than red, participants should search for the red target, and vice versa. Whereas group performance was far from optimal, some observers switched to the easier target as the distractors changed colour (although they still failed to switch at an optimal point); some always searched for the same colour and avoided switching; and some frequently switched between targets, but not in a way that related to the distractor colours. The flexible adjustment of attention to colour and the differences between individuals in this ability highlight the limitations of salience and guidance alone in accounting for visual search performance.

In another study, observers had to find multiple targets while ignoring distractors in a search array. The targets were defined by either one (e.g., colour) or two features (e.g., colour and shape) [81,82]. When the target was defined by a single feature, observers frequently switched between target types (e.g., picking a yellow target, then red, then yellow again). When observers searched for a target defined by two features (shape and colour), most of them tended to pick the same target type in long runs (e.g., all yellow squares) until they they exhausted search and then moved on to search for the other type of the target (e.g., red triangles). Notably, four out of 16 participants, termed "super-foragers", continued to use the same strategy in the two conditions, but with no apparent cost to performance. In other words, a minority of people seem able to hold two target templates in mind simultaneously. In an earlier demonstration of individual differences in search strategy, Araujo and colleagues found that the majority of participants chose inefficient search strategies when planning a saccade to one of the two possible target locations [83]. In this experiment, the target was a randomly-oriented letter T located in one of two clusters of letter Ls. The probability of the target occurring in one of the clusters was 80% and was signalled to the observers by the different luminance levels of the two clusters. The distance between each cluster of letters and central fixation varied between the trials. Because the display was presented for only 500 msc, the best performance would be achieved by inspecting the high probability location first. Yet, the cue signalling the higher probability was only used by one out of the six participants. The other five observers preferred to inspect the closer cluster first (to varying degrees) and then the remaining cluster, even though the display had been removed before the second saccade arrived.

One might be tempted to conclude that some people are simply better at search than others, but our recent work suggested that something more complex underlies these differences [84]. We tested the same set of 64 participants with the split-half [78], adaptive choice [79] and foraging [82] paradigms. Although each of these tasks has been shown to have a good test-retest reliability ($r \approx 0.7$–0.9), we found a surprising lack of correlations across tasks. Not only was there a wide range of differences in how people approached a visual search task, but these individual differences depended critically on task structure. Given the range of performance we observed in these three tasks, it is possible that the interaction between these two factors (individual differences and task structure) was responsible for far more of the variation we see in human search behaviour than has been explained by visual salience and guided search. The important question that remains to be understood is not whether people are optimal searchers or not, but rather why some people are optimal and under what circumstances. Given that this person–circumstance interaction accounted for the majority of variance in these three different search tasks, understanding it will not only help us build more powerful models of visual search itself, but it also has the potential to facilitate efficient search in industrial and social circumstances (two particularly high-stakes examples are security and health care).

Individual differences in search can also have importance in patient populations. We observed a wide range of individual differences in a series of experiments looking at how eye movement strategies of healthy people are affected by simulated visual deficits [85]. Visual information in one hemifield was removed on-line while participants searched a display of lines for the target (line tilted 45° to the right) embedded in an array of homogeneously- or heterogeneously-oriented lines, similar to those shown in Figure 2, except the arrays were not split into two halves; instead, distractor orientations were sampled from the same range across the array (either homogeneous or heterogeneous). Making eye movements to the initially-sighted side did not harm performance for a heterogeneous background, but for the homogeneous background, the logic was similar as for the split/half arrays described above: a target on the visible half of the display could already be detected, without the need for any eye movements. The effective strategy was therefore to make large eye movements to the blind side, to reveal the part of the display that was currently hidden. We found that participants continued to use the inefficient strategy of making eye movement to the sighted side even when the search was very easy and the target could be easily ascertained to not be present in the periphery. When we exposed participants to the simulated visual deficits over five days, with financial incentives for performance improvements, the majority of participants became gradually more efficient [86], but those who were inefficient at the beginning of the experiment were still far from optimal, while individual differences in strategy were very stable over a week of repeated exposure to the task. Variability in performance between patients with visual field deficits has often been attributed to factors such as the site and extent of the lesion and the age at the onset (see for example [87,88]). One gap in patient studies that our research on individual differences highlighted is the knowledge of premorbid strategies. We cannot reasonably assume that the patients will show optimal performance post-lesion if their premorbid strategy was poor, and we have shown that these differences are large and persistent in healthy populations.

Variation in abilities has been extensively explored elsewhere, for instance in the process of comparing shapes [89] and in the face recognition literature, where it has been shown that observers vary on a spectrum from developmental prosopagnosia (profound inability to recognise familiar faces) to super-recognisers (observers that are significantly better than average at recognising faces [90]). Studies of twins suggest that a genetic component might drive the differences, and the ability to recognise faces is independent of more general aspects of cognition such as attention and intelligence [91]. Much can be learned about a particular skill by understanding the full range of individual differences, and visual search research, like many other areas of psychology, has not yet realized the full potential of this approach.

4. Conclusions

Visual search continues to be a useful model task and a rich source of data for understanding a wide range of perceptual, cognitive, and motor skills. Much has already been learned about perception and attention from studying eye movements and attention during visual search tasks. Much remains to be understood about strategy, bias, and decision, and visual search is poised to provide insights into these domains, as well. In our conclusions, therefore, we offer four inter-related recommendations for future research into visual search to tackle, to move the field toward a more complete understanding of this complex, but important task.

1. The focus of our experiments and analyses should not only be to explain average patterns, but also to account for variance. The large sources of variance, relative to smaller ones, will be the more powerful predictors of search performance. If we can understand and control these, the smaller sources of variance will be easier to tackle. Individual differences and variability within conditions should not be hidden away in averaged data, but made a central part of our models and theories.

2. A related suggestion is to be cautious in interpreting measures of central tendency, such as means and medians. Given the large range of individual differences we have observed in most of our own search data, and the spurious conclusions we could have reached if we relied on average patterns alone, we think it is important to consider carefully whether a measure of central tendency is, in fact, a good representation of a particular set of data. That is, is the mean (or median) pattern similar to most of the trial and individual level results? If a particular summary statistic is not an accurate or adequate representation of most of your data, do not report it. Instead, show the full range of results so other researchers can understand how variable a given behaviour is within and between individuals. This variance does not indicate a failed manipulation or "noisy data"; instead, consider that it contains essential information, without which we cannot fully understand visual search performance.

3. Based on observations of independent sources of variance across different tasks [84,89], it is clearly important to address directly the question of how confidently we can apply conclusions from one search task to related and unrelated tasks and contexts. For example, we often assume that visual primitives like line segments and Gabor patches will scale up to more complex scenes and objects or that basic phenomena like attentional capture or inhibition of return will be easy to observe in real-world situations. In fact, it is difficult to find straightforward instances in the literature where these basic effects have clearly generalized from the laboratory to more

complex real-world situations. It is important to note that the context-specificity of a given effect is not an indication that it is trivial or unimportant. Instead, it is an important source of data for understanding the constraints and boundary conditions for patterns of results that can be reliably produced in the laboratory. Directly measuring how particular manipulations and interventions affect search in a variety of situations can be a fruitful source of insight into these effects.

4. We all have a tendency to stay within the bounds of familiar theories and models. Looking outside the vision and attention literature can lead to many new useful ideas and explanations, especially in visual search, which is a rich and complex task. Our understanding can be enriched from insights and models from other fields such as decision-making, learning, human factors and individual differences.

Author Contributions: Writing, original draft preparation, A.D.F.C., A.N. and A.R.H.; writing, review and editing, A.D.F.C., A.N. and A.R.H.

Funding: This research was funded by the James S. McDonnell Foundation (scholar award to A.R.H.) and the Education and Social Research Council of the U.K. (PhD studentship to A.N.)

Conflicts of Interest: The authors declare no conflict of interest. The funders had no role in the design of the study; in the collection, analyses, or interpretation of data; in the writing of the manuscript; nor in the decision to publish the results.

References

1. Palmer, J.; Verghese, P.; Pavel, M. The psychophysics of visual search. *Vis. Res.* **2000**, *40*, 1227–1268. [CrossRef]
2. Eckstein, M.P.; Thomas, J.P.; Palmer, J.; Shimozaki, S.S. A signal detection model predicts the effects of set size on visual search accuracy for feature, conjunction, triple conjunction, and disjunction displays. *Percept. Psychophys.* **2000**, *62*, 425–451. [CrossRef] [PubMed]
3. Parr, T.; Friston, K.J. Attention or salience? *Curr. Opin. Psychol.* **2019**, *29*, 1–5. [CrossRef] [PubMed]
4. Anderson, B.A. Neurobiology of value-driven attention. *Curr. Opin. Psychol.* **2018**. [CrossRef] [PubMed]
5. Geng, J.J.; Witkowski, P. Template-to-distractor distinctiveness regulates visual search efficiency. *Curr. Opin. Psychol.* **2019**, *29*, 119–125. [CrossRef] [PubMed]
6. Theeuwes, J. Goal-Driven, Stimulus-Driven and History-Driven selection. *Curr. Opin. Psychol.* **2019**, *29*, 97–101. [CrossRef] [PubMed]
7. Wolfe, J.M.; Utochkin, I.S. What is a preattentive feature? *Curr. Opin. Psychol.* **2019**, *29*, 19–26. [CrossRef] [PubMed]
8. Gaspelin, N.; Luck, S.J. Inhibition as a potential resolution to the attentional capture debate. *Curr. Opin. Psychol.* **2019**. [CrossRef]
9. Itti, L.; Koch, C.; Niebur, E. A model of saliency-based visual attention for rapid scene analysis. *IEEE Trans. Pattern Anal. Mach. Intell.* **1998**, *20*, 1254–1259. [CrossRef]
10. Itti, L.; Koch, C. A saliency-based search mechanism for overt and covert shifts of visual attention. *Vis. Res.* **2000**, *40*, 1489–1506. [CrossRef]
11. Bylinskii, Z.; Judd, T.; Oliva, A.; Torralba, A.; Durand, F. What do different evaluation metrics tell us about saliency models? *arXiv* **2016**, arXiv:1604.03605.
12. Bylinskii, Z.; Judd, T.; Borji, A.; Itti, L.; Durand, F.; Oliva, A.; Torralba, A. MIT Saliency Benchmark. Available online: http://saliency.mit.edu/ (accessed on 10 September 2019).
13. Kong, P.; Mancas, M.; Thuon, N.; Kheang, S.; Gosselin, B. Do Deep-Learning Saliency Models Really Model Saliency? In Proceedings of the 2018 25th IEEE International Conference on Image Processing (ICIP), Athens, Greece, 7–10 October 2018; IEEE: Piscataway, NJ, USA, 2018; pp. 2331–2335.
14. Kümmerer, M.; Wallis, T.S.; Gatys, L.A.; Bethge, M. Understanding low-and high-level contributions to fixation prediction. In Proceedings of the 2017 IEEE International Conference on Computer Vision, Venice, Italy, 22–29 October 2017; pp. 4799–4808.
15. Walther, D.; Koch, C. Modeling attention to salient proto-objects. *Neural Netw.* **2006**, *19*, 1395–1407. [CrossRef] [PubMed]

16. Nuthmann, A.; Henderson, J.M. Object-based attentional selection in scene viewing. *J. Vis.* **2010**, *10*, 20. [CrossRef] [PubMed]

17. Cadieu, C.F.; Hong, H.; Yamins, D.L.; Pinto, N.; Ardila, D.; Solomon, E.A.; Majaj, N.J.; DiCarlo, J.J. Deep neural networks rival the representation of primate IT cortex for core visual object recognition. *PLoS Comput. Biol.* **2014**, *10*, e1003963. [CrossRef] [PubMed]

18. Wolfe, J.M.; Cave, K.R.; Franzel, S.L. Guided search: An alternative to the feature integration model for visual search. *J. Exp. Psychol. Hum. Percept. Perform.* **1989**, *15*, 419. [CrossRef] [PubMed]

19. Wolfe, J.; Cain, M.; Ehinger, K.; Drew, T. Guided Search 5.0: Meeting the challenge of hybrid search and multiple-target foraging. *J. Vis.* **2015**, *15*, 1106. [CrossRef]

20. Wolfe, J.M.; Gray, W. Guided search 4.0. In *Integrated Models of Cognitive Systems*; Oxford University Press: New York, NY, USA, 2007; pp. 99–119.

21. Treisman, A.M.; Gelade, G. A feature-integration theory of attention. *Cogn. Psychol.* **1980**, *12*, 97–136. [CrossRef]

22. Remington, R.W.; Johnston, J.C.; Yantis, S. Involuntary attentional capture by abrupt onsets. *Percept. Psychophys.* **1992**, *51*, 279–290. [CrossRef] [PubMed]

23. Corbetta, M.; Shulman, G.L. Control of goal-directed and stimulus-driven attention in the brain. *Nat. Rev. Neurosci.* **2002**, *3*, 201. [CrossRef]

24. Bacon, W.F.; Egeth, H.E. Overriding stimulus-driven attentional capture. *Percept. Psychophys.* **1994**, *55*, 485–496. [CrossRef]

25. Duncan, J.; Humphreys, G.W. Visual search and stimulus similarity. *Psychol. Rev.* **1989**, *96*, 433. [CrossRef] [PubMed]

26. Lavie, N. Distracted and confused?: Selective attention under load. *Trends Cogn. Sci.* **2005**, *9*, 75–82. [CrossRef] [PubMed]

27. Folk, C.L.; Remington, R.W.; Johnston, J.C. Involuntary covert orienting is contingent on attentional control settings. *J. Exp. Psychol. Hum. Percept. Perform.* **1992**, *18*, 1030. [CrossRef] [PubMed]

28. von Mühlenen, A.; Lleras, A. No-onset looming motion guides spatial attention. *J. Exp. Psychol. Hum. Percept. Perform.* **2007**, *33*, 1297. [CrossRef] [PubMed]

29. Cosman, J.D.; Vecera, S.P. Attentional capture under high perceptual load. *Psychon. Bull. Rev.* **2010**, *17*, 815–820. [CrossRef] [PubMed]

30. Di Lollo, V.; Enns, J.T.; Rensink, R.A. Competition for consciousness among visual events: The psychophysics of reentrant visual processes. *J. Exp. Psychol. Gen.* **2000**, *129*, 481. [CrossRef] [PubMed]

31. Awh, E.; Belopolsky, A.V.; Theeuwes, J. Top-down versus bottom-up attentional control: A failed theoretical dichotomy. *Trends Cogn. Sci.* **2012**, *16*, 437–443. [CrossRef] [PubMed]

32. Anderson, B.A.; Laurent, P.A.; Yantis, S. Value-driven attentional capture. *Proc. Natl. Acad. Sci. USA* **2011**, *108*, 10367–10371. [CrossRef]

33. Hulleman, J.; Olivers, C.N. On the brink: The demise of the item in visual search moves closer. *Behav. Brain Sci.* **2017**, *40*. [CrossRef]

34. Zelinsky, G.J. A theory of eye movements during target acquisition. *Psychol. Rev.* **2008**, *115*, 787. [CrossRef]

35. Findlay, J.M. Global visual processing for saccadic eye movements. *Vis. Res.* **1982**, *22*, 1033–1045. [CrossRef]

36. Zelinsky, G.J.; Sheinberg, D.L. Eye movements during parallel–serial visual search. *J. Exp. Psychol. Hum. Percept. Perform.* **1997**, *23*, 244. [CrossRef] [PubMed]

37. Klein, R.; Farrell, M. Search performance without eye movements. *Percept. Psychophys.* **1989**, *46*, 476–482. [CrossRef]

38. Boot, W.R.; Kramer, A.F.; Becic, E.; Wiegmann, D.A.; Kubose, T. Detecting transient changes in dynamic displays: The more you look, the less you see. *Hum. Factors* **2006**, *48*, 759–773. [CrossRef] [PubMed]

39. Najemnik, J.; Geisler, W.S. Optimal eye movement strategies in visual search. *Nature* **2005**, *434*, 387. [CrossRef] [PubMed]

40. Najemnik, J.; Geisler, W.S. Eye movement statistics in humans are consistent with an optimal search strategy. *J. Vis.* **2008**, *8*, 4. [CrossRef]

41. Hoppe, D.; Rothkopf, C.A. Multi-step planning of eye movements in visual search. *Sci. Rep.* **2019**, *9*, 144. [CrossRef]

42. Clarke, A.D.; Green, P.; Chantler, M.J.; Hunt, A.R. Human search for a target on a textured background is consistent with a stochastic model. *J. Vis.* **2016**, *16*, 4. [CrossRef]

43. Boccignone, G.; Ferraro, M. Modelling gaze shift as a constrained random walk. *Phys. A Stat. Mech. Its Appl.* **2004**, *331*, 207–218. [CrossRef]
44. Tatler, B.W. The central fixation bias in scene viewing: Selecting an optimal viewing position independently of motor biases and image feature distributions. *J. Vis.* **2007**, *7*, 1–17. [CrossRef]
45. Clarke, A.D.; Tatler, B.W. Deriving an appropriate baseline for describing fixation behaviour. *Vis. Res.* **2014**, *102*, 41–51. [CrossRef] [PubMed]
46. Tatler, B.W.; Vincent, B.T. The prominence of behavioural biases in eye guidance. *Vis. Cognit.* **2009**, *17*, 1029–1054. [CrossRef]
47. Gilchrist, I.D.; Harvey, M. Evidence for a systematic component within scan paths in visual search. *Vis. Cognit.* **2006**, *14*, 704–715. [CrossRef]
48. Parkhurst, D.; Law, K.; Niebur, E. Modeling the role of salience in the allocation of overt visual attention. *Vis. Res.* **2002**, *42*, 107–123. [CrossRef]
49. Zelinsky, G.J. Using eye saccades to assess the selectivity of search movements. *Vis. Res.* **1996**, *36*, 2177–2187. [CrossRef]
50. Nuthmann, A.; Matthias, E. Time course of pseudoneglect in scene viewing. *Cortex* **2014**, *52*, 113–119. [CrossRef] [PubMed]
51. Foulsham, T.; Gray, A.; Nasiopoulos, E.; Kingstone, A. Leftward biases in picture scanning and line bisection: A gaze-contingent window study. *Vis. Res.* **2013**, *78*, 14–25. [CrossRef] [PubMed]
52. Over, E.; Hooge, I.; Vlaskamp, B.; Erkelens, C. Coarse-to-fine eye movement strategy in visual search. *Vis. Res.* **2007**, *47*, 2272–2280. [CrossRef]
53. Klein, R.M. Inhibition of return. *Trends Cogn. Sci.* **2000**, *4*, 138–147. [CrossRef]
54. Sumner, P. Inhibition versus attentional momentum in cortical and collicular mechanisms of IOR. *Cognit. Neuropsychol.* **2006**, *23*, 1035–1048. [CrossRef]
55. MacInnes, W.J.; Hunt, A.R.; Hilchey, M.D.; Klein, R.M. Driving forces in free visual search: An ethology. *Atten. Percept. Psychophys.* **2014**, *76*, 280–295. [CrossRef] [PubMed]
56. Clarke, A.D.; Stainer, M.J.; Tatler, B.W.; Hunt, A.R. The saccadic flow baseline: Accounting for image-independent biases in fixation behaviour. *J. Vis.* **2017**, *17*, 12. [CrossRef] [PubMed]
57. Dickinson, A. Actions and habits: The development of behavioural autonomy. *Philos. Trans. R. Soc. Lond. Ser. B Biol. Sci.* **1985**, *308*, 67–78. [CrossRef]
58. Morvan, C.; Maloney, L.T. Human visual search does not maximize the post-saccadic probability of identifying targets. *PLoS Comput. Biol.* **2012**, *8*, e1002342. [CrossRef] [PubMed]
59. Clarke, A.D.; Hunt, A.R. Failure of intuition when choosing whether to invest in a single goal or split resources between two goals. *Psychol. Sci.* **2016**, *27*, 64–74. [CrossRef] [PubMed]
60. Hunt, A.R.; von Mühlenen, A.; Kingstone, A. The time course of attentional and oculomotor capture reveals a common cause. *J. Exp. Psychol. Hum. Percept. Perform.* **2007**, *33*, 271. [CrossRef]
61. Kowler, E. *Eye Movements and Their Role in Visual and Cognitive Processes*; Elsevier: Amsterdam, The Netherlands, 1990.
62. Platt, M.L.; Glimcher, P.W. Neural correlates of decision variables in parietal cortex. *Nature* **1999**, *400*, 233. [CrossRef] [PubMed]
63. Schall, J.D. Neural basis of deciding, choosing and acting. *Nat. Rev. Neurosci.* **2001**, *2*, 33. [CrossRef] [PubMed]
64. DeMiguel, V.; Garlappi, L.; Uppal, R. How inefficient are simple asset allocation strategies. *Rev. Financ. Stud.* **2009**, *22*, 1915–1953. [CrossRef]
65. Maier, N.R. The behaviour mechanisms concerned with problem solving. *Psychol. Rev.* **1940**, *47*, 43. [CrossRef]
66. Krechevsky, I. Brain mechanisms and "hypotheses". *J. Comp. Psychol.* **1935**, *19*, 425. [CrossRef]
67. Krechevsky, I. Brain mechanisms and variability: II. Variability where no learning is involved. *J. Comp. Psychol.* **1937**, *23*, 139. [CrossRef]
68. Dukewich, K.R.; Klein, R.M. Finding the target in search tasks using detection, localization, and identification responses. *Can. J. Exp. Psychol. Can. Psychol. Exp.* **2009**, *63*, 1. [CrossRef] [PubMed]
69. Wolfe, J.M.; Palmer, E.M.; Horowitz, T.S. Reaction time distributions constrain models of visual search. *Vis. Res.* **2010**, *50*, 1304–1311. [CrossRef] [PubMed]

70. Tatler, B.W.; Brockmole, J.R.; Carpenter, R.H. LATEST: A model of saccadic decisions in space and time. *Psychol. Rev.* **2017**, *124*, 267. [CrossRef] [PubMed]
71. Charnov, E.L. Optimal foraging, the marginal value theorem. *Theor. Popul. Biol.* **1976**, *9*, 129–136. [CrossRef]
72. McNair, J.N. Optimal giving-up times and the marginal value theorem. *Am. Nat.* **1982**, *119*, 511–529. [CrossRef]
73. Smith, A.D.; Hood, B.M.; Gilchrist, I.D. Visual search and foraging compared in a large-scale search task. *Cogn. Process.* **2008**, *9*, 121–126. [CrossRef]
74. Gilchrist, I.D.; North, A.; Hood, B. Is visual search really like foraging? *Perception* **2001**, *30*, 1459–1464. [CrossRef]
75. Wolfe, J.M. When is it time to move to the next raspberry bush? Foraging rules in human visual search. *J. Vis.* **2013**, *13*, 10. [CrossRef]
76. Chun, M.M.; Wolfe, J.M. Just say no: How are visual searches terminated when there is no target present? *Cogn. Psychol.* **1996**, *30*, 39–78. [CrossRef] [PubMed]
77. Hedge, C.; Powell, G.; Sumner, P. The reliability paradox: Why robust cognitive tasks do not produce reliable individual differences. *Behav. Res. Methods* **2018**, *50*, 1166–1186. [CrossRef] [PubMed]
78. Nowakowska, A.; Clarke, A.D.; Hunt, A.R. Human visual search behaviour is far from ideal. *Proc. R. Soc. B* **2017**, *284*, 20162767. [CrossRef] [PubMed]
79. Irons, J.L.; Leber, A.B. Choosing attentional control settings in a dynamically changing environment. *Atten. Percept. Psychophys.* **2016**, *78*, 2031–2048. [CrossRef] [PubMed]
80. Irons, J.L.; Leber, A.B. Characterizing individual variation in the strategic use of attentional control. *J. Exp. Psychol. Hum. Percept. Perform.* **2018**, *44*, 1637. [CrossRef] [PubMed]
81. Jóhannesson, Ó.I.; Thornton, I.M.; Smith, I.J.; Chetverikov, A.; Kristjánsson, A. Visual foraging with fingers and eye gaze. *i-Perception* **2016**, *7*, 2041669516637279. [CrossRef] [PubMed]
82. Kristjánsson, Á.; Jóhannesson, Ó.I.; Thornton, I.M. Common attentional constraints in visual foraging. *PLoS ONE* **2014**, *9*, e100752. [CrossRef]
83. Araujo, C.; Kowler, E.; Pavel, M. Eye movements during visual search: The costs of choosing the optimal path. *Vis. Res.* **2001**, *41*, 3613–3625. [CrossRef]
84. Clarke, A.; Irons, J.; James, W.; Leber, A.B.; Hunt, A.R. Stable individual differences in strategies within, but not between, visual search tasks. *PsyArXiv* **2019**. [CrossRef]
85. Nowakowska, A.; Clarke, A.D.; Sahraie, A.; Hunt, A.R. Inefficient search strategies in simulated hemianopia. *J. Exp. Psychol. Hum. Percept. Perform.* **2016**, *42*, 1858. [CrossRef]
86. Nowakowska, A.; Clarke, A.D.; Sahraie, A.; Hunt, A.R. Practice-related changes in eye movement strategy in healthy adults with simulated hemianopia. *Neuropsychologia* **2019**, *128*, 232–240. [CrossRef] [PubMed]
87. Zihl, J. Oculomotor scanning performance in subjects with homonymous visual field disorders. *Vis. Impair. Res.* **1999**, *1*, 23–31. [CrossRef]
88. Tant, M.; Cornelissen, F.W.; Kooijman, A.C.; Brouwer, W.H. Hemianopic visual field defects elicit hemianopic scanning. *Vis. Res.* **2002**, *42*, 1339–1348. [CrossRef]
89. Cooper, L.A. Individual differences in visual comparison processes. *Percept. Psychophys.* **1976**, *19*, 433–444. [CrossRef]
90. Russell, R.; Duchaine, B.; Nakayama, K. Super-recognizers: People with extraordinary face recognition ability. *Psychon. Bull. Rev.* **2009**, *16*, 252–257. [CrossRef] [PubMed]
91. Zhu, Q.; Song, Y.; Hu, S.; Li, X.; Tian, M.; Zhen, Z.; Dong, Q.; Kanwisher, N.; Liu, J. Heritability of the specific cognitive ability of face perception. *Curr. Biol.* **2010**, *20*, 137–142. [CrossRef]

Review

The Changing Landscape: High-Level Influences on Eye Movement Guidance in Scenes

Carrick C. Williams [1],* and Monica S. Castelhano [2]

[1]   Department of Psychology, California State University San Marcos, San Marcos, CA 92069, USA
[2]   Department of Psychology, Queen's University, Kingston, ON K7L 3N6, Canada
*   Correspondence: cawilliams@csusm.edu

Received: 31 March 2019; Accepted: 24 June 2019; Published: 28 June 2019

Abstract: The use of eye movements to explore scene processing has exploded over the last decade. Eye movements provide distinct advantages when examining scene processing because they are both fast and spatially measurable. By using eye movements, researchers have investigated many questions about scene processing. Our review will focus on research performed in the last decade examining: (1) attention and eye movements; (2) where you look; (3) influence of task; (4) memory and scene representations; and (5) dynamic scenes and eye movements. Although typically addressed as separate issues, we argue that these distinctions are now holding back research progress. Instead, it is time to examine the intersections of these seemingly separate influences and examine the intersectionality of how these influences interact to more completely understand what eye movements can tell us about scene processing.

Keywords: eye movements; scenes; attention

---

1. Introduction

Visual scenes are ubiquitous but complex concepts. Scenes are typically defined as any view of the natural world [1–6]. However, we will adopt a narrower definition often used by researchers (e.g., [3]), where a scene is defined as a human-scaled view of an environment that is made up of space-defining surfaces and larger elements [7], containing a number of smaller objects arranged in specific spatial locations [8–10], and forming a coherent semantic concept [1–3]. Each component of this definition can contribute individually to the understanding of the visual world [1].

Breaking down the definition, the surfaces and larger elements that define the shape of the space of an environment have been shown to be crucial for the initial perceptual processing of scenes (i.e., spatial envelope [11–13]). This space, as defined by these larger elements, is the basic building block of the scene (e.g., floors, walls, and ceiling for indoor scenes). The shape and position of these elements have been linked to processes involved in navigation and visual search, and we will explore this in more detail below.

The arrangement of the smaller elements of a scene are relative to the space and position of larger space-defining elements. In addition, the arrangement follows the rules of physical objects (e.g., gravity), and this arrangement can affect both search and identification processing of individual elements [14,15]. Further, the position of these elements relative to the observer may have differing effects on processing [16,17].

The third component of the definition, semantic coherence, introduces the importance of prior knowledge. For example, knowing that the scene is a kitchen provides information that affects how the scene is represented based on the likelihood of objects present, functions performed in that space, and the likely spatial arrangement of certain elements [14,18,19].

The final aspect of the definition is that scenes are human-scaled environments. Considering that this is the way our real-world environment is normally viewed, it is reasonable to think that these views would be most familiar, easiest to process, and most relevant for considering different factors having an effect on processing. Although researchers can examine other perspectives, such as those that are either too small (e.g., the world from the view of a cat) or too large (e.g., a satellite image of a city), there are visual and structural properties that are available at the human scale that would not be available with other views.

The above definition points to the complexity of trying to encompass all aspects of real-world scenes in a single definition. However, we do acknowledge that this definition is strongly linked with the stimuli used to depict scenes in labs (i.e., photographs, line drawings, computer generated images) and does not reflect the changing scales and embedded nature of scene perception that occurs when we are perceiving, processing, and navigating a real-world environment. We will come back to this limitation later and discuss potential ways that future research can address it. Although the issues are complex, how we view and interpret scenes is important in our understanding of human visual behavior. The question then is how to assess the visual evaluation of scenes. One can approach the problem multiple ways; we will focus on the use of eye movements to examine how scenes are processed and understood. Rayner and Pollatsek [5] reviewed the extant literature on eye movements and scene perception in the early 1990s. Critically, although most of their research focuses on reading, Rayner and Pollatsek laid out some basic questions that eye movements could answer regarding scene processing. They point to the tension (that still exists) between the processing of varied and complex visual information from scenes and the effect of task on what type of information is acquired and processed. Their review covered what was known at that time about scenes and eye movements, but in the last 30 years, the study of eye movements and scene perception has exploded. Researchers from around the globe have since explored the effects of attention on understanding and remembering scenes as well as the effects of scene knowledge and memory on the deployment of attention.

However, to gain traction on the complex processes involved in scene processing, researchers have made simplifying assumptions about scenes. For instance, scene context is known to have a strong influence on behavior, but the exact nature of that influence is rarely specified, even though it could arise from the various sources as described in the definition above. The field as a whole has begun to recognize that these simplifying assumptions may limit the explanatory power of theories and as such, limit our knowledge of scene processing to a relatively narrow case of visual processing.

The present review will examine the current state of research on scene processing and its development over the last 10 years with a focus on eye movements. Although our review will necessarily be selective, we hope to highlight not only how some of the simplifying assumptions made in the past may have run their course but also how, to move forward, research in eye movements and scene perception will need to embrace the intersectionality of how different influences and sources interact.

In this review, we examine the processing of real-world scene information through eye movements to explore the timeline of processes in more fine-grained units and explore what processes are critical in scene viewing. The eye movement record of a person viewing a scene allows researchers to examine: (1) attention and eye movements; (2) where you look; (3) influence of task; (4) memory and scene representations; and (5) dynamic scenes and eye movements. We address each point in turn.

Why Use Eye Movements to Study Scenes?

Although eye movements provide an excellent record of the location and timing of where the eyes are pointed, this record would be meaningless unless we could tie the eye movement record to mental processes. Early demonstrations showed that eye movements reflected different ways information was processed [20,21], and researchers have consistently since found that eye movements are tied to cognitive processing and attention [22–24]. When the eyes move, attention precedes the eyes to the intended location and remains at that location for some time before moving to another location (potentially the location of the next saccadic target). Although not absolute (attention can and does move to locations that the eyes do not travel to), the location of a fixation is known to be an attended location. By taking advantage of the various fixated locations, researchers can get a better understanding of why those spatial locations were attended and as well as the order of those selections. Based on the definition of a scene provided above, eye movements can provide many advantages that other measurements cannot:

1. **Eye movements are natural.** Because of the structure of the eye, people naturally move their eyes to point the location of the highest acuity in the retina (i.e., the fovea, ~2° of visual angle) at what they wish to "look" at (see Kowler [25] for a more detailed description of the visual field). To compensate for this limited area of high acuity, people rotate the eyes to focus light from different physical locations onto the fovea. Importantly, in contrast to cognitive tasks that require the experimenter to train the participant in how to respond correctly, researchers do not teach participants how to move their eyes. In fact, it takes effort and monitoring by the researcher if the goal is to have participants not move their eyes. Most people are unaware that the eyes move a number of times per second (~3 eye movements per second [26]). Although not completely implicit, the relative ease of eye movements and their relative "invisibility" makes them an ideal tool for non-invasively observing behavior. In addition, the measurement of eye movements has become relatively easier in the last two decades with the costs of eye trackers falling and the ease of use of these devises increasing. Overall, eye movements provide a low-cost way to unobtrusively observe natural behavior.

2. **Eye movements are fast.** Eye movements and fixations operate on a time scale that allows researchers to have greater precision in their measurements. Saccadic eye movements generally take less than 50 ms (frequently much less) to rotate the eyes from pointing at one part of the visual world to pointing at another part of the visual world. Once the eyes have rotated to point to the new location, they pause or fixate for a brief amount of time (e.g., 100–400 ms). While the eye is in motion, visual processing from the eyes is limited through saccadic suppression [26], so of cognitive interest is when the eye is relatively still (such as during a fixation) and visual information is acquired. Borrowing from reading research, scene processing has been measured using different fixation measures based on duration, number/count, and location. However, aggregate fixation measures that define processing across different temporal windows have proven to be especially useful. For instance, gaze duration (the sum of the fixation durations on a region of interest from the first fixation in the region to when the eyes leave that region) can give an indication of the time to initially process and recognize an object. Subsequent fixations (second gaze duration or total time) would indicate that additional information gathering was needed or that a checking/confirming process was necessary.

3. **Eye movements operate across a spatial dimension.** Unlike temporal measures that are inherently unidimensional (e.g., reaction time), measuring where the eyes are directed allows researchers to determine areas of a stimulus that the participant is currently prioritizing. The spatially distributed information of eye movements allows researchers to have a direct measure of prioritization of information available to the observer, examine commonalities in prioritization across individuals, and allow for other interesting spatially aggregate measures. For instance, the proportion of the image fixated and the scan path length can each indicate the extent of exploratory vs. focused behavior. Some tasks encourage greater exploration of the scene (e.g., memorization), whereas others constrain that exploration (e.g., visual search). Further, scan path allows for a direct measure of efficiency of the eye movements as it can be used to create a ratio of the distance taken to reach a critical region to the shortest distance possible. Thus, with all 360° of possible prioritization for the next fixation, the spatial dimension allows for a rich set of measures that reflect different types of processing.

4. **Eye movements operate across a temporal dimension.** Because the eyes have to move from one location to the next in a serial manner, eye movement data also provide a temporal record of processing in addition to the spatial record. This information allows researchers to identify the order that scene features are processed, potentially indicating their relative importance to the task.

In addition, fixations typically last only a few hundred milliseconds, which is much shorter than many complex tasks (e.g., search) take to complete. The serial fixation record can be examined to determine, at a more fine-grained time scale, the processing that was occurring at each point in the trial rather than on a global scale (i.e., reaction time).

All of these features make eye movements an especially useful tool for studying the processing that occurs in scenes. The ability to create a record of the spatial locations visited and the timing of those visits with a measurement that is practically implicit to the participant allows researchers to probe scene processing in ways that other methods do not possess. As mentioned above, the fact that fixation location is linked to attentional processing is critical to the use of eye movements as a cognitive measure. The following section examines research into this connection with regard to scene processing.

2. Attention and Eye Movements

Researchers have demonstrated that there is a tight link between eye movements and attention [27–30]. From this, researchers have also established that attention is tightly linked with the planning and execution of an eye movement to a new location in the environment. For instance, in a seminal study, Hoffman and Subramaniam [23] demonstrated that attending to one location while saccading to another did not improve performance at the attended location, but instead improvement was seen at the location of the saccade landing point. In recent years, researchers have highlighted exceptions to this assumed link between attention and saccade targets [27–29]. For instance, Golomb et al. [31–33] revealed facilitation effects at both retinotopic and spatiotopic coordinates. They argue that a lingering retinotopic trace persists after the eyes have moved, and it persists for a small window of time with the updated spatiotopically relevant area. This research highlights not only the connection between attention and eye movements but also the temporal shifts of processing within a fixation as the current fixation shifts to a new location.

Interestingly, the demonstration of the connection between eye movements and attention also presents an interesting conundrum for the interpretation of fixation durations. As stated above, the decision of where to move the eye is inherently a part of measurements that reflect when to move the eyes. Thus, the link between eye movement and attentional processes are not a straightforward causal relationship, as many early studies posited [34]. These more recent studies do not dispute the link between attention and eye movements, but rather highlight how information processing at different positions relative to the current eye position are updated over time and introduce a more fractured view of the role of attention relative to eye movements.

Although it is true that stimulus properties are often researched as the main driving force influencing where we look, there also seems to be a shift in the understanding of how stimulus properties drive eye movement location. Other influences on the direction of attention and eye movements have recently come to the fore. For instance, internal tendencies for how information is acquired seems to have an independent effect on eye movement guidance [35–38]. As an example, there is the central bias effect [36], where the eyes have a bias to look to the center of the image. Although initial eye movements seem to be positioned centrally, this effect dissipates with time [35,39], but other factors influence it as well. For instance, Bindeman [35] found that both the center of the scene and the screen itself play a role in the bias, as only early eye movements were directed to the center. Furthermore, Rothkegel et al. [39] had participants initially fixate on a scene to one side and found that the initial fixation voluntarily made by participants was not affected by the scene context because it was independent of the location of the most informative regions. Inherent in the interpretation of the central bias is the notion (either explicitly or implicitly) that the attentional window starts wide and then focuses on scene details. This default mode of processing is ubiquitous regardless of stimuli details. We assume attention and eye movements interact in the selection, planning, and execution of a new voluntary eye movement. It is how we select the information to be scrutinized that we now turn to.

3. Where You Look

By far, the largest number of research studies on eye movements and scenes have examined the question of where we look. The driving force behind this question is of course that certain types of information are prioritized for further inspection or that certain characteristics of the environment tend to capture attention (or draw attention) to them. Although there is still some focus on how stimulus properties capture attention, more recent studies have introduced a number of new ways of thinking about and categorizing the type of visual properties that receive further scrutiny through eye movement planning. We examine these different influences in turn, beginning with the omnipresent properties of the stimulus itself, and then exploring other factors.

3.1. Influence of Stimulus Properties

Traditionally, researchers have examined basic and higher-order image features and how they drive the eyes in a bottom-up manner. Computer vision has been incredibly influential in how researchers theorize about how eye movements are guided, most predominantly through computational models of saliency [40,41]. Visual saliency and saliency maps try to define areas that "stand out" from the background as potential points of interest. When looking at images, the eyes rarely go to large homogenous areas such as a blue sky or a blank wall [20,21,42]. Saliency calculations attempt to find the areas of the image based on the low-level features that can be extracted from the image itself. Saliency maps highlight the coordinates of the points that stand out and allow for a ranking of importance within the image. Low-level features such as color, orientation, and intensity [40,43], as well as second-order features such as intersections and edges [44,45] have been found to affect eye movement planning. Many researchers have explored the combined and separate contributions of low-level features to eye movement guidance (e.g., color [46]), but there has been movement away from a purely bottom-up approach.

Although saliency has inspired a number of theoretical models and research studies, over the past decade, the limits of saliency as an explanatory tool has become more pronounced [47–49]. First, inherent in models of saliency is the notion that information selection is passive and based solely on the properties of the image regardless of the individual's intent. However, across many studies, it should be noted individuals are actively seeking visual input, regardless of task [49,50]. Second, there is an overall movement away from classifying influences as either purely top-down or bottom-up [49,51]. Instead, researchers have begun to examine different sources of information (e.g., immediate history with a task) and how those sources are combined and interact. For instance, many recent models are finding ways to incorporate higher-level information such as meaning and objects into how scene information is selected. There are many computational models that have since been proposed to better represent higher-order information, but review of that work is beyond the scope of the current paper. However, we will consider the different approaches for considering higher-order information and its influence on eye movement guidance.

3.2. Meaning or Object as the Unit of Selection

Rather than treating information as either purely top-down or bottom-up sources, researchers are finding that the combination of these factors best explain eye movement planning. One method has been to identify objects (rather than low-level features) as the unit of selection for eye movement planning [52–55]. This stems from early demonstrations that observers tend to prefer to focus on objects rather than background elements [20,21]. In these cases, objects are defined as meaningful entities that are visually distinct from the background. For instance, Stoll et al. [55] found that the preferred landing position on an object was centered on it in relation to the object's boundaries. Unlike the low-level features that posit that local edges (changes in contrast) may attract attention, an object-centered approach conveys that the center of mass within those edges are the targets of eye movements (e.g., [56]). Indeed, Pereira and Castelhano [8] found that fixations were directed at groups of objects within a larger scene context during search, and they concluded that object content provided specific information about where to aim fixations.

The issue with using objects in calculating where to attend is the fact that it is often difficult to define what an object is from the image itself. The complexity of figure-ground separation is only magnified with multiple objects at multiple depths in scenes. Without defining objects, researchers have sought to update saliency models with other higher-level features. For instance, proto-objects can be used as the unit of selection [57–61]. Although the definitions vary across studies, there is a general consensus that proto-objects are defined as fragments of a feature-similar visual space or are pre-attentive structures with limited spatial and temporal coherence. These signals of objects (rather than relying on the separation of whole object representations) serve to circumvent the question of how you know the identity of what is there before you know that something is there (e.g., [62]). The proto-object approach also circumvents the distinction between purely bottom-up, low-level features and top-down priorities. For instance, Wischnewski et al. [60] argue that having proto-objects as the unit of incoming information allows for models to incorporate different aspects of the visual scenes that may not otherwise be possible (such as temporal-spatial changes over time). Thus, proto-objects allow for further integration of different sources of information without having to define individual objects per se.

Other methods have been to establish the ground-truth of informativeness or meaningfulness of scene regions [63,64]. For instance, Henderson and colleagues had a separate group of participants rate the meaningfulness of small regions of the image. Using these ratings, the allocation of eye movements to different regions of the scenes were predicted from their level of meaningfulness as derived from these rating studies. The concept of meaning in this case is somewhat related to the proto-object properties mentioned above, in that it allows for object parts or high-level features to be the unit of analysis. In addition, the combination of presenting participants with isolated regions of the scene and using their ability to interpret that information (to varying degrees) results in a combination of low-level features traditionally used in saliency map, with a top-down, high-level interpretation of those features.

3.3. Semantic Integrity within the Larger Scene Context

The influence of overall scene semantics on selection is typically examined by contrasting eye movements to semantically congruent and semantically incongruent objects within a scene context. Any differences in fixating congruent and incongruent objects would be the result of the scene semantics because incongruency, especially, is defined by the object-scene relationship. Although there are multiple ways in which semantic incongruency could influence eye movements, researchers have concentrated on two main questions: whether inconsistent objects attract attention to themselves from a distance and how inconsistent objects affect processing, once attended. We will examine each question in turn.

First examined by Loftus and Mackworth [65], the question of whether semantically inconsistent objects automatically attract attention has been studied for decades. Despite being intuitively appealing, subsequent studies produced mixed results [66–69], and this inconsistency has continued in more recent studies [14,70–73]. For instance, Võ and Henderson [71] found that objects that were inconsistent with the scene context (e.g., a computer printer on a stove in the kitchen) did not attract initial fixations, suggesting that participants were not immediately drawn to these objects. On the other hand, Lapointe and Milliken [73] found that there was a tendency for inconsistent objects to be detected more quickly during a change detection task. The difference in patterns of results illustrate important interactions between stimulus properties (e.g., the size of the critical object) and task (e.g., visual search vs. change detection), highlighting another instance where different types of scene properties interact.

The mixed results seem to be associated with object size. For instance, it is unclear across studies whether the object's identity can be extracted from parafoveal or peripheral information. To the extent that object size is constrained by the limited availability of information peripherally, a decrease in that object's ability to draw attention is seen [74–76]. This question links to the question of when, during the execution of an eye movement, object identity is extracted. Is it parafoveally just prior to fixation or only once the details have been directly fixated and processed? We will examine this question in more detail below.

With regards to the influence of scene semantics, is the question of how semantic inconsistencies affect processing once objects are fixated. This question is less controversial than the first in that it has been well established that inconsistent or incongruent objects do lead to longer scrutiny and overall longer processing times [66,67,77]. More recent studies have demonstrated similar patterns of results [14,70,78], although there have been some exceptions [73,79]. We can also examine the converse effects of scene semantics and examine how consistent semantic information can positively affect eye movement guidance [6,67,80,81]. One technique to examine the influence of scene information on eye movement guidance is the Flash Preview-Moving Window (FPMW) paradigm [80]. The FPMW paradigm has participants briefly view a scene image preview (250 ms, which is too brief to execute an eye movement). Following the brief scene preview, a target label is presented that the participant needs to search for in a scene. The search scene is then presented, but the participant's view of the scene is limited to a ~4° radius window centered on current gaze position. Because the window is locked to gaze, the participant views the scene as if viewing it through a paper tube. In this way, rather than relying on the immediately available visual information extracted from the periphery, planning of eye movements outside the window would require observers to rely on the representation of the scene from that initial preview. By manipulating the relationships of the preview to the search scene, researchers have explored several aspects of how scene representations affect eye movement guidance. For instance, researchers have shown that the scene semantic category did little to improve search performance [70,80], that specific details about the scene seems to be important [8,14,80], details about the target help [18,70], and that extraction of useful information occurs quite quickly [82]. More recently, researchers have also examined the effects of domain expertise and interestingly found that when radiologists viewed chest x-rays, the previews provided much smaller benefit than would be expected based on search in scenes [83].

As was suggested by previous research, studies in the past decade have found that consistent objects lead to more efficient search performance [8,14,70,84–86]. Researchers posit that the semantic relatedness of the object not only to the scene context, but also to other objects in the scenes, led to faster search. For instance, Hwang et al. [84] used annotated photo images (from LabelMe Database [87]) to examine the contribution of semantically related objects to the guidance of eye movements. They found that there was a tendency for the next fixation to be programmed to an object that was semantically similar to the currently fixated object. Further, Mack and Eckstein [86] found that when semantically related objects were placed in close proximity, search was much faster. In a recent development, Võ, and colleagues [9] proposed a key role for certain larger objects (anchor objects) that are associated with other objects (e.g., stove and a pot). They demonstrated that the presence of anchor object led to more effective guidance during search. Because anchor objects are typically large scene elements, they fit with the earlier studies as they can be identified in the parafovea or periphery providing semantic guidance within the scene [84,88,89]. In addition, this information can also provide spatial information that could aid gaze guidance, which we will discuss next.

3.4. Influences of Spatial Associations

Scene shape has been closely linked to scene categories [7,10–13,90,91] and has been used to explain how eye movements are guided during visual search [6,8,14,81,92,93]. That overall scene structure has an influence on search and object processing has been known for some time [4,80,81,94,95]. For instance, Castelhano and Heaven's [14] study included a factorial manipulation of scene semantics and spatial positioning of target objects. Rather than finding that spatial consistency was only useful within semantically consistent scenes, they found that both semantic and spatial information influence search independently and additively.

Researchers examining when objects are out of place in a scene (i.e., in a spatially inconsistent location), have found that visual search performance is slowed [14,15,18,72,96]. For instance, Hillstrom et al. [96] found that when target objects' location switched from a spatially consistent location to an improbable (mug on floor) or an impossible (mug in the air) location between the preview and search scene, performance significantly worsened. Furthermore, Castelhano and Witherspoon [18] found a strong link between the target object's function and its spatial location in the larger scene context. They found that when the functions of novel objects were learned, participants were able to locate them much more quickly than when only the visual features of the target object were known. Further research has shown the link between action, function, and spatial organization is thought to be strongly linked in scene representations [19].

More recently, Castelhano et al. [17,97,98] have argued for the importance of scene surfaces in guiding attention during visual search. The Surface Guidance Framework posits that attention is directed to surfaces in the scene most associated with the target object. For example, (1) upper (e.g., ceiling, upper walls), (2) middle (e.g., countertops, tabletops, desktops, stovetops), and (3) lower regions (e.g., floor, lower walls), are associated with specific objects: (1) upper (e.g., painting, wall clock), (2) middle (e.g., blender, book), and (3) lower (e.g., garbage bin, shoes). By dividing the scene into relevant and irrelevant surfaces, we can define target-relevant and target-irrelevant regions for any scene-object combination. This, in turn, allows for the examination of how previous knowledge about the scene context and its association affects processing across regions. For instance, Pereira and Castelhano [97] found that suddenly onset objects were more likely to capture attention when they appeared upon a target-relevant than a target-irrelevant surface. Thus, how attention is deployed is closely tied to the scene structure, where surfaces can act as a larger object-based region across which attention is allocated [99,100].

Interestingly, researchers have also begun to inquire about differences in processing across the spatial depth of a scene. For instance, recent studies have shown qualitatively different processing of spaces closer to the observer [16,17,101–103]. For instance, Castelhano and Fernandes [17] found that foreground information (from the center of the total scene depth to the position of the observer within the scene) had a great initial influence on initial scene perception than background information. Furthermore, Bonner and Epstein [101] have found that activity in the occipital place area (OPA) was linked to perceiving potential paths for movement in immediate surroundings. In addition, Josephs and Konkle [16] have found that the spaces that are reachable are represented qualitatively differently than objects and larger scene spaces. Given the qualitative differences in processing across depth, it stands to reason that information closer in depth may have different utility than information farther away and thus, may differently affect eye movement guidance and visual search in a scene. Indeed, in a recent study, Man and Castelhano [103] found a consistent effect of scene depth, where targets placed closer in space (in the foreground of the scene) were found faster and with fewer fixations than those placed in the background, regardless of semantic association and regardless of target size. Thus, across studies information is processed differently across scene depth.

Across the different influences on how eye movements are guided, one commonality is in how these influences are assessed. Much of the research on how high-level information is prioritized or captures attention is attained through tasks such as visual search, change detection, or free viewing. To some extent, the commonalities across tasks point back to the notion discussed by Rayner and Pollatsek [5] that when trying to understand scene processing, task may be irrelevant as the complex nature of the scene processing is required regardless of how the observer is processing that information. However, we also know from a number of studies that task plays a crucial role in how information is processed. We turn to the question of the influence of task next.

4. Effects of Task

Although the contributions of the elements discussed above have formed a large part of the research on eye movements and scenes in the last decade, the importance of task and task-relevance of scene information has become more apparent. Buswell [20] and Yarbus [21] both described the importance of the task that the observer is performing. Rayner and Pollatsek [5] raised the issue of task in scene viewing and identified its many challenges. In contrast to reading, scene perception is not simply one task; search, memorization, or "free viewing" can all be tasks performed on scenes. Each task can alter the eye movement pattern dramatically because different aspects of the scene are relevant to each task. One of the most famous of these demonstrations was performed by Yarbus [21]. In his study, Yarbus examined the eye movements of a participant when looking at a single stimulus, the painting "An Unexpected Visitor," with different tasks. Yarbus found that the locations of fixations varied with the task (see also [104]). Castelhano et al. [50] replicated the finding that the locations of fixations varied with the viewing task, but they also found that some measures, like fixation durations, were task invariant. The finding that the task can dramatically alter what is viewed indicates that it can interact with other aspects of scene processing.

The question of the task is critical when interpreting eye movements on scenes. Across the scene processing literature, many types of tasks have been employed. For instance, to encourage participants to scan over the spatial extent of the entire scene, tasks that direct the participant minimally are typically used, such as free viewing [20,105,106], memorization [50,107], or aesthetic judgments [108,109]. These less focused tasks encourage participants to scan a large proportion of the scene as participants are uncertain of the importance of any particular detail. Buswell [20] demonstrated that when participants were instructed to "look at the pictures in their normal manner" (p. 136) fixations were widespread. However, this breadth is not of uniform density and tends to be focused on parts of the scenes containing meaningful objects [42,50,63,110]. In contrast to less focused tasks, tasks that involve focused processing, such as search and change detection, require participants to be more directed in their viewing in line with a specific goal. Buswell [20] demonstrated that when participants searched for a person in a window in the same scene that they had freely viewed previously, the fixations were concentrated on possible locations where the target could be. In general, the extent of the eye movements executed under search instructions tend to be more focused on possible areas of the scene that the object can occur [6]. Because search tasks necessitate that some aspects are relevant and some are not, they allow researchers to manipulate the relative importance of physical features or meaning through the specification of the search target. For example, Peacock et al. [64] had participants look for bright patches in a scene (a physical feature), but they found that even though the search task did not encourage processing of the meaning of the objects in the scene, meaningfulness continued to affect where people looked in the scene. In addition, when the search target is absent, participants will search for an extended period of time, examining progressively more of the image and allowing for a more direct comparison of the scene processing between search and memorization (e.g., [111]).

Consistent with the theme of the current review, it is impossible to divorce viewing task from the type of information obtained while viewing a scene. The viewer's task acts as a filter that highlights aspects of the scene that are consistent with the current goal. In some instances, the goal is relatively undefined like free viewing, which leads to fixation patterns that vary widely. On the other hand, focused goals like search lead to a more constrained viewing pattern. Regardless of the specific pattern, it is impossible to examine eye movements in scene processing without considering the task.

5. Influence of Scene Representations in Memory

As stated above, fixations on an image provide a record of the objects and areas that are attended to in the image. Because of this connection, the fixation record provides a reliable indication of aspects of a scene that could be encoded into memory [112]. In addition, the serial nature of eye movements allows researchers to examine the order of encoding information into memory. Given these two facets, the influence of eye movements on memory could actually be bi-directional: in one direction, a previously encountered scene could influence the fixation patterns and in the other direction, the eye movement pattern is tied to the scene representation in memory.

5.1. What Is Remembered of a Scene from a Fixation?

Fixation location provides a useful marker for what is attended in a scene and what can be processed with the highest acuity. It stands to reason that the memory for scenes would similarly be tied to fixation location. Although there have been claims that little to no memory exists of scenes [113,114], more recently the general consensus is that scene information can be retained reasonably well [115]. Thus, it is not a question of the visual representation's existence, but the degree to which visual details are stored.

As highlighted earlier, the question of task is critical when examining eye movements. This question is just as critical when examining the interaction of eye movements and memory. What one is doing when fixating an object or part of a scene influences the quality and robustness of the memory retained. Some tasks encourage the participant to try explicitly to encode the information that is presented [116,117]. In other tasks, like visual search or aesthetic judgments, memory is retained incidentally [92,118,119]. Incidental encoding may be tested with an explicit memory test, such as a two alternative forced choice task (e.g., which of these hats did you see in a previously presented image? [111]), but encoding scene details was not the focus of the task. Finally, researchers also employ the relatively undefined task of "free viewing," which may fall in between the explicit and incidental nature of encoding because the task itself does not provide an instruction of how to process the scene. Thus, participants are left to their own discretion as to how to process the image.

With regard to memory for scene representations, in most viewing tasks encoding is incidental in nature. Tasks such as navigating, searching, or judging the environment are far more common and scene representations are formed even without explicit instructions. For instance, Castelhano and Henderson [111] examined memory for objects in scenes that were acquired either incidentally or intentionally. They found that objects that were directly fixated were remembered better than chance regardless of the manner in which they were encoded. More recent studies have extended this finding to incidentally viewed parts of scenes and found that increased viewing of an area of the scene led to better memory performance [120]. Thus, studies have demonstrated that even without an intention to remember, fixated locations tended to be encoded.

The memory for objects that are encountered when viewing a scene are heavily influenced by both the fixation on the object and the task being performed. Draschkow and Võ [121] had participants perform a multistage search through the same environment where participants were told to find items to pack for a trip. They examined memory for objects that were relevant and those that were irrelevant to the task. They found that relevant objects to the packing task were remembered better than irrelevant objects. However, although greater fixation time on objects before they were targets did predict faster subsequent finding, the memory advantage of the relevant objects was not related to fixation time (see also [122,123]). This finding demonstrates a more complex interaction of the task and memory than simply one based on fixation time.

Another paradigm used to assess memory for scenes and target objects is the contextual cuing paradigm [124]. In this paradigm, after a target is found in the first block of trials, the viewer uses the knowledge of that scene to move the eyes efficiently to the previously found target in subsequent searches [125–127]. In this paradigm, rather than explicitly testing memory, representations of scenes and objects are shown through a decrease in the time to fixate a previously located target object. The speed up from one presentation to the next demonstrates that memory existed from the previous encounter that can be used when viewing the next presentation.

Although repeating the exact scene can be informative, it is also possible to demonstrate more generalized learning of specific contingencies across different scenes. Brockmole and Võ [128] found that people could learn and thus locate a target letter more quickly, when the target letter always appeared on a specific object in a scene (e.g., the letter appeared on a pillow across multiple bedroom exemplars). This learning and speeded fixation was even evident when a more general association was needed between a category of scenes and the target location. Clearly, memory is affected by the locations of fixations on the scene. However, this relationship is not an absolute in that simply knowing the amount of time a part of the scene is fixated is not a perfect predictor of memory. Instead, the relationship of fixation to memory is affected by the task performed and the role of individual object details in the task. Without considering both components, the relationship between fixation and visual memory can be opaque.

5.2. How Does Memory of the Scene Influence Current Fixations?

One would assume that having previously fixated an object in a scene would make it easier to find that object later, but that advantage may be limited [78,129]. Võ and Wolfe [119] had participants repeatedly search the same scene for different targets, while their eyes were tracked. Critically, because the same scene was being searched several times, the target object of the current search task had been a distractor on previous searches. Võ and Wolfe found that even though the current target had been viewed on previous trials, it did not significantly reduce the time to find that current target. Although the shift from distractor to target did not improve search performance, once the object had been found as a target, subsequent searches for that target object were facilitated. On the other hand, Hollingworth [130] attempted to replicate these results with more participants and found that there was indeed facilitation of having viewed distractor objects on subsequent searches for distractor-turned-target objects. Similar to the previous study [78], Hollingworth found that searching for a previously located target produced a much larger facilitation of search. These results indicate that the way in which an object is processed influences memory, but there is a general facilitation for previously processing an object, regardless of the type of processing.

Another means by which memory can influence eye movements is when an undetected change occurs in the scene. Memory for the previous object information within the scene can be demonstrated by longer subsequent fixations on a changed object compared to when the object has not changed [131–133]. For instance, Võ, Zwickel, and Schneider [131] showed that when an object changed location from one presentation to the next, gaze durations on the changed object were lengthened, even when that change was not explicitly detected. This implicit detection of a change via fixations demonstrates that eye movements can be more sensitive to some cognitive processes than explicit response.

It is clear that what is fixated can affect scene memory and vice versa. However, what about the pattern of eye movements themselves? Do people have to reenact the eye movements they made when learning the scene to remember the scene [56,134–136]? This scan path idea of Noton and Stark [134] argued that the memory for a scene would be improved if the eye movement pattern were repeated. However, subsequent research has indicated that in its strongest sense, the pattern of eye movements during retrieval does not have to match the pattern at encoding in order to recognize the scene. For example, Foulsham and Kingstone [56] found that there was no advantage of viewing one's own pattern of eye movements on memory of a scene fixating compared to viewing someone else's eye movement pattern. However, more recent studies have attempted to show that under limited circumstances, eye movements can be repeated between an initial viewing and a subsequent retrieval. Some of the stronger evidence for this claim comes from studies of "looking at nothing" studies. Johansson and Johansson [135] found that memory for objects was improved if people fixate where the object had previously been located (although the screen was blank at the time of retrieval) compared to when they fixate another location. This type of memory advantage extends to imagined scenes as well [136]. Thus, looking where something used to be appears to provide a boost to memory. In addition, it seems that there can be overlap of portions of the scan path from one viewing to the next [137,138]. Wynn et al. [137] found that fixations near the start and end of a search sequence were similar between viewings of the same scene in a change detection task. However, the fixations between the initial fixations on the scene and the final fixations on the scene did not match between the two views. Given the lack of similarity for most of the scan path and the heavy influence of the change detection task, these results provide limited support for scan path recapitulation as part of the retrieval process. In a similar vein, Bochynska and Laeng [138] compared a free viewing task (i.e., allowing participants to move their eyes) with a restricted viewing (i.e., participants could not move their eyes outside a central box) and found that memory for a sequence was better with free viewing compared to the restricted viewing. Although evidence that being able to move the eyes is better than not moving the eyes (see [139] for a similar finding with face stimuli), it is not strong evidence that eye gaze had to follow the same pattern to retrieve information. In general, although the execution of eye movements is important to remembering scenes, the recapitulation of the eye movements appears to add limited explanatory power.

Both the impact of memory on eye movements and the impact of eye movements on memory for the scenes clearly demonstrates that scene processing is an interactive process. In the first encounter with a scene, what is fixated and for how long influences what is stored in memory. At the other end, prior experience with a scene influences what is selected to be viewed and for how long. The fact that information can flow both directions indicates the strong connection between attention and memory. Although the idea is not new—William James pointed to the function of attention and the stream of consciousness in a similar way [140]—the ability to use eye movements as an observable method to measure this connection has greatly helped our understanding of the interaction.

As mentioned above, much of the research on scene perception has been limited to static scenes depicted on computer monitors in various formats. However, there has been significant progress in the research of scenes that more closely mirrors the experience of an observer when embedded in a scene. Next, we examine how representations differ as information changes over time and how eye movements differ when reflecting dynamic scene processing.

6. Dynamic Scenes and Eye Movements

A vast majority of the work examining eye movements and scene perception has used static images projected on a computer monitor. Limiting the scope of the research to static images makes sense when one considers the complexity of the stimuli. However, static images are a specific case of viewing compared to how the real world is processed. Dynamic environments can involve changes caused by the individual moving through or interacting with the environment [141–144] or by watching a dynamic scene unfold [145–147]. Although the studying of eye movements involved in the perception of dynamic scenes creates some technical challenges, the investigation of the how the eyes move in natural, dynamic scenes allows researchers to have a better understanding of scenes more generally.

Even though scenes can be dynamic in two ways (as mentioned above), most of the work in the last decade has focused on dynamic scenes that unfold over time before the viewer. The first question that arises in examining these dynamic scenes is the differences in eye movement patterns between static and dynamic scenes. Although they may contain similar information at different levels of analysis, dynamic scenes' additional motion cues do alter the fixation patterns. The basic motion of dynamic scenes leads to larger differences between static and dynamic scenes than between different types of dynamic scenes [145]. Further, Mital et al. [146] examined the visual factors that predict eye movements while watching dynamic images. They found that motion within the dynamic scene was the most predictive of where the eyes would fixate. The differences between static and dynamic scenes even extend to the common laboratory findings, such as oculomotor capture, that occur regularly in static images, but they may be absent or altered in dynamic scenes [148].

Previous findings have also indicated that dynamic scenes' motion cues lead to a higher degree of "attentional synchrony," where people tend to look at the same location at the same time [38,145,149,150]. For instance, Dorr et al. [145] found that viewing professional movies resulted in fixation patterns that tend to cluster in the same regions of any individual shot. Similarly, Tseng et al. [38] attempted to disentangle the contribution of various viewing factors (e.g., saliency, photographer bias) on the production of the center bias in dynamic scene viewing and found that the tendency of points of high interest and salience to be located in the center of the image was the strongest contributor to the center bias. These common factors can be described as the "tyranny of film," which has been shown to be a dominant factor in where people look in films [150]. These studies indicate that there are strong influences that encourage the viewer to cluster their fixations in a similar location, especially when the scene is changing over time. The question that remains is whether this is due to simple stimulus-based factors or is it a result of higher-level information being associated with motion. Interestingly, more naturally filmed scenes as well as static scenes are not processed with the same type of central focus that is seen with professionally filmed scenes [143].

However, viewing task seems to affect the similarity of fixation patterns between groups [151,152]. Smith and Mital [151] found that the highest level of cross-participant similarity in eye movements was associated with dynamic scenes during a free viewing task. This high degree of similarity indicates that fixation location is at least in part due to endogenous and task-related factors. Further, Foulsham and Kingstone [152] compared where people fixate while virtually walking around scenes on a computer screen. The comparison indicated that even when taking into account sequential information, eye movements on static scenes were not a better predictor of the location of real-world fixations than a model that simply relied on a central bias. Together, these results warrant caution when attempting to generalize eye movements from a static image to those in the real world [145].

The study of dynamic scenes is an increasingly promising area for the study of eye movements and scenes. Findings that eye movements in dynamic scenes do not correspond to those in static scenes leads to a need to develop new methodologies and possible new theoretical frameworks. Static scenes have provided a good starting point for researchers by providing more control over the stimuli. However, as eye movements change when the same information is presented dynamically, researchers need to consider how other factors could interact with the motion inherent in dynamic scenes. For instance, the task given the person or the effect of the immediately preceding history of an unfolding scene could both influence the fixation locations even with the motion of dynamic scenes. Moreover, of course, the interaction of the observer with the environment, such as when the observer is allowed to navigate in the real world, will require increasingly complex theoretical approaches. One tractable method of investigating dynamic scenes may be addressed through immersive environments (i.e., Virtual Reality) as they will allow researchers to have control over both types of dynamic scenes.

7. Conclusions

When Rayner and Pollatsek [5] reviewed the literature on eye movements and scene perception, they felt compelled to justify the use of eye movements as a way to study the cognitive processing of scenes. At that time, studies were demonstrating that scene representations could be formed with extremely brief presentations that were too short to allow for planning and executing eye movements [153]. As a result, researchers posited that measuring eye movements would not be particularly informative for scene perception because sufficient information was garnered without them. However, the explosion of eye movement research in scenes since that review, and especially in the last decade, clearly indicates that eye movement measures are indispensable as a tool for understanding scene processing.

We began this review by examining a number of influences that have motivated research in scene processing and eye movements. We reviewed research examining: (1) attention and eye movements; (2) where you look; (3) influence of task; (4) memory and scene representations; and (5) dynamic scenes. The research in each of these areas has advanced dramatically in the last decade and led to considerable improvement in our understanding of the visual processes involved in scene perception. However, what is not clear is where to go from here. Do we continue to drill down on each of these influences and possibly other factors that may have an effect or is there an alternative approach?

Although we have organized and discussed the five influences separately, we also made note of the different ways in which these influences could interact. Here, we hope to highlight that the future advancement of the field will be in examining these interactions directly. For example, although one can argue about the relative contributions of top-down and bottom-up processes, the way in which different sources of information (scene gist knowledge, expected layout, probability of location, and meaning, among others) contribute to processing may depend on other critical factors such as task, history/memory, and type of scene (embedded 3D or viewing 2D). This would require researchers to view these combinations not as a set of additive, independent contributions that are typically referred to as top-down effects, but rather begin to describe how certain critical factors and influences interact and lead to qualitatively different types of processing compared to other combinations. Rather than a list or a hierarchy, researchers need to move toward an understanding of how the landscape of "top-down" factors is shaped. For example, Awh et al. [51] posits that experience affects performance in a way that is different from previous homogeneous views of top-down influences. Defined as a form of selection history, they found an impact of the previous trials' selections on the current trial on performance. The approach has begun the process of breaking down what we mean by top-down influences by including the unique contribution of experience.

The future of using eye movements to study scenes appears to be at the intersections of the areas we have described. By examining the intersectionality (processing that is unique to the specific combination) of these influences, we can begin to ask more complex questions. This argument is similar to that of Wertheimer (1923/1938) who, in comparing Gestalt Psychology to Structuralism [154], stated "I stand at the window and see a house, trees, sky. Theoretically I might say there were 327 brightnesses and nuances of colour. Do I have '327'? No. I have sky, house, and trees" ([155], p. 71). The recognition of the Gestalt Psychologists was that in attempting to break down the scene, the actual goal of understanding scenes was lost. A more holistic approach was needed. With respect to eye movements and scenes, a substantial amount of research has attempted to break down scenes into their components to understand the nature of viewing them. However, it will take looking beyond the components to the intersections, in order to gain a better understanding of the scene perception and eye movements.

Author Contributions: C.C.W. and M.S.C. contributed equally to writing this review. Each author wrote portions of the first draft, and both authors edited subsequent drafts.

Funding: This work was partially supported by funding from the Natural Sciences and Engineering Research Council of Canada, Canadian Foundation for Innovation, and Ontario Ministry of Research and Innovation to MSC.

Conflicts of Interest: The authors declare no conflict of interest.

References

1. Castelhano, M.S.; Henderson, J.M.M. The influence of color on the perception of scene gist. *J. Exp. Psychol. Hum. Percept. Perform.* **2008**, *34*, 660–675. [CrossRef] [PubMed]
2. Henderson, J.M. Human gaze control during real-world scene perception. *Trends Cogn. Sci.* **2003**, *7*, 498–504. [CrossRef] [PubMed]
3. Henderson, J.M.; Hollingworth, A. High-level scene perception. *Annu. Rev. Psychol.* **1999**, *50*, 243–271. [CrossRef]
4. Oliva, A. Gist of the scene. In *Neurobiology of Attention*; Itti, L., Rees, G., Tsotsos, J.K., Eds.; Academic Press: Cambridge, MA, USA, 2005; pp. 251–256.
5. Rayner, K.; Pollatsek, A. Eye movements and scene perception. *Can. J. Psychol.* **1992**, *46*, 342. [CrossRef] [PubMed]
6. Torralba, A.; Oliva, A.; Castelhano, M.S.; Henderson, J.M. Contextual guidance of eye movements and attention in real-world scenes: The role of global features in object search. *Psychol. Rev.* **2006**, *113*, 766–786. [CrossRef]
7. Oliva, A.; Torralba, A. The role of context in object recognition. *Trends Cogn. Sci.* **2007**, *11*, 520–527. [CrossRef]
8. Pereira, E.J.; Castelhano, M.S. Peripheral guidance in scenes: The interaction of scene context and object content. *J. Exp. Psychol. Hum. Percept. Perform.* **2014**, *40*, 2056–2072. [CrossRef]
9. Võ, M.L.-H.; Boettcher, S.E.; Draschkow, D. Reading scenes: How scene grammar guides attention and aids perception in real-world environments. *Curr. Opin. Psychol.* **2019**, *29*, 205–210. [CrossRef]
10. Intraub, H. Rethinking scene perception: A multisource model. *Psychol. Learn. Motiv.* **2010**, *52*, 231–264.
11. Greene, M.R.; Oliva, A. The briefest of glances: The time course of natural scene understanding. *Psychol. Sci.* **2009**, *20*, 464–472. [CrossRef]

12. Greene, M.R.; Oliva, A. Recognition of natural scenes from global properties: Seeing the forest without representing the trees. *Cogn. Psychol.* **2009**, *58*, 137–176. [CrossRef] [PubMed]
13. Oliva, A.; Torralba, A. Modeling the shape of the scene: A holistic representation of the spatial envelope. *Int. J. Comput. Vis.* **2001**, *42*, 145–175. [CrossRef]
14. Castelhano, M.S.; Heaven, C. Scene context influences without scene gist: Eye movements guided by spatial associations in visual search. *Psychon. Bull. Rev.* **2011**, *18*, 890–896. [CrossRef]
15. Malcolm, G.L.; Henderson, J.M. Combining top-down processes to guide eye movements during real-world scene search. *J. Vis.* **2010**, *10*, 4. [CrossRef] [PubMed]
16. Josephs, E.L.; Konkle, T. Perceptual dissociations among views of objects, scenes, and reachable spaces. *J. Exp. Psychol. Hum. Percept. Perform.* **2019**, *45*, 715–728. [CrossRef] [PubMed]
17. Castelhano, M.S.; Fernandes, S. The Foreground Bias: Initial scene representations dominated by foreground information. *J. Vis.* **2018**, *18*, 1240. [CrossRef]
18. Castelhano, M.S.; Witherspoon, R.L. How you use it matters: Object function guides attention during visual search in scenes. *Psychol. Sci.* **2016**, *27*, 606–621. [CrossRef]
19. Greene, M.R.; Baldassano, C.; Esteva, A.; Beck, D.M.; Fei-Fei, L. Visual scenes are categorized by function. *J. Exp. Psychol. Gen.* **2016**, *145*, 82–94. [CrossRef] [PubMed]
20. Buswell, G. *How People Look at Pictures: A Study of the Psychology and Perception in Art*; Univ. Chicago Press: Oxford, UK, 1935.
21. Yarbus, A.L. *Eye Movements and Vision*; Springer: Boston, MA, USA, 1967.
22. Deubel, H.; Schneider, W.X. Saccade target selection and object recognition: Evidence for a common attentional mechanism. *Vis. Res.* **1996**, *36*, 1827–1837. [CrossRef]
23. Hoffman, J.E.; Subramaniam, B. The role of visual attention in saccadic eye movements. *Percept. Psychophys.* **1995**, *57*, 787–795. [CrossRef] [PubMed]
24. Rayner, K.; McConkie, G.W.; Ehrlich, S. Eye movements and integrating information across fixations. *J. Exp. Psychol. Hum. Percept. Perform.* **1978**, *4*, 529–544. [CrossRef] [PubMed]
25. Kowler, E. Eye movements: The past 25 years. *Vis. Res.* **2011**, *51*, 1457–1483. [CrossRef] [PubMed]
26. Rayner, K. Eye movements and attention in reading, scene perception, and visual search. *Q. J. Exp. Psychol.* **2009**, *62*, 1457–1506. [CrossRef] [PubMed]
27. Casarotti, M.; Lisi, M.; Umiltà, C.; Zorzi, M. Paying attention through eye movements: A computational investigation of the premotor theory of spatial attention. *J. Cogn. Neurosci.* **2012**, *24*, 1519–1531. [CrossRef] [PubMed]
28. Cavanagh, P.; Hunt, A.R.; Afraz, A.; Rolfs, M. Visual stability based on remapping of attention pointers. *Trends Cogn. Sci.* **2010**, *14*, 147–153. [CrossRef] [PubMed]
29. Rolfs, M.; Jonikaitis, D.; Deubel, H.; Cavanagh, P. Predictive remapping of attention across eye movements. *Nat. Neurosci.* **2011**, *14*, 252–256. [CrossRef] [PubMed]
30. Zhao, M.; Gersch, T.M.; Schnitzer, B.S.; Dosher, B.A.; Kowler, E. Eye movements and attention: The role of pre-saccadic shifts of attention in perception, memory and the control of saccades. *Vis. Res.* **2012**, *74*, 40–60. [CrossRef] [PubMed]
31. Golomb, J.D.; Marino, A.C.; Chun, M.M.; Mazer, J.A. Attention doesn't slide: Spatiotopic updating after eye movements instantiates a new, discrete attentional locus. *Atten. Percept. Psychophys.* **2011**, *73*, 7–14. [CrossRef] [PubMed]
32. Golomb, J.D.; Nguyen-Phuc, A.Y.; Mazer, J.A.; McCarthy, G.; Chun, M.M. Attentional facilitation throughout human visual cortex lingers in retinotopic coordinates after eye movements. *J. Neurosci.* **2010**, *30*, 10493–10506. [CrossRef] [PubMed]
33. Golomb, J.D.; Pulido, V.Z.; Albrecht, A.R.; Chun, M.M.; Mazer, J.A. Robustness of the retinotopic attentional trace after eye movements. *J. Vis.* **2010**, *10*, 1–12. [CrossRef] [PubMed]
34. Rizzolatti, G.; Riggio, L.; Dascola, I.; Umiltá, C. Reorienting attention across the horizontal and vertical meridians: Evidence in favor of a premotor theory of attention. *Neuropsychologia* **1987**, *25*, 31–40. [CrossRef]
35. Bindemann, M. Scene and screen center bias early eye movements in scene viewing. *Vis. Res.* **2010**, *50*, 2577–2587. [CrossRef] [PubMed]
36. Tatler, B.W. The central fixation bias in scene viewing: Selecting an optimal viewing position independently of motor biases and image feature distributions. *J. Vis.* **2007**, *7*, 4. [CrossRef] [PubMed]

37. Tatler, B.W.; Vincent, B.T. The prominence of behavioural biases in eye guidance. *Vis. Cogn.* **2009**, *17*, 1029–1054. [CrossRef]

38. Tseng, P.H.; Carmi, R.; Cameron, I.G.M.; Munoz, D.P.; Itti, L. Quantifying center bias of observers in free viewing of dynamic natural scenes. *J. Vis.* **2009**, *9*, 4. [CrossRef] [PubMed]

39. Rothkegel, L.O.M.; Trukenbrod, H.A.; Schütt, H.H.; Wichmann, F.A.; Engbert, R. Temporal evolution of the central fixation bias in scene viewing. *J. Vis.* **2017**, *17*, 3. [CrossRef]

40. Itti, L.; Koch, C. A saliency-based search mechanism for overt and covert shifts of visual attention. *Vis. Res.* **2000**, *40*, 1489–1506. [CrossRef]

41. Itti, L.; Koch, C.; Niebur, E. A model of saliency-based visual attention for rapid scene analysis. *IEEE Trans. Pattern Anal. Mach. Intell.* **1998**, *20*, 1254–1259. [CrossRef]

42. Mackworth, N.H.; Morandi, A.J. The gaze selects informative details within pictures. *Percept. Psychophys.* **1967**, *2*, 547–552. [CrossRef]

43. Borji, A.; Sihite, D.N.; Itti, L. Quantitative analysis of human-model agreement in visual saliency modeling: A comparative study. *IEEE Trans. Image Process.* **2013**, *22*, 55–69. [CrossRef]

44. Erdem, E.; Erdem, A. Visual saliency estimation by nonlinearly integrating features using region covariances. *J. Vis.* **2013**, *13*, 11. [CrossRef] [PubMed]

45. Frey, H.-P.; König, P.; Einhäuser, W. The role of first- and second-order stimulus features for human overt attention. *Percept. Psychophys.* **2007**, *69*, 153–161. [CrossRef] [PubMed]

46. Nuthmann, A.; Malcolm, G.L. Eye guidance during real-world scene search: The role color plays in central and peripheral vision. *J. Vis.* **2016**, *16*, 3. [CrossRef] [PubMed]

47. Bruce, N.D.B.; Wloka, C.; Frosst, N.; Rahman, S. On computational modeling of visual saliency: Examining what's right, and what's left. *Vis. Res.* **2015**, *116*, 95–112. [CrossRef] [PubMed]

48. Henderson, J.M.; Brockmole, J.R.; Castelhano, M.S.; Mack, M. Visual saliency does not account for eye movements during visual search in real-world scenes. In *Eye Movements: A Window on Mind and Brain*; van Gompel, R., Fischer, M., Murray, W.S., Hill, R.L., Eds.; Elsevier: Oxford, UK, 2007; pp. 537–562.

49. Tatler, B.; Hayhoe, M.; Land, M.; Ballard, D. Eye guidance in natural vision: Reinterpreting salience. *J. Vis.* **2011**, *11*, 5. [CrossRef] [PubMed]

50. Castelhano, M.S.; Mack, M.L.; Henderson, J.M. Viewing task influences eye movement control during active scene perception. *J. Vis.* **2009**, *9*, 6. [CrossRef] [PubMed]

51. Awh, E.; Belopolsky, A.V.; Theeuwes, J. Top-down versus bottom-up attentional control: A failed theoretical dichotomy. *Trends Cogn. Sci.* **2012**, *16*, 437–443. [CrossRef] [PubMed]

52. Bindemann, M.; Scheepers, C.; Burton, A.M. Viewpoint and center of gravity affect eye movements to human faces. *J. Vis.* **2009**, *9*, 7. [CrossRef]

53. Henderson, J.M.; Malcolm, G.L.; Schandl, C. Searching in the dark: Cognitive relevance drives attention in real-world scenes. *Psychon. Bull. Rev.* **2009**, *16*, 850–856. [CrossRef]

54. Nuthmann, A.; Henderson, J.M. Object-based attentional selection in scene viewing. *J. Vis.* **2010**, *10*, 20. [CrossRef]

55. Stoll, J.; Thrun, M.; Nuthmann, A.; Einhäuser, W. Overt attention in natural scenes: Objects dominate features. *Vis. Res.* **2015**, *107*, 36–48. [CrossRef] [PubMed]

56. Foulsham, T.; Kingstone, A. Fixation-dependent memory for natural scenes: An experimental test of scanpath theory. *J. Exp. Psychol.* **2013**, *142*, 41–56. [CrossRef] [PubMed]

57. Driver, J.; Davis, G.; Russell, C.; Turatto, M.; Freeman, E. Segmentation, attention and phenomenal visual objects. *Cognition* **2001**, *80*, 61–95. [CrossRef]

58. Henderson, J.; Chanceaux, M.; Smith, T.J. The influence of clutter on real-world scene search: Evidence from search efficiency and eye movements. *J. Vis.* **2009**, *9*, 32. [CrossRef] [PubMed]

59. Walther, D.; Koch, C. Modeling attention to salient proto-objects. *Neural Netw.* **2006**, *19*, 1395–1407. [CrossRef] [PubMed]

60. Wischnewski, M.; Belardinelli, A.; Schneider, W.X.; Steil, J.J. Where to look next? Combining static and dynamic proto-objects in a TVA-based model of visual attention. *Cognit. Comput.* **2010**, *2*, 326–343. [CrossRef]

61. Zelinsky, G.J.; Yu, C.-P. Clutter perception is invariant to image size. *Vis. Res.* **2015**, *116*, 142–151. [CrossRef]

62. Grill-Spector, K.; Kanwisher, N. Visual recognition. *Psychol. Sci.* **2005**, *16*, 152–160. [CrossRef]

63. Henderson, J.M.; Hayes, T.R. Meaning-based guidance of attention in scenes as revealed by meaning maps. *Nat. Hum. Behav.* **2017**, *1*, 743–747. [CrossRef]

64. Peacock, C.E.; Hayes, T.R.; Henderson, J.M. Meaning guides attention during scene viewing, even when it is irrelevant. *Atten. Percept. Psychophys.* **2019**, *81*, 20–34. [CrossRef]

65. Loftus, G.R.; Mackworth, N.H. Cognitive determinants of fixation location during picture viewing. *J. Exp. Psychol. Hum. Percept. Perform.* **1978**, *4*, 565–572. [CrossRef] [PubMed]

66. De Graef, P.; Christiaens, D.; D'Ydewalle, G.G.; De Graef, P.; Christiaens, D.; D'Ydewalle, G.G. Perceptual effects of scene context on object identification. *Psychol. Res.* **1990**, *52*, 317–329. [CrossRef] [PubMed]

67. Henderson, J.M.; Weeks, P.A.; Hollingworth, A. The effects of semantic consistency on eye movements during complex scene viewing. *J. Exp. Psychol. Hum. Percept. Perform.* **1999**, *25*, 210–228. [CrossRef]

68. Becker, M.W.; Pashler, H.; Lubin, J. Object-intrinsic oddities draw early saccades. *J. Exp. Psychol. Hum. Percept. Perform.* **2007**, *33*, 20–30. [CrossRef]

69. Underwood, G.; Humphreys, L.; Cross, E. Congruency, saliency and gist in the inspection of objects in natural scenes. *Eye Mov.* **2007**, *IV-VII*, 567–579. [CrossRef]

70. Castelhano, M.S.; Heaven, C. The relative contribution of scene context and target features to visual search in scenes. *Atten. Percept. Psychophys.* **2010**, *72*, 1283–1297. [CrossRef] [PubMed]

71. Võ, M.L.-H.; Henderson, J.M. Object-scene inconsistencies do not capture gaze: Evidence from the flash-preview moving-window paradigm. *Atten. Percept. Psychophys.* **2011**, *73*, 1742–1753. [CrossRef]

72. Võ, M.L.-H.; Henderson, J.M. Does gravity matter? Effects of semantic and syntactic inconsistencies on the allocation of attention during scene perception. *J. Vis.* **2009**, *9*, 24. [CrossRef]

73. LaPointe, M.R.P.; Milliken, B. Semantically incongruent objects attract eye gaze when viewing scenes for change. *Vis. Cogn.* **2016**, *24*, 63–77. [CrossRef]

74. De Groot, F.; Huettig, F.; Olivers, C.N.L. When meaning matters: The temporal dynamics of semantic influences on visual attention. *J. Exp. Psychol. Hum. Percept. Perform.* **2016**, *42*, 180–196. [CrossRef]

75. Spotorno, S.; Tatler, B.W.; Faure, S. Semantic consistency versus perceptual salience in visual scenes: Findings from change detection. *Acta Psychol. (Amst)* **2013**, *142*, 168–176. [CrossRef] [PubMed]

76. Vo, M.L.-H.; Wolfe, J.M. Differential electrophysiological signatures of semantic and syntactic scene processing. *Psychol. Sci.* **2013**, *24*, 1816–1823. [CrossRef] [PubMed]

77. Friedman, A. Framing pictures: The role of knowledge in automatized encoding and memory for gist. *J. Exp. Psychol. Gen.* **1979**, *108*, 316–355. [CrossRef] [PubMed]

78. Võ, M.L.-H.; Wolfe, J.M. The interplay of episodic and semantic memory in guiding repeated search in scenes. *Cognition* **2013**, *126*, 198–212. [CrossRef] [PubMed]

79. LaPointe, M.R.P.; Lupianez, J.; Milliken, B. Context congruency effects in change detection: Opposing effects on detection and identification. *Vis. Cogn.* **2013**, *21*, 99–122. [CrossRef]

80. Castelhano, M.S.; Henderson, J.M.J.M. Initial scene representations facilitate eye movement guidance in visual search. *J. Exp. Psychol. Hum. Percept. Perform.* **2007**, *33*, 753–763. [CrossRef] [PubMed]

81. Neider, M.B.; Zelinsky, G.J. Scene context guides eye movements during visual search. *Vis. Res.* **2006**, *46*, 614–621. [CrossRef]

82. Võ, M.L.-H.; Henderson, J.M. The time course of initial scene processing for eye movement guidance in natural scene search. *J. Vis.* **2010**, *10*, 14. [CrossRef]

83. Litchfield, D.; Donovan, T. Worth a quick look? Initial scene previews can guide eye movements as a function of domain-specific expertise but can also have unforeseen costs. *J. Exp. Psychol. Hum. Percept. Perform.* **2016**, *42*, 982–994. [CrossRef]

84. Hwang, A.D.; Wang, H.-C.; Pomplun, M. Semantic guidance of eye movements in real-world scenes. *Vis. Res.* **2011**, *51*, 1192–1205. [CrossRef]

85. Malcolm, G.L.; Henderson, J.M. The effects of target template specificity on visual search in real-world scenes: Evidence from eye movements. *J. Vis.* **2009**, *9*, 8. [CrossRef] [PubMed]

86. Mack, S.C.; Eckstein, M.P. Object co-occurrence serves as a contextual cue to guide and facilitate visual search in a natural viewing environment. *J. Vis.* **2011**, *11*, 9. [CrossRef] [PubMed]

87. Russell, B.C.; Torralba, A.; Murphy, K.P.; Freeman, W.T. Labelme: A database and web-based tool for image annotation. *Int. J. Comput.* **2008**, *77*, 157–173. [CrossRef]

88. Nuthmann, A. How do the regions of the visual field contribute to object search in real-world scenes? Evidence from eye movements. *J. Exp. Psychol. Hum. Percept. Perform.* **2014**, *40*, 342–360. [CrossRef] [PubMed]

89. Nuthmann, A. Fixation durations in scene viewing: Modeling the effects of local image features, oculomotor parameters, and task. *Psychon. Bull. Rev.* **2017**, *24*, 370–392. [CrossRef] [PubMed]

90. Walther, D.B.; Chai, B.; Caddigan, E.; Beck, D.M.; Fei-Fei, L. Simple line drawings suffice for functional MRI decoding of natural scene categories. *Proc. Natl. Acad. Sci. USA* **2011**, *108*, 9661–9666. [CrossRef]

91. O'Connell, T.P.; Walther, D.B. Dissociation of salience-driven and content-driven spatial attention to scene category with predictive decoding of gaze patterns. *J. Vis.* **2015**, *15*, 20. [CrossRef]

92. Hollingworth, A. Two forms of scene memory guide visual search: Memory for scene context and memory for the binding of target object to scene location. *Vis. Cogn.* **2009**, *17*, 273–291. [CrossRef]

93. Wolfe, J.M.; Võ, M.L.-H.; Evans, K.K.; Greene, M.R. Visual search in scenes involves selective and nonselective pathways. *Trends Cogn. Sci.* **2011**, *15*, 77–84. [CrossRef]

94. Biederman, I.; Glass, A.L.; Stacy, E.W. Searching for objects in real-world scenes. *J. Exp. Psychol.* **1973**, *97*, 22–27. [CrossRef]

95. Sanocki, T.; Epstein, W. Priming spatial layout of scenes. *Psychol. Sci.* **1997**, *8*, 374–378. [CrossRef]

96. Hillstrom, A.P.; Segabinazi, J.D.; Godwin, H.J.; Liversedge, S.P.; Benson, V. Cat and mouse search: The influence of scene and object analysis on eye movements when targets change locations during search. *Philos. Trans. R. Soc. Lond. B Biol. Sci.* **2017**, *372*, 20160106. [CrossRef] [PubMed]

97. Castelhano, M.S.; Pereira, E.J. Searching through the clutter: Using surface guidance framework to explore set size effects in scenes. *Manuscr. Prog.* **2019**, under review.

98. Pereira, E.J.; Castelhano, M.S. Attentional capture is contingent on scene region: Using surface guidance framework to explore attentional mechanisms during search. *Psychon. Bull. Rev.* **2019**, 1–9. [CrossRef] [PubMed]

99. Malcolm, G.L.; Shomstein, S. Object-based attention in real-world scenes. *J. Exp. Psychol. Gen.* **2015**, *144*, 257–263. [CrossRef] [PubMed]

100. Vatterott, D.B.; Vecera, S.P. The attentional window configures to object and surface boundaries. *Vis. Cogn.* **2015**, *23*, 561–576. [CrossRef]

101. Bonner, M.F.; Epstein, R.A. Coding of navigational affordances in the human visual system. *Proc. Natl. Acad. Sci. USA* **2017**, *114*, 4793–4798. [CrossRef]

102. Bonner, M.F.; Epstein, R.A. Computational mechanisms underlying cortical responses to the affordance properties of visual scenes. *PLoS Comput. Biol.* **2018**, *14*, e1006111. [CrossRef]

103. Man, L.L.Y.; Castelhano, M.S. Across the planes: Differing impacts of foreground and background information on visual search in scenes. *J. Vis.* **2018**, *18*, 384. [CrossRef]

104. DeAngelus, M.; Pelz, J.B. Top-down control of eye movements: Yarbus revisited. *Vis. Cogn.* **2009**, *17*, 790–811. [CrossRef]

105. Pannasch, S.; Schulz, J.; Velichkovsky, B.M. On the control of visual fixation durations in free viewing of complex images. *Atten. Percept. Psychophys.* **2011**, *73*, 1120–1132. [CrossRef] [PubMed]

106. Pannasch, S.; Velichkovsky, B.M. Distractor effect and saccade amplitudes: Further evidence on different modes of processing in free exploration of visual images. *Vis. Cogn.* **2009**, *17*, 1109–1131. [CrossRef]

107. Henderson, J.M.; Shinkareva, S.V.; Wang, J.; Luke, S.G.; Olejarczyk, J. Predicting cognitive state from eye movements. *PLoS ONE* **2013**, *8*, e64937.

108. Franke, I.S.; Pannasch, S.; Helmert, J.R.; Rieger, R.; Groh, R.; Velichkovsky, B.M. Towards attention-centered interfaces. *ACM Trans. Multimed. Comput. Commun. Appl.* **2008**, *4*, 1–13. [CrossRef]

109. Choe, K.W.; Kardan, O.; Kotabe, H.P.; Henderson, J.M.; Berman, M.G. To search or to like: Mapping fixations to differentiate two forms of incidental scene memory. *J. Vis.* **2017**, *17*, 8. [CrossRef] [PubMed]

110. Antes, J.R. The time course of picture viewing. *J. Exp. Psychol.* **1974**, *103*, 62–70. [CrossRef] [PubMed]

111. Castelhano, M.S.; Henderson, J.M. Incidental visual memory for objects in scenes. *Vis. Cogn.* **2005**, *12*, 1017–1040. [CrossRef]

112. Hollingworth, A. Scene and position specificity in visual memory for objects. *J. Exp. Psychol. Learn. Mem. Cogn.* **2006**, *32*, 58. [CrossRef]

113. Rensink, R.A. The dynamic representation of scenes. *Vis. Cogn.* **2000**, *7*, 17–42. [CrossRef]

114. Rensink, R.A.; O'Regan, J.K.; Clark, J.J. To see or not to see: The need for attention to perceive changes in scenes. *Psychol. Sci.* **1997**, *8*, 368–373. [CrossRef]

115. Konkle, T.; Brady, T.F.; Alvarez, G.A.; Oliva, A. Scene memory is more detailed than you think: The role of categories in visual long-term memory. *Psychol. Sci.* **2010**, *21*, 1551–1556. [CrossRef] [PubMed]

116. Kardan, O.; Berman, M.G.; Yourganov, G.; Schmidt, J.; Henderson, J.M. Classifying mental states from eye movements during scene viewing. *J. Exp. Psychol. Hum. Percept. Perform.* **2015**, *41*, 1502–1514. [CrossRef] [PubMed]

117. Mills, M.; Hollingworth, A.; Van der Stigchel, S.; Hoffman, L.; Dodd, M.D. Examining the influence of task set on eye movements and fixations. *J. Vis.* **2011**, *11*, 17. [CrossRef] [PubMed]

118. Subramanian, R.; Shankar, D.; Sebe, N.; Melcher, D. Emotion modulates eye movement patterns and subsequent memory for the gist and details of movie scenes. *J. Vis.* **2014**, *14*, 31. [CrossRef]

119. Võ, M.L.H.; Wolfe, J.M. When does repeated search in scenes involve memory? Looking at versus looking for objects in scenes. *J. Exp. Psychol. Hum. Percept. Perform.* **2012**, *38*, 23–41. [CrossRef] [PubMed]

120. Olejarczyk, J.H.; Luke, S.G.; Henderson, J.M. Incidental memory for parts of scenes from eye movements. *Vis. Cogn.* **2014**, *22*, 975–995. [CrossRef]

121. Draschkow, D.; Võ, M.L.-H. Of "what" and "where" in a natural search task: Active object handling supports object location memory beyond the object's identity. *Atten. Percept. Psychophys.* **2016**, *78*, 1574–1584. [CrossRef]

122. Võ, M.L.; Wolfe, J.M. The role of memory for visual search in scenes. *Ann. N. Y. Acad. Sci.* **2015**, *1339*, 72–81.

123. Williams, C.C. Incidental and intentional visual memory: What memories are and are not affected by encoding tasks? *Vis. Cogn.* **2010**, *18*, 1348–1367. [CrossRef]

124. Chun, M.M.; Jiang, Y. Contextual cueing: Implicit learning and memory of visual context guides spatial attention. *Cogn. Psychol.* **1998**, *36*, 28–71. [CrossRef]

125. Castelhano, M.S.; Fernandes, S.; Theriault, J. Examining the hierarchical nature of scene representations in memory. *J. Exp. Psychol. Learn. Mem. Cogn.* **2018**. [CrossRef] [PubMed]

126. Brockmole, J.R.; Castelhano, M.S.; Henderson, J.M. Contextual cueing in naturalistic scenes: Global and local contexts. *J. Exp. Psychol. Learn. Mem. Cogn.* **2006**, *32*, 699–706. [CrossRef] [PubMed]

127. Brockmole, J.R.; Henderson, J.M. Using real-world scenes as contextual cues for search. *Vis. Cogn.* **2006**, *13*, 99–108. [CrossRef]

128. Brockmole, J.R.; Võ, M.L.-H. Semantic memory for contextual regularities within and across scene categories: Evidence from eye movements. *Atten. Percept. Psychophys.* **2010**, *72*, 1803–1813. [CrossRef] [PubMed]

129. Josephs, E.L.; Draschkow, D.; Wolfe, J.M.; Võ, M.L.-H. Gist in time: Scene semantics and structure enhance recall of searched objects. *Acta Psychol. (Amst)* **2016**, *169*, 100–108. [CrossRef] [PubMed]

130. Hollingworth, A. Task specificity and the influence of memory on visual search: Comment on Võ and Wolfe (2012). *J. Exp. Psychol.* **2012**, *38*, 1596–1603. [CrossRef] [PubMed]

131. Võ, M.L.-H.; Zwickel, J.; Schneider, W.X. Has someone moved my plate? The immediate and persistent effects of object location changes on gaze allocation during natural scene viewing. *Atten. Percept. Psychophys.* **2010**, *72*, 1251–1255. [CrossRef] [PubMed]

132. Hollingworth, A.; Williams, C.C.; Henderson, J.M. To see and remember: Visually specific information is retained in memory from previously attended objects in natural scenes. *Psychon. Bull. Rev.* **2001**, *8*, 761–768. [CrossRef] [PubMed]

133. Henderson, J.M.; Williams, C.C.; Castelhano, M.S.; Falk, R.J. Eye movements and picture processing during recognition. *Percept. Psychophys.* **2003**, *65*, 725–734. [CrossRef] [PubMed]

134. Noton, D.; Stark, L. Scanpaths in saccadic eye movements while viewing and recognizing patterns. *Vis. Res.* **1971**, *11*, 929–942. [CrossRef]

135. Johansson, R.; Johansson, M. Look here, eye movements play a functional role in memory retrieval. *Psychol. Sci.* **2014**, *25*, 236–242. [CrossRef] [PubMed]

136. Laeng, B.; Bloem, I.M.; D'Ascenzo, S.; Tommasi, L. Scrutinizing visual images: The role of gaze in mental imagery and memory. *Cognition* **2014**, *131*, 263–283. [CrossRef] [PubMed]

137. Wynn, J.S.; Bone, M.B.; Dragan, M.C.; Hoffman, K.L.; Buchsbaum, B.R.; Ryan, J.D. Selective scanpath repetition during memory-guided visual search. *Vis. Cogn.* **2016**, *24*, 15–37. [CrossRef] [PubMed]

138. Bochynska, A.; Laeng, B. Tracking down the path of memory: Eye scanpaths facilitate retrieval of visuospatial information. *Cogn. Process.* **2015**, *16*, 159–163. [CrossRef] [PubMed]

139. Henderson, J.M.; Williams, C.C.; Falk, R.J. Eye movements are functional during face learning. *Mem. Cognit.* **2005**, *33*, 98–106. [CrossRef] [PubMed]

140. James, W. *The Principles of Psychology*; Henry Holt and Company: New York, NY, USA, 1890.

141. Hayhoe, M. Vision using routines: A functional account of vision. *Vis. Cogn.* **2000**, *7*, 43–64. [CrossRef]

142. Hayhoe, M.; Ballard, D. Eye movements in natural behavior. *Trends Cogn. Sci.* **2005**, *9*, 188–194. [CrossRef] [PubMed]
143. Land, M.F.; McLeod, P. From eye movements to actions: How batsmen hit the ball. *Nat. Neurosci.* **2000**, *3*, 1340–1345. [CrossRef] [PubMed]
144. Schneider, E.; Villgrattner, T.; Vockeroth, J.; Bartl, K.; Kohlbecher, S.; Bardins, S.; Ulbrich, H.; Brandt, T. EyeSeeCam: An eye movement-driven head camera for the examination of natural visual exploration. *Ann. N. Y. Acad. Sci.* **2009**, *1164*, 461–467. [CrossRef]
145. Dorr, M.; Martinetz, T.; Gegenfurtner, K.R.; Barth, E. Variability of eye movements when viewing dynamic natural scenes. *J. Vis.* **2010**, *10*, 28. [CrossRef]
146. Mital, P.K.; Smith, T.J.; Hill, R.L.; Henderson, J.M. Clustering of gaze during dynamic scene viewing is predicted by motion. *Cognit. Comput.* **2011**, *3*, 5–24. [CrossRef]
147. Marius 't Hart, B.; Vockeroth, J.; Schumann, F.; Bartl, K.; Schneider, E.; König, P.; Einhäuser, W. Gaze allocation in natural stimuli: Comparing free exploration to head-fixed viewing conditions. *Vis. Cogn.* **2009**, *17*, 1132–1158. [CrossRef]
148. Hinde, S.J.; Smith, T.J.; Gilchrist, I.D. In search of oculomotor capture during film viewing: Implications for the balance of top-down and bottom-up control in the saccadic system. *Vis. Res.* **2017**, *134*, 7–17. [CrossRef] [PubMed]
149. Goldstein, R.B.; Woods, R.L.; Peli, E. Where people look when watching movies: Do all viewers look at the same place? *Comput. Biol. Med.* **2007**, *37*, 957–964. [CrossRef] [PubMed]
150. Loschky, L.C.; Larson, A.M.; Magliano, J.P.; Smith, T.J. What would jaws do? The tyranny of film and the relationship between gaze and higher-level narrative film comprehension. *PLoS ONE* **2015**, *10*, e0142474. [CrossRef] [PubMed]
151. Smith, T.J.; Mital, P.K. Attentional synchrony and the influence of viewing task on gaze behavior in static and dynamic scenes. *J. Vis.* **2013**, *13*, 16. [CrossRef] [PubMed]
152. Foulsham, T.; Kingstone, A. Are fixations in static natural scenes a useful predictor of attention in the real world? *Can. J. Exp. Psychol. Can. Psychol. Exp.* **2017**, *71*, 172–181. [CrossRef]
153. Potter, M.C. Short-term conceptual memory for pictures. *J. Exp. Psychol. Hum. Learn. Mem.* **1976**, *2*, 509–522. [CrossRef]
154. Titchener, E.B. The postulates of a structural psychology. *Philos. Rev.* **1898**, *7*, 449. [CrossRef]
155. Wertheimer, M. Untersuchungen zur lehre von der gestalt. II. *Psychol. Res.* **1923**, *4*, 301–350. [CrossRef]

Review

Meaning and Attentional Guidance in Scenes: A Review of the Meaning Map Approach

John M. Henderson [1,2,*], Taylor R. Hayes [1], Candace E. Peacock [1,2] and Gwendolyn Rehrig [2]

[1]   Center for Mind and Brain, 267 Cousteau Place, University of California, Davis, CA 95618, USA;
      trhayes@ucdavis.edu (T.R.H.); cepeacock@ucdavis.edu (C.E.P.)
[2]   Department of Psychology, University of California, Davis, CA 95618, USA; glrehrig@ucdavis.edu
*   Correspondence: johnhenderson@ucdavis.edu; Tel.: +1-530-754-455

Received: 27 February 2019; Accepted: 7 May 2019; Published: 10 May 2019

Abstract: Perception of a complex visual scene requires that important regions be prioritized and attentionally selected for processing. What is the basis for this selection? Although much research has focused on image salience as an important factor guiding attention, relatively little work has focused on semantic salience. To address this imbalance, we have recently developed a new method for measuring, representing, and evaluating the role of meaning in scenes. In this method, the spatial distribution of semantic features in a scene is represented as a meaning map. Meaning maps are generated from crowd-sourced responses given by naïve subjects who rate the meaningfulness of a large number of scene patches drawn from each scene. Meaning maps are coded in the same format as traditional image saliency maps, and therefore both types of maps can be directly evaluated against each other and against maps of the spatial distribution of attention derived from viewers' eye fixations. In this review we describe our work focusing on comparing the influences of meaning and image salience on attentional guidance in real-world scenes across a variety of viewing tasks that we have investigated, including memorization, aesthetic judgment, scene description, and saliency search and judgment. Overall, we have found that both meaning and salience predict the spatial distribution of attention in a scene, but that when the correlation between meaning and salience is statistically controlled, only meaning uniquely accounts for variance in attention.

Keywords: attention; scene perception; eye movements

1. Introduction

The world contains an enormous amount of visual information, but human vision and visual cognition are severely limited in their processing capacity: Only a small fraction of the latent information can be analyzed at any given moment. Efficient visual cognition therefore requires selecting the information that is most relevant at the present moment for understanding and acting on the world. The primary way in which this selection takes place in natural vision is with overt attention via eye movements [1–9], as shown in Figure 1. Close or direct fixation of the scene region containing relevant information is typically necessary to perceive its visual details, to unambiguously determine its identity and meaning, and to encode it into short- and long-term memory. That is, what we see and understand about the world is determined by where we look [10].

Given the importance of eye movements for perception and cognition, a critical issue concerns understanding the representations and processes that guide the eyes through a visual scene in real time to support perception, cognition, and behavior [11]. Models based on image salience have provided an influential approach to eye movement guidance in scene perception. For static images, these models propose that attention is controlled by contrasts in primitive image features such as luminance, color, and edge orientation [12–14]. A key concept is the saliency map, which is generated by salience over the primitive features. Attention is then assumed to be captured or "pulled" to the most visually salient scene regions represented by the saliency map [15–21]. Because the primitive features are semantically uninterpreted, scene regions are prioritized for attentional selection based on image properties alone. The appeal of this type of image salience model is that visual salience is both neurobiologically plausible in the sense that the visual system is known to compute the assumed primitive features, and computationally tractable in the sense that working models have been implemented that generate image salience from these features [4].

Figure 1. Scan pattern of a single viewer freely viewing a real-world scene.

We note that "salience" has different interpretations in different literatures, and so we want to be clear about which interpretation we are using. In vision science and attention research, the term is typically reserved for models in the Koch and Ullman tradition based on the idea that the human visual system computes difference maps from basic image features that are then combined and used to guide attention [19]. On the other hand, in computer vision and image processing, models that predict attention regardless of the underlying processes are also sometimes referred to as saliency models. This difference in usage can lead to confusion. For our purposes here, we specifically focus on image salience in the Koch and Ullman tradition, and to eliminate ambiguity, we use the terms "image salience" and "salience" to refer to that concept.

In contrast to models based on image salience, cognitive guidance models emphasize the important role of scene semantics in directing attention in scenes. In this view, attention is "pushed" by the cognitive system to scene regions that are semantically informative and cognitively relevant [10]. Cognitive guidance is consistent with evidence suggesting that viewers attend to semantically informative regions of a scene [5,6,22–25], as well as scene regions that are task-relevant [6,26–34]. For example, according to the cognitive relevance model [35,36], the attentional priority assigned to a scene region is based on its inherent meaning (e.g., "cup") as well as its meaning with respect to the scene (e.g., "a cup in an office") and the goals of the viewer (e.g., "I am looking for something I can put water into"). In the cognitive relevance model, the scene image is of course important:

It is the input that enables these higher-level semantic processes and interpretations. Additionally, the scene image is the input for forming a map of potential attention targets. However, for this model, the critical hypothesis is that the image-based parse generates a "flat" (that is, unranked) landscape of potential targets rather than a landscape ranked by image salience. Attentional priority is then based on the detected and predicted informativeness (i.e., meaning) of the scene regions and objects in that landscape [3,4].

Until recently, it has been difficult to compare directly the influence of image salience and meaning on attentional guidance in scenes, because to do so requires representing both of them in a format that allows for comparable quantitative predictions of attentional distribution over the scene. Saliency map models naturally provide this type of prediction [17,18,20,21,30,37]. Unfortunately, it is far more difficult to create a computational model of meaning than it is to create a computational model of image salience, a likely reason that saliency models have been so popular [4,10]. Given this difficulty, studies of meaning-based models of attention in scenes have typically focused on manipulations of one or at most a small number of specific scene regions or objects [22,38–41]. However, these types of manipulations do not allow a direct comparison of image salience and semantic informativeness across the entire scene.

The issue we have recently pursued, then, is this: How can we generate and represent the spatial distribution of semantic informativeness over a scene in a format that supports direct comparison with a saliency map? To address this issue, we introduced a new approach based on meaning maps [42]. Meaning maps were inspired by two classic scene viewing studies [23,24]. The central idea of a meaning map is that it represents the spatial distribution of semantic informativeness over a scene in the same format as a saliency map represents the spatial distribution of image salience. To create meaning maps, we use crowd-sourced responses given by large numbers of naïve subjects who rate the meaningfulness of scene patches. Specifically, photographs of real environments are divided into dense arrays of objectively defined circular overlapping patches at two spatial scales (Figure 2). The two scales and numbers of patches are chosen based on simulations showing that we can recover ground truth visual properties of scenes from them [42]. Large numbers of workers on Mechanical Turk each rate a randomly selected subset of individually presented patches taken from the set of scenes to be rated. We then construct meaning maps for each scene by averaging these ratings by pixel over patches and raters and smoothing the results (Figure 3). Like image salience, meaning is spatially distributed non-uniformly across scenes, with some scene regions relatively rich in semantic informativeness and other regions relatively sparse. Meaning maps represent this spatial distribution pixel by pixel, and so offer a foundation for directly comparing the relative roles of meaning and image salience on attentional guidance.

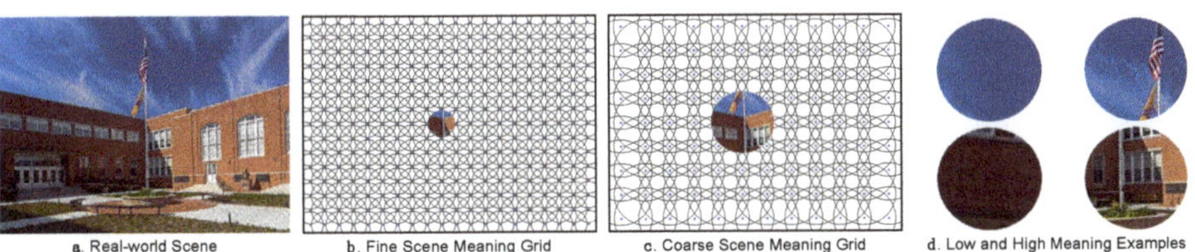

Figure 2. (**a**) A real-world scene; (**b**) fine scale, and (**c**) coarse scale patches from the patch grids; (**d**) examples of patches rated low (left column) and high (right column) in meaning.

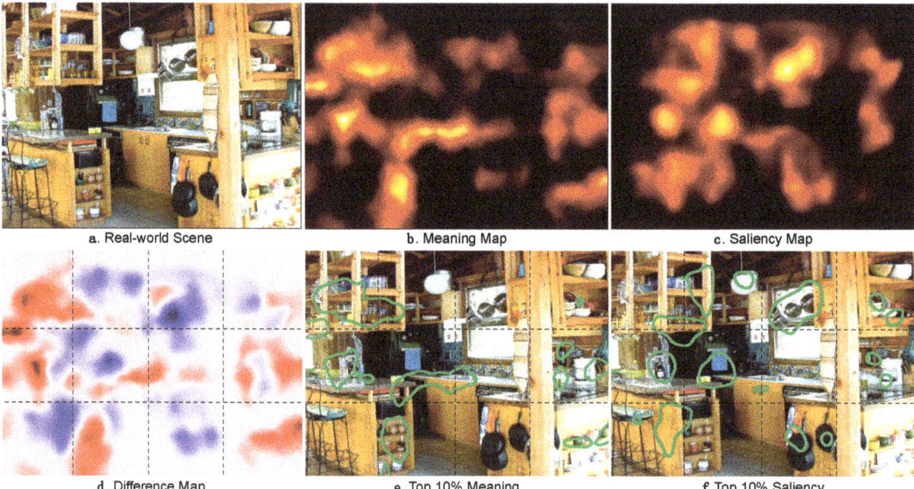

Figure 3. (**a**) Real-world scene; (**b**) scene's meaning map; (**c**) saliency map; (**d**) difference map showing regions that are more meaningful (red) and salient (blue); (**e**) regions in the top 10% of meaning values; (**f**) regions in the top 10% of saliency values.

In sum, meaning maps provide a conceptual analog of saliency maps by representing the spatial distribution of semantic features associated with informativeness across a scene. Meaning maps generate predictions concerning attentional guidance that can be tested using the same methods that have been used to test predictions from saliency maps. This allows contrasting predictions from semantic informativeness and image salience to be directly compared [42].

2. Review of Results

Meaning maps and saliency maps serve as predictions of how attention will be guided through scenes. The key empirical question is how well these predictions accord with observed distributions of attention produced by people viewing scenes.

In a first study directed toward this question, we asked subjects to view a set of scenes for 12 s each while their eye movements were recorded [42]. Subjects viewed the scenes to prepare for memory and aesthetic judgment questions that were presented after viewing. We operationalized attention maps as fixation density maps that reflected the spatial distribution of eye fixations across the scene. Importantly, the attention maps represented attention in the same format and on the same scale as the meaning and saliency maps. We then contrasted the degree to which the spatial distribution of meaning and salience predicted viewers' attention over the scenes. The results showed that both meaning and salience predicted attention, but that when the association between meaning and salience was statistically controlled with semi-partial correlations, only meaning uniquely accounted for attentional variance. Furthermore, the influence of meaning was observed both at the very beginning of scene viewing and throughout the trial. Figure 4 summarizes the data from this study. This result was observed in both scene memorization and aesthetic judgment viewing tasks. Given the strong observed correlation between meaning and salience, and the finding that only meaning accounted for variance in attention when that correlation was controlled, we concluded that the existing data are consistent with a theory in which meaning is the main factor guiding attention through scenes.

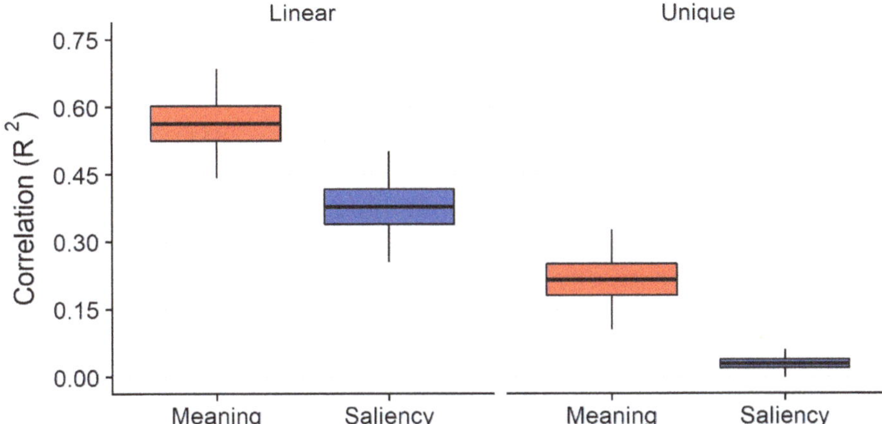

Figure 4. Squared linear correlation and semi-partial (unique) correlation for meaning and image salience across 40 scenes from a scene memorization task in Henderson and Hayes (2017) [42]. Box plots show the grand mean (black horizontal line), 95% confidence intervals (colored box), and one standard deviation (black vertical line).

In our initial study, we focused on the spatial distribution of attention by measuring the locations of fixations [42]. However, fixations also differ by duration, and fixation duration is known to reflect ongoing visual and cognitive activity [43–49]. Therefore, in a reanalysis of the data from Henderson and Hayes (2017), we examined the relative influences of semantic features and image salience on the distribution of attention taking into account attentional dwell time at each location [42,50]. This was accomplished by creating attention maps that weighted each fixation by its duration. Again, the question was whether these duration-weighted attention maps would be best accounted for by meaning maps or saliency maps. Using the same analysis methods with these duration-weighted attention maps, we replicated all of the basic data patterns that we observed in the original study. Specifically, we found that both meaning and salience were associated with attention, but that when the correlation between meaning and salience was statistically controlled, only meaning accounted for variance in attention. Once again, the influence of meaning was observed both at the beginning of scene viewing and throughout the trial. Therefore, whether the measure of attention is based on location only or includes dwell time, the answer with respect to meaning and image salience is the same [50].

In our initial experiments demonstrating the advantage of meaning maps over saliency maps, subjects viewed scenes in order to prepare for memory and aesthetic judgment questions that were presented after viewing [42,50]. In those tasks, the responses were off-line with respect to scene perception, so viewers may not have guided attention as quickly or under as much control as they might in a task requiring real-time responses during viewing. Therefore, in the next study we investigated how well meaning and image salience account for attention when the subject is actively engaged with and responding to the scene continuously in real time, and the guidance of attention is directly relevant to the real-time task. For this purpose, we used two scene description tasks [51].

We drew on evidence that language production is incremental in the sense that speakers interweave planning and speaking instead of planning the entire production and then executing the plan [52]. That is, speakers typically plan and produce small units of speech (words and phrases) that are tied to each scene region that they fixate. Scene description therefore allows us to examine the relative influences of semantic information and image salience under conditions in which on-line changes in attention to specific scene regions are functional and necessary. We used this basic paradigm in two experiments. In one experiment, subjects described what someone might do in each scene, and in a second experiment, they simply described each scene. In both experiments, subjects were asked to begin their description when the scene appeared and to continue speaking for 30 s of scene presentation. Their eyes were tracked throughout each trial. The main result was that, once again, both meaning and salience were associated with the spatial distribution of attention, but when the correlation between meaning and salience was statistically controlled, only meaning accounted for variance in attention. This basic result was seen in both experiments. Therefore, once again we found no evidence for a unique influence of image salience on attention that could not be explained by the relationship between image salience and meaning, whereas the unique influence of meaning could not be attributed to salience.

So far, across all four tasks we have described (memorization, aesthetic judgment, scene description, and action description), meaning was highly relevant. Perhaps for tasks in which image salience is necessary and meaning is irrelevant, salience would better predict attention. To test this hypothesis, in a final set of experiments we examined the role of meaning and salience in two experiments using tasks for which meaning was completely irrelevant and saliency was critical: a brightness rating task in which participants rated each scene for its overall brightness, and a brightness search task in which participants counted the number of bright patches in each scene [53]. If meaning was used to guide attention in the previous tasks because those tasks emphasized the semantic content of the scenes, then the relationship between meaning and attention should no longer hold in the tasks that focus on the image itself. On the other hand, if the use of meaning to guide attention is a fundamental property of the operation of the attention system when faced with real-world scenes, then we should continue to see a relationship between meaning and attention even if meaning is irrelevant. Using the same methods as the previous studies, the striking finding was that in both the brightness rating and brightness search tasks, the results were very similar to the prior experiments: When the correlation between meaning and salience was controlled, only meaning uniquely accounted for significant variance in attention. These results showed that the relationship between meaning and attention is not restricted to viewing tasks that require the viewer to analyze meaning. The results support theories in which scene semantics play a central role in attentional guidance in scenes.

3. Discussion

We have reviewed the meaning map approach and shown how we have used it to investigate the relative roles of semantic informativeness and image salience on attentional guidance. Our results strongly suggest a fundamental and mandatory role for meaning in attentional guidance in real-world scenes. These results are consistent with a growing understanding that both overt and covert visual attention are often under the influence of the meaning of the visual stimulus, even when that meaning seems irrelevant to the task. Examples of this type of semantic effect are influences of object meaning on eye movements in the visual world paradigm [54], and influences of semantic object relationships on covert attention [55]. Furthermore, we found that the observed relationship between meaning and attention was about as strong as it could be given the variability in attention maps across subjects [42]. In the remainder of this section we consider and discuss some additional related issues.

First, although it has sometimes been proposed that image salience and semantic content are likely to be correlated in scenes, it has been difficult to test this hypothesis directly. The meaning map approach provides such a method. Further, as we have shown across several studies, that correlation is robust. An important implication of this finding is that previous results demonstrating a relationship between saliency maps and attention cannot be taken as evidence for a functional role of salience in guiding attention. Indeed, as we have reviewed above, essentially the entire relationship between image salience and attention can be attributed to the association between image salience and semantic content.

Second, it is important to be clear that meaning maps are not a theory of scene semantics. They are simply a method for tapping into and representing human judgments concerning the relative informativeness of scene regions continuously over space. Meaning maps provide an operational definition of the spatial distribution of meaning that can be quantified, but they do not offer direct insight into the nature of scene semantics or how meaning is represented in the mind and brain.

That said, it may be that the meaning map approach can be used as a tool for beginning to get a handle on the nature of scene semantics. For example, the type of scene meaning we have investigated so far has been based on ratings of scene patches that were presented to raters independently of the scenes from which the patches were taken. These experiments therefore focus on the role of what we call scene-intrinsic context-free meaning on attention [51]. We might collect ratings using other types of questions that could provide insight into other types of semantic representations. For example, we could compare context-free to contextualized meaning in which the degree of meaning associated with a scene region is based on its relationship to the scene as a whole rather than on its own. Similarly, we might consider goal-related meaning, in which the meaning of a scene region is based on its relationship to the viewer's goals rather than intrinsic to the scene itself. Using judiciously chosen ratings, we might be able to unravel how different types of meaning are related to each other and to performance over different perceptual and cognitive tasks. The meaning map approach provides a method for pursuing these questions. Relatedly, the meaning maps we have investigated so far based on context-free ratings may not be the type of meaning most associated with attention, and we may therefore be underestimating the relationship between semantic features and attention.

One advantage of saliency maps over meaning maps is that the former are image computable: They can be derived automatically and without human intervention from computational models. In comparison, meaning maps are not image computable and require human raters. For this reason, one might suggest that saliency models are to be preferred as an explanation of human attention. From an engineering perspective, this view has merit. However, from a vision and cognitive science perspective, it does not. In our view, the interesting psychological claim of the image salience hypothesis is that human attention in real-world scenes is driven by contrasts in basic image features. This claim has been supported by a large number of experiments showing a correlation between saliency maps and human fixations. The alternative hypothesis we are pursuing is that image salience effects are actually disguised meaning effects, because image features and semantic features in scenes are correlated [42]. Indeed, this is what we find, with very little if any unique variance accounted for by salience once the variance accounted for by semantic features is controlled. This comparison is not one of a computational model versus human ratings, but one of two competing psychological theories. That is, we are concerned with psychological principles here, not modeling. There is no logical requirement that testing this (or any) psychological hypothesis requires image computable semantic features. Furthermore, until we have a computational model of the entire semantic system (which is clearly a long way off), there is no other way to go about comparing salience to semantics. Saliency models have been influential because they have been the only game in town [4,10]. Meaning maps provide an alternative game, but their creation does require human judgment. From our perspective, this approach is similar to the approach that uses human labeling to parse and label objects in scenes as in the labelMe database [56] to produce object ground truth. Of course, it would be very interesting to use the ground truth represented by meaning maps to try to train a model to find meaningful regions (and indeed we are pursuing this idea), but that is not necessary for testing the theories at stake.

Relatedly, it has been argued that Koch and Ullman-inspired saliency models like GBVS are no longer state of the art, but instead have been replaced by a newer class of models based on deep neural networks (DNNs) that have recently been found to predict human attention quite well [57]. Given this, one might ask why we should take standard saliency models as the baseline for comparison to meaning. In our view, although these DNN models are impressive from the perspective of pushing the boundaries of deep learning and big data, it is not clear at this point how much they have to say about active biological vision. For example, DNN models are trained on fixations over one set of scenes and then predict fixations on another set of scenes. It is clear that humans do not learn where to fixate based on supervised learning from the fixations of others or by ingesting large amounts of external fixation data. It is also unclear whether the mechanisms used by DNNs to predict attention operate on the same principles that are used by the human brain. For these reasons, although we watch developments in this field with great interest, we are not yet sure how their successes and failures should be interpreted from the perspective of human attentional processes.

Author Contributions: J.H. wrote the first draft and all authors edited and contributed to later drafts. All authors contributed to the conceptualization, implementation, and analysis of subsets of the experiments reviewed here.

Funding: This research was supported by the National Eye Institute of the National Institutes of Health under award number R01EY027792. The content is solely the responsibility of the authors and does not necessarily represent the official views of the National Institutes of Health.

Acknowledgments: We thank the members of the UC Davis visual cognition group, and Carrick Williams and two anonymous reviewers, for their helpful feedback on this work. We also thank Brad Wyble, Tom Wallis, and others for a stimulating Twitter discussion of the issues raised here.

Conflicts of Interest: The authors declare no conflicts of interest.

References

1. Land, M.F.; Hayhoe, M.M. In what ways do eye movements contribute to everyday activities? *Vis. Res.* **2001**, *41*, 3559–3565. [CrossRef]
2. Hayhoe, M.M.; Ballard, D. Eye movements in natural behavior. *Trends Cogn. Sci.* **2005**, *9*, 188–194. [CrossRef]
3. Henderson, J.M. Human gaze control during real-world scene perception. *Trends Cogn. Sci.* **2003**, *7*, 498–504. [CrossRef] [PubMed]
4. Henderson, J.M. Gaze Control as Prediction. *Trends Cogn. Sci.* **2017**, *21*, 15–23. [CrossRef]
5. Buswell, G.T. *How People Look at Pictures*; University of Chicago Press: Chicago, IL, USA, 1935.
6. Yarbus, A.L. *Eye Movements and Vision*; Plenum Press: New York, NY, USA, 1967; ISBN 0306302985.
7. Henderson, J.M.; Hollingworth, A. High-level scene perception. *Ann. Rev. Psychol.* **1999**, *50*, 243–271. [CrossRef] [PubMed]
8. Rayner, K. The 35th Sir Frederick Bartlett Lecture: Eye movements and attention in reading, scene perception, and visual search. *Q. J. Exp. Psychol.* **2009**, *62*, 1457–1506. [CrossRef]
9. Liversedge, S.P.; Findlay, J.M. Saccadic eye movements and cognition. *Trends Cogn. Sci.* **2000**, *4*, 6–14. [CrossRef]
10. Henderson, J.M. Regarding scenes. *Curr. Dir. Psychol. Sci.* **2007**, *16*, 219–222. [CrossRef]
11. Henderson, J.M. Eye movements and scene perception. In *The Oxford Handbook of Eye Movements*; Liversedge, S.P., Gilchrist, I.D., Everling, S., Eds.; Oxford University Press: New York, NY, USA, 2011; pp. 593–606.
12. Treisman, A.M.; Gelade, G. A Feature-Integration Theory of Attention. *Cogn. Psychol.* **1980**, *12*, 97–136. [CrossRef]
13. Wolfe, J.M. Guided Search 2.0. A revised model of visual search. *Psychon. Bull. Rev.* **1994**, *1*, 202–238. [CrossRef] [PubMed]
14. Wolfe, J.M.; Horowitz, T.S. Five factors that guide attention in visual search. *Nat. Hum. Behav.* **2017**, *1*, 1–8. [CrossRef]
15. Borji, A.; Parks, D.; Itti, L. Complementary effects of gaze direction and early saliency in guiding fixations during free viewing. *J. Vis.* **2014**, *14*, 3. [CrossRef]
16. Borji, A.; Sihite, D.N.; Itti, L. Quantitative analysis of human-model agreement in visual saliency modeling: A comparative study. *IEEE Trans. Image Proc.* **2013**, *22*, 55–69. [CrossRef]
17. Harel, J.; Koch, C.; Perona, P. Graph-Based Visual Saliency. *Adv. Neural Inf. Proc. Syst.* **2006**, 1–8.
18. Itti, L.; Koch, C. Computational modelling of visual attention. *Nat. Rev. Neurosci.* **2001**, *2*, 194–203. [CrossRef]
19. Koch, C.; Ullman, S. Shifts in Selective Visual Attention: Towards the Underlying Neural Circuitry. *Hum. Neurobiol.* **1985**, *4*, 219–227.
20. Itti, L.; Koch, C.; Niebur, E. A model of saliency-based visual attention for rapid scene analysis. *IEEE Trans. Pattern Anal. Mach. Intell.* **1998**, *20*, 1254–1259. [CrossRef]
21. Parkhurst, D.; Law, K.; Niebur, E. Modeling the role of salience in the allocation of overt visual attention. *Vis. Res.* **2002**, *42*, 107–123. [CrossRef]
22. Loftus, G.R.; Mackworth, N.H. Cognitive determinants of fixation location during picture viewing. *J. Exp. Psychol.* **1978**, *4*, 565–572. [CrossRef]
23. Antes, J.R. The time course of picture viewing. *J. Exp. Psychol.* **1974**, *103*, 62–70. [CrossRef]

24. Mackworth, N.H.; Morandi, A.J. The gaze selects informative details within pictures. *Percept. Psychophy.* **1967**, *2*, 547–552. [CrossRef]
25. Wu, C.C.; Wick, F.A.; Pomplun, M. Guidance of visual attention by semantic information in real-world scenes. *Front. Psychol.* **2014**, *5*, 1–13. [CrossRef] [PubMed]
26. Tatler, B.W.; Hayhoe, M.M.; Land, M.F.; Ballard, D.H. Eye guidance in natural vision: Reinterpreting salience. *J. Vis.* **2011**, *11*, 5. [CrossRef] [PubMed]
27. Rothkopf, C.A.; Ballard, D.H.; Hayhoe, M.M. Task and context determine where you look. *J. Vis.* **2007**, *7*, 16.1-20. [CrossRef] [PubMed]
28. Hayhoe, M.M.; Ballard, D. Modeling Task Control of Eye Movements Minireview. *Curr. Biol.* **2014**, *24*, R622–R628. [CrossRef]
29. Einhäuser, W.; Rutishauser, U.; Koch, C. Task-demands can immediately reverse the effects of sensory-driven saliency in complex visual stimuli. *J. Vis.* **2008**, *8*, 2.1-19. [CrossRef] [PubMed]
30. Torralba, A.; Oliva, A.; Castelhano, M.S.; Henderson, J.M. Contextual guidance of eye movements and attention in real-world scenes: The role of global features in object search. *Psychol. Rev.* **2006**, *113*, 766–786. [CrossRef] [PubMed]
31. Castelhano, M.S.; Mack, M.L.; Henderson, J.M. Viewing task influences eye movement control during active scene perception. *J. Vis.* **2009**, *9*, 6.1-15. [CrossRef] [PubMed]
32. Neider, M.B.; Zelinsky, G. Scene context guides eye movements during visual search. *Vis. Res.* **2006**, *46*, 614–621. [CrossRef] [PubMed]
33. Turano, K.A.; Geruschat, D.R.; Baker, F.H. Oculomotor strategies for the direction of gaze tested with a real-world activity. *Vis. Res.* **2003**, *43*, 333–346. [CrossRef]
34. Foulsham, T.; Underwood, G. How does the purpose of inspection influence the potency of visual salience in scene perception? *Perception* **2007**, *36*, 1123–1138. [CrossRef]
35. Henderson, J.M.; Brockmole, J.R.; Castelhano, M.S.; Mack, M. Visual saliency does not account for eye movements during visual search in real-world scenes. In *Eye Movements: A Window on Mind and Brain*; Van Gompel, R.P.G., Fischer, M.H., Murray, W.S., Hill, R.L., Eds.; Elsevier Ltd.: Oxford, UK, 2007; pp. 537–562, ISBN 9780080449807.
36. Henderson, J.M.; Malcolm, G.L.; Schandl, C. Searching in the dark: Cognitive relevance drives attention in real-world scenes. *Psychon. Bull. Rev.* **2009**, *16*, 850–856. [CrossRef]
37. Borji, A.; Itti, L. State-of-the-art in visual attention modeling. *IEEE Trans. Pattern Anal. Mach. Intell.* **2013**, *35*, 185–207. [CrossRef]
38. De Graef, P.; Christiaens, D.; d'Ydewalle, G. Perceptual effects of scene context on object identification. *Psychol. Res.* **1990**, *52*, 317–329. [CrossRef]
39. Henderson, J.M.; Weeks, P.A., Jr.; Hollingworth, A. The effects of semantic consistency on eye movements during complex scene viewing. *J. Exp. Psychol.* **1999**, *25*, 210–228. [CrossRef]
40. Võ, M.L.H.; Henderson, J.M. Does gravity matter? Effects of semantic and syntactic inconsistencies on the allocation of attention during scene perception. *J. Vis.* **2009**, *9*, 1–15. [CrossRef]
41. Brockmole, J.R.; Henderson, J.M. Prioritizing new objects for eye fixation in real-world scenes: Effects of object-scene consistency. *Vis. Cogn.* **2008**, *16*, 375–390. [CrossRef]
42. Henderson, J.M.; Hayes, T.R. Meaning-based guidance of attention in scenes as revealed by meaning maps. *Nat. Hum. Behav.* **2017**, *1*, 743–747. [CrossRef]
43. Henderson, J.M.; Pierce, G.L. Eye movements during scene viewing: Evidence for mixed control of fixation durations. *Psychon. Bull. Rev.* **2008**, *15*, 566–573. [CrossRef]
44. Nuthmann, A.; Smith, T.J.; Engbert, R.; Henderson, J.M. CRISP: A computational model of fixation durations in scene viewing. *Psychol. Rev.* **2010**, *117*, 382–405. [CrossRef]
45. Henderson, J.M.; Smith, T.J. How are eye fixation durations controlled during scene viewing? Further evidence from a scene onset delay paradigm. *Vis. Cogn.* **2009**, *17*, 1055–1082. [CrossRef]
46. Glaholt, M.G.; Reingold, E.M. Direct control of fixation times in scene viewing: Evidence from analysis of the distribution of first fixation duration. *Vis. Cogn.* **2012**, *20*, 605–626. [CrossRef]
47. Henderson, J.M.; Nuthmann, A.; Luke, S.G. Eye movement control during scene viewing: Immediate effects of scene luminance on fixation durations. *J. Exp. Psychol.* **2013**, *39*, 318–322. [CrossRef]
48. Van Diepen, P.; Ruelens, L.; d'Ydewalle, G. Brief foveal masking during scene perception. *Acta Psychol.* **1999**, *101*, 91–103. [CrossRef]

49. Luke, S.G.; Nuthmann, A.; Henderson, J.M. Eye movement control in scene viewing and reading: Evidence from the stimulus onset delay paradigm. *J. Exp. Psychol.* **2013**, *39*, 10–15. [CrossRef] [PubMed]
50. Henderson, J.M.; Hayes, T.R. Meaning guides attention in real-world scene images: Evidence from eye movements and meaning maps. *J. Vis.* **2018**, *18*. [CrossRef] [PubMed]
51. Henderson, J.M.; Hayes, T.R.; Rehrig, G.; Ferreira, F. Meaning Guides Attention during Real-World Scene Description. *Sci. Rep.* **2018**, *8*, 1–9. [CrossRef]
52. Ferreira, F.; Swets, B. How incremental is language production? Evidence from the production of utterances requiring the computation of arithmetic sums. *J. Mem. Lang.* **2002**, *46*, 57–84. [CrossRef]
53. Peacock, C.E.; Hayes, T.R.; Henderson, J.M. Meaning guides attention during scene viewing, even when it is irrelevant. *Atten. Percept. Psychophy.* **2019**, *81*, 20–34. [CrossRef]
54. Huettig, F.; McQueen, J.M. The tug of war between phonological, semantic and shape information in language-mediated visual search. *J. Mem. Lang.* **2007**, *57*, 460–482. [CrossRef]
55. Shomstein, S.; Malcolm, G.L.; Nah, J.C. Intrusive Effects of Task-Irrelevant Information on Visual Selective Attention: Semantics and Size. *Curr. Opin. Psychol.* **2019**. [CrossRef] [PubMed]
56. Russell, B.C.; Torralba, A.; Murphy, K.P.; Freeman, W.T. 2008 LabelMe. *Int. J. Comput. Vis.* **2008**, *77*, 157–173. [CrossRef]
57. Kummerer, M.; Wallis, T.S.A.; Gatys, L.A.; Bethge, M. Understanding Low- and High-Level Contributions to Fixation Prediction. In Proceedings of the 2017 IEEE International Conference on Computer Vision (ICCV), Venice, Italy, 22–29 October 2017; pp. 4799–4808. [CrossRef]

Review

Eye Movements in Medical Image Perception: A Selective Review of Past, Present and Future

Chia-Chien Wu [1,2,*] and Jeremy M. Wolfe [1,2,3]

1   Visual Attention Lab, Department of Surgery, Brigham & Women's Hospital, 65 Landsdowne St, Cambridge, MA 02139, USA; jwolfe@bwh.harvard.edu
2   Department of Radiology, Harvard Medical School, Boston, MA 02115, USA
3   Department of Ophthalmology, Harvard Medical School, Boston, MA 02115, USA
*   Correspondence: cchienwu@gmail.com

Received: 1 April 2019; Accepted: 18 June 2019; Published: 20 June 2019

Abstract: The eye movements of experts, reading medical images, have been studied for many years. Unlike topics such as face perception, medical image perception research needs to cope with substantial, qualitative changes in the stimuli under study due to dramatic advances in medical imaging technology. For example, little is known about how radiologists search through 3D volumes of image data because they simply did not exist when earlier eye tracking studies were performed. Moreover, improvements in the affordability and portability of modern eye trackers make other, new studies practical. Here, we review some uses of eye movements in the study of medical image perception with an emphasis on newer work. We ask how basic research on scene perception relates to studies of medical 'scenes' and we discuss how tracking experts' eyes may provide useful insights for medical education and screening efficiency.

Keywords: visual search; eye movements; medical image perception

1. Introduction

Detection and diagnosis in medicine are frequently based on analysis of medical images. Clinicians of various specializations consume a vast volume of medical images each day. They perform remarkable tasks with these images but they are not perfect. For instance, though more than two million new cases of breast cancer and lung cancer were diagnosed worldwide in 2018 according to the report from World Cancer Research Fund (worldwide cancer data, https://www.wcrf.org/dietandcancer/cancer-trends/worldwide-cancer-data), we know that many cancers are not discovered, even though they may be visible in the image (e.g., [1–4]). Though the acceptance of routine cancer screening has risen and the imaging technology has continued to advance, false negative and false positive rates continue to be higher than we would expect or desire [5,6]. Measuring the eye movements of experts as they perform medical image perception tasks is one way to identify possible weak spots in the processes of medical image perception, raising the possibility of interventions that could improve performance. Eye tracking can also be used to assess the effectiveness of those interventions. Finally, from the vantage point of the basic science of perception, expert performance with medical images can give us insight into the processes of more ordinary acts of scene perception.

One of the interesting aspects of medical image perception research is that the stimuli keep changing. Fifty years ago, questions about medical image perception would have been questions about static 2D, achromatic x-ray images, presented on film. Today, technologies, like CT, create a set of virtual slices through some volume of the body (e.g., the chest) and produce a 3D volume of image data to be examined [7]. They could look at nuclear medicine images (e.g., positron emission tomography—PET) or ultrasound [8], where a 3D dataset might be rendered as a rotating figure. Furthermore, many images are in color today [9]. The dataset might be 4-dimensional, with a time-varying fourth dimension as in

CT angiography where a contrast agent is injected into the bloodstream and tracked in 3D as it sweeps the heart, brain, etc. [10]. Thus, there are three spatial dimensions and the images evolve over time [11], creating 4D datasets.

Each advance in technology requires (or should require) a new set of psychophysical studies because each new modality presents new opportunities and challenges for the perceptual and cognitive capabilities of the observers. What we learned about search strategies or patterns of errors in 2D films may be only loosely applicable to newer forms of imaging. It is nearly impossible to develop better viewing strategies without understanding how human perception works in these new modalities.

Though the technology changes, many of the basic perceptual issues do not. Kundel, Nodine and their colleagues and students have worked for many years on a set of issues that remain relevant today. We will organize this brief review starting with the scanpaths that can be measured during visual search, since the sequences of eye movements are the basic data that is collected in eye movement studies in medical image research. Once scanpaths are collected, they can be aggregated in various ways to address other questions such as the extent of the "useful field of view" (e.g., How much of the image can be processed around the current point of fixation?) and the nature of search errors (e.g., Was the missed cancer fixated during search?). The topic of "satisfaction of search" is an extension of the topic of search errors. Finally, we will discuss what people can perceive in a single glimpse—the 'gist' that can be extracted when there is no scanpath at all.

For other important topics in the field of eye movements in medical image perception (e.g., medical training and education), there are other useful reviews: For example, [12–14].

2. Scanpaths: Searching in Scenes and Medical Images

The sequence of saccades and fixations made when an observer views an image is known as the "scanpath" [15,16]. It has long been used as a clue to what observers are doing when they are looking at something. Essentially all eye movement studies begin by recording a set of scanpaths. Scanpaths tend to be substantially different each time an image is viewed. Thus, for many of the studies described below, scanpaths are aggregated in ways that lose the precision of space and time that a single scanpath would possess. For example, it is often useful to measure the likelihood that a specific location in the image was fixated in the scan paths of many observers. In such an analysis, temporal order will be sacrificed to create a spatial map of fixations.

Yarbus [17] famously held that the scanpath could reveal what an observer was thinking about when viewing a scene, though that may not be as straightforward as he thought [18] (also see [19–21]). In the absence of eye tracking, people are surprisingly ineffective at knowing which part of an image they have looked at. We [22,23] investigated how well people could recall their own fixations after a brief period of scene search. In Võ et al [22]., observers were asked to perform a change detection task. They viewed a scene for 3 s and then saw a new version of the scene and were asked what had changed. On 25% of trials, after viewing the first scene for three seconds, observers were asked to click on 12 locations that they thought they had fixated. Humans make voluntary eye movements about 4 times per second; hence, 12 fixations in 3 s. People were not random, but the results show that observers' memories for where they had looked in the scene were no better than their guesses about where someone else might have looked in the same scene. That is, you might know that it would be reasonable to look at the coffee mug on the desk while viewing an office scene, but you have no privileged access to whether you actually looked at the mug. Kok et al. [23] went on to demonstrate that online feedback has only a marginal effect on memory for the scanpath. They used a gaze-contingent window during the search to highlight where observers were looking and it was still difficult for those observers to maintain a representation of where they had looked once the search was done.

Obviously, the failure to remember where you have looked before could contribute to errors in medical image reading because radiologists may also have poor representations of which areas they examined in an image. This could be particularly true for 3D volumes of image data like CT and MR. What have you really "looked at", once you have scrolled through a sequence of images covering the volume of the chest, for example (see Figure 1)? This is closely related to the question of the UFOV, which is discussed below. With eye tracking, it would be possible to give feedback to a radiologist who is completing a search: For example, an eye-tracking computer might tell the observer, "You may have a good reason, but do you know you have not looked at *this* entire region?" or "You spent a lot of time looking at this spot. You did not label it as abnormal but it clearly grabbed your attention. Do you want to reconsider before you move on?" Kundel, Nodine, & Krupinski [24] found that giving feedback of this second sort to radiologists had a positive impact, and explorations of this type of intervention continue [25]. However, in other contexts, being told where you have fixated may not be that useful [26,27].

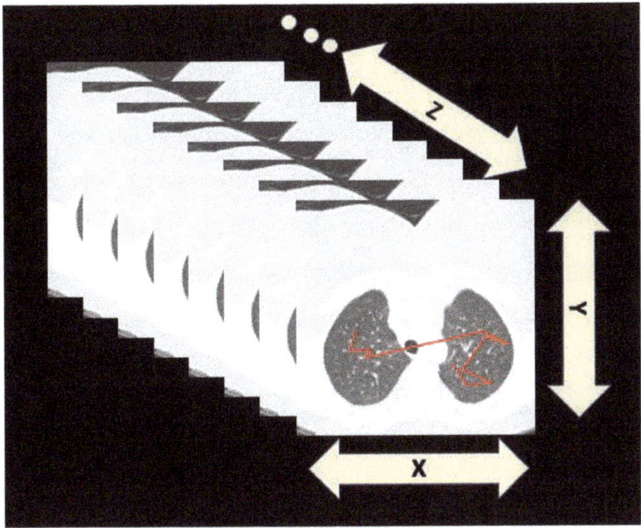

Figure 1. Eye tracking in 3D stacks of images means keeping track of the eye's position in the XY plane and the depth (Z) of the currently displayed image.

The increasing use of techniques like computed tomography (CT) has converted the measuring of the scanpath from a 2D to a 3D problem. Many modern imaging technologies create 3D volumes of image data. These are often rendered into 'stacks' of 2D images. The reader typically scrolls back and forth through the stack while examining the currently visible 2D image (see Figure 1).

Thus, the eyes move in the XY plane while movement in Z, depth in the stack, is typically controlled by the observer through the workstation.

There has been a limited amount of research into search through 3D volumes of image data [28–33]. These 3D volumes of data represent an increase in the information/images that observers need to process. They also, necessarily, change observers' eye movements patterns from a 2D search strategy in X&Y dimensions to a 3D search in X, Y, & Z. Drew et al. [34] had 24 radiologists search for lung nodules in stacks of images drawn from patients undergoing testing for lung cancer. As shown in Figure 2, to visualize the data, they color-coded the slices so that each quadrant had its own hue. Then they plotted that hue (a coarse measure of XY position) as a function of time in the search and the Z dimension, the slice in depth. They reported that radiologists could be coarsely split into two groups: "drillers" who moved rapidly in depth while staying in a relatively constant spot in the XY plane. In contrast, scanners moved slowly in depth while looking much more widely in the current XY plane.

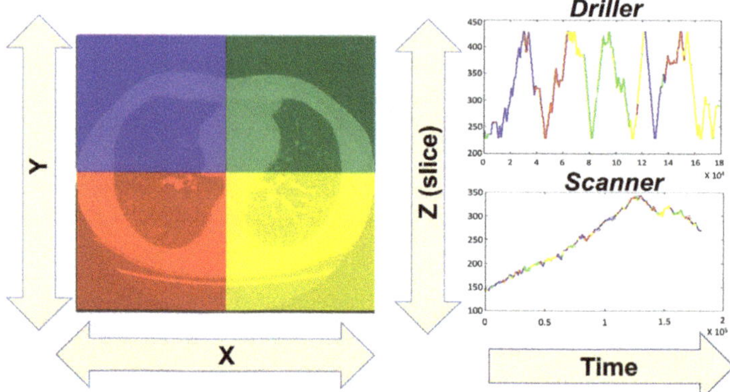

Figure 2. To visualize an approximation of the 3D scanpath through the lungs, position in the XY plane is coarsely color-coded into four quadrants (left-hand image). Depth (the slice in the stack) is plotted as a function of time-on-task on the right, with colors indicating the XY position. The terms "driller" and "scanner" are explained in the main text.

We know that viewing lung CT images as an observer-controlled 'movie' is superior to viewing slices in a 2D array [35] but we don't yet have enough data to know if we should be recommending 'driller' or 'scanner' strategies. Nevertheless, this type of scanpath research is an example of how eye movement recording can change our understanding of how the behavior of radiologists (or other experts) responds to changes in technology.

Once revealed, drilling and scanning seem like rather natural categories of scanpaths for 3D volumes of images. However, different technologies and different tasks may produce different oculomotor approaches. In breast imaging, the analog to the set of slices produced by CT is produced by digital breast tomosynthesis (DBT) [36], a somewhat different x-ray technology that also produces a stack of slices. At least one manufacturer explicitly trains radiologists to "drill" (albeit, not using that term). Nevertheless, when Aizenman et al. [28] measured scanpaths of radiologists reading DBT stacks, they found that the XYZ paths did not conform to the driller or scanner pattern from lung CT. Readers did tend to move rapidly back and forth through the depth of the breast, consistent with drilling, but they did not restrict themselves to one part of the breast in any rigorous manner. They seemed to scan and drill at the same time ("scrilling"?).

In thinking about possible differences between scanning in lung CT and breast tomosynthesis, it is worth noting that the different screening modalities are often used for different diagnostic purposes. For example, DBT is often used as a secondary diagnostic aid in addition to traditional mammography. But other 3D volumes of images, such as lung CT, serve as the primary screening tool. These different diagnostic tasks probably lead to very different scanpaths. Thus, it is important to search for general rules and also to check for those rules in multiple specific cases.

Returning to lung CT, if one looks more closely at either drilling or scanning behavior, one can see that readers are toggling quickly, back and forth between images. They may be looking for any items that might be nodules to see if they pop in and out of visibility as the observer toggles between a relatively few slices in the 3D stack. Other features, like vessels, snake through the image over many slices and would move rather than vanishing as the viewer moves a short distance through the stack. This shows one benefit of "toggling" between two (or more) nearly identical images, looking for change. Looking for change in this way is most effective when the toggled images are largely identical. Interestingly, there may be benefits even when images are not identical, as would be the case with a pair of mammograms of the same breast, taken at two different times. Drew et al. [37] asked radiologists to compare positive cases of mammograms with the negative prior exams acquired 2-3 years earlier from the same patients. Radiologists read the current and prior stimuli either in Side-By-Side mode or in Toggle mode. In Toggle mode, the radiologist could alternate between current and prior images at the same location on the screen. In Side-By-Side mode, current and prior images were visible at the same time on the screen. Drew et al. found that toggling produced a substantial improvement in time (~15%) and a small improvement in accuracy (~6%). The time benefit seems to result from the reduction in the number of required eye movements. In side-by-side viewing, readers made many saccades between the two images. Even though saccades are "cheap", enough of them can add up to a real cost in time. The possible benefits of toggling for accuracy deserve further study.

3. The Useful Field of View (UFOV)

Scanpaths raise an interesting problem. It is self-evident that observers move their eyes around an extended scene in order to see it better. But what exactly does that mean? It is clear that, as you fixate on this word *here, right now,* you can see the whole screen or page in front of you, but if your task is to read letters, your useful field of view (UFOV) is much smaller. You can only actually do that task in a narrow area around the current point of fixation. Some constraints on the UFOV are based on fundamental visual properties. Turning to Figure 3, if you fixate on the "1", you can read the letter "A" but if you fixate on "2", the decline in acuity away from the point of fixation will probably mean that you cannot read the letter "P". If you fixate on 3, you will find that it is hard to resolve the letter "C" in the middle of the string "DCT", even though it is the same size as the clearly visible "A" in Line 1 above. In Line 3, the problem is crowding [38,39], a reduction in our ability to process/recognize one object if it is surrounded by others.

Figure 3. Acuity and crowding limit the UFOV.

Beyond these basic visual limits, there is also an important attentional limit on the UFOV (e.g., [40]). This is illustrated in Figure 4.

Figure 4. Fixate on the 'x' at the center and find the incorrect number.

If you fixate on the 'x' at the center of this clock face, you should note that you can read each number in turn. If, however, you need to determine which number is out of sequence, you may find that a single fixation of a quarter to a third of a second is not adequate to do the task. If you were lucky, the '2' in the 7 o'clock position was inside your UFOV for this task at this moment. If not, if you only had one fixation, you would have missed it.

The same constraints that shape the UFOV in tasks like those illustrated in Figures 3 and 4, will also apply in medical image search. When radiologists search a breast or a lung for a possibly cancerous mass, they will have a UFOV specific to the current image and task (e.g., the UFOV will be larger if the current target is larger). That clinical UFOV, like any other, will be affected by acuity, crowding and attentional factors. Thus, having an estimate of the UFOV can help us to understand whether the radiologist did or did not "look" at the whole image.

Sanders [41] was an early pioneer in this area, investigating what we are calling the UFOV though he preferred the term "functional visual field" (FVF) [42,43]. Sanders divided the visual field into three attentional areas: the stationary field, in which people can process information without moving their eyes; the eye field, in which eye movements but not head movements are required to sample the information; and the head field, in which head movements would be necessary. He found that, in a target detection task, observers barely made any eye movements when the target was presented within 30 degrees. What we have learned, over the subsequent 50 years of attention research, is that measuring the UFOV for almost any task will involve more factors than those most extensively discussed by Sanders [41]. The size of UFOV will interact with the type of images, the properties of the target and its surroundings, and with human visual and attentional capacities. In a study of searching for low contrast lung nodules, Kundel and colleagues [44] reported that "The visual field size that is most effective in detecting nodules during search has a radius of 3.5 degrees visual angle. Nodule detection may be limited by basic neurologic constraints on human scanning performance". In a separate study, Carmody et al. [45] asked observers to look for a nodule in chest x-ray films. The images were presented only for 300 msec to simulate a single fixation. They found that detection rates dropped by one-half when the nodules were presented at 5 degrees from the fixation. Apparently, it is hard to establish a reliable UFOV measure, even if the task is restricted to something as straight-forward as search for nodules in chest x-ray. Thus, any estimate, based on a scanpath, of how much of an image was examined should be treated with caution. It will be based on assumptions about the size of UFOV. That said, statements about *relative* coverage are more convincing. Unless there is some reason that the UFOV might change between conditions, it is reasonable to use scanpath data to say that observers looked at more of a scene under condition A than under condition B.

UFOV questions certainly do not become simpler when the stimuli are 3D volumetric images, such as lung CT. In one of very few studies, Ebner et al. [46] have reported that the nodule detection rate on chest CT was highly correlated not only with the size of UFOV, as measured by the nodule eccentricity ("transverse distance" in their paper), but with the nodule size and local lung complexity. There are essentially no data on the UFOV in DBT to say nothing of other tasks such as CT angiography where complex 3D image data changes over time. If it was important to know what percentage of an image set was examined in some specific task, UFOV measures would need to be made for that task. It is probably more useful to ask slightly different questions, based on the scanpath. For example, when should we be concerned that a reader is not looking at enough of an image?

Thus, with the cautions described above, it is possible, given some UFOV assumptions, to use scanpaths to make statements about how much of an image or a case has been examined. However, it is by no means clear how much coverage is enough. The unthinking answer is that we would hope that the radiologist would look at the "whole image" but that is clearly incorrect. Suppose you walk into the kitchen to look for a pepper grinder that is, in fact, not present. How much of the kitchen should you look at before declaring the object to be absent (akin to declaring the breast to be normal)? Looking at "everything" would be foolish. You, as a kitchen expert, know that the pepper grinder is never on the floor, even if it could be. On the very rare occasion that it is on the floor, you might fail to find it but, most of the time, not looking at "everything" is sensible. Similarly, an expert radiologist will know that there are some parts of an image that do not require fixation. Indeed, one of the oculomotor hallmarks of developing expertise is a tendency to look at less of the image [47,48]. We can see the utility of learning what not to look at in a study by Rubin et al. [49] where they found the radiologists on average search only 26% of the lung parenchyma yet encompass 75% of nodules in their search volume". This applies beyond radiology; for example, in Dermatology [50]. Other oculomotor metrics also change as expertise develops [51] but the pruning back of the scanpath is an important and general sign of expertise.

Returning to the idea of using eye movement feedback, discussed above, if the scanpath is to be used to warn the expert that some areas of the image are unexamined, those warnings should acknowledge that some areas simply do not need to be examined. Thus, it would be foolish to build a system that would insist that a reader fixate every part of an image before allowing the reader to move on to the next case and it might be clever if an artificial intelligence (AI) system could learn where in the image/scene a target might be and where it could never be.

4. Scanpath Signatures of Errors in Medical Image Perception

The motivation for eye movement feedback of the sort described above would be to prevent false negative (miss) errors in search. That implies that readers miss findings when they do not look at them. Not looking at a target is certainly one reason that the target might be missed, but it is not the only reason. In medical image perception, Kundel and colleagues developed a useful taxonomy of false negative errors, based on eye movement recording. For example, Kundel, Nodine & Carmody [52] recorded and analyzed eye movements from clinicians who were searching chest x-rays for lung nodules. They argued that clinicians' eye movements could be used to distinguish three types of errors: Search, recognition, and decision errors. In various works, the terminology has varied somewhat but the core idea has remained about the same. In Figure 5, we will use those terms to describe the taxonomy. The scanpaths in Figure 5 are invented for purposes of illustration. Suppose that the yellow arrow is pointing to a finding in the breast that should be reported as suspicious. Kundel and colleagues argued that there were three different ways that this target might be missed. In a search error, the target is never fixated. A recognition error is said to occur when the eyes fixate on the target briefly and then move on, with no indication that the reader noted anything of interest. Finally, multiple and/or long fixations on a target indicate a decision error, if the reader still fails to report the finding. This pattern indicates that, implicitly or explicitly, the reader knew that this spot deserved attention but then the reader made the wrong decision and did not mark the spot as abnormal. In their lung nodule study, Kundel, Nodine & Carmody found that clinicians made about 30% search errors, 25% recognition and 45% decision errors. Similar proportions are found in a variety of studies (e.g., [47,53,54]).

How does this taxonomy of errors apply to 3D volumes of image data? The short answer is that the correct studies have not been done, but there are some hints. A major reason for moving to a 3D modality (e.g., lung CT) over a 2D one (chest x-ray) is that findings that are ambiguous in 2D are clarified by 3D. A downside of the move is that the increase in images leads to an increase in the time per case and to pressure to move quickly through the images. One might suspect that these factors would decrease the proportion of decision errors while increasing the proportion of search errors in which the target was never fixated. Drew et al. [34] found the signs of such a shift and a report by [49] that readers only search 26% of the lung in lung CT also points in that direction. The topic deserves further study. A more complete review of the work on specific types of search errors can be found in [12].

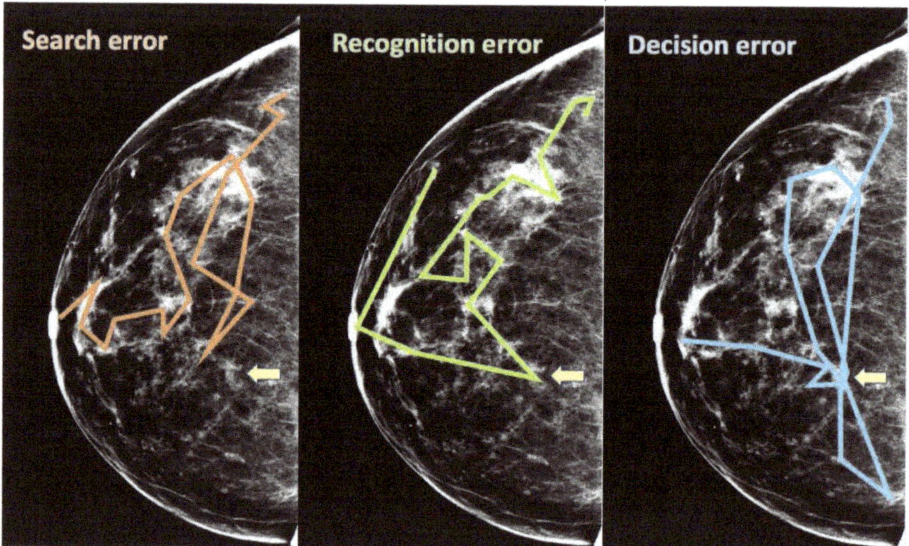

Figure 5. Three types of false negative (miss) errors, as proposed by Kundel and colleagues.

5. Incidental Findings and Satisfaction of Search Errors

Two, possibly related varieties of errors deserve some further mention. "Incidental findings" are findings that may be clinically significant but are not the primary target of the clinician's search. A lung nodule would be a primary target in a search for signs of lung cancer. The same lung nodule would be an incidental finding if it was noticed in the course of an exam to determine if the patient had pneumonia [55]. Radiologists are typically expected to report incidental findings [56]. There is considerable debate about reporting and management of incidental findings because many of them turns out not to require any action. Raising them to attention can cause unnecessary worry and unnecessary medical care [57–60]. On the other hand, not reporting a finding that turns out to be clinically significant is a potent source of malpractice suits [61]. It is important to note that radiologists are trained to detect incidental findings. They would not typically stumble on a finding by chance. They know what they are looking for and, indeed, may have specific search strategies designed to detect problems that are not the specific focus of the case.

Medical image perception researchers do not need to solve the issue around the management of incidental findings. We can focus on reducing the number of targets that are not found at all. Clinicians cannot successfully manage what they don't see. What kind of errors are missed incidental findings? In the eye tracking study that produced the driller/scanner data as shown in Figure 2, an image of a gorilla was inserted into the final case. It spanned five slices in the stack of CT images and was easily detectable in a variety of control conditions. Nevertheless, 20 of 24 radiologists failed to notice it [62]. The gorilla was chosen as a stimulus because it is the iconic stimulus [63] for studying the phenomenon of "inattentional blindness" [64]. The experiment showed that expertise does not immunize the expert against inattentional blindness, even when the stimuli are the subject of the expertise. When radiologists are looking for a small, round, white nodule, they are likely to miss a big visible gorilla right in front of their eyes because attention will be guided to the wrong set of basic color and shape features for gorilla detection [65].

Since the observers were being eye tracked, it was possible to apply the Kundel taxonomy to these miss errors (understanding that there are only 20 data points in total here, from the 20 miss trials). Observers spent nearly 6 s on average looking at the five slices that contained the gorilla. On average, they fixated on the gorilla itself for an average of 329 msec. Thus, most of these do not appear to be search errors. These appear to be recognition errors where the eyes landed on the gorilla but the strange identity of the item did not register with the observer. It seems quite unlikely that these could be decision errors. Of course, this does not mean that *all* misses or even most incidental finding errors are recognition errors. However, the result does make the point that an expert can look at something very odd in an image and, if looking for something else, fail to note the oddity.

As multiple radiologists have pointed out, gorillas are not an ideal model for incidental findings. When a radiologist misses a nodule while looking for pneumonia, she knows that lung cancer is something that can plausibly happen in a lung. Gorillas do not happen in lungs. In more recent work, we have developed a lab analog of incidental findings in which non-expert observers reliably miss over 30% of targets that they know that they are looking for [66]. Observers are looking for three specific images and, at the same time, for any members of any of three broad categories, like animals, hats, or fruit. They find the specific targets with ease but make large numbers of errors on the categorical targets. We don't yet know if observers fixate the targets that they fail to detect because the eye tracking experiments have not been done. The results of the gorilla study would imply that we will find that observers can fixate on an elephant and still fail to report having found an animal.

Satisfaction of Search (SoS) errors are a class of false negative errors that are somewhat similar to incidental finding errors in the sense that, in both cases, one target, either the one you are searching for or the one has been found, interferes with the detection of another. In the SoS case, finding one target in an image makes it less likely that observers will report a second target compared to cases where only the second target is present [67]. The name comes from the original account of the source of these errors. It was initially proposed that, having found one target, observers were "satisfied" and, thus, abandoned the search too soon, before finding the second target. Subsequent research suggests that this theory is not correct [68–70] but the name has persisted, though Cain et al. [71] have tried to get the field to shift to "subsequent search misses (SSM)".

There are two groups who have conducted the most extensive work on SoS: Berbaum and his colleagues (reviewed in [72,73]) and Cain, Adamo, Mitroff and colleagues [71]. Here, we want to focus on what eye movements can tell us about SoS errors. Kundel, Nodine, and their co-workers concluded that most SoS errors were recognition errors. The observers fixated on the missed items but only relatively briefly. If they spent a longer time, they tended to find the target [70]. Berbaum et al. [74] found that the type of task made a difference in x-ray studies of the abdomen. For some classes of radiologic exam, they found that the SoS errors tended to be search errors where the reader did not look in the right place. For other tasks, like Samuel et al. [70], they found a large proportion of recognition errors. On the other hand, in an eye tracking study of chest x-rays Berbaum et al. [75] found very few search errors, 35% recognition errors, and 58% decision errors. In that study, the readers apparently looked at the targets for some time and decided not to call them targets. Using a very different task, search for low contrast T's among L's, with a non-expert population of observers, Cain et al. [71] found search errors to be the largest category (37.8%), with the second largest category, at 24.3%, being a new type of error that they called "resource depletion errors". They define resource depletion errors as errors that arise when the first target depletes working memory resources that could be used to find the second target. Recognition and Decision errors together account for only 20% of errors in their study. Clearly, SoS (or SSM) errors are not the product of a unitary mechanism in search. As shown in [71], there is a taxonomy of these errors, as there is a taxonomy of errors in search more generally.

6. Scene Gist and Medical Scene Gist

There are other emerging uses of eye tracking in medical image perception. For example, Drew and colleagues have been using eye tracking to address the effects of interruption on radiologists [76,77]. However, we want to finish this brief review with some consideration of medical image searches that involve little or no oculomotor activity. To quote Kundel [78], "Clearly much of what happens in perception precedes exhaustive visual scanning of the image." An important part of the Kundel-Nodine model of search in medical images is a "holistic" stage lasting about a second [79–81] during which the observer might be processing the whole image without needing to move the eyes. Much of the basis for proposing this holistic stage comes from eye tracking studies that often show the eyes of experts moving to the target almost immediately [82,83].

Basic research in visual attention would divide these holistic effects into more than one component. "Covert attention" can be shifted more rapidly and more frequently than the eyes [84–86] so attention might have reached a target before the eyes had a chance to move. The first eye movement might be simply confirming what had been found. Second, a set of basic features guides the deployment of attention and the eyes [87,88]. Thus, in a search for a lung nodule, attention will be guided to a location that contained the small, white, and round features which are characteristic of a nodule. Again, the first eye movement might go to a target because the target's features provided successful guidance about where to best deploy the eyes.

Kundel and Nodine [79] investigated whether there is any useful information available before the eyes move by asking radiologists to evaluate X-rays with only a 200 msec exposure to the images. With these particular stimuli, expert performance was nearly perfect with unlimited viewing. Surprisingly, with just a 200 msec glimpse of the image, their classification accuracy was still about 70%. In their following study, Carmody, Nodine and Kundel [89] systematically varied the exposure duration to test how detection performance changes over the first half-second of exposure. They found that, across the different level of image visibilities, the performance reached an asymptote after 240 msec. Moreover, even for the least visible cases, there was still substantial information available in the first quarter of the second. This global analysis not only occurs in lung screening, but also in the screening of mammography [90]. The Kundel-Nodine group argues that a major part of the development of expertise is a growing ability to do this holistic processing. They invoke this holistic stage to explain why experts make fewer eye movements.

Interestingly, there is an aspect of this holistic stage of processing that does not serve to direct the eyes to the target. Evans et al. [91] showed radiologists a brief flash of a mammogram for durations of from 250 msec to 2 s and asked them to classify this case as normal or abnormal (Would you call back this patient?). They found that radiologists could perform at above chance levels at all stimulus durations, even those that did not permit a voluntary eye movement. Importantly, this awareness of abnormality did not appear to be based on visible features of a lesion because when radiologists were also asked to place a localizing mark on the most suspicious location on a white outline mask of the breast after the brief presentation of stimulus, their performance was not better than chance. This was true regardless of their rated degree of confidence that the presented image was abnormal. Readers were not simply getting lucky and fixating a lesion on a subset of trials. In a subsequent study, Evans et al. [92] showed that this global "gist signal" was not based on a break in the normal asymmetry between breasts nor was it a proxy for breast density, a known risk factor for breast cancer. In fact, radiologists were able to discriminate between normal and abnormal at above chance levels when the "abnormal" images were the images from the breast contralateral to the breast with overt signs of cancer. Since there is no lesion presented in the image, the observed performance cannot be due to a lucky fixation on a mass. Something about the texture of the breast tissue is abnormal and experts have become sensitive to that signal. Brennan et al. [93] repeated the experiment but with the "priors", the mammograms acquired three years before the women developed overt signs of cancer. Even though there were no localized lesions in these priors, radiologists could still detect this gist signal at an above chance level. A signal that is available years before the cancer develops could be a useful imaging biomarker of cancer risk.

7. Conclusions

In this brief review, we have tried to show the usefulness of eye tracking for understanding how experts perform tasks like those involved in clinical radiology. A similar story could be told about many other expert domains. Scanpaths tell a story. The story may not be as clear as optimistic interpretations of Yarbus' work might suggest, but the sequence of eye movements and the placement of fixations relative to targets tell us a lot about the processes of expert visual search. Eye movement recordings are of particular interest in analyzing errors and, we may hope, in testing efforts to reduce those errors. It is notable how many of the basic issues in this field were outlined and studied by Kundel, Nodine, and their group (as well as by other labs and earlier researchers). However, their path breaking work is not the end of the story. As long as the advances in medical imaging create new stimuli to improve medical interpretation, there are always new scientific questions that need to be investigated.

Funding: This research was funded by NIH grant number: NCI: CA207490; NEI: EY017001.

Conflicts of Interest: The authors declare no conflict of interest.

References

1. Boyer, B.; Hauret, L.; Bellaiche, R.; Graf, C.; Bourcier, B.; Fichet, G. Retrospectively detectable carcinomas: Review of the literature. *J. Radiol.* **2004**, *85*, 2071–2078. [CrossRef]
2. Hoff, S.R.; Abrahamsen, A.-L.; Samset, J.H.; Vigeland, E.; Klepp, O.R.; Hofvind, S. Breast Cancer: Missed Interval and Screening-detected Cancer at Full-Field Digital Mammography and Screen-Film Mammography,Äî Results from a Retrospective Review. *Radiology* **2012**, *264*, 378–386. [CrossRef] [PubMed]
3. Martin, J.E.; Moskowitz, M.; Milbrath, J.R. Breast cancer missed by mammography. *AJR Am. J. Roentgenol.* **1979**, *132*, 737–739. [CrossRef] [PubMed]
4. Pisano, E.D.; Gatsonis, C.; Hendrick, E.; Yaffe, M.; Baum, J.K.; Acharyya, S.; Conant, E.F.; Fajardo, L.L.; Bassett, L.; D'orsi, C.; et al. Diagnostic Performance of Digital versus Film Mammography for Breast-Cancer Screening. *N. Engl. J. Med.* **2005**, *353*, 1773–1783. [CrossRef] [PubMed]
5. Le, M.T.; Mothersill, C.E.; Seymour, C.B.; Mcneill, F.E. Is the false-positive rate inmammography in North America too high? *Br. J. Radiol.* **2016**, *89*. [CrossRef] [PubMed]
6. Seely, J.M.; Alhassan, T. Screening for breast cancer in 2018—What should we be doing today? *Curr. Oncol.* **2018**. [CrossRef] [PubMed]
7. Hedlund, L.W.; Anderson, R.F.; Goulding, P.L.; Beck, J.W.; Effmann, E.L.; Putman, C.E. Two methods for isolating the lung area for a CT scan for density information. *Radiology* **1982**, *144*, 353–357. [CrossRef] [PubMed]
8. Kotsianos-Hermle, D.; Wirth, S.; Fischer, T.; Hiltawsky, K.M.; Reiser, M. First clinical use of a standardized three-dimensional ultrasound for breast imaging. *Eur. J. Radiol.* **2009**, *71*, 102–108. [CrossRef] [PubMed]
9. Celebi, M.E.; Schaefer, G. *Color Medical Image Analysis*; Emre Celebi, M., Schaefer, G., Eds.; Springer: Dordrecht, The Netherlands, 2013.
10. Moscariello, A.; Takx, R.A.; Schoepf, U.J.; Renker, M.; Zwerner, P.L.; O'Brien, T.X.; Allmendinger, T.; Vogt, S.; Schmidt, B.; Savino, G.; et al. Coronary CT angiography: Image quality, diagnostic accuracy, and potential for radiation dose reduction using a novel iterative image reconstruction technique,Äîcomparison with traditional filtered back projection. *Eur. Radiol.* **2011**, *21*, 2130–2138. [CrossRef]
11. Eddleman, C.S.; Jeong, H.J.; Hurley, M.C.; Zuehlsdorff, S.; Dabus, G.; Getch, C.G.; Batjer, H.H.; Bendok, B.R.; Carroll, T.J. 4D radial acquisition contrast-enhanced MR angiography and intracranial arteriovenous malformations: Quickly approaching digital subtraction angiography. *Stroke* **2009**, *40*, 2749–2753. [CrossRef] [PubMed]
12. Brunye, T.T.; Drew, T.; Weaver, D.L.; Elmore, J.G. A Review of Eye Tracking for Understanding and Improving Diagnostic Interpretation. *Cogn. Res. Princ. Implic. (CRPI)* **2019**, *4*, 7. [CrossRef] [PubMed]
13. Van der Gijp, A.; Ravesloot, C.J.; Jarodzka, H.; van der Schaaf, M.F.; van der Schaaf, I.C.; van Schaik, J.P.; Ten Cate, T.J. How visual search relates to visual diagnostic performance: A narrative systematic review of eye-tracking research in radiology. *Adv. Health Sci. Educ. Theory Pract.* **2017**, *22*, 765–787. [CrossRef] [PubMed]
14. Krupinski, E.A. Current Perspectives in Medical Image Perception. *Atten. Percept. Psychophys.* **2010**, *72*, 1205–1217. [CrossRef] [PubMed]
15. Noton, D.; Stark, L. Scanpaths in eye movements during pattern perception. *Science* **1971**, *171*, 308–311. [CrossRef]
16. Noton, D.; Stark, L. Scanpaths in saccadic eye movements while viewing and recognizing patterns. *Vis. Res.* **1971**, *11*, 929–942. [CrossRef]

17. Yarbus, A.L. *Eye Movements and Vision*; Plenum: New York, NY, USA, 1967.
18. Greene, M.R.; Liu, T.; Wolfe, J.M. Reconsidering Yarbus: Pattern classification cannot predict observer's task from scan paths. *Vis. Res.* **2012**, *62*, 1–8. [CrossRef]
19. Bahle, B.; Mills, M.; Dodd, M.D. Human Classifier: Observers can deduce task solely from eye movements. *Atten. Percept. Psychophys.* **2017**, *79*, 1415–1425. [CrossRef]
20. Damiano, C.; Wilder, J.; Walther, D.B. Mid-level feature contributions to category-specific gaze guidance. *Atten. Percept. Psychophys.* **2019**, *81*, 35–46. [CrossRef]
21. Kardan, O.; Berman, M.G.; Yourganov, G.; Schmidt, J.; Henderson, J.M. Classifying mental states from eye movements during scene viewing. *J. Exp. Psychol. Hum. Percept. Perform.* **2015**, *41*, 1502–1514. [CrossRef]
22. Võ, M.L.H.; Aizenman, A.M.; Wolfe, J.M. You think you know where you looked? You better look again. *J. Exp. Psychol. Hum. Percept. Perform.* **2016**, *42*, 1477–1481. [CrossRef]
23. Kok, E.M.; Aizenman, A.M.; Võ, M.L.H.; Wolfe, J.M. Even if I showed you where you looked, remembering where you just looked is hard. *J. Vis.* **2017**, *17*, 2. [CrossRef] [PubMed]
24. Kundel, H.L.; Nodine, C.F.; Krupinski, E.A. Computer-displayed eye position as a visual aid to pulmonary nodule interpretation. *Investig. Radiol.* **1990**, *25*, 890–896. [CrossRef]
25. Donovan, T.; Manning, D.J.; Crawford, T. Performance changes in lung nodule detection following perceptual feedback of eye movements. *Proc. SPIE* **2008**, *6917*. [CrossRef]
26. Drew, T.; Williams, L.H. Simple eye-movement feedback during visual search is not helpful. *Cogn. Res. Princ. Implic.* **2017**, *2*, 44. [CrossRef] [PubMed]
27. Peltier, C.; Becker, M.W. Eye movement feedback fails to improve visual search performance. *Cogn. Res. Princ. Implic.* **2017**, *2*, 47. [CrossRef]
28. Aizenman, A.; Drew, T.; Ehinger, K.A.; Georgian-Smith, D.; Wolfe, J.M. Comparing search patterns in digital breast tomosynthesis and full-field digital mammography: An eye tracking study. *J. Med. Imaging* **2017**, *4*, 045501. [CrossRef]
29. Den Boer, L.; van der Schaaf, M.F.; Vincken, K.L.; Mol, C.P.; Stuijfzand, B.G.; van der Gijp, A. Volumetric Image Interpretation in Radiology: Scroll Behavior and Cognitive Processes. *Adv. Health Sci. Educ.* **2018**, *23*, 783–802. [CrossRef]
30. D'Ardenne, N.M.; Nishikawa, R.M.; Wu, C.C.; Wolfe, J.M. Occulomotor Behavior of Radiologists Reading Digital Breast Tomosynthesis (DBT). In Proceedings of the SPIE Medical Imaging, San Diego, CA, USA, 6–21 February 2019.
31. Mercan, E.; Shapiro, L.G.; Brunyé, T.T.; Weaver, D.L.; Elmore, J.G. Characterizing Diagnostic Search Patterns in Digital Breast Pathology: Scanners and Drillers. *J. Digit. Imaging* **2018**, *31*, 32–41. [CrossRef]
32. Timberg, P.; Lång, K.; Nyström, M.; Holmqvist, K.; Wagner, P.; Förnvik, D.; Tingberg, A.; Zackrisson, S. Investigation of viewing procedures for interpretation of breast tomosynthesis image volumes: A detection-task study with eye tracking. *Eur. Radiol.* **2013**, *23*, 997–1005. [CrossRef]
33. Venjakob, A.C.; Mello-Thoms, C.R. Review of prospects and challenges of eye tracking in volumetric imaging. *J. Med. Imaging* **2015**, *3*. [CrossRef]
34. Drew, T.; Vo, M.L.-H.; Olwal, A.; Jacobson, F.; Seltzer, S.E.; Wolfe, J.M. Scanners and drillers: Characterizing expert visual search through volumetric images. *J. Vis.* **2013**, *13*. [CrossRef] [PubMed]
35. Seltzer, S.E.; Judy, P.F.; Adams, D.F.; Jacobson, F.L.; Stark, P.; Kikinis, R.; Swensson, R.G.; Hooton, S.; Head, B.; Feldman, U. Spiral CT of the chest: Comparison of cine and film-based viewing. *Radiology* **1995**, *197*, 73–78. [CrossRef] [PubMed]
36. Baker, J.A.; Lo, J.Y. Breast Tomosynthesis: State-of-the-Art and Review of the Literature. *Acad. Radiol.* **2011**, *18*, 1298–1310. [CrossRef] [PubMed]
37. Drew, T.; Aizenman, A.M.; Thompson, M.B.; Kovacs, M.D.; Trambert, M.; Reicher, M.A.; Wolfe, J.M. Image toggling saves time in mammography. *J. Med. Imaging* **2016**, *3*, 011003. [CrossRef] [PubMed]
38. Levi, D.M. Crowding-An essential bottleneck for object recognition: A mini-review. *Vis. Res.* **2008**, *48*, 635–654. [CrossRef]
39. Manassi, M.; Whitney, D. Multi-level Crowding and the Paradox of Object Recognition in Clutter. *Curr. Biol.* **2018**, *28*, R127–R133. [CrossRef]
40. Hulleman, J.; Olivers, C.N.L. The impending demise of the item in visual search. *Behav. Brain Sci.* **2017**, *40*, e132. [CrossRef]

41. Sanders, A.F. Some aspects of the selective process in the functional visual field. *Ergonomics* **1970**, *13*, 101–117. [CrossRef]

42. Ikeda, M.; Takeuchi, T. Influence of foveal load on the functional visual field. *Percept. Psychophys.* **1975**, *18*, 255–260. [CrossRef]

43. Sanders, A.F.; Houtmans, M.J.M. Perceptual modes in the functional visual field. *Acta Psychol.* **1985**, *58*, 251–261. [CrossRef]

44. Kundel, H.L.; Nodine, C.F.; Thickman, D.; Toto, L. Searching for lung nodules. A comparison of human performance with random and systematic scanning models. *Investig. Radiol.* **1987**, *22*, 417–422. [CrossRef]

45. Carmody, D.P.; Nodine, C.F.; Kundel, H.L. An analysis of perceptual and cognitive factors in radiographic interpretation. *Perception* **1980**, *9*, 339–344. [CrossRef] [PubMed]

46. Ebner, L.; Tall, M.; Choudhury, K.R.; Ly, D.L.; Roos, J.E.; Napel, S.; Rubin, G.D. Variations in the functional visual field for detection of lung nodules on chest computed tomography: Impact of nodule size, distance, and local lung complexity: Impact. *Med. Phys.* **2017**, *44*, 3483–3490. [CrossRef] [PubMed]

47. Krupinski, E.A. Visual scanning patterns of radiologists searching mammograms. *Acad. Radiol.* **1996**, *3*, 137–144. [CrossRef]

48. Kelly, B.S.; Rainford, L.A.; Darcy, S.P.; Kavanagh, E.C.; Toomey, R.J. The Development of Expertise in Radiology: In Chest Radiograph Interpretation, "Expert" Search Pattern May Predate "Expert" Levels of Diagnostic Accuracy for Pneumothorax Identification. *Radiology* **2016**, *280*, 252–260. [CrossRef] [PubMed]

49. Rubin, G.D.; Roos, J.E.; Tall, M.; Harrawood, B.; Bag, S.; Ly, D.L.; Seaman, D.M.; Hurwitz, L.M.; Napel, S.; Roy Choudhury, K. Characterizing Search, Recognition, and Decision in the Detection of Lung Nodules on CT Scans: Elucidation with Eye Tracking. *Radiology* **2015**, *274*, 276–286. [CrossRef] [PubMed]

50. Dreiseitl, S.; Pivec, M.; Binder, M. Differences in examination characteristics of pigmented skin lesions: Results of an eye tracking study. *Artif. Intell. Med.* **2012**, *54*, 201–205. [CrossRef] [PubMed]

51. Bertram, R.; Kaakinen, J.; Bensch, F.; Helle, L.; Lantto, E.; Niemi, P.; Lundbom, N. Eye Movements of Radiologists Reflect Expertise in CT Study Interpretation: A Potential Tool to Measure Resident Development. *Radiology* **2016**, *281*, 805–815. [CrossRef]

52. Kundel, H.L.; Nodine, C.F.; Carmody, D. Visual scanning, pattern recognition and decision-making in pulmonary nodule detection. *Investig. Radiol.* **1978**, *13*, 175–181. [CrossRef]

53. Hu, C.H.; Kundel, H.L.; Nodine, C.F.; Krupinski, E.A.; Toto, L.C. Searching for bone fractures: A comparison with pulmonary nodule search. *Acad. Radiol.* **1994**, *1*, 25–32. [CrossRef]

54. Kundel, H.L.; Nodine, C.F.; Krupinski, E.A. Searching for lung nodules. Visual dwell indicates locations of false-positive and false-negative decisions. *Investig. Radiol.* **1989**, *24*, 472–478. [CrossRef]

55. Beigelman-Aubry, C.; Hill, C.; Grenier, P.A. Management of an incidentally discovered pulmonary nodule. *Eur. Radiol.* **2007**, *17*, 449–466. [CrossRef] [PubMed]

56. Lumbreras, B.; Donat, L.; Hernández-Aguado, I. Incidental findings in imaging diagnostic tests: A systematic review. *Br. J. Radiol.* **2010**, *83*, 276–289. [CrossRef] [PubMed]

57. Heller, R.E. Counterpoint: A Missed Lung Nodule Is a Significant Miss. *J. Am. Coll. Radiol.* **2017**, *14*, 1552–1553. [CrossRef] [PubMed]

58. Oren, O.; Kebebew, E.; Ioannidis, J.P.A. Curbing Unnecessary and Wasted Diagnostic Imaging. *JAMA* **2019**, *321*, 245–246. [CrossRef]

59. Pandharipande, P.V.; Herts, B.R.; Gore, R.M.; Mayo-Smith, W.W.; Harvey, H.B.; Megibow, A.J.; Berland, L.L. Authors' Reply. *J. Am. Coll. Radiol.* **2016**, *13*, 1025–1027. [CrossRef] [PubMed]

60. Pandharipande, P.V.; Herts, B.R.; Gore, R.M.; Mayo-Smith, W.W.; Harvey, H.B.; Megibow, A.J.; Berland, L.L. Rethinking Normal: Benefits and Risks of Not Reporting Harmless Incidental Findings. *J. Am. Coll. Radiol.* **2016**, *13*, 764–767. [CrossRef] [PubMed]

61. Clayton, E.W.; Haga, S.; Kuszler, P.; Bane, E.; Shutske, K.; Burke, W. Managing incidental genomic findings: Legal obligations of clinicians. *Genet. Med.* **2013**, *15*, 624–629. [CrossRef]

62. Drew, T.; Vo, M.L.-H.; Wolfe, J.M. The Invisible Gorilla Strikes Again: Sustained Inattentional Blindness in Expert Observers. *Psychol. Sci.* **2013**, *24*, 1848–1853. [CrossRef] [PubMed]

63. Simons, D.J.; Chabris, C.F. Gorillas in our midst: Sustained inattentional blindness for dynamic events. *Perception* **1999**, *28*, 1059–1074. [CrossRef] [PubMed]

64. Mack, A.; Rock, I. *Inattentional Blindness*; MIT Press: Cambridge, MA, USA, 1998.

65. Most, S.B.; Simons, D.J.; Scholl, B.J.; Jimenez, R.; Clifford, E.; Chabris, C.F. How not to be seen: The contribution of similarity and selective ignoring to sustained inattentional blindness. *Psychol. Sci.* **2001**, *12*, 9–17. [CrossRef] [PubMed]

66. Wolfe, J.M.; Alaoui-Soce, A.; Schill, H. How did I miss that? Developing mixed hybrid visual search as a 'model system' for incidental finding errors in radiology. *Cogn. Res. Princ. Implic. (CRPI)* **2017**, *2*, 35. [CrossRef] [PubMed]

67. Tuddenham, W.J. Visual search, image organization, and reader error in roentgen diagnosis. Studies of the psycho-physiology of roentgen image perception. *Radiology* **1962**, *78*, 694–704. [CrossRef] [PubMed]

68. Berbaum, K.S.; Franken, E.A., Jr.; Dorfman, D.D.; Rooholamini, S.A.; Coffman, C.E.; Cornell, S.H.; Cragg, A.H.; Galvin, J.R.; Honda, H.; Kao, S.C. Time course of satisfaction of search. *Investig. Radiol.* **1991**, *26*, 640–648. [CrossRef]

69. Berbaum, K.S.; Franken, E.A., Jr.; Dorfman, D.D.; Rooholamini, S.A.; Kathol, M.H.; Barloon, T.J.; Behlke, F.M.; Sato, Y.U.T.A.K.A.; Lu, C.H.; El-Khoury, G.Y.; et al. Satisfaction of search in diagnostic radiology. *Investig. Radiol.* **1990**, *25*, 133–140. [CrossRef]

70. Samuel, S.; Kundel, H.L.; Nodine, C.F.; Toto, L.C. Mechanism of satisfaction of search: Eye position recordings in the reading of chest radiographs. *Radiology* **1995**, *194*, 895–902. [CrossRef] [PubMed]

71. Cain, M.S.; Adamo, S.H.; Mitroff, S.R. A taxonomy of errors in multiple-target visual search. *Vis. Cogn.* **2013**, *21*, 899–921. [CrossRef]

72. Berbaum, K.S.; Franken, E.A.; Caldwell, R.T.; Shartz, K. Satisfaction of search in traditional radiographic imaging. In *The Handbook of Medical Image Perception and Techniques*; Krupinski, E.A., Samei, E., Eds.; Cambridge U Press: Cambridge, UK, 2010; pp. 107–138.

73. Berbaum, K.S.; Franken, E.A.; Caldwell, R.T.; Shartz, K.; Madsen, M. Satisfaction of search in radiology. In *The Handbook of Medical Image Perception and Techniques*, 2nd ed.; Samei, E., Krupinski, E.A., Eds.; Cambridge U Press: Cambridge, UK, 2019; pp. 121–166.

74. Berbaum, K.S.; Franken, E.A., Jr.; Dorfman, D.D.; Miller, E.M.; Krupinski, E.A.; Kreinbring, K.; Caldwell, R.T.; Lu, C.H. Cause of satisfaction of search effects in contrast studies of the abdomen. *Acad. Radiol.* **1996**, *3*, 815–826. [CrossRef]

75. Berbaum, K.S.; Franken, E.A., Jr.; Dorfman, D.D.; Miller, E.M.; Caldwell, R.T.; Kuehn, D.M.; Berbaum, M.L. Role of faulty visual search in the satisfaction of search effect in chest radiography. *Acad. Radiol.* **1998**, *5*, 9–19. [CrossRef]

76. Drew, T.; Williams, L.H.; Aldred, B.; Heilbrun, M.E.; Minoshima, S. Quantifying the costs of interruption during diagnostic radiology interpretation using mobile eye-tracking glasses. *J. Med. Imaging* **2018**, *5*, 031406. [CrossRef]

77. Williams, L.H.; Drew, T. Distraction in diagnostic radiology: How is search through volumetric medical images affected by interruptions? *Cogn. Res. Princ. Implic.* **2017**, *2*, 12. [CrossRef] [PubMed]

78. Kundel, H.L. How to minimize perceptual error and maximize expertise in medical imaging. *Proc. SPIE* **2007**, *6515*. [CrossRef]

79. Kundel, H.L.; Nodine, C.F. Interpreting chest radiographs without visual search. *Radiology* **1975**, *116*, 527–532. [CrossRef] [PubMed]

80. Kundel, H.L.; Nodine, C.F.; Conant, E.F.; Weinstein, S.P. Holistic component of image perception in mammogram interpretation: Gaze-tracking study. *Radiology* **2007**, *242*, 396–402. [CrossRef] [PubMed]

81. Kundel, H.L.; Nodine, C.F.; Krupinski, E.A.; Mello-Thoms, C. Using gaze-tracking data and mixture distribution analysis to support a holistic model for the detection of cancers on mammograms. *Acad. Radiol.* **2008**, *15*, 881–886. [CrossRef] [PubMed]

82. Nodine, C.F.; Mello-Thoms, C. The role of expertise in radiologic image interpretation. In *The Handbook of Medical Image Perception and Techniques*; Krupinski, E.A., Samei, E., Eds.; Cambridge U Press: Cambridge, UK, 2010; pp. 139–156.

83. Nodine, C.F.; Mello-Thoms, C. Acquiring expertise in radiologic image interpretation. In *The Handbook of Medical Image Perception and Techniques*, 2nd ed.; Samei, E., Krupinski, E.A., Eds.; Cambridge U Press: Cambridge, UK, 2019; pp. 139–156.

84. Posner, M.I. Orienting of attention. *Quart. J. Exp. Psychol.* **1980**, *32*, 3–25. [CrossRef]

85. Posner, M.I.; Cohen, Y. Components of attention. In *Attention and Performance X*; Bouma, H., Bouwhuis, D.G., Eds.; Erlbaum: Hillside, NJ, USA, 1984; pp. 55–66.

86. Taylor, T.L.; Klein, R.M. On the causes and effects of inhibition of return. *Psychon. Bull. Rev.* **1999**, *5*, 625–643. [CrossRef]

87. Wolfe, J.M.; Horowitz, T.S. What attributes guide the deployment of visual attention and how do they do it? *Nat. Rev. Neurosci.* **2004**, *5*, 495–501. [CrossRef]

88. Wolfe, J.M.; Horowitz, T.S. Five factors that guide attention in visual search. *Nat. Hum. Behav.* **2017**, *1*, 0058. [CrossRef]

89. Carmody, D.P.; Nodine, C.F.; Kundel, H.L. Finding lung nodules with and without comparative visual scanning. *Percept. Psychophys.* **1981**, *29*, 594–598. [CrossRef]

90. Nodine, C.F.; Mello-Thoms, C.; Kundel, H.L.; Weinstein, S.P. Time course of perception and decision making during mammographic interpretation. *AJR Am. J. Roentgenol.* **2002**, *179*, 917–923. [CrossRef] [PubMed]

91. Evans, K.K.; Birdwell, R.L.; Wolfe, J.M. If You Don't Find It Often, You Often Don't Find It: Why Some Cancers Are Missed in Breast Cancer Screening. *PLoS ONE* **2013**, *8*, e64366. [CrossRef] [PubMed]

92. Evans, K.; Haygood, T.M.; Cooper, J.; Culpan, A.-M.; Wolfe, J.M. A half-second glimpse often lets radiologists identify breast cancer cases even when viewing the mammogram of the opposite breast. *Proc. Natl. Acad. Sci. USA* **2016**, *113*, 10292–10297. [CrossRef] [PubMed]

93. Brennan, P.C.; Gandomkar, Z.; Ekpo, E.U.; Tapia, K.; Trieu, P.D.; Lewis, S.J.; Wolfe, J.M.; Evans, K.K. Radiologists can detect the 'gist' of breast cancer before any overt signs of cancer appear. *Sci. Rep.* **2018**, *8*, 8717. [CrossRef] [PubMed]

Review

Using Eye Movements to Understand how Security Screeners Search for Threats in X-ray Baggage

Nick Donnelly [1], Alex Muhl-Richardson [2], Hayward J. Godwin [3,*] and Kyle R. Cave [4]

[1]   Department of Psychology, Liverpool Hope University, Liverpool L16 9JD, UK; donneln@hope.ac.uk
[2]   Department of Psychology, University of Cambridge, Cambridge CB2 3EB, UK; am2662@cam.ac.uk
[3]   Psychology, University of Southampton, Southampton SO17 1BJ, UK
[4]   Department of Psychological and Brain Sciences, University of Massachusetts
      Amherst, Amherst, MA 01003, USA; kcave@psych.umass.edu
*   Correspondence: hg102@soton.ac.uk

Received: 28 February 2019; Accepted: 1 June 2019; Published: 4 June 2019

Abstract: There has been an increasing drive to understand failures in searches for weapons and explosives in X-ray baggage screening. Tracking eye movements during the search has produced new insights into the guidance of attention during the search, and the identification of targets once they are fixated. Here, we review the eye-movement literature that has emerged on this front over the last fifteen years, including a discussion of the problems that real-world searchers face when trying to detect targets that could do serious harm to people and infrastructure.

Keywords: visual search; eye movements; X-ray images; security screening

---

1. Introduction

The job of an airport screener is to stop a wide range of prohibited items being taken on board aircraft. Some of these prohibited items are threats (e.g., guns, knives, explosives) and some are restricted for other reasons (e.g., liquids, narcotics, etc.). Airport screeners search for the presence of these prohibited items in two or three-dimensional images derived from X-raying screening trays. Their job is, therefore, one that requires the inspection of visual images for the presence of prohibited items.

While the ability to detect prohibited items in X-ray images of baggage has been explored for some decades, it came on to the radar of experimental psychologists following the terrorist attacks on the World Trade Center in New York in 2001. From this time onwards, researchers sought to better understand the perceptual and cognitive challenges associated with airport baggage screening. In this review, we report some key data from experimental studies in which eye movements have been recorded in order to better understand how, when, and why errors are made during the search. The results have been informative from both a theoretical and practical perspective.

2. What Are the Images Used in Baggage Screening?

Compared to most real-world search stimuli and medical images, baggage images have less structure. The contents and arrangement of items in bags and of bags on screening trays are difficult to predict. This leaves the screener with few expectations about where these objects should appear within the image, and thus makes decision making more uncertain than in most other search tasks.

Baggage images are typically, but not always, colored (or more correctly *pseudo-colored*) using a standard color mapping (see Figure 1). The artificial color mapping depicts the density of the objects in the image, and also provides some information about object materials. Interpreting density is relatively straightforward: areas with no material or with very low density appear as white, with higher density regions appearing as darker and more saturated colors. If a region is so dense that no X-rays pass through it, it is depicted as black. Interpreting X-ray images is tricky because these arbitrary color mappings produce a visual world that is very different from what we are accustomed to.

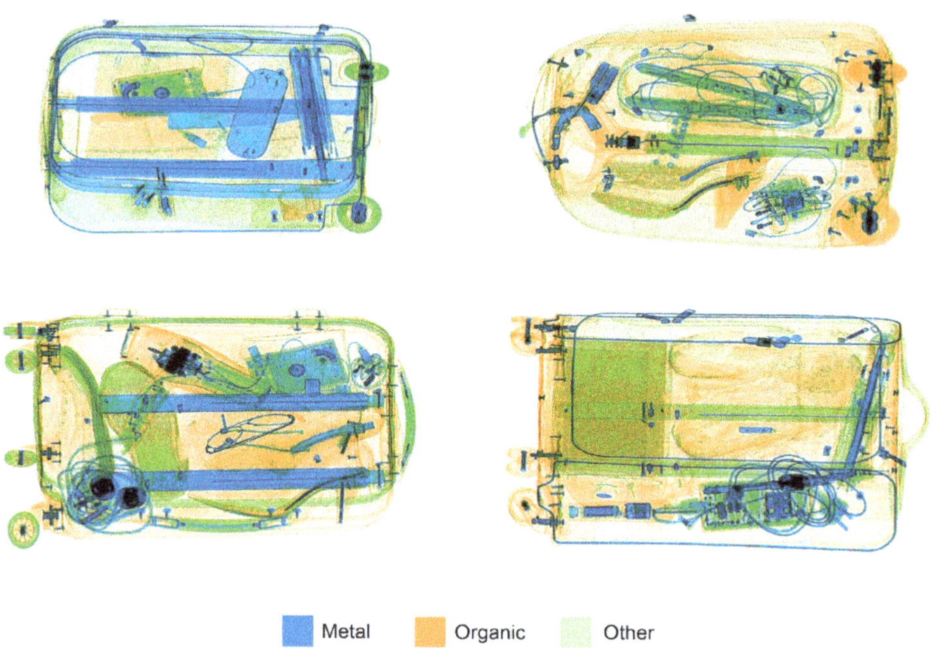

Figure 1. Example X-ray images of bags, two contain threats (top left—knife; bottom right–pistol); images from the CaSePIX image library [1]. Reproduced with permission from the copyright holder Dr Greg Davis.

The depiction of information about the materials making up the objects is more complex. The X-rays that have passed through the objects can be used to infer an atomic number, which is then used to put each part of the image into one of three categories. Those regions with the lowest atomic numbers will include most explosives, but will of course include many non-threatening organic materials as well, such as foods and fabrics. These regions of the image are assigned an orange hue in the standard color mapping. Regions with the highest atomic numbers include metal weapons such as guns and knives, but also many harmless objects, and appear as blue. Atomic numbers in between these ranges are often shown as green, and can generally be regarded as less threatening because they are not metal and are not explosives.

The challenge to the interpretation of X-ray images is increased because all but the densest objects have a certain amount of transparency. Once X-rays have been emitted from the X-ray sources and passed through one object, they may pass through other objects before striking the collector and being registered. Thus, one part of the image often depicts information about multiple objects. This overlap introduces the additional problem of segmenting the different objects from one another.

There is another aspect of object overlap in X-ray images that is very different from our usual experience. When an orange, green, or blue hue is assigned to each part of the image, it is based on an *average* of the atomic numbers of the different materials that the X-ray beam has passed through. Thus, if an organic object with atomic number in the low range overlaps with a metal object with atomic number in the high range, the region of overlap in the image may be depicted as belonging to the middle range.

Clearly, interpreting overlapping objects in X-ray images requires careful consideration. Adding to the challenge is that these searches are often conducted in a busy work environment with time pressure from long lines of passengers who are concerned about missing a flight. Thus, it is important for us to find ways to make the task easier and to increase accuracy and efficiency. This is particularly important in light of the fact that, although search times (and limits) vary from airport to airport and from country to country, screeners often will take around only six seconds to complete their searches [2].

Recent innovations to security screening include dual view where screeners make their decisions from interpreting two orthogonal views (and sometimes a photographic image of the tray as well) [3]. Computed tomography (CT) scanners have also become available as an alternative to more traditional X-ray machines. An apparent advantage of CT scanners is that they can generate a 3D image of a tray which can be manipulated on-screen, including advancing through 2D slices of the full 3D volume [4]. The challenge to threat detection in baggage search that these conditions create is altered but not fundamentally changed by recent innovations aimed at supporting image interpretation by providing richer imagery.

While these advanced imaging systems add information that should help in the detection of prohibited items, they may also create new challenges. We note two of these challenges here as we think they are important, despite being largely unexplored. First, search and interpretation from viewing two orthogonal views requires the ability to make eye movements to the same or related locations on the different images showing orthogonal views. As far as we are aware there is only a single unpublished study on the success and utility of making these eye movements with respect to baggage screening [5]. Second, radiological imaging requires searching for targets in slices of a 3D volume, but what is being imaged in medical imaging is of a known structure and complexity, and what is being identified are variations from a norm. The risk to baggage screening of using this technology is that features consistent with an early emerging target may be 'tracked' through slices, leading to misses from targets appearing later in the sequence of slices. We know of only a single study that has explored an analogue of this situation [6].

3. What Are the Decisions Made by Baggage Screeners?

While we refer to baggage search throughout this review, in line with the existing literature on this topic, in the real world of baggage screening, the images are more correctly defined as the contents of *screening trays*. Screening trays can contain bags, laptops, and other items that are routinely X-rayed such as keys, wallets and mobile telephones. The clarification is an important one to make. It is important to be clear that the visual search and threat identification challenge that these different situations deliver for screeners changes somewhat unpredictably on a tray-by-tray basis. The important point is that some decisions can be made without the need for extensive searching at all, but on the basis that the gist of the tray indicates the presence of very little or nothing that comes close to resembling a target. We define gist here in a very limited way as providing an estimation of the clutter of the objects within a tray. Virtually empty trays may allow rejection on the basis that the limited material present can easily be discounted as resembling a target. We do not mean to imply gist emerging from object co-occurrence, configurations or scene backgrounds [7].

A screener can either "clear" a bag and let it pass, or "reject" it. A bag may be rejected because a prohibited object(s) can be seen and readily identified in an image. However, a bag may also be "rejected" simply because the screener cannot be certain that prohibited objects(s) are not present in an image, probably because there are many objects overlapping one another. In either case, the decision to reject bags usually leads to further investigation through hand searching.

4. The Security Search Task

Having considered the images from which 'clear' and 'reject' decisions are made and the nature of those decisions, we now turn to eye movements and the security search task. The color coding of atomic density allied to the uncertainties created by contents, the arrangement of contents, and viewpoint mean that color is often used to guide attention to possible targets. Fortunately, we know quite a lot about how people search for color and the eye movement behavior associated with this search. In addition, studies have explored the costs associated with simultaneously searching for more than one color, as is the case when searching for, for example, the orange indicating possible plastic explosive and the blue/black indicating possible wire, guns or knives.

There are a number of prominent models of visual search [8–10], and fewer such models of visual search and eye movement behavior [11,12]. Despite their differences, the models of search share the common idea that, during the search, our limited resources need to be directed to given objects in order to determine whether those objects are targets [13]. In the classic Guided Search model, which has been revised several times since its inception [8,14–16], attention is directed towards target-similar objects, one at a time, until a target is found, or until the searcher decides to quit. The decision to terminate search can depend on a number of factors, including the time spent searching and errors made on previous trials [17]. Models of eye movements and visual search share much the same overall architecture, though they do generally focus more on the mechanisms that govern when and where to move the eyes during a search [11,18,19].

The recording of eye movements during the search opened up new avenues for understanding the information processing that takes place during the search. Many early studies of eye movements and the search of complex images involved radiographic image screening. These studies adapted and developed the language of guidance in visual search to focus on the fractionation of what takes place in terms of information-processing during visual search [20,21]. Put simply, by recording eye movement behavior, researchers were able to determine when, how and why targets were missed; this was, and still is, a question of vital importance when it comes to real-world search tasks in which missing a target can have severe consequences. These studies developed a new framework for understanding eye movements during the search by focusing on failures of *guidance* and failures of *decision-making*.

Failures of guidance were measured using three eye movement metrics: namely, the probability that a target was fixated during a search, the speed at which a target was fixated during the search, and the number of nontargets fixated. Connecting with classic models of search, such as Guided Search [8], under this view, an easier search task can be conceptualised as one wherein the target is directly and rapidly fixated. As a search task becomes more difficult, there is a longer delay before the target is fixated, as more and more nontargets are fixated, and the target is less likely to be fixated at all. As a direct consequence of this reduction in the efficacy of guidance, targets in more difficult search tasks can then be missed purely because of a failure in guidance. An early eye tracking study by McCarley and colleagues [22] found that after practice, subjects fixated fewer nontargets, but were no more likely to fixate the target. We will discuss what factors make search tasks more difficult in the context of baggage screening in more detail below.

Even if a target is directly fixated, this is not a guarantee that it will be detected. Such failures of decision-making in search can also be measured using two eye movement metrics: the time between fixating the target and identifying it as a target, and the probability of detecting a target after fixating it. When a target is more complex or difficult to recognise, this increases the time required to identify it, and, worse still, reduces the probability that it will be detected even after being directly fixated. The study mentioned earlier by McCarley et al. [22] found that search practice substantially improved target identification, although some of that improvement was tied to specific target images.

With this basic framework now set out, we will turn to outlining several ways in which exactly these forms of error have been studied extensively in the context of tasks inspired by airport X-ray baggage screening, beginning with the study of target templates in baggage search.

An understanding of the time taken to search baggage comes from studies of attentional control and guidance. A number of different theories of attentional control [8,10] are built on the assumption that early visual processing provides limited types of information that can be used to guide attentional selection. In difficult searches like baggage search, this attentional guidance is based on a mental representation with information about the target to be found, sometimes called the "target template".

Once attentional guidance has determined which portions of the input should be selected, those portions are then processed more fully to recognize objects and interpret their configurations relative to one another. In general, the information stored in the target template seems to be constrained by the rather severe limitations that apply to visual working memory. Despite those limits, visual search can be very flexible, as demonstrated by experiments by Wolfe and colleagues [23,24] in which participants are asked to search for as many as 100 different targets. These searches probably do not benefit from the accurate guidance that is possible when searching for a single color or orientation.

In bags that are largely empty, the guidance of attention may be so precise that attention is quickly directed to the search target, or the absence of any target can be quickly determined. Guidance may instead be imprecise, perhaps because it has not been possible to develop an effective target template (although even target templates that inaccurately specify target features afford guidance proportional to their target similarity [25]). In addition, it may also be the case that a bag is so densely packed with multiple metal and organic objects that focused attention is required to segment and identify and possible objects. Many baggage searches will fall somewhere in between these easy and difficult extremes. In general, attention should be guided toward the dark blue and black regions that might be metal weapons, and to the orange regions that might be explosives. By preventing attention to the light and green regions, guidance can limit the amount of time spent in attentional processing.

Attentional guidance is fairly accurate when a target is known to have a specific color [26]. Information about the target color can be used to form a mental representation or template to guide attention toward items with similar colors, avoiding most of the other colors. Security search is somewhat more complicated than this simple single-color search, however: the threat targets can be either black, blue, or orange, and the regions that can be ruled out based on color can be either green or low-saturation colors close to white. Performance in searches for multiple targets is often worse in the search for two targets than in the search for a single target, although the nature of the cost varies depending on the task. Sometimes, accuracy is lower or response times are longer in dual-target search, as shown by Menneer and colleagues [27] in searches for abstract target stimuli defined by color, orientation, or shape. Menneer et al. [28] also found dual-target costs in accuracy for searches among X-ray images of objects, as long as the two targets differed from one another in color. Eye-tracking studies, both with abstract shapes [29] and with X-ray images [26] have helped to illustrate one source of the dual-target cost. In those experiments, participants who were searching for two types of targets with different colors made a number of fixations to distractors that had colors different from either target.

The drop in search efficiency between 1-color search and 2-color search was explored with a set of very basic abstract stimuli by Stroud et al. [30]. The target was distinguished from distractors by a difficult shape discrimination, but it could nonetheless be found quickly if its color was known in advance. In 1-color search, fixation rates to colors that were very different from the target were low, demonstrating effective color guidance. However, when the target could appear in either of two very different colors, there were many more fixations to colors that were very different from the distractor. Dual-target search was much slower because participants were spending time fixating distractors that should have been excluded by color guidance. Search guidance could be degraded further by expanding the set of possible target colors to eight, so that the range of possible colors covered half of the possible hues [31]. The implication of this finding for baggage screening is clear: the simultaneous search for more than one target reduces guided eye movements to targets and increases unguided eye movements to things that are very unlikely to be targets. The time limited aspect of baggage screening makes the increased number of unguided fixations in multiple target versus single target search a significant problem when considered in the context of the job.

While color may be the best cue to guide attention in baggage search, it is not a reliable cue to the presence of prohibited items, as many non-prohibited items are also coded in orange and blue. Participants in these studies have the option to employ color guidance or not in these particular tasks because the target can often be identified by shape. The Stroud et al. study shows that if many of the items with the target color also have a shape that makes them a distractor, then participants are much less likely to guide their eye movements by color, but if most of the items with the target colors are actual targets, then there will be a stronger tendency to direct fixations to the target colors and away from colors that were never targets [31]. This pattern suggests that participants have some high-level control over how guidance is used in search: if they expect color to lead them to a target on a high proportion of trials, then they will employ color guidance more often. For baggage screeners to make use of color to guide their eye movements, they may well need to resist this tendency, and instead decide to use color guidance, despite the low likelihood that what is fixated will be a prohibited item.

Other experiments have demonstrated flexibility in search guidance that may be informative to understanding how to manage searching for multiple targets. In an eye-movement study by Beck, Hollingworth, and Luck [32], participants could either search for two target colors simultaneously, or search first for one color and then switch to the other, depending on the instructions. When left to choose their search method on their own, participants seem to do something in between purely simultaneous and purely successive search: they often switch between fixating one target color and fixating the other, but at a given moment during the search they seem to favour one target over the other [33].

In summary, when a search task, like baggage search, requires holding two colors in working memory, participants are more likely to perform the task without guiding attention by color, even though guidance may make their search more efficient. This occurs whether both colors are search targets, or one is held for another task [34]. When guidance is not used, there may be many fixations to distractors that could be avoided, and the search may take much longer to complete.

5. Identification of Complex, Overlapping Transparent Objects

As noted above, one of the key problems associated with searching through baggage X-rays is the fact that they contain a wide array of varied overlapping and transparent objects. The manner in which these objects combine creates a uniquely difficult task wherein objects can be difficult to identify (see reference [35] for an excellent illustration of this point).

Applying what is known regarding human search behavior to the basic properties of the images that screeners search through is a difficult task, mainly because the vast majority of visual search experiments have utilised displays wherein objects do not overlap with one another. Studies that have examined the deleterious effect of searching through cluttered, overlapping displays [36,37] find that search is impaired for these difficult images.

More recently, we conducted a series of experiments [38] that focused on the problem of overlap and transparency in search and also recorded the eye movement behavior of participants as they searched the displays. We found that, for opaque displays, target detection rates fell substantially and reaction times increased when object overlap increased. We also found that perceptual selection and perceptual identification errors increased substantially as overlap increased. Moving to transparent displays, the same basic pattern emerged, with the important difference that reaction times were even longer for higher levels of overlap than in the opaque displays. Crucially, this was likely to be the result of the fact that, in transparent images, the complexity of the displays *obscures* but does not *remove* information as it does in opaque displays. The cost associated with perceptual identification in transparent images led to long verification times because of the need to examine multiple different possibilities of grouping and object identity before a final decision could be reached.

Despite the apparent concerns regarding the effects of overlapping images in complex displays, there is reason to be confident that there may be ways to assist those searching these difficult images. In the same set of studies [38], we also found that presenting the objects on separate three-dimensional depth planes aids search and ameliorates the effects of overlap, suggesting that future research would benefit from further understanding the benefits of depth in aiding in the segmentation and identification of complex, overlapping objects during the search.

6. The Problem of Low Prevalence

It is fortunate that targets in X-ray baggage are very rare indeed, but the relative scarcity of targets can influence the likelihood of detecting those targets once they finally do appear. One line of studies focuses on the prevalence of targets, which is defined as the proportion of trials on which a target is presented. These studies demonstrate that low target prevalence, e.g., 2%, results in a reduction in the target detection rates compared to higher prevalence levels, e.g., 50% [39–41].

Early studies of this prevalence effect used behavioral measures, including response times, response accuracy, and signal detection theory measures [42]. The general finding was that reductions in target prevalence resulted in a shift in the response criterion such that participant responses became more conservative and less likely to respond that a target was present. By this we mean that their hit rate and false alarm rate were reduced and target-absent responses were sped up relative to when target prevalence was higher [41,43–47].

Later and more recent studies of eye movement behavior in conditions of varied prevalence have helped to elaborate on how, when and why rare targets are missed when prevalence is low. For example, a reduction in target prevalence increases failures of perceptual selection, whereby searchers fixate fewer objects in each trial as prevalence is reduced [48]. It is worth noting that efforts to overcome this shortcoming by providing feedback in relation to where fixations have and have not been made are unlikely to help [49,50].

Moreover, the failure of perceptual selection is added to by the fact that targets are fixated more slowly when target prevalence is low [48]. Moreover, a reduction in target prevalence also increases the likelihood of perceptual identification errors, with participants not just being slower to identify targets after fixating them, but also less likely to detect targets after fixating them [48,51,52].

It is worth noting that some of the challenges caused by the low prevalence of actual threat items are mitigated in the real world of baggage screening by the inclusion of other prohibited items that must be searched for. Even without the inclusion of other prohibited items that must be searched for, the detection of threat items may be aided by the inclusion of Threat Image Projection (TiP) items amongst baggage images. TiP items are fictitious guns, knives and IEDs that are pseudo-randomly presented in the context of whole bags to screeners. TiP items are always of low prevalence - one published study reported on TiP detection rates using data from real screeners performing their jobs in situ [53] where the prevalence rate was set at 4%. That low prevalence rate is, however, higher than the actual incidence of these kinds of threats.

Although target prevalence remains an issue for baggage screeners, they are detecting prohibited items, including threat items, more often than might be supposed. Importantly, with the inclusion of prohibited items beyond threats, they are doing so at a higher rate than envisaged in some experimental analogues of baggage screening where target prevalence is manipulated.

7. Differences in Individual Screeners

So far we have considered how studying eye movements has informed us of the challenges associated with finding and identifying prohibited items during baggage search. In doing so it seems to us that key eye movement metrics reflect different stages of processing and decision making associated with baggage search. In this final section, we consider how these metrics might be influenced by individual differences.

As a population, baggage screeners vary in their age, experience and training, in addition to their cognitive abilities and affective characteristics. Differences in screener performance are certainly contributed to by individual differences in basic perceptual processing (e.g., references [35,54]). There is evidence for reliable individual differences over time in sensitivity and some evidence of modest age-related decline that has been attributed to reducing efficiency of perceptual and cognitive abilities that cannot be overcome by years of performing the screening task [54,55].

More importantly, and perhaps not surprisingly, differences in screener performance are contributed to by training focussed on developing robust templates of prohibited items [35]. A question that emerges is whether training improves search guidance to possible prohibited items or verification for identification? While there is likely to be some improvement in both, there is good evidence that the effect of training is more striking on processes associated with verification than guidance [22,56]. The improvement in target detection that comes with training is more striking with respect to IEDs than guns, knives and other kinds of prohibited items [57]. It seems that training mostly (though not exclusively) improves the robustness of the templates that allow screeners to more readily match what is seen to what they know, and improvement is greatest when there is most to learn.

A pattern emerges of good screeners being defined as having (1) an ability to make a fast and accurate 'clear' decision based on the gist derived from the image; (2) excellent guidance of attention to potential prohibited items; (3) speed in verifying or rejecting item identity based on having robust target templates to match items against; while also ensuring that they are (4) sufficiently exhaustive in search to ensure all possible targets are investigated and all viable interpretations of items are tested against target templates.

While these are distinct issues that could be explored in future studies, the current literature allows only a coarser analysis, which we consider in terms of three broad categories. The first is how working memory capacity and attentional control might influence search and guidance. The second is how factors related to conservatism in decision-making influence the thoroughness of search and the time given to verify target presence and absence (and how this relates to the first category). Finally, we consider a third broad category of individual differences that may relate specifically to search through slices of a 3D data volume.

Our goal in exploring these issues is simple. Prior evidence clearly shows that baggage screening performance is associated with basic perceptual skills and task knowledge. Beyond that, is there any evidence of systematic relationships between cognitive and affective factors and eye movement behavior when searching complex images then we might be able to use to inform the selection of baggage screeners? We ask this question in light of the four characteristics of good screening that were outlined above and in the spirit of a hypothesis worthy of exploration. We do not have a view of the relative importance of the individual differences that we discuss for baggage screening, especially since they may jointly influence the performance of any given individual.

8. Working Memory Capacity and Attentional Control

While working memory capacity (WMC) does not predict performance in very simple searches, it does in complex search tasks, including those that demand sustained high levels of attentional control [58–60]. We previously discussed how target templates are held or processed in visual working memory (e.g. references [61,62]). It follows from this that effective dual-target search will typically involve more WMC (even in searches which require long-term memory storage [63], WM is still needed for encoding, retrieving and maintaining templates [64]).

It is also conceivable that holding two targets in WM, or one target in conjunction with an item from a simultaneous memory task, may require some sort of memory organization or segregation to minimise interference. Maintaining this WM segregation may require more resources than merely searching for a single target. This is consistent with both the study noted earlier which showed that adding extra WM load can interfere with search guidance and lead to more unguided fixations [34] and with evidence that when search distractors match a color held in WM, saccades are slower and less accurate [65]. Greater WMC must therefore be advantageous in dual-target search tasks that involve the guidance of eye movements to two targets [66].

Extra resources will also help sustain high levels of attentional control and avoid erroneous eye movements due to the conflicts that must be resolved as different targets try to pull attention in different directions [67]. Furthermore, maintaining attentional control is likely to be particularly challenging when a dual-target search is coupled with low levels of target prevalence [68,69]. It is beyond the scope of the current review to discuss the relationship between WM and attention at any length (see reference [70] for more detail), but we can say that WMC predicts attentional control in an antisaccade task and, in the same study, high WMC participants exhibited less of a performance cost when switching from an anti- to a prosaccade task [71]. Similar costs may be observed when two target templates are active successively and guidance shifts from one target to another over time; not only will extra resources be needed to coordinate the switch, but residual information from previous task sets, including target templates, will also increase the likelihood of eye movements to irrelevant items [72–75].

As individuals search, some of the resources that could have been allocated to dual-target guidance may instead be engaged in other cognitive tasks; they may be monitoring how well they are doing in the task, or speculating on the motivation for the experiment, or making plans for the rest of the day [76]. Participants may also be motivated to hold some resources in reserve due to a built-in aversion to high levels of resource utilization that ensures resource availability for unanticipated future tasks. In summary, while it not been tested directly, it follows from these behavioral and eye movement laboratory studies that good baggage screening performance would be associated with high WMC and good attentional control.

9. Setting Conservative Decision Thresholds

While low target prevalence makes search and target verification more challenging, good baggage screeners will be relatively resistant to these effects; they will tend to ensure all locations that might contain a target are searched and that those locations are processed until all possible interpretations of image features are considered before deciding to clear or reject. Doing so equates to good baggage screeners setting conservative decision thresholds for terminating their inspection of individual fixations during the search and for continuing fixations before ending search, including when no target is found. The ability to set conservative decision thresholds will be influenced by training and a number of studies have examined the performance of trained screeners [55,77]. In these studies, screener performance is often characterized in terms of Signal Detection Theory measures that reflect the relationship between screeners' hit rates and false alarm rates [78].

Apart from training, the setting of decision thresholds will also be influenced by the individual tendency to be 'satisfied' with search. 'Satisfaction of search' was a term used primarily in the context of radiographic screening to describe failures to detect subsequent targets following the detection of a first [79]. Different accounts of this effect exist, but it is likely that it is not related to satisfaction at all, but rather a combination of a bias towards finding subsequent targets that are similar to an initial target (perceptual set bias) and the depletion of cognitive resources associated with finding an initial target [80,81]. This phenomenon has since been renamed 'subsequent search misses' to reflect the contribution of these other factors [81]. Individual tendencies linked to satisfaction are an important source of individual variance in the setting of quitting thresholds and response criteria during the search [82–84]. Consistent with the notion of resource depletion, recent evidence links individual differences in WMC to errors of both perceptual selection (failure to fixate targets) and perceptual identification (failure of verification) during the search [85]. In a follow-up study, the effect of WMC on perceptual selection errors was attributed to quitting thresholds that were not adequately conservative; that is to say, searchers quit before they had fixated targets [68].

There are a number of psychological measures that tap into individual tendencies for satisfaction during the search. One of these is the 'Maximization Scale' [86], a 13-item personality scale that assesses the degree to which an individual is a 'maximizer', who will strive for the best outcome, or a 'satisficer', who will accept outcomes that are good enough. Maximization is also conceptually linked to perfectionism and it has been demonstrated that perfectionists who engage with X-ray search and object identification tasks perform these tasks more accurately and faster than other individuals [87]. Attention to detail, as assessed by a subscale of the Autism Quotient, predicts accuracy in X-ray baggage search and has been the basis of a recent scale developed specifically to assess aptitude for X-ray baggage screening, the XRIndex [88].

One component of trait anxiety, Intolerance of Uncertainty (IU), may also lead to a tendency to settle for poorer performance in a search. IU represents the extent to which individuals experience (and can cope with) worry about uncertain future events [89,90] and can bias decision making towards minimizing uncertainty. IU positively predict false alarm rates in a complex search task where participants searched for low prevalence color targets. In this task the color targets appeared in arrays of colored squares whose appearance changed dynamically over time through an ordered color space [6]. The increased false alarms reflect a decision to prematurely classify as targets items that are still distractors (and have yet to become targets).

Compared to simple laboratory search tasks, the complexities of X-ray search may increase the likelihood of individuals who tend to be easily satisfied, or those with greater IU, to terminate searches prematurely (i.e. lowering quitting thresholds). We suspect that there are some contingencies between the thresholds for terminating fixations both to possible targets and in target verification, WMC, maximisation and IU. At the very least, it is an area worthy of further research.

10. Searching through Three-Dimensional Volumes

Earlier in this review we discussed how baggage search is merely altered, and not fundamentally changed, by advances in screening technology. One of the ways in which screening is altered by CT scanning technology is that screeners are able to search through sequential 2D slices of a 3D volume. Interacting with images in this way introduces a dynamic element into the search task, whereby the image to be searched changes dynamically as screeners move through the slices. Each 2D slice is also related to the slices either side of it in 3D space, such that moving through slices will cause objects to emerge over time in a predictable way. This technology is relatively new in baggage screening and the implications of it for the eye movements made during the search are unclear (though see reference [91] for a discussion of related issues in medical imaging).

A priori it seems to us that searching for prohibited items with a technology that allows search to have a predictive aspect across 'slices' brings a risk that search can become unduly focussed on these locations. From a different paradigm, there is evidence that individual differences in WMC and IU influence the number of fixations made when monitoring dynamically changing color displays for the onset of targets [6]. For individuals with low WMC, those with high IU made fewer eye movements relative to those with low IU. This issue requires much further investigation. Nevertheless, this result emphasizes the importance of measuring eye movements as a tool to help develop our understanding some of the psychological challenges that new ways of the data visualizations bring.

11. General Discussion and Summary

We have known for a long time that security searches of baggage can be very difficult. Security screeners start at a disadvantage when searching through X-ray images, because those images differ in fundamental ways from real-world images. The objects do not appear in the colors that we associate with them, and overlapping objects produce a very unnatural sort of transparency. After years of research, we now have a better understanding of some of the other factors can make searches especially difficult.

Some factors, such as the large numbers of objects in some bags and the degree to which those objects overlap one another in the image, may be addressable by developing new technology that can enhance images, perhaps by rotating them or presenting them in depth. Eye tracking has been a useful tool in assessing technologies and measuring how searchers use them, and as both the display technologies and the experimental methods advance, we can expect eye tracking studies to be even more valuable in assessing future technological advances.

Other factors that complicate X-ray security searches include limitations on cognitive mechanisms for attention and object recognition. Eye tracking data have allowed us to better understand some of these limitations, such as the prevalence effect and the dual-target cost. It may be possible to partly or fully overcome some of these limitations by developing new training methods. For instance, it may be possible to train subjects to search thoroughly even though targets rarely appear, or to guide their search effectively using multiple target templates. Once those methods are developed, additional eye tracking studies will allow us to measure their benefits.

If some of these limitations cannot be eliminated by training, it may still be possible to improve security searches by devising ways to identify those individuals with the most effective perceptual and attentional mechanisms for this type of search. The studies reviewed above suggest that testing for WMC, maximization and tolerance to uncertainty may help to find better screeners. Research on individual differences that are relevant to security search is just getting off the ground, so there is probably lots to learn here.

The eye tracking studies reviewed here have given us a better understanding of the many factors that make security searches difficult, and in some cases they are helping to find methods of improving performance. These are, of course, important contributions to the safety and well-being of large numbers of people, but there have been other benefits to emerge from this research as well. In addition to the practical improvements in detecting threats, this research has contributed to our understanding of visual search and the mental processes that control attention and object identification. For instance, one set of studies shows that decisions to stop search are not driven solely by the stimuli but are influenced by internal factors that have their origins outside of the attentional system. Another set of studies shows that subjects do not effectively guide their attention when searching for two colors if they can instead identify the target with a single shape discrimination, even though this method makes the search much less efficient. These findings open up new questions about the basic nature of attentional control that go beyond security search. We can expect that future research in this area will lead to further advances in our theoretical understanding of visual cognition, while also enhancing our abilities to detect threats.

Funding: This research was funded by a Defence and Security Accelerator grant awarded to Nick Donnelly and Hayward J. Godwin (ref: ACC6000166). Kyle R. Cave was supported by sabbatical visit funding from the University of Massachusetts Amherst and Liverpool Hope University.

Conflicts of Interest: The authors declare no conflict of interest.

References

1.	Davis, G. CaSePIX X-ray Image Library. University of Cambridge: Cambridge, UK, Unpublished work. 2017.
2.	Harris, D.H. How to Really Improve Airport Security. *Ergon. Des.* **2002**, *10*, 17–22. [CrossRef]
3.	Franzel, T.; Schmidt, U.; Roth, S. Object Detection In Multi-view X-Ray Images. In Proceedings of the Joint DAGM (German Association for Pattern Recognition) and OAGM Symposium, Graz, Austria, 28–31 August 2012; Pinz, A., Pock, T., Bischof, H., Leberl, F., Eds.; Springer: Berlin/Heidelberg, Germany, 2012; pp. 144–154.
4.	Hättenschwiler, N.; Mendes, M.; Schwaninger, A. Detecting Bombs in X-Ray Images of Hold Baggage: 2D Versus 3D Imaging. *Hum. Factors* **2019**, *61*, 305–321. [CrossRef] [PubMed]
5.	Koehler, K.; Eckstein, M.P. Beyond Scene Gist: Objects Guide Search More Than Scene Background. *J. Exp. Psychol. Hum. Percept. Perform.* **2017**, *43*, 1177–1193. [CrossRef] [PubMed]
6.	Wolfe, J.M.; Cave, K.R.; Franzel, S.L. Guided Search: An Alternative To The Feature Integration Model For Visual Search. *J. Exp. Psychol. Hum. Percept. Perform.* **1989**, *15*, 419–433. [CrossRef] [PubMed]
7.	Duncan, J.; Humphreys, G. Beyond The Search Surface: Visual Search And Attentional Engagement. *J. Exp. Psychol.* **1992**, *18*, 578–588. [CrossRef]
8.	Treisman, A.; Gelade, G. A Feature-Integration Theory Of Attention. *Cogn. Psychol.* **1980**, *12*, 97–136. [CrossRef]
9.	Zelinsky, G.J. A Theory of Eye Movements During Target Acquisition. *Psychol. Rev.* **2008**, *115*, 787–835. [CrossRef] [PubMed]
10.	Hulleman, J.; Olivers, C.N.L. The Impending Demise Of The Item In Visual Search. *Behav. Brain Sci.* **2017**, *40*, e132. [CrossRef] [PubMed]
11.	Eckstein, M.P. Visual Search: A Retrospective. *J. Vis.* **2011**, *11*, 14. [CrossRef]

12. Wolfe, J.M. Guided Search 4.0: Current Progress With A Model Of Visual Search. In *Integrated Models of Cognitive Systems*; Gray, W., Ed.; Oxford University Press: New York, NY, USA, 2007; pp. 99–120.

13. Wolfe, J.M. Guided Search 2.0 A Revised Model of Visual Search. *Psychon. Bull. Rev.* **1994**, *1*, 202–238. [CrossRef]

14. Wolfe, J.M.; Gancarz, G. Guided Search 3.0. In *Basic and Clinical Applications of Vision Science*; Lakshminarayanan, V., Ed.; Springer: Dordrecht, the Netherlands, 1997.

15. Chun, M.M.; Wolfe, J.M. Just Say No: How Are Visual Searches Terminated When There Is No Target Present? *Cogn. Psychol.* **1996**, *30*, 39–78. [CrossRef] [PubMed]

16. Findlay, J.M.; Walker, R. A Model Of Saccade Generation Based On Parallel Processing And Competitive Inhibition. *Behav. Brain Sci.* **1999**, *22*, 661–674; discussion 674–721. [CrossRef] [PubMed]

17. Tatler, B.W.; Brockmole, J.R.; Carpenter, R.H.S. LATEST: A Model Of Saccadic Secisions In Space And Time. *Psychol. Rev.* **2017**, *124*, 267–300. [CrossRef] [PubMed]

18. Kundel, H.L.; La Follette, P.S. Visual Search Patterns And Experience With Radiological Images. *Radiology* **1972**, *103*, 523–528. [CrossRef] [PubMed]

19. Nodine, C.F.; Kundel, H.L. Using Eye Movements To Study Visual Search And To Improve Tumor Detection. *RadioGraphics* **2013**, *7*, 1241–1250. [CrossRef]

20. Mccarley, J.S.; Kramer, A.F.; Wickens, C.D.; Vidoni, E.D.; Boot, W.R. Visual Skills In Airport-Security Screening. *Psychol. Sci.* **2004**, *15*, 302–306. [CrossRef] [PubMed]

21. Wolfe, J.M. Saved By A Log: How Do Humans Perform Hybrid Visual And Memory Search? *Psychol. Sci.* **2012**, *23*, 698–703. [CrossRef]

22. Wolfe, J.M.; Aizenman, A.M.; Boettcher, S.E.P.; Cain, M.S. Hybrid Foraging Search: Searching For Multiple Instances Of Multiple Types Of Target. *Vis. Res.* **2016**, *119*, 50–59. [CrossRef]

23. Hout, M.C.; Goldinger, S.D. Target Templates: The Precision Of Mental Representations Affects Attentional Guidance And Decision-Making In Visual Search. *Atten. Percept. Psychophys.* **2015**, *77*, 128–149. [CrossRef]

24. Menneer, T.; Stroud, M.J.; Cave, K.R.; Li, X.; Godwin, H.J.; Liversedge, S.P.; Donnelly, N. Search For Two Categories Of Target Produces Fewer Fixations To Target-Color Items. *J. Exp. Psychol. Appl.* **2012**, *18*, 404–418. [CrossRef]

25. Menneer, T.; Phillips, L.; Donnelly, N.; Barrett, D.J.K.; Cave, K.R. Search Efficiency For Multiple Targets. *Cogn. Technol.* **2004**, *9*, 22–25.

26. Menneer, T.; Cave, K.R.; Donnelly, N. The Cost Of Search For Multiple Targets: Effects Of Practice And Target Similarity. *J. Exp. Psychol. Appl.* **2009**, *15*, 125–139. [CrossRef] [PubMed]

27. Stroud, M.J.; Menneer, T.; Cave, K.R.; Donnelly, N.; Rayner, K. Search For Multiple Targets Of Different Colors: Misguided Eye Movements Reveal A Reduction Of Color Selectivity. *Appl. Cogn. Psychol.* **2011**, *25*, 971–982. [CrossRef]

28. Stroud, M.J.; Menneer, T.; Cave, K.R.; Donnelly, N. Using The Dual-Target Cost To Explore The Nature Of Search Target Representations. *J. Exp. Psychol. Hum. Percept. Perform.* **2012**, *38*, 113–122. [CrossRef] [PubMed]

29. Stroud, M.J.; Menneer, T.; Kaplan, E.; Cave, K.R.; Donnelly, N. We Can Guide Search By A Set Of Colors, But Are Reluctant To Do It. *Atten. Percept. Psychophys.* **2019**, *81*, 377–406. [CrossRef] [PubMed]

30. Beck, V.M.; Hollingworth, A.; Luck, S.J. Simultaneous Control Of Attention By Multiple Working Memory Representations. *Psychol. Sci.* **2012**, *23*, 887–898. [CrossRef] [PubMed]

31. Cave, K.R.; Menneer, T.; Nomani, M.S.; Stroud, M.J.; Donnelly, N. Dual Target Search Is Neither Purely Simultaneous Nor Purely Successive. *Q. J. Exp. Psychol.* **2018**, *71*, 169–178. [CrossRef]

32. Menneer, T.; Cave, K.R.; Kaplan, E.; Stroud, M.J.; Chang, J.; Donnelly, N. The Relationship Between Working Memory And The Dual-Target Cost In Visual Search Guidance. *J. Exp. Psychol. Hum. Percept. Perform.* **2019**. [CrossRef] [PubMed]

33. Hardmeier, D.; Hofer, F.; Schwaninger, A. The X-Ray Object Recognition Test (X-Ray ORT)—A Reliable And Valid Instrument For Measuring Visual Abilities Needed In X-Ray Screening. In Proceedings of the 39th Annual 2005 International Carnahan Conference On Security Technology, Las Palmas, Spain, 11–14 October 2005; pp. 189–192.

34. Verghese, P.; Mckee, S.P. Visual Search In Clutter. *Vis. Res.* **2004**, *44*, 1217–1225. [CrossRef] [PubMed]

35. Adamo, S.H.; Cain, M.S.; Mitroff, S.R. Targets Need Their Own Personal Space: Effects Of Clutter On Multiple-Target Search Accuracy. *Perception* **2015**, *44*, 1203–1214. [CrossRef]

36. Godwin, H.J.; Menneer, T.; Liversedge, S.P.; Cave, K.R.; Holliman, N.S.; Donnelly, N. Adding Depth To Overlapping Displays Can Improve Visual Search Performance. *J. Exp. Psychol. Hum. Percept. Perform.* **2017**, *43*, 1532–1549. [CrossRef] [PubMed]
37. Mitroff, S.R.; Biggs, A.T. The Ultra-Rare-Item Effect: Visual Search For Exceedingly Rare Items Is Highly Susceptible To Error. *Psychol. Sci.* **2014**, *25*, 284–289. [CrossRef] [PubMed]
38. Schwark, J.D.; Macdonald, J.; Sandry, J.; Dolgov, I. Prevalence-Based Decisions Undermine Visual Search. *Vis. Cogn.* **2013**, *21*, 541–568. [CrossRef]
39. Menneer, T.; Donnelly, N.; Godwin, H.J.; Cave, K.R. High Or Low Target Prevalence Increases The Dual-Target Cost In Visual Search. *J. Exp. Psychol. Appl.* **2010**, *16*, 133–144. [CrossRef] [PubMed]
40. Green, D.M.; Swets, J.A. *Signal Detection Theory And Psychophysics*; Wiley: New York, NY, USA, 1966.
41. Fleck, M.S.; Mitroff, S.R. Rare Targets Are Rarely Missed In Correctable Search. *Psychol. Sci.* **2007**, *18*, 943–947. [CrossRef] [PubMed]
42. Godwin, H.J.; Menneer, T.; Cave, K.R.; Helman, S.; Way, R.L.; Donnelly, N. The Impact Of Relative Prevalence On Dual-Target Search For Threat Items From Airport X-Ray Screening. *Acta Psychol.* **2010**, *134*, 79–84. [CrossRef]
43. Godwin, H.J.; Menneer, T.; Cave, K.R.; Donnelly, N. Dual-Target Search For High And Low Prevalence X-Ray Threat Targets. *Vis. Cogn.* **2010**, *18*, 1439–1463. [CrossRef]
44. Ishibashi, K.; Kita, S.; Wolfe, J.M. The Effects Of Local Prevalence And Explicit Expectations On Search Termination Times. *Atten. Percept. Psychophys.* **2012**, *74*, 115–123. [CrossRef]
45. Wolfe, J.M.; Horowitz, T.S.; Kenner, N. Rare Items Often Missed In Visual Searches. *Nature* **2005**, *435*, 439–440. [CrossRef]
46. Godwin, H.J.; Menneer, T.; Riggs, C.A.; Cave, K.R.; Donnelly, N. Perceptual Failures In The Selection And Identification Of Low-Prevalence Targets In Relative Prevalence Visual Search. *Atten. Percept. Psychophys.* **2015**, *77*, 150–159. [CrossRef]
47. Peltier, C.; Becker, M.W. Eye Movement Feedback Fails To Improve Visual Search Performance. *Cogn. Res. Princ. Implic.* **2017**, *2*, 47. [CrossRef] [PubMed]
48. Drew, T.; Williams, L.H. Simple Eye-Movement Feedback During Visual Search Is Not Helpful. *Cogn. Res. Princ. Implic.* **2017**, *2*, 44. [CrossRef]
49. Godwin, H.J.; Menneer, T.; Riggs, C.A.; Taunton, D.; Cave, K.R.; Donnelly, N. Understanding The Contribution Of Target Repetition And Target Expectation To The Emergence Of The Prevalence Effect In Visual Search. *Psychon. Bull. Rev.* **2016**, *23*, 809–816. [CrossRef] [PubMed]
50. Godwin, H.J.; Menneer, T.; Cave, K.R.; Thaibsyah, M.; Donnelly, N. The Effects Of Increasing Target Prevalence On Information Processing During Visual Search. *Psychon. Bull. Rev.* **2015**, *22*, 469–475. [CrossRef] [PubMed]
51. Meuter, R.F.I.; Lacherez, P.F. When And Why Threats Go Undetected: Impacts Of Event Rate And Shift Length On Threat Detection Accuracy During Airport Baggage Screening. *Hum. Factors* **2015**, *58*, 218–228. [CrossRef] [PubMed]
52. Schwaninger, A.; Hardmeier, D.; Hofer, F. Measuring Visual Abilities And Visual Knowledge Of Aviation Security Screeners. *IEEE ICCST Proc.* **2004**, *38*, 258–264.
53. Koller, S.M.; Drury, C.G.; Schwaninger, A. Change Of Search Time And Non-Search Time In X-Ray Baggage Screening Due To Training. *Ergonomics* **2009**, *52*, 644–656. [CrossRef] [PubMed]
54. Schwaninger, A.; Hardmeier, D.; Riegelnig, J.; Martin, M. Use It And Still Lose It? *Geropsych* **2010**, *23*, 169–175. [CrossRef]
55. Bleckley, M.K.; Durso, F.T.; Crutchfield, J.M.; Engle, R.W.; Khanna, M.M. Individual Differences In Working Memory Capacity Predict Visual Attention Allocation. *Psychon. Bull. Rev.* **2003**, *10*, 884–889. [CrossRef]
56. Halbherr, T.; Schwaninger, A.; Budgell, G.R.; Wales, A. Airport Security Screener Competency: A Cross-Sectional And Longitudinal Analysis. *Int. J. Aviat. Psychol.* **2013**, *23*, 113–129. [CrossRef]
57. Poole, B.J.; Kane, M.J. Working-Memory Capacity Predicts The Executive Control Of Visual Search Among Distractors: The Influences Of Sustained And Selective Attention. *Q. J. Exp. Psychol.* **2009**, *62*, 1430–1454. [CrossRef] [PubMed]
58. Sobel, K.V.; Gerrie, M.P.; Poole, B.J.; Kane, M.J. Individual Differences In Working Memory Capacity And Visual Search: The Roles Of Top-Down And Bottom-Up Processing. *Psychon. Bull. Rev.* **2007**, *14*, 840–845. [CrossRef] [PubMed]

59. Gunseli, E.; Olivers, C.N.L.; Meeter, M. Effects Of Search Difficulty On The Selection, Maintenance, and Learning Of Attentional Templates. *J. Cogn. Neurosci.* **2014**, *26*, 2042–2054. [CrossRef]

60. Olivers, C.N.L.; Peters, J.; Houtkamp, R.; Roelfsema, P.R. Different States In Visual Working Memory: When It Guides Attention And When It Does Not. *Trends Cogn. Sci.* **2011**, *15*, 327–334. [CrossRef] [PubMed]

61. Hollingworth, A.; Luck, S.J. The Role Of Visual Working Memory In The Control Of Gaze During Visual Search. *Atten. Percept. Psychophys.* **2009**, *71*, 936–949. [CrossRef] [PubMed]

62. Barrett, D.J.K.; Zobay, O. Attentional Control Via Parallel Target-Templates In Dual-Target Search. *Plos ONE* **2014**, *9*, E86848. [CrossRef]

63. Meier, M.E.; Kane, M.J. Attentional Control And Working Memory Capacity. In *The Wiley Handbook Of Cognitive Control*; Wiley: Chichester, West Sussex, UK, 2017; pp. 50–63.

64. Peltier, C.; Becker, M.W. Individual Differences Predict Low Prevalence Visual Search Performance. *Cogn. Res. Princ. Implic.* **2017**, *2*, 5. [CrossRef]

65. Helton, W.S.; Russell, P.N. Feature Absence-Presence And Two Theories Of Lapses Of Sustained Attention. *Psychol. Res.* **2011**, *75*, 384–392. [CrossRef]

66. Friedman, N.P.; Miyake, A. Unity And Diversity Of Executive Functions: Individual Differences As A Window On Cognitive Structure. *Cortex* **2017**, *86*, 186–204. [CrossRef]

67. Kane, M.J.; Bleckley, M.K.; Conway, A.R.A.; Engle, R.W. A Controlled-Attention View Of Working-Memory Capacity. *J. Exp. Psychol. Gen.* **2001**, *130*, 169–183. [CrossRef]

68. Meiran, N. Modeling Cognitive Control In Task-Switching. *Psychol. Res.* **2000**, *63*, 234–249. [CrossRef]

69. Meiran, N.; Chorev, Z.; Sapir, A. Component Processes In Task Switching. *Cogn. Psychol.* **2000**, *41*, 211–253. [CrossRef]

70. Pashler, H. Task Switching And Multitask Performance. In *Attention And Performance XVIII: Control Of Cognitive Processes*; Monsell, S., Driver, J., Eds.; MIT Press: Cambridge, MA, USA, 2000; pp. 277–307. ISBN 0262133679.

71. Adamo, S.H.; Cain, M.S.; Mitroff, S.R. An Individual Differences Approach To Multiple-Target Visual Search Errors: How Search Errors Relate To Different Characteristics Of Attention. *Vis. Res.* **2017**, *141*, 258–265. [CrossRef]

72. Kiyonaga, A.; Egner, T. Working Memory As Internal Attention: Toward An Integrative Account Of Internal And External Selection Processes. *Psychon. Bull. Rev.* **2013**, *20*, 228–242. [CrossRef]

73. Wolfe, J.M.; Brunelli, D.N.; Rubinstein, J.; Horowitz, T.S. Prevalence Effects In Newly Trained Airport Checkpoint Screeners: Trained Observers Miss Rare Targets, Too. *J. Vis.* **2013**, *13*, 33. [CrossRef]

74. Sterchi, Y.; Hättenschwiler, N.; Schwaninger, A. Detection Measures For Visual Inspection Of X-Ray Images Of Passenger Baggage. *Atten. Percept. Psychophys.* **2019**. [CrossRef]

75. Tuddenham, W.J. Visual Search, Image Organization, and Reader Error In Roentgen Diagnosis. *Radiology* **1962**, *78*, 694–704. [CrossRef]

76. Cain, M.S.; Mitroff, S.R. Memory For Found Targets Interferes With Subsequent Performance In Multiple-Target Visual Search. *J. Exp. Psychol. Hum. Percept. Perform.* **2013**, *39*, 1398–1408. [CrossRef]

77. Cain, M.S.; Adamo, S.H.; Mitroff, S.R. A Taxonomy Of Errors In Multiple-Target Visual Search. *Vis. Cogn.* **2013**, *21*, 899–921. [CrossRef]

78. Berbaum, K.S.; Schartz, K.M.; Caldwell, R.T.; Madsen, M.T.; Thompson, B.H.; Mullan, B.F.; Ellingson, A.N.; Franken, E.A. Satisfaction Of Search From Detection Of Pulmonary Nodules In Computed Tomography Of The Chest. *Acad. Radiol.* **2013**, *20*, 194–201. [CrossRef]

79. Berbaum, K.S.; Krupinski, E.A.; Schartz, K.M.; Caldwell, R.T.; Madsen, M.T.; Hur, S.; Laroia, A.T.; Thompson, B.H.; Mullan, B.F.; Franken, E.A. Satisfaction Of Search In Chest Radiography 2015. *Acad. Radiol.* **2015**, *22*, 1457–1465. [CrossRef]

80. Adamo, S.H.; Cain, M.S.; Mitroff, S.R. Satisfaction At Last: Evidence For The "Satisfaction" Account For Multiple-Target Search Errors. In Proceedings of the Medical Imaging 2018: Image Perception, Observer Performance, and Technology Assessment, Houston, TX, USA, 7 March 2018.

81. Peltier, C.; Becker, M.W. Decision Processes In Visual Search As A Function Of Target Prevalence. *J. Exp. Psychol. Hum. Percept. Perform.* **2016**, *42*, 1466–1476. [CrossRef]

82. Schwartz, B.; Ward, A.; Monterosso, J.; Lyubomirsky, S.; White, K.; Lehman, D.R. Maximizing Versus Satisficing: Happiness Is A Matter Of Choice. *J. Personal. Soc. Psychol.* **2002**, *83*, 1178–1197. [CrossRef]

83. Onefater, R.A.; Kramer, M.R.; Mitroff, S.R. Perfection And Satisfaction: A Motivational Predictor Of Cognitive Abilities. In Proceedings of the Annual Workshop On Object Perception, Attention, and Memory, Vancouver, BC, USA, 8–9 November 2017.

84. Rusconi, E.; Ferri, F.; Viding, E.; Mitchener-Nissen, T. Xrindex: A Brief Screening Tool For Individual Differences In Security Threat Detection In X-Ray Images. *Front. Hum. Neurosci.* **2015**, *9*, 1–18. [CrossRef]

85. Birrell, J.; Meares, K.; Wilkinson, A.; Freeston, M.H. Toward A Definition Of Intolerance Of Uncertainty: A Review Of Factor Analytical Studies Of The Intolerance Of Uncertainty Scale. *Clin. Psychol. Rev.* **2011**, *31*, 1198–1208. [CrossRef]

86. Buhr, K.; Dugas, M.J. Investigating The Construct Validity Of Intolerance Of Uncertainty And Its Unique Relationship With Worry. *J. Anxiety Disord.* **2006**, *20*, 222–236. [CrossRef]

87. Muhl-Richardson, A.; Godwin, H.J.; Garner, M.; Hadwin, J.A.; Liversedge, S.P.; Donnelly, N. Individual Differences In Search And Monitoring For Color Targets In Dynamic Visual Displays. *J. Exp. Psychol. Appl.* **2018**, *24*, 564–577. [CrossRef]

88. Venjakob, A.C.; Mello-Thoms, C.R. Review Of Prospects And Challenges Of Eye Tracking In Volumetric Imaging. *J. Med. Imaging* **2015**, *3*, 011002. [CrossRef]

89. Godwin, H.J.; Donnelly, N.; Liversedge, S.P. Eye Movement Behavior Of Airport X-Ray Screeners When Using Different Dual-View Display Systems. **2009**. Unpublished Report.

90. Drew, T.; Boettcher, S.; Wolfe, J.M. Searching While Loaded: Visual Working Memory Does Not Interfere With Hybrid Search Efficiency But Hybrid Search Uses Working Memory Capacity. *Psychon. Bull. Rev.* **2016**, *23*, 201–212. [CrossRef]

91. Williams, L.H.; Drew, T. Working Memory Capacity Predicts Search Accuracy For Novel As Well As Repeated Targets. *Vis. Cogn.* **2018**, *26*, 463–474. [CrossRef]

Review

Eye Behavior During Multiple Object Tracking and Multiple Identity Tracking

Jukka Hyönä [1,*], Jie Li [2,3] and Lauri Oksama [4]

[1]   Department of Psychology, University of Turku, FI-20014 Turku, Finland
[2]   Institutes of Psychological Sciences, Hangzhou Normal University, Hangzhou 311121, China
[3]   School of Psychology, Beijing Sport University, Beijing 100084, China
[4]   Finnish Defence Research Agency, Human Performance Division, P.O. Box 5, FI-04401 Järvenpää, Finland
*   Correspondence: hyona@utu.fi

Received: 28 February 2019; Accepted: 23 July 2019; Published: 31 July 2019

Abstract: We review all published eye-tracking studies to date that have used eye movements to examine multiple object (MOT) or multiple identity tracking (MIT). In both tasks, observers dynamically track multiple moving objects. In MOT the objects are identical, whereas in MIT they have distinct identities. In MOT, observers prefer to fixate on blank space, which is often the center of gravity formed by the moving targets (centroid). In contrast, in MIT observers have a strong preference for the target-switching strategy, presumably to refresh and maintain identity-location bindings for the targets. To account for the qualitative differences between MOT and MIT, two mechanisms have been posited, a position tracking (MOT) and an identity tracking (MOT & MIT) mechanism. Eye-tracking studies of MOT have also demonstrated that observers execute rescue saccades toward targets in danger of becoming occluded or are about to change direction after a collision. Crowding attracts the eyes close to it in order to increase visual acuity for the crowded objects to prevent target loss. It is suggested that future studies should concentrate more on MIT, as MIT more closely resembles tracking in the real world.

Keywords: eye movements; multiple object tracking; multiple identity tracking; dynamic attention

---

1. Introduction

In many real-life visual tasks, people are required to keep track of multiple moving objects. This is true, for example, of car drivers maneuvering a vehicle across a busy intersection where the driver has to keep track of other vehicles approaching the intersection and of pedestrians crossing the street. This is also true of a football player making the decision of whom to pass the ball. When making that decision under stringent time constraints, he or she needs to be aware of both where his/her own teammates are as well as the players of the opponent team are located and moving toward. In the human factors' literature, such awareness is called situation awareness [1]. It is genuinely dynamic in nature in that the visual environment (e.g., the intersection or the football field) is constantly changing, so that situation awareness has to be updated accordingly in order to adequately represent the moment-to-moment fluctuations in the relevant task environments.

The experimental research on the dynamic tracking performance has mimicked the real-life visual environments using two types of laboratory tasks: multiple object tracking (MOT; for a review, see [2]) and multiple identity tracking (MIT; [3]). In both tasks, the moving objects are presented on a computer screen. The pioneering study, [4] introduced the MOT task. In MOT, the objects to be tracked are visually identical (e.g., black circles) so that they can be differentiated from each other only by their spatiotemporal properties. In the beginning of the task, a subset of the objects is designated as the targets, the remaining objects are distracters. The MOT task is depicted in Figure 1. The seminal work [4] demonstrated that people can track about four identical objects. Later it was shown [3] that there are significant individual differences in the MOT performance.

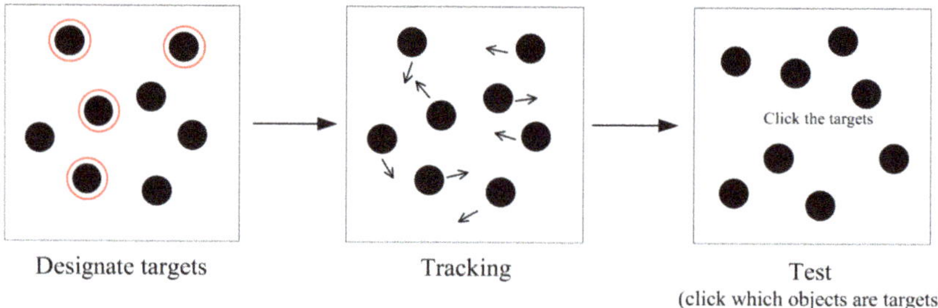

Figure 1. In the multiple object tracking (MOT) task, the targets to be tracked are first designated, for example, by drawing a red circle around the targets. After the target designation, the target and distracter items (all identical to each other) move for a few seconds. For the test phase, all objects stop moving and the participant is asked to click on the targets.

The multiple identity tracking task was developed by Oksama and Hyönä [3,5]. It resembles more closely real-life visual environments in that the to-be-tracked objects all have distinct identities (like in traffic or sports). The MIT task is depicted in Figure 2. It was originally designed to mimic the visual environment air-traffic controllers or fighter pilots operate on. Air-traffic controllers monitor the incoming and outgoing aircraft depicted on a big screen by call signals that identify each aircraft. In the laboratory experiments the objects to be tracked have been, for example, line drawings of common objects or faces [5], cartoon characters [6] or words [7]. Horowitz et al. [6] demonstrated that tracking of multiple distinct identities is more difficult than tracking the whereabouts of identical objects. Perhaps not more than two objects with distinct identities can be successfully tracked.

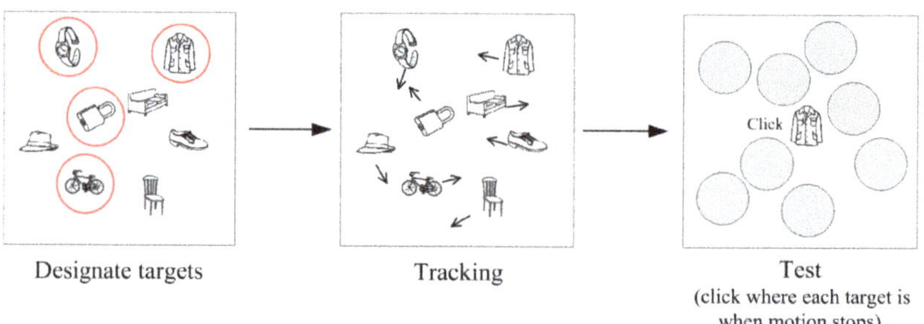

Figure 2. In the multiple identity tracking (MIT) task, all objects have distinct identities; in this example they are line drawings of common objects. In the initial target designation stage, the targets to be tracked are shown, for example, by drawing a red circle around them. Then all objects move for a few seconds. In the test phase, all objects stop moving and the targets are probed, for example, one by one by asking the participant to locate where each target was positioned at the time the movement was terminated.

The majority of MOT and MIT studies have examined the tracking performance by using performance accuracy as the dependent measure. However, there are some studies that have registered observers' eye movements during the tracking performance. The eye-tracking studies of MOT and MIT are very useful, as they provide direct evidence regarding the allocation of overt attention during tracking. Thus, it may be used, for example, to study the extent to which observers overtly switch attention between the targets to perform the task. In this review paper, we go over all the eye-tracking studies conducted on MOT and MIT. We first review the MOT studied followed by the

MIT studies and direct comparisons of the two tasks. We conclude the review by suggesting directions for future research.

2. Eye Behavior during MOT

2.1. Where Are Observers Looking at during MOT?

In this section, we review evidence on where observers look at during MOT. We first go over the studies providing evidence for the view that the center-looking strategy is the predominant strategy in MOT. Subsequently, we review evidence demonstrating that the target-looking strategy is in fact as dominant as the center-looking strategy. A summary of the eye movement studies of MOT is provided in Table 1.

Table 1. Observed scanning strategies in the eye-tracking studies of multiple object tracking (MOT).

Study	Stimulus Size (°)	Set-Size	Speed (°/s)	Analysis Method	Scanning Strategy
Fehd & Seiffert [8]	2.1	1	15	Shortest distance rule	Target: 96%
Follow-up	3				Centroid: 66%; Target: 9%
	3				Centroid: 42%; Target: 11%
	4				Centroid: 42%; Target 9%
	5				Centroid: 42%; Target 8%
Fehd & Seiffert [9]				AoI: 5° in diameter	
Experiment 1	2.1	4	3, 6, 12, 18, 24		Centroid:~25%; Target: ~10%
Experiment 2	0.06–0.3	4	3, 6, 8, 12, 24		Centroid: 34%; Target: 8%
Experiment 3	1.8	3	12		Centroid: 43%; Target 13%
Zelinsky & Neider [10]	0.5–1.1	1	1.1	Shortest distance rule	Target: 94%
		2			Centroid: 47%; Target: 37%
		3			Centroid: 39%; Target: 42%
		4			Centroid: 24%; Target: 52%
Zelinsky & Todor [11]	0.5–1.1	2–4	Not reported	"Rescue saccade": landing 1° from a target	Anticipatory rescue saccades were initiated to occluded targets
Huff et al. [12]	1.3–2.2	3		AoI: same size as the objects	
Experiment 1			2		Centroid:~7%; Target: ~10%
			4		Centroid:~7.5%; Target:~8.5%
			6		Centroid:~10%; Target:~8%
Experiment 2			4		Centroid:~11%; Target:~8%
			10		Centroid:~12.5%; Target:~5.5%
Vater et al. [13]	1	4	6, 9, 12	AoI: 5° in diameter	Centroid: 30%; Target: 11%
Vater et al. [14]	1	4	6	Gaze-vector distances	Gaze was closer to centroid than to target regardless of target changes
Vater et al. [15]	1	4	6	Relative gaze distance to the targets	Gaze was closer to crowded than uncrowded targets. Anticipatory saccades were initiated to targets colliding with border.
Lukavský [16]	1	4	5	AoI: 1° for centroid, 2° for objects; Normalized scanpath saliency	Centroid: 7.7%; Target: 12.6%; Anticrowding point: 12.2%; Target eccentricity minimizing point: 9%
Dechterenko & Lukavsky [17]	1	4	Adaptive	Normalized scanpath saliency	The model accounting for the crowding effect yielded the best performance.
Lukavský & Děchtěrenko [18]	1	4	5	Local maximum of similarity	Gaze position lagged by approximately 110 ms behind the scene content.
Děchtěrenko, Lukavský, & Holmqvist [19]	1	4	5	Correlation distance	Scan patterns in flipped trials differed only slightly from those of the original trials.
Oksama & Hyönä [20]					
Experiment 1	2.1	2–5	2.6, 6.3, 10.3	AoI: 3.4°	Blank area: 48%; Target: 21%; Centroid: 7%; Distracter: 4%
Experiment 3	2.1	2–5	2.6, 6.3, 10.3	AoI: 3.4°	Blank area: 48%; Target: 24%; Centroid: 7%; Distracter: 5%

2.1.1. Center-Looking Strategy

The Central Areas between Targets are Frequently Gazed at during MOT

An intriguing question is where observers are looking at when tracking multiple targets moving around in the visual field. In the seminal eye-tracking study of MOT, Fehd and Seiffert [8] provide evidence for the view that observers predominately look at the central area between the targets, presumably for grouping multiple targets into one single virtual object to track [21].

In the main experiment of the study, eye movements were registered when observers tracked 1 or 3 target dots out of 8 dots, which randomly moved about on the screen for 3 s with a speed of 15°/s. The trials in which the observers correctly tracked the targets were included in eye movement analyses. The 8 dots and the center of the targets (i.e., the centroid of the triangle formed by the 3 targets, which is the intersection of the medians) were defined for each frame as the competing areas of interest (AoIs). A location competition analysis with the shortest-distance rule was used to determine the location to which the gaze was closest. Each AoI was assigned a weight of zero at the beginning of the trial, and after each frame the AoI closest to the gaze position received an increase in its weight while the remaining AoIs received a decrease in weight. The AoI with the highest weight value on a given frame was considered the winner for that frame. The time each AoI was the winner across all frames was summed up, and then the average percentage of the time that the gaze was directed towards each AoI was calculated.

Their results showed that in the 1-target trials, the vast majority of the gazes (96.0%) were closest to the target dot. This is understandable as observers did not need to distribute their attention across multiple targets. More importantly, when tracking 3 targets, 65.7% of the time the gaze was closest to the center of the targets, while 8.5% of the gaze time was on each target dot, and only 1.0–2.2% on each distracter. The researchers replicated the finding in a follow-up experiment, which showed that participants looked significantly more at the center (41.6–42.4%) than at each of the targets when tracking 3 (10.7%), 4 (9.3%), and 5 (8.1%) targets in an array of 10 dots. Fehd and Seiffert [8] conclude that, "when multiple objects are tracked, more time is spent looking towards the center of the target array than at each target individually" (p. 206).

Such a center-looking strategy was replicated or partially replicated by subsequent eye-tracking MOT studies [9,10,13]. For example, Zelinsky and Neider [10] adopted an experimental setting somewhat different from the basic MOT paradigm, as they mimicked real-life dynamic visual environments by presenting on the screen moving sharks that moved underwater. Underwater scene is free of contextual constraints, as sharks can freely select and change their movement trajectory. In terms of features, sharks are more complex objects than dots, thus better approximating the kind of objects tracked in real life. In the experiment, 9 identical sharks moved about for 20 s in a 3D underwater scene (i.e., movement in depth was also rendered possible). Target set-size varied from 1 to 4. When the movement stopped, one of the sharks was probed by drawing a red circle around it, and the participant had to decide whether or not it was a target. The results showed that when tracking just one target, observers spent 94% of the time gazing at the single target; when tracking 2, 3, or 4 targets, observers spent 47%, 39%, 24% of the gaze time at the center.

Center-Looking Includes Centroid-Looking, Anticrowding, and More

Whereas, there is a consensus that observers frequently look at the central areas between multiple targets, it is more controversial which specific central locations observers are looking at during MOT and what strategies their eye behavior represents.

Fehd and Seiffert [8] propose that center-looking represents centroid-looking. Observers look at the centroid of the virtual polygon formed by the targets, which serves the function of mentally grouping multiple targets into one virtual object to track. In their analyses, they compared two types of central locations: the centroid of the virtual object for grouping, and the central points for minimizing the eccentricities of the targets (either the maximum eccentricity of any one target, or the average eccentricity of all the targets). The authors note that the eccentricity minimizing points are very close to the centroid, so it may be difficult to tell them apart, and hence they analyzed the frames where eccentricity minimizing points differed from the centroid by more than one degree. The result showed that the average distance to the gaze was larger from the eccentricity minimizing points than from the centroid (4.8–5.4° vs. 4.1°). Moreover, in the location competition analysis the eccentricity minimizing points were winners less often than the centroid (28.4–28.6% vs. 65.7%). Fehd and Seiffert [8] also compared the centroid with the central location calculated by averaging the coordinates of the targets. It is the same location when there are 3 targets but not when there are 4 or 5 targets. They did not describe explicitly how the centroid was defined when there were 4 or 5 targets. Presumably, it was defined as the geometric center of the polygon formed by the targets. Their results showed that the location of the average target coordinates did not correspond to the gaze position as closely as the centroid. Overall, Fehd and Seiffert [8] considered the centroid-looking strategy to fit best to the data and being a more parsimonious explanation, as it presumably reflects a grouping strategy whereby the targets are grouped to form a single virtual object.

However, the dominant centroid-looking strategy may originate from the analysis methods used for determining to which AoI each gaze position belongs. Only the seminal studies using the shortest-distance rule found the majority of the gazes being closest to the centroid [8,10]. With this rule, gaze was defined as either looking at the centroid or one of the objects depending on which the distance was shorter. Thus, many gazes on central areas were categorized as looking at the centroid, even though they were fairly distant from it. For instance, as mentioned above, in the Fehd and Seiffert's [8] study, the average distance between gaze and centroid was 4.1°. Subsequent studies used AoIs of different size to define whether gaze is at the centroid or on individual objects; they found that the centroid-looking strategy accounted for about 4–30% of the total gaze time [9,12,13,16,20], in comparison to 40–66% found in the seminal studies using the shortest-distance rule [8,10]. In particular, using AoIs of 3.4° in size, Oksama and Hyönä [20] found 7% of gaze time on the centroid, in contrast to 21% on individual targets, 4% on distracters, 4% on the screen center, and about half of the time (48%) on blank area that was neither the centroid nor the screen center, implying that center-looking entails more than merely centroid-looking.

Lukavský [16] compared five different types of central locations representing five types of center-looking strategies during MOT. Three of them were the centroid, the eccentricity minimizing point for the targets, and the screen center; the other two were the eccentricity minimizing point for all objects, and the anticrowding point, which minimizes the ratio between each target's distance from the gaze point and the distance from every distracter. It is noteworthy that the computation of the last two central locations takes into account both the locations of the targets and the distracters. Lukavský [16] compared the adherence of the observed data to the five models of the center-looking strategy, and found highest adherence for the anticrowding point, closely followed by the target eccentricity minimizing point and the centroid, whereas the adherence of the all-object eccentricity minimizing point was lower, and that of the screen center was lowest. Moreover, observers spent a larger percentage of time gazing at the anticrowding point than the target eccentricity minimizing point or the centroid (12.2% vs. 9.0% vs. 7.7%). Therefore, Lukavský [16] suggests that an important function of the center-looking strategy is to mitigate the effect of crowding and thus reduce the danger of confusing targets with surrounding distracters.

2.1.2. Target-Looking

Individual Targets are Frequently Gazed at during MOT

Besides the central areas, the individual targets are frequently looked at during MOT as well, whereas the non-targets are seldom looked at. In fact, the total gaze time on all the targets is usually comparable to or larger than the gaze time on central areas [8,10,12,16]. For instance, even though Fehd and Seiffert [8] emphasized the center-looking strategy, the target-looking strategy also featured prominently in their data. While the center-looking strategy accounted for 41.6–42.4% of the gaze time, each of the targets was gazed at for 10.7%, 9.3%, and 8.1% of the time when tracking 3, 4, and 5 targets. It means that the target-looking accounted approximately for about 32.1%, 37.2%, and 40.5% (i.e., 10.7% × 3, 9.3% × 4, and 8.1% × 5) of the gaze time in total, which is comparable to the center-looking time. The Zelinsky and Neider [10] study also observed pronounced target-looking. When there were 2, 3, 4 targets, the time spent on targets accounted for 37%, 42%, 52%, which is comparable to or even greater than that spent on the center (47%, 39%, and 24%, for 2, 3, and 4 targets, respectively).

The percentages of target-looking and center-looking both varied with the size of AoIs used in the eye-tracking MOT studies, yet it seems that target-looking is always comparable in frequency or even higher than centroid-looking. For instance, Huff, Papenmeier, Jahn and Hesse [12] used AoIs of 1.3° to 2.2° in size, which was of the same size as the objects. The results showed about 10% of the gaze time on the centroid, and about 5–10% on each of the 3 targets on average, resulting in about 15–30% spent on the targets in total. In Oksama and Hyönä's [20] study, AoIs of 3.4° were used, while the size of the objects was around 2°. The results showed 7% of the gaze time on the centroid, and 21% on all the targets in total. The number of targets varied from 2 to 5, so that the average time on each target was around 4–10%. In Lukavský's [16] study, the centroid was defined by an AoI of 1°, equal to the size of the moving objects, whereas the AoIs for objects were 2° (this is not explicitly stated in the paper but implied by "AOIs defined by targets and distracters are four times larger"). The results showed 7.7% of the gaze time on the centroid, and 3.1% of the gaze time on each of the 4 targets on average, resulting in about 12.6% on the targets in total. Fehd and Seiffert [9] used two different sizes of AoIs in their Experiment 2. The size of the objects was 0.06–0.3°. They first used AoIs of 0.6°. The results showed very small percentages of gaze overlap with the AoIs. Gaze overlapped with the centroid 4.0% of the time, and 0.9% and 0.3% with each of the 4 targets and 6 distracters, resulting in 3.6% on targets and 1.8% on distracters. They then increased the AoI size to 5°. Gaze overlapped with the centroid for 34.3%, and 7.8% and 3.1% for each of the targets and distracters, resulting in 31.2% and 18.6% of gaze time on the targets and distracters, respectively. The percentages were similar to those in their Experiment 1 with 5° AoIs, which were about 25% on the centroid, 40% on the targets, and 18% on the distracters. In sum, across the different studies, as the AoI size varied between 0.6–5°, the range of the average percentage of gaze time on each target was about 1–10%, resulting in the total target-looking time of 4–40%, while the centroid-looking was about 4–34%.

It has been demonstrated that observers frequently switch their gaze back and forth between the center and the targets rather than switching from target to target. Fehd and Seiffert [9] showed that in each trial, there were 5.9 center-to-target switches on average, which was significantly more than target-to-target switches (1.9). The results were replicated by Vater, Kredel, and Hossner [13], who found 5.5 center–target switches per trial in comparison with 1.3 target–target switches. The Oksama and Hyönä [20] study provided support for another possible strategy, namely the look-at-one-target-strategy. The results showed that regardless the number of the targets that needed to be tracked (varied from 2 to 5), the number of targets gazed at during tracking remained constant, about $1\frac{1}{2}$ targets. It implies that observers may choose one target to be overtly tracked by the eyes, while the other targets are tracked peripherally in a covert fashion. "Such a strategy seems both psychologically and computationally more plausible than the continuous computation of the centroid dynamically delineated by the moving targets, which appears computationally quite demanding" [20]. It should be noted, however, that in their study the look-at-one-target-strategy was less frequent than the look-at-blank-space strategy.

Gaze to Individual Targets Lags behind the Targets Rather Than Extrapolate Target Motion

Another interesting question about target-looking is that when observers look at targets, whether the eyes extrapolate target motion and land on a position where the target is expected to go to. Lukavský and Děchtěrenko [18] examined this question yet failed to find evidence in support of motion extrapolation in MOT. Instead, they observed that in all the tested conditions the gaze lagged behind the targets. In their study, the stimuli comprised 8 gray disks of 1° of visual angle in size that moved in a constant speed of 5°/s. Four targets among 4 distracters were tracked for 10 s. Importantly, they presented forward motion trajectories, repeated for 4 times, and backward motion trajectories that were the forward trajectories played back in reverse order. The logic is that if there is no motion prediction, gaze behavior would be highly similar between the forward and backward trajectories. In order to make this comparison possible, the eye movement data were reversed for the backward trajectories. The procedure also allowed determining the temporal lag or lead time of gaze behavior in relation to object motion.

Using the logic described above, they [18] observed in Experiment 1 (20 participants) an average lag of 114 ms in gaze behavior with respect to object motion, with all participants demonstrating a lag (from 52 ms to 173 ms). In Experiment 2 (27 participants), they manipulated target set-size (2 or 4) to investigate whether track load affects motion extrapolation. It was found that the gaze lagged behind object motion a bit less (a non-significant difference) with low (2 targets) than high (4 targets) track load (93 ms vs. 108 ms). In other words, an easy tracking condition did not lead to a significantly shorter lag. Yet, it should be noted that object motion was not very predictable. In Experiment 3, the predictability of object motion was manipulated. In Experiment 3a (23 participants), the trajectories used in Experiment 2 (baseline) were pitted against more predictable trajectories, while in Experiment 3b (28 participants) the baseline condition was compared against a less predictable (i.e., more chaotic) condition. It was observed that more predictable trajectories did not reliably shorten the gaze lag (although there was a tendency in that direction), but the more chaotic condition reliably delayed the lag by 32 ms (102 ms vs. 134 ms). It is suggested that the 100-ms lag may reflect oculomotor limits. Generally speaking, the study [18] failed to find evidence in support of motion extrapolation in MOT, as in all tested conditions gaze lagged behind the targets.

2.1.3. Summary: Both Center-Looking and Target-Looking during MOT

In sum, during MOT, observers frequently look at the central areas between targets. The central locations mainly include the centroid of the virtual polygon formed by the targets, the anticrowding point, and the eccentricity minimizing point. The individual targets are frequently looked at as well, while non-targets are seldom looked at. The total time of target-looking is comparable to center-looking. Observers frequently switch the gaze back and forth between the center and the targets, manifesting a center-target-switching strategy; yet they may also overtly follow one target by gazing at it, while tracking the other targets peripherally in a covert fashion. When looking at the targets, the gaze does not extrapolate their motion trajectories but instead lags behind.

It is noteworthy that the studies differ in the way eye movement data were analyzed. In most studies (e.g., [8,10]) raw data (i.e., x and y coordinates) are assigned to AOIs (for an open-source automated tool for conducting such analyses, see [22]), while in some studies (e.g., [20]) the raw data are first parsed into fixations and saccades, after which fixations are assigned to the AOIs. The downside of the former analysis procedure is that also saccades are signed to AOIs despite the fact that no visual information processing is carried out during saccades due to saccadic suppression. This may not be considered a serious problem, as saccades are rapid and thus consume relatively little time. On the other hand, the downside of the latter analysis procedure is that it cannot differentiate smooth pursuit movements from fixations. Yet, in our data, we seldom observe smooth pursuits when multiple objects are to be tracked. The situation is dramatically different when only one target is tracked; observers are likely to track the single target with smooth pursuit eye movements (see [8,10]).

2.2. What Influences Eye Behavior during MOT?

2.2.1. Number of Target Objects

To date, the results are inconsistent regarding how eye behavior during MOT varies as a function of the number of targets. Clearly, when tracking one single target, observers spend almost all the gaze time (about 95%) on the target [8,10]. When tracking more than one target, some studies show that as the number of targets increases, the total target-looking time increases while the average looking time on each target decreases. For instance, Zelinsky and Neider [10] showed that as the number of targets increased from 2, 3, to 4, the total target-looking time increased from 37%, 42%, to 52%, while the average looking time on each target dropped from 18%, 14%, to 13%, respectively. Fehd and Seiffert [8] found a similar trend; by showing that as the number of targets increased from 3, 4, to 5, the total target-looking time increased from 32.1%, 37.2%, to 40.5%, while the average looking time on each target dropped from 10.7%, 9.3%, to 8.1%, respectively. Zelinsky and Neider [10] suggest that the increase in total target-looking time is due to observers shifting from the centroid-looking strategy to the target-switching strategy when there are more targets to track. This argument was supported by the finding that as the number of targets increased from 2, 3, to 4, the centroid-looking time decreased from 47%, 39%, to 24%. Moreover, Zelinsky and Todor [11] found that the gaze-target distance was generally larger for tracking 2 rather than 3 or 4 targets, perhaps reflecting a more pronounced centroid-looking strategy in set-size 2. In contrast, the Fehd and Seiffert [8] study suggests that the centroid-looking strategy remains constant regardless of the number of targets; as the number of targets varied from 3 to 5, the centroid-looking time remained at 41.6–42.4%.

In contrast to the seminal eye-tracking studies of MOT, Oksama and Hyönä [20] found that as the number of targets increased from 2 to 5, almost all the measures of eye behavior remained constant, including the number of eye visits to targets, the number of visited targets, average fixation duration, the number of fixations, pupil size, and the number of blinks. The results provide strong support for parallel tracking models of MOT. The discrepancy between the studies may originate from the variation in motion speed, as discussed below.

2.2.2. Motion Speed

Fehd and Seiffert [9] argue that regardless of target speed, the prevalence of the centroid-looking strategy remains unchanged. They examined the possibility that observers adopt the centroid-looking strategy only when individual targets are moving too fast to be followed with the eyes. The authors varied the motion speed from slow to very fast (3, 6, 12, 18, or 24°/s); yet, they found no effect of speed in the percentage of gaze time on the centroid, targets, and distracters. Moreover, the gaze time on the centroid was always longer than that on each target and distracter regardless of speed. On the other hand, the dwell time decreased as the speed increased, showing that "At the slower speeds, participants viewed the center for prolonged periods of time, while at higher speeds they made quick glances to the center" (p. 5). Similarly, Vater et al. [13] found no effect of speed in the percentage of gaze time on the centroid, target, and distracter, when motion speed was varied from 6°/s, 9°/s, to 12°/s.

In contrast, Huff et al. [12] provided evidence for the view that the relative importance of centroid-looking increases with object speed. They manipulated object speed to be 2, 4, or 6°/s. The results showed that when the speed was slow (2°/s), the percentage of gaze time on the centroid was even lower than that on a target (about 7% vs. 10%). As the object speed increased, the centroid-looking strategy became more frequent while the target-looking preference decreased. Huff et al. [12] speculate that the centroid-looking preference reflects its benefits over the target-switching strategy. The centroid's velocity and degree of movement is lower than those of targets, with the difference being particularly pronounced at higher speed. This presumably explains why the centroid-looking preference is increased with higher object speeds.

Oksama and Hyönä [20] suggests that as the motion speed increases from low to high, observers switch from a more serial tracking to a more parallel tracking, manifested in decreased fixations on individual targets and increased fixations on the central areas between targets. The authors varied the speed from 2.6°/s (slow), 6.3°/s (medium), to 10.7°/s (fast), while varying the number of targets from 2 to 5. The results showed that as the object speed increased, the participants spent less time looking at the targets and more on the center. Moreover, when the speed was slow, the number of fixations and target visits increased while the fixation duration tended to decrease as the number of targets increased. On the other hand, with a medium speed, the number of fixations and target visits as well as the fixation duration remained constant as the number of targets increased. Finally, with a high speed, the trend was that the number of fixations and target visits decreased and the fixation duration increased as a function of the number of targets. The results indicate that at slow speed observers adopt a more serial tracking, switching the eyes from one target to another, while as the speed increases they adopt a more parallel tracking strategy.

The discrepancy in the results reviewed above may originate from the differences in the tested object speeds. Thus, perhaps Huff et al. [12] found increased centroid-looking associated with an increase in motion speed, because the speed in their study varied from low to medium (2, 4, and 6°/s), whereas Fehd and Seiffert [9] and Vater et al. [13] found constant centroid-looking as a function of motion speed, because the speed in these studies varied mostly from medium to high.

Research has also shown that as object speed increases, observers make fewer saccades, presumably to reduce the risks of losing targets during saccadic eye movements. When objects are moving slowly, even if some information is lost due to saccadic suppression, it can still be easily recovered after the saccade, whereas recovery may be impossible at high motion speed, since the locations of the objects may have significantly shifted position during a saccade [17].

In sum, observers frequently switch the gaze from target to target when the targets are moving slowly, showing a pattern of tracking the targets in a serial fashion, whereas they exhibit more center-looking at medium to high object speeds, demonstrating parallel tracking of targets.

2.2.3. Stimulus Size

In principal, the decrease in stimulus size leads to reduced peripheral resolution of the targets, so that observers may fixate on the individual targets more often and hence the centroid-looking strategy may be abandoned. Fehd and Seiffert [9] examined this possibility in their Experiment 2. Squares were used as stimuli. They varied in size from 1 to 5 pixels on a side, subtending 0.06° to 0.3° of visual angle, so that the largest stimulus was seven times smaller than the stimuli used in their Experiment 1. When using AoIs of the same size as in Experiment 1 (5° in diameter), the results revealed the same centroid-looking strategy as in Experiment 1, manifested in 34.3% of the gaze time on the centroid and 7.8% on each separate target. When using AoIs double the size of the stimuli, which were significantly smaller than in Experiment 1, the centroid was still gazed at more than each individual target, although the percentages were much smaller than in Experiment 1, 4.0% on the centroid and 0.9% on the targets. Furthermore, as size decreased from 0.3° to 0.12°, the percentage of gaze time on the centroid decreased from 5.0% to 3.2%, while the gaze time on targets increased from 0.9% to 1.3%, suggesting a small trend for decreased centroid-looking and increased target-looking with decreased stimulus size. Overall, the study showed that target-looking increases and centroid-looking still prevails when stimulus size is small, demonstrating that the centroid-looking strategy "provides some value to tracking objects beyond the use of peripheral vision" (p. 7).

2.2.4. Crowding and Collision

During MOT, crowding and collision commonly co-occur and may influence the tracking performance and eye behavior. Crowding is established when the objects move close to one another so that it is more difficult to differentiate the targets from the distracters. At the same time, the probability for object collisions is increased, which may also impact tracking. Vater, Kredel, and Hossner [15] disentangled the effects of crowding and collision on eye movements during tracking. Collision was manipulated by having the targets collide with the bordering frame in one condition during a critical period of 0.5 s, which was compared to a no collision condition. Crowding was also manipulated by two conditions. In the crowding condition, during the critical period three of the four targets formed a group where each target was closely surrounded by a distracter (the fourth target was not); in the no-crowding condition, the group of three targets was not closely surrounded by distracters.

The authors measured the distance between the gaze and the targets, as well as the number of saccades initiated to the targets in different conditions. Saccades to the targets were initiated much more frequently prior to the collision (64%) than after the collision (9%). These anticipatory saccades are assumed to be made to update changes in motion trajectories after the collision against the boundary. On the other hand, crowding did not affect the probability of initiating a saccade; yet, the gaze was located closer to the target group than to the isolated target in the crowding than in the no-crowding condition. The result is consistent with the study of [11], who found that the gaze-target distance decreased as the target-other-object distance became smaller. Similarly, the modelling studies of Lukavský and colleagues [16,17] observed that the anticrowding model is the best model for explaining the gaze locations during tracking. This is presumably due to attempts to increase visual acuity for targets in order to be better able to keep them separate from the distracters. In sum, crowding draws the eyes toward the crowded area, while collision induces anticipatory saccades to the target.

2.2.5. Occlusion

Zelinsky and Todor [11] examined eye movements in a MOT task where objects occlude each other. The experiment was otherwise identical to their earlier study [10] except that in half of the trials at least one target, usually more than one, was involved in an occlusion. Observers tracked 2, 3 or 4 targets out of 9 objects for a period of 20 s. The main goal of the study was to assess the premise that observers "look at targets in order to prevent occlusion-related track losses" [11] (p. 4). In order to do that the authors examined the number of saccades programmed to occluded targets during the occlusion period (800 ms around the occlusion). These saccades were named as rescue saccades, as they are presumably carried out to prevent from losing track of occluded targets.

The results showed that rescue saccades occurred in 45% of the occlusion periods, but only in 8% of the time before an occlusion period of the same duration. The rescue saccades were anticipatory in nature, as they frequently (33.7%) appeared within the time bin of 800 ms prior to occlusion. Interestingly, the probability of rescue saccades was at the same level (42.3%) in near-occlusion events where a target came close to another object (closer than 1° distance) and then moved away without being occluded, suggesting that close distance may signal forthcoming occlusion and hence triggers anticipatory saccades. In addition, the authors observed no difference in the frequency of rescue saccades between the target-target occlusions (or near-occlusions) and the target-distracter occlusions. This suggests that the process is driven by the possibility of confusion regardless of whether a target would be confused with a distracter or another target. The finding that the incidence of rescue saccades was no more likely with distracters than with targets makes sense by considering that in real-life tracking tasks, unlike in MOT, it is equally detrimental to confuse a target with another target as with a distracter. This is because in real-life tracking tasks the targets have distinct identities that have to be kept separate from each other. Thus, the predisposition of the eye movement system may be to automatically initiate saccades to targets in proximity to other objects. In sum, similarly to collisions, target occlusion induces anticipatory saccades to the targets.

2.2.6. Abrupt Viewpoint Changes

Huff et al. [12] examined the effects of abrupt viewpoint changes in MOT on observers' eye movement patterns. Changing abruptly the viewpoint resembles a situation where a person watches a football match on TV where the game is presented as sequential shots from different camera angles. In Experiment 1, eight white spheres (a size of 1.3 to 2.2°) moved on a floor plane resembling a checkerboard. Three spheres were designated as the targets. Objects moved for 5 s in straight lines in randomly chosen directions with a constant speed of 2, 4 or 6°/s. In half of the trials, after 3 s the camera viewpoint changed abruptly by 20° either to the left or right. At the end of the trial when the objects stopped moving, the participant had to click on the targets with a computer mouse. In Experiment 2, the authors [12] made three changes to the procedure. First, the timing of the abrupt viewpoint change was varied in order to prevent anticipation. Second, object speed was increased to 10°/s; this speed was compared to the medium speed condition of Experiment 1 (4°/s). Finally, participants were asked to report after each trial whether or not they were able to track all the targets.

The analyses showed that after the abrupt viewpoint change, there was a drop-in gaze time on targets. The drop was temporary, as after 500 ms following the viewpoint change, gaze time on targets returned to the level it was prior to the change. The drop-in gaze time was smaller and non-significant for centroid looking. The first saccade after the change was executed faster to the centroid than to a target. It suggests that a new centroid was readily recalculated. In the 500-ms time bin after the viewpoint change, there were also more saccades to the centroid than to the target. Moreover, these effects of viewpoint change on gaze behavior existed only when the motion speed was relatively slow (2–6°/s) but disappeared when the speed was fast (10°/s). Huff et al. [12] speculate that the centroid-looking preference reflects its benefits over the target-switching strategy. The viewpoint change affects less the location of the centroid than the targets, which is assumed to explain why the gaze on the centroid was less influenced by viewpoint change. Moreover, the centroid's velocity and degree of movement is lower than those of targets. This difference is particularly pronounced with higher speeds. This presumably explains why the centroid-looking preference is increased with higher objects speeds.

In sum, abrupt viewpoint changes do not substantially affect centroid-looking, as saccades are quickly made to the centroid after the change; meanwhile, target-looking temporarily drops after the viewpoint change.

2.2.7. Trajectory Repetition and Flipping

Lukavský [16] investigated whether the repetition of motion trajectory would increase similarity in eye behavior. The repetition effects were studied by presenting the odd-numbered trials in the MOT experiments as unique trials, while the even-numbered trials reappeared once in each of the 4 blocks. Similarity in scanpaths was analyzed with a measure of normalized scan path saliency [23]. It was found that repeated trials produced more similar scanpaths than unique trials. This finding was replicated by using mean gaze distance between frames as the dependent measure. Observers were to some extent aware of the fact that some trials were repeated. In a post-experiment recognition test they were presented a subset of the trials and asked to judge whether or not they had seen them during the experiment. Repeated trials were recognized somewhat better than unique trials (47% vs. 39%).

To prevent the contamination of observers noticing the repetition and consciously learning the trials, Děchtěrenko, Lukavský, and Holmqvist [19] geometrically flipped the trajectories along the *x*- or *y*-axes in the repeated trials. In addition, the trajectories shared common segments for 6 s while differed for the rest (2 s) of the motion. Using the measure of the correlation distance metric, the authors showed that even though the scan patterns in the flipped trials differed significantly from those in the original trials, the difference was small (as little as a 13% increase of overall distance). In sum, eye behavior generally remains constant when the trajectories of object motion are repeated or flipped.

2.3. What Are the Functions of Eye Behavior during MOT?

In this section, we outline possible functions of eye movements carried out during MOT. The currently available research suggests that they may serve the function of grouping targets together, counteracting detrimental effects of crowding, occlusion and collision, as well as detecting changes in targets.

2.3.1. Centroid-Looking for Grouping

The idea put forth in the seminal eye movement studies of MOT was that the main function of eye behavior during MOT is to help group multiple targets as a single virtual polygon by gazing at the centroid of the polygon [9,10,21]. This idea was supported by the studies investigating the relationship between eye behavior and tracking performance. Zelinsky and Neider [10] computed correlations between the adopted eye movement strategy and tracking accuracy. The correlations demonstrated that increased time looking at the centroid was associated with better performance, whereas increasing time to look at the targets led to poorer performance.

Fehd and Seiffert [9] instructed the participants of their Experiment 3 to track the targets with three types of looking strategy: free-looking, center–looking, and target-looking. In the first experimental block the participants were free to use any strategy they wished, after which they were instructed to use the latter two strategies in different blocks. The center-looking instructions emphasized that the participants should "keep their gaze near the center point of the target group or near a target", and "when they looked at one of the targets, to look back at the center before looking at another target." The target-looking instructions emphasized that the participants should "keep their gaze near a target", and "when they looked away from one target, to be sure to look at another target." The results showed that the free-looking strategy resulted in better tracking accuracy (83%) than the center–looking (77%) or the target-looking (57%) strategy, while the center–looking strategy was reliably better than the target-looking strategy. Moreover, most of the participants preferred the center–looking strategy to the target-looking strategy, as during the free-looking block the center–looking strategy was used more frequently than the target-looking strategy (40% vs. 13%). It is concluded that "instructed center-looking is beneficial to tracking but not as helpful as allowing participants to move their eyes naturally" [9] (p. 9).

2.3.2. Target-Looking and Center-Looking for Resolving Crowding

Target-looking is helpful for extracting high-resolution information of the targets in order to enhance the distinction between the targets and the distracters in crowding situations. Zelinsky and Todor [11] reported that people tend to look closer to the targets that are in the proximity of other objects. In their analysis, the gaze-to-target distance was plotted as a function of target-to-other-object (the other object could be a distracter or another target) distance. It was found that the gaze-target distance decreased as the target-other-object distance became smaller. For target set-size 2 and 3 this occurred primarily for short target-object distances, whereas for target set-size 4 the decline was linear. Moreover, the gaze-target distance was generally larger for set-size 2 than set-size 3 or 4. The results indicate that observers gaze closer to individual targets to resolve local crowding clusters. Similarly, Vater et al. [15] showed that "gaze is located closer to targets when they are crowded, as would be expected to reduce negative crowding effects by utilizing the higher spatial acuity of foveal vision" (p. 1).

On the other hand, one important function of center-looking is to gaze at the anticrowding point between the targets, which minimizes the ratio between each target's distance from the gaze point and distance from every distracter, so as to reduce crowding globally. Such an anticrowding model shows highest adherence to the observed eye movement data, which even outperformed the model of centroid-looking [16,17].

2.3.3. Detecting Changes in Object Form and Motion by Using Peripheral Vision

Vater et al. [13] investigated in Experiment 2 the detection of form and motion changes in peripheral vision during MOT. Motion change was operationalized as a sudden stop of target movement for 0.5 s, while form change was operationalized as a target changing for 0.5 s from square to diamond (square rotated by 45°). Assuming that the participants looked at the centroid at the time of the change (a specific manipulation was arranged to create a static centroid for the critical period), the change took place 15° into the periphery from the fixation point. The participants were asked to press a button as soon as they detected a change; at the end of the trial, a number was projected to each object, and the participants were required to name the number of the changed target. If no change occurred, they were to recall all four targets. Eye behavior was analyzed for trials where the target was correctly detected and the gaze was farther than 5° away from the to-be-changed target at the time the change was initiated. The results demonstrated significantly better detection of motion than form changes. Participants were also faster to do so, as indicated by the higher percentage in the motion than form change condition of button presses prior to fixating the changed target. On the other hand, a saccade to the changed target was initiated faster in the form than motion change condition, indicating that observers are more inclined to use foveal vision to inspect form than motion change. The average saccadic reaction times were above 0.5 s, which means that typically the saccade reached the target after the termination of the momentary (0.5 s) change. In other words, the changes were detected with the help of peripheral vision. The trials during which the critical saccade reached the target before the change was terminated did not result in better change detection than the trials where the saccade reached the target after the change had ended. This is taken as further proof that detection was achieved by peripheral vision.

As a direct follow-up, Vater, Kredel and Hossner [14] manipulated the peripheral location of the change so that it took place either relatively near (5–10°) or farther away (15–20°) from the centroid. The results of Experiment 1 showed that eccentricity influenced the change detection success; changes were detected more poorly in the far than near eccentricities. Eccentricity also affected the detection speed for form changes but not for motion changes; form changes were slower to detect at far eccentricities. In Experiment 2 [15], two types of temporary changes in target motion needed to be detected: a temporary stop (similarly to Experiment 1) and a slowdown of motion. Motion changes were detected better and faster at near than far eccentricities. Moreover, change detection was faster in the stop than in the slowdown condition. In sum, it is concluded (see also [13]) that "peripheral vision is more capable of detecting motion than form changes due to its high motion sensitivity and low acuity" [14] (p. 912).

2.3.4. Saccades Induced by Changes and Forthcoming Occlusion and Collision

As mentioned earlier, Zelinsky and Todor [11] found that rescue saccades are made to targets that are about to be occluded. Furthermore, these saccades are anticipatory in nature in that saccades are made before the occlusion. The results demonstrate the instrumental nature of eye movements during MOT in the form of rescue saccades that are a crucial tool in selectively processing "targets that are in danger of being lost" due to occlusion [11] (p. 10). Yet, it is not known from this study whether tracking accuracy was actually improved by the execution of rescue saccades.

Similarly, Vater et al. [15] showed that anticipatory saccades are made to targets that are about to collide with the border frame. In their study, a target-stop-detection task was adopted as the secondary task. The results showed that a change in target motion was detected poorer in the collision than the no-collision condition. The novel finding relevant to the present section is that it was confined to cases where an anticipatory saccade was executed toward the to-be-colliding targets. This suggests that saccades disrupt the continuous flow of visual information and thus impair change detection. Moreover, as mentioned in the previous section, a saccade could also be elicited by target changes. Typically, the saccade reached the target after the termination of the momentary change, and the saccade did not benefit change detection even in cases when it reached the target before the change had ended [13].

In sum, the results reviewed in this section showed that during tracking saccades are initiated to changed targets and targets that are about to be occluded or collided to each other. With good reasons, these saccades can be assumed to be beneficial to tracking, as indicated by their name (rescue saccades). Yet they may disrupt information processing due to saccadic suppression. Such eye behavior is likely to be automatically induced by lower-level visual processing, despite not necessarily being useful or being even detrimental to the tracking performance. Clearly, more research needs to be conducted on this issue before firmer conclusions can be made on its role in MOT.

2.4. Summary of MOT Studies: A Processing Architecture for Modeling Eye Movements during MOT

In sum, during MOT eyes frequently gaze at the central area between targets for grouping and anticrowding. Moreover, the eyes are located closer to crowded targets to extract high-resolution location information. High motion speed and abrupt viewpoint change lead to more gazes performed towards the center of the targets, presumably for anchoring the reference frame and tracking the targets in parallel. Repeated or flipped trajectories result in more similar eye behavior than unique trajectories, whereas the eye gaze at the targets does not extrapolate the target trajectories but lags behind the targets. Changes in target form and motion, as well as forthcoming occlusion and collision, are perceived in the peripheral vision and induce saccadic eye movements towards the relevant targets. Even though they may be assumed to benefit tracking, they may actually disrupt object tracking due to saccadic suppression.

We suggest that eye movements during MOT should be understood in relation to the processes of object tracking, which contain different levels of processing. We propose that for successful tracking, the process at the most elementary level is to perceive the targets' locations at each time point. The second processing level is to perceive the target objects at different locations and time points as a coherent group of moving targets, while the third level is to continuously allocate attention to the targets while suppressing the non-targets. The various factors mentioned in the prior sections may exert an impact on different levels of processing and thus influence the eye behavior in different ways. Crowding and stimulus size may mainly affect the first level of processing. On the other hand, motion speed, viewpoint change, trajectory repetition, occlusion, and collision may mainly affect the second level. Finally, the number of targets may mainly affect the third level of processing. Thus, eye behavior observed during tracking is likely to reflect a combination of processing at different levels.

The key for the success at the first level of processing is to enhance the resolution of the perceived target locations, so as to maximize the distinction between the target and distracter locations. Small stimulus size may make it difficult to perceive each target location. Thus, the eyes may land closer to targets to perceive the target locations. Moreover, close distance between objects (i.e., crowding) may lead to confusion between the target and distracter locations; in such cases the use of foveal vision is beneficial for resolving the crowding effect. Thus, the eyes are likely to gaze at the areas where targets are crowded by other objects.

The linking process at the second level can be influenced by various factors. High object speed leads to increased distance between target locations perceived at different time points, while viewpoint-change disrupts the reference frame of the whole scene. In such situations, moving the eyes may further interrupt the processing of object locations, and hence the preferred eye behavior is to maintain the gaze at a central location between the targets. If target trajectories are familiar via repetition, they may facilitate the second-level processing. On the other hand, occlusions and collisions increase the uncertainty of target locations before and after such events. In real-life visual environments, occlusion and collision may hint at the possibility of occurrence of potentially threatening events, so that saccades are automatically programmed toward such events for extracting detailed information with the foveal vision. Whether such rescue saccades are helpful or disruptive to tracking requires further research.

At the third level of processing, cognitive resources are severely limited, so that only a few objects can be simultaneously attended and maintained. In addition, as the number of objects increases, the cognitive resources distributed to each object decreases. Thus, the effect of the number of targets mainly reflects the capacity limitation during MOT. One way to overcome the limitation is to group multiple objects into one single virtual object. Thus, observers are inclined to look at the centroid of multiple targets so as to facilitate the grouping process.

Eye behavior at the lower processing levels is likely to be bottom-up, automatically induced by the computation of the visual scene, whereas eye behavior serving the high-level processes may be more subject to top-down control. Thus, observers are more likely to become aware of their eye behavior at this level. In sum, to achieve good tracking performance in MOT, the gaze should be optimally anchored to the crowded targets, while monitoring the objects with peripheral vision. Saccades towards targets should be initiated only when the risk of losing track of the targets is greater at the current gaze position than the saccade-related cost [16].

3. Eye Behavior during MIT

As described in the Introduction, MIT differs from MOT in that not only the target position but also the target identity needs to be dynamically updated. Arguably, MIT more closely resembles real-life tracking tasks than MOT in the sense that being aware of the target identity is typically relevant to the task at hand. For example, a parent tracking the whereabouts of his/her children on a crowded beach is highly motivated to discriminate his/her own children from the other bathers.

As MIT has generally been studied to a less extent than MOT, there is only a handful of eye-tracking studies of MIT. They are reviewed next (for their summary, see Table 2). Similarly, to the section on MOT, we first review studies examining where observers look at during MIT, followed by a review of the factors influencing eye movements in MIT and the possible functions of eye movements in MIT.

Table 2. Observed scanning strategies in eye-tracking studies of multiple identity tracking (MIT).

Study	Stimuli	Set-Size	Speed (deg/s)	Analysis Method	Scanning Strategy
Doran, Hoffman & Scholl [24]	lines that varied in size	3	2	AoI: 1 deg	Target: ~17%; Centroid: ~7%; Distracter: ~5%
Oksama & Hyönä [20]					
Experiment 2	line drawings 1.9 × 1.8 deg	2–5	2.6, 6.3, 10.3	AoI: 3.4 deg	Target: 53%; Blank area: 25%; Centroid: 2%; Distracter: 2%
Experiment 3	line drawings 1.9 × 1.8 deg	2–5	2.6, 6.3, 10.3	AoI: 3.4 deg	Target: 52%; Blank area: 24%; Centroid: 2%; Distracter: 2%
Li, Oksama & Hyönä [25]	Landolt rings, 1.6 deg	3	8.6	AoI: 2 deg	
Experiment 1					Target: 73%; Centroid: 6%
Experiment 2					Target: 66%; Centroid: 6%
Li, Oksama & Hyönä [26]	faces (1.7–2.3 deg), color discs (2 deg), line drawings (2 × 2 deg)	3,4	4.5	AoI: 2.5 deg	
All-Present					Target: 82%; Distracter: 6%
None-Present					Target: 76%
Wu & Wolfe [27]	hidden animals (3 × 3 deg)	3–5	6	AoI: 4 deg	Target: ~35%; Blank area: ~35%; Centroid: ~20%; Distracter: ~8%

3.1. Where Are Observers Looking at during MIT?

The eye-tracking study of Doran, Hoffman and Scholl [24] is conceived by the authors as a MOT study. However, as they registered observers' eye movements when they tracked moving lines of different length and orientation, we review it as an MIT study. The targets namely have unique identities defined by their length and orientation. Yet, it isn't a straightforward MIT study either, because the lines constantly changed their length.

In Experiment 1 of Doran et al. [24], ten observers tracked three target lines out of six for 20 s (Experiment 2 is not reviewed, as the observers were not allowed to move their eyes during tracking). The lines constantly changed length, orientation and velocity with the maximum velocity of 2°/s. After the movement stopped, they were to click with the mouse on the targets. The targets were tracked with 84% accuracy. For the eye movement analysis, four AoIs were determined: centroid, line center, line end, and line location other than center or end. A fixation was assigned to the nearest AoI provided that it was no further than 1° away from it. Line center and end were delineated as AoIs, as they were the two possible probe locations.

Observers spent more time fixating on the targets (~17%) than the centroid (~7%) or the distracters (~5%); the percentages are approximated on the basis of Figure 6 in Doran et al. [24]. Most target fixations were on positions not occupied by the probes (~30%). Doran et al. [24] expected to find evidence for a centroid-looking strategy. They suggest several possibilities for the failure to do so: (a) the stimuli used by them do not have a single locus of attention, (b) the targets often appear in close proximity to each other thus requiring good visual acuity to tell them apart, and/or (c) the probe detection task "may have biased observers' gaze toward individual objects" [24] (p. 594). In sum, although Doran et al. did not observe any prevalent scanning strategy, the target-switching strategy was more often used than the centroid-looking strategy.

Oksama and Hyönä [20] examined eye behavior during MIT using stimuli that were more comparable to MOT stimuli than the lines used by Doran et al. [24]. In Experiment 2 they registered eye movements during MIT. Observers tracked 2–5 targets among 6 moving objects. The stimuli were line-drawings of real-life objects (e.g., flower, coat). They subtended a maximum visual angle of $1.9 \times 1.8°$. The same speed conditions (2.6, 6.3, and 10.3°/s) were used as in their Experiment 1 in which eye movements were recorded during MOT (see above). After the movement stopped, all objects were masked, one of the targets was probed, after which the participants were to click on the probed target on a new screen where all 6 objects were displayed.

For the eye fixation analyses, AoIs (3.4° in diameter) were determined for targets, distracters, centroid, and screen center. Oksama and Hyönä [20] found that observers spent the majority of trial time fixating on targets (53%), but very little time on the centroid (1.5%) or distracters (1.7%). The reminder of the time was spent on blank area (25%) outside any AoI. Thus, the prevalent strategy was the target-switching strategy, whereas the centroid-looking strategy frequently found in MOT studies was completely absent.

Li, Oksama and Hyönä [25] used Landolt rings as stimuli in their MIT study. Landolt rings are rings with a gap in some part of the ring. In the study, rings with different identity were created by having the gap appear in different compass orientations. In both experiments of Li et al. [25], observers tracked 3 rings among a total of 8 rings. Their diameter was 1.6°. The rings moved in random directions at the speed of 8.6°/s. Object motion lasted for 3–6 s. After the movement stopped, all objects were masked and one target was probed by presenting it on the screen center. Participants were to click on the probed target or a "Not a target" response option, if (s)he thought it was not a target. Data of 22 participants were included in the analyses of Experiment 1 and those of 25 participants in the analyses of Experiment 2. For the analysis of eye behavior, fixations were assigned to the closest object using 1-degree bins (0–1°, 1–2°, 2–3°, etc.).

Experiment 1 of Li et al. [25] showed that 73% of all fixations landed no farther than 2° from the closest target, whereas only 6% fixations landed close to the centroid. In other words, the target-switching strategy was highly prevalent. Similarly, in Experiment 2, 66% of the fixations were located no further than 2° from a target. The centroid was seldom (6%) looked at.

Li, Oksama and Hyönä [26] investigated how different identities that varied in their visual resolution are tracked. They had participants track faces (high resolution), line drawings (medium resolution) and color disks (low resolution). A pretest demonstrated that these stimuli are identified in the parafovea and periphery with variable success. In the pretest, the stimuli were presented individually 2.5, 5 and 7.5° away from the fixation point. With color discs the identification rate was near ceiling for all these eccentricities, whereas a significant drop as a function of eccentricity was observed for faces. Identification of line drawings was at ceiling for the two nearest eccentricities but dropped in the 7.5° eccentricity.

In Experiment 1, Li et al. [26] examined the tracking of colored, ellipse-shaped facial images ($1.7° \times 2.3°$). Set-size was varied between participants; 20 participants tracked 3 faces among 3 distracter faces, while another 19 participants tracked 4 faces among 4 distracter faces. Faces moved randomly with a speed of 4.5°/s for a period of 4 to 8 s. After the movement stopped, all faces were masked and each target was probed by presenting it on the screen center one by one. The participants were required to click on the probed targets. In Experiment 2 [26], tracking of color discs (2° in diameter) was examined. Each disc appeared in one of nine possible colors. Twenty-one observers tracked 3 target discs among 3 distracter discs; another 19 participants tracked 4 target discs among 4 distracter discs. Experiment 2 was otherwise comparable to Experiment 1. Experiment 3 was comparable to Experiments 1 and 2 apart from the stimuli, which were line drawings ($2° \times 2°$) of common objects. Twenty participants tracked 3 targets among 3 distracters, while another 20 participants tracked

4 targets among 4 distracters. In all three experiments, an AoI of 2.5° around the moving objects was used.

The study of Li et al. [26] established a strong preference for the target-switching strategy; 82% of fixations fell on targets and only 6% on distracters. The target-switching strategy prevailed for all object identities, even though it was somewhat weaker for color discs (73%) than line drawings (85%) or faces (89%). In other words, tracking during MIT appears to be inherently serial and not only limited to high-resolution stimuli.

In sum, the available evidence strongly and consistently suggests that the default eye movement strategy in MIT is the target-switching strategy. The prevalence of the target-switching strategy has been established with different kinds of stimuli.

3.2. What Factors Influence Eye Behavior during MIT?

In this section, we review studies that have examined effects of target set-size, target speed and target type on the eye behavior during MIT.

3.2.1. Effects of Set-Size and Speed

Oksama and Hyönä [20] showed that the number of target visits increased as a function of set-size (2–5 targets), while the fixation time of each visit decreased as a function of set-size. An analogous finding was observed by Li et al. [26] for fixation time when comparing the tracking of 3 versus 4 targets. Oksama and Hyönä [20] also observed a decrease of target visits as a function of speed as well as an interaction between set-size and speed. The interaction suggested that the set-size effect was observable for slow and medium speed but not so much for fast object speed. The number of updated targets (i.e., targets visited with the eyes at least once) showed that target visits closely corresponded with the set-size, indicating that all targets were visited with the eyes at least once. The only exception was set-size 5 combined with fast speed; in this condition participants updated about 4 targets. This finding suggests that with an increase in target speed, observers may not be able to track with their eyes every target, particularly when there are several targets to be tracked. The decrease in target fixation time as a function of set-size indicates that observers move faster their eyes between targets when there are more targets to be tracked.

Oksama and Hyönä [20] also observed that pupil dilated as a function of set-size and speed. Moreover, these effects interacted suggesting that the set-size effect was stronger in the fast than slow speed condition. As pupil size presumably reflects, among other things, attentional effort, these findings suggest that MIT becomes increasingly attentionally demanding with an increase in target set-size and target speed. Finally, the number of blinks was also reduced as a function of set-size and speed. The decrease in blink rate as a function of increase in set-size and speed indicates that observers opt for maximal visual sampling when the task becomes demanding.

3.2.2. Effects of Target Type

Li et al. [25] conducted two experiments using Landolt rings as stimuli. In Experiment 1, the difficulty of perceiving the identity was varied by manipulating the gap width. The width of the narrow gap was 0.05° and that of the wide gap 0.2°. In Experiment 2, the visibility of the gap was kept constant (the wide gap condition of Experiment 1 was employed), but attentional demands were varied by manipulating the similarity in gap orientation among the targets. The gaps in two rings with a similar gap orientation differed by 20–40°, while the third ring had a gap orientation that differed by more than 80°.

Experiment 1 of Li et al. [25] showed that the probability of fixating on a target at least once during tracking was greater for narrow than wide gaps, which suggests that target fixations are particularly needed for refreshing identity-location bindings for less perceivable identities. In Experiment 2 more fixations were found on attentionally more demanding targets (a similar gap orientation) than less demanding targets (a dissimilar gap orientation). In contrast, fixation duration was shorter on attentionally more demanding than less demanding targets. This trade-off between fixation frequency and duration "may reflect observers' efforts in keeping the identity of the two similar targets distinct from each other" (p. 620).

As already mentioned above, Li et al. [26] found that the target-switching strategy prevailed for all tested object identities, color discs, line drawings and faces, although it was somewhat weaker for color discs. Color discs differ from faces and line drawings in that they are more readily perceivable in peripheral vision.

To sum up, the currently available evidence suggests that the target-switching strategy is the dominant eye movement strategy in MIT regardless of target type. It is even more dominant when a high-resolution representation needs to be constructed for the target identities to tell them apart.

3.3. What Functions Do Eye Movements Serve during MIT?

Above, we have reviewed evidence demonstrating the prevalence of the target-switching strategy in MIT. In this section, we review the evidence on the functions of target fixations. The evidence suggests that target fixations serve the purpose of establishing and refreshing identity-location bindings (Section 3.3.1), enhancing the tracking performance (Section 3.3.3) and clustering targets in conflict detection (Section 3.3.4). The present evidence for coupling of the fixation target and the attentional target (Section 3.3.2) suggests that the tightness depends on the type of identity to be tracked.

3.3.1. Establishing and Refreshing Identity-Location Bindings

Li et al. [26] investigated the role of target fixations by employing the gaze-contingent display change paradigm [28,29] to manipulate the availability of the moving objects. This paradigm makes possible to manipulate what is presented in the visual field contingent on where the observer looks at from moment to moment. Four presentation conditions were used: (a) all objects present, (b) only the fixated object visually available (all other objects were replaced with placeholders), (c) all but the fixated object present (once a fixation is initiated on a target, it is masked by a placeholder), and (d) none of the objects present during tracking. In the All-Present condition, observers can utilize both their foveal and peripheral vision for tracking; in the Fovea-Present condition, they can only use foveal vision; in the Periphery-Present condition they can only use peripheral vision; finally, in the None-Present condition tracking is carried out solely with the help of the visuospatial working memory. If only the foveated target is tracked at each moment, the Fovea-Present and All-Present conditions should result in equally good tracking accuracy, whereas the Periphery-Present condition should impair the performance. On the other hand, if multiple identities are tracked simultaneously, the All-Present condition should produce the best performance, followed by the Periphery-Present conditions, as in these conditions multiple identities are simultaneously available.

In Experiment 1 of Li et al. [26], faces were used as stimuli. The results showed that tracking accuracy was practically identical between the Fovea-Present and All-Present conditions (69.5% vs. 69.3%), which produced higher accuracy than the Periphery-Present and None-Present conditions that produced identical accuracy (62.3%). The pattern of results is completely consistent with the view that faces are tracked serially one at a time. Having all targets simultaneously available did not improve the performance from the situation where only the fixated target was available. Moreover, making the foveated target unavailable led to an equally poor performance as when no targets were available.

In Experiment 2 of Li et al. [26], color discs were used as stimuli. With set-size 3 tracking accuracy was near ceiling and it did not differ between the display conditions. With set-size 4, the All-Present and Periphery-Present conditions produced the best performance (86.3% vs. 84.5%) that was significantly better than the accuracy in the Fovea-Present (77.6%) and None-Present (71.7%) conditions. The pattern of results is consistent with the view that color discs are tracked in parallel. The tracking became better with the increase in the number of visually available targets (0, 1, and 3 in the None-Present, Fovea-Present and Periphery-Present conditions, respectively).

In Experiment 3 of Li et al. [26], black-and-white line drawings of common objects were used as stimuli. The All-Present condition produced the best tracking accuracy (89%), followed by the Fovea-Present and Periphery-Present conditions that produced practically equal (86%) accuracy, which was better than for the None-Present condition (82%). The pattern of results suggests that tracking was not completely serial, as seeing targets in the periphery resulted in better performance than seeing no targets. It was not completely parallel either, because seeing multiple targets in the periphery did not result in better performance than seeing just one at fovea.

Taken together, Li et al. [26] conclude that "the performance accuracy results indicate that the manner of tracking multiple objects varies in the serial-parallel continuum according to the identifiability of the objects" (p. 270). When object identities are readily identifiable in peripheral vision, as is the case with color discs, tracking is parallel, whereas poor peripheral identifiability of object identities, as is the case with facial images, leads to serial tracking. In other words, target fixations are necessary for refreshing identity-location bindings for high-resolution stimuli, but they are not required for tracking low-resolution images.

Interestingly, observers frequently fixated target locations even when there was nothing to see. This became apparent in the None-Present condition where only the placeholders were visible during tracking and identity tracking needed to be performed by the help of visual-spatial short-term memory. Li et al. [26] found that 75% of dwell time was spent fixating on the target placeholders. The frequency of target visits was not affected by identity type (faces, color discs, line drawings). The frequent visits to placeholders bears resemblance to the "looking-at-nothing phenomenon" [30–32]; looking at a location previously occupied by an object may active its memory representation [33].

The Periphery-Present condition resembled the None-Present condition in that once a target area was fixated, there was no identity information to be seen in the fovea. Nevertheless, similarly to the None-Present condition, observers made frequent fixations on the placeholders. Moreover, they more frequently visited the placeholders of faces and line drawings than those of color discs. In other words, the above findings reflect observers' intention of sampling visual information for target identities.

To sum up, the study of Li et al. [26] demonstrated that in MIT fixations on targets serve the purpose of (a) establishing identity-location bindings for high-resolution stimuli and (b) refreshing identity-location bindings in general. The former function has to do with the need of perceiving the identity of high-resolution stimuli with the foveal vision, while the latter function reflects memory updating.

3.3.2. Coupling of Attention and Fixation

Three studies have examined the degree to which the attentional target and the fixation target are coupled during MIT. Doran et al. [24] studied tracking of moving lines that constantly changed length, orientation and velocity. Tracking was combined with a secondary task of responding to probes (small grey circles) presented for 213 ms at random intervals either on line center or end. Participants pressed a button as soon as they detected a probe. They were instructed to prioritize tracking over probe detection. Probes were more readily detected when appearing on the line center than end (the so-called concentration effect). This concentration effect was not observed when the fixation happened to be very close (less than 4°) to the probe on the target, probably due to the benefits of high visual acuity for all probes. On the other hand, it was established for farther fixation-probe distances. The concentration effect suggests that attention is more readily concentrated on object centers. Yet, the eye movement data showed that this concentration effect was not coupled with a corresponding concentration of fixations on the object center. Thus, attention and fixation were not tightly coupled.

Li et al. [25] came to a different conclusion regarding the coupling of attention and fixation. They showed that duration of fixations varied as a function of the distance from a target; it was longest when it was no farther than 1° from a target and became shorter as the distance increased up to 4°. The few fixations landing on the centroid did not show such a relationship; instead fixation duration remained stable regardless of its distance to the centroid. The preference for staying fixated on the targets is taken as evidence that these fixations reflect the process of establishing and refreshing the identity-location binding for that target. In Experiment 1, they also found that the probability of fixating on a target at least once during the tracking interval was greater for narrow than wide gaps (Landolt rings were used as stimuli). In Experiment 2, more fixations were found to land on attentionally more demanding targets (a similar gap orientation) than less demanding targets (a dissimilar gap orientation). These results suggest that target fixations are particularly needed for refreshing identity-location bindings for less perceivable and distinguishable identities. Moreover, they suggest a close coupling between the fixation target and the attentional target.

Finally, the gaze-contingent display change study of Li et al. [26] provided evidence for the view that the coupling of the fixation and attention target depends on the type of identity to be tracked. As reviewed above in more detail, the overall pattern of their results showed that when tracking high-resolution stimuli, the fixation and attention targets are tightly coupled in that observers appear to strongly focus their attention to the fixated target. On the other hand, with low-resolution stimuli tracking is more parallel, meaning that observers simultaneously attend to more than one target. In other words, the attentional target and the fixation target are decoupled.

To sum up this section, the currently available evidence suggests that the tightness of coupling between attention and fixation during MIT is modified by the type of identity to be tracked. When the targets are readily perceivable in the visual periphery, as is the case with color discs and dot probes, attention and fixation may be decoupled. On the other hand, with high-resolution stimuli requiring foveal vision to be identified attention and fixation are tightly coupled.

3.3.3. Enhancing the Tracking Performance

Li et al. [25] examined the tracking accuracy as a function of the recency of target fixation. They observed that the recency of target fixation was linearly related to tracking accuracy with most recently fixated targets producing the best tracking accuracy and temporarily more distantly fixated targets having poorer accuracy. This linear trend was slightly steeper for narrow than wide gaps (Landolt rings were used as stimuli). Moreover, the farther away the last fixation was from the probed target, the poorer its tracking accuracy was. This trend was established only for the narrow gap rings. These results show that fixations on targets benefit tracking; on the other hand, when a target is not recently fixated, its identity-location binding is outdated and thus in danger of becoming lost.

Li et al. [26] replicated the recency effect [25]; targets fixated just before they were probed were associated with better tracking accuracy than previously fixated targets. This held true for faces in all presentation conditions and for line drawings except for the None-Present condition. In other words, the benefit of target fixation was particularly prominent when the targets required high-resolution information to be identified. These findings suggest that eye visits to targets benefit the tracking of high-resolution targets (e.g., faces) but not necessarily low-resolution targets (e.g., color discs).

3.3.4. Detecting Target Conflicts

Landry, Sheridan and Yufik [34] registered eye movements in a tracking task where observers' task was also to detect possible conflicts between targets. The examined tracking task was designed to resemble a task that air traffic controllers are exposed to. "The task was to identify targets predicted to conflict (defined as targets at the same "altitude" that pass within a particular distance on the display of one another)" (p. 93). When a conflict was identified, the participants (14 observers inexperienced in air traffic control) were asked to click on the targets in conflict. As the secondary task, they were to click on new targets appearing on the screen as well as on old targets just before they departed from the screen. Two conflict conditions were created: a demanding (5–7 conflicts) and a less demanding (1–3 conflicts) condition. The number of targets on the screen remained constant (14). Although we review this study as an MIT study, it is not clear how the target identities were marked, as this information is not provided. As the task was to resemble that of air traffic controllers, it is possible that alphabetic call signals were used as target identities.

Conflict detection among the untrained participants was low, yet the false alarm rate was also low. The eye-tracking data showed that the number of fixations on targets detected as conflicts were not significantly higher than on targets not selected as conflicts. Moreover, there were no more transitions between target pairs detected as conflicts than those not detected. Thus, these results do not support the view that eye fixations would be instrumental in detecting conflicting target trajectories. However, Landry et al. [34] also studied possible clustering of aircraft based on gaze transitions between targets. This was examined by the Virtual Associative Network (VAN) model developed by the authors. VAN represents a unified network of moving objects encompassing the entire visual scene as well as a "dynamic network partitioning into cohesive and externally weakly coupled clusters" (p. 93). Their eye movement analysis revealed that the probability of being within the same cluster was highest (50%) for target pairs correctly detected as being in conflict with each other, and it was lower (30%) for missed conflicts. "This indicates that the ability to detect a conflict may be affected by the ability to group the conflicting targets within a cluster" (p. 99).

In sum, there is suggestive evidence, based on a single study, that eye movements are used to cluster conflicting targets together when detecting conflicting target trajectories.

3.4. Summary of the Eye-Tracking Studies of MIT

All eye movement studies of MIT demonstrated a preference for using a target-switching strategy. In most studies, the preference was very strong so that up to 80% of the tracking time was spent on fixating targets.

Constantly fixating targets is not an epiphenomenon in MIT, but target fixations benefit tracking. Most recently fixated targets are associated with better tracking accuracy than more distantly fixated. The target-switching strategy is particularly relevant for high-resolution targets whose identities cannot be perceived peripherally. Target fixations not only serve the purpose of visual sampling and updating of identity information, but they are also in the service of working memory. Even when objects move hidden behind a placeholder, the majority of fixations fall on targets. Fixating the positions of hidden identities boosts their activation in working memory (cf. the looking-at-nothing phenomenon).

4. Comparison Eye Behavior in MOT and MIT

Three recent studies [20,27,35] have directly compared eye behavior during MOT and MIT. In this section, we review them one by one.

4.1. Oksama and Hyönä (2016)

In Experiment 3 of Oksama and Hyönä [20], performance differences in MOT and MIT were examined by within-participant comparisons. The experimental procedure was identical to that of their Experiment 1 and 2 (see above) with the following exceptions. The MOT stimuli were identical line drawings of a lobster and only half of the trials of Experiment 1 and 2 were presented. MOT and MIT were performed as separate blocks. Data from twelve observers were included in the analyses.

Tracking accuracy was 90% or better except for set-size 5 in MIT where it went down to approximately 78%. When carrying out the MOT task, observers spent most of their time (48%) looking at a blank space that was neither the centroid nor the screen center, but less time on targets (24%). In contrast, during MIT observers spent most of the time fixating targets (52%) and less time fixating blank space (26%). During MIT, the number of target visits and the number of updated targets increased linearly as a function of set-size, whereas in MOT it remained constant. On average, targets were visited in MIT at the rate of 1.1 Hz, while in MOT the rate was half of that (0.5 Hz). Pupil size increased as a function of set-size, it did so more deeply during MIT than MOT. On the other hand, blink rate decreased as a function of set-size. In sum, participants sampled the dynamic display much more frequently in MIT than in MOT. Moreover, MIT was attentionally more demanding than MOT, as indexed by the pupil size. This was despite the fact that MOT included 10 moving objects (including distracters), whereas MIT included 6 objects. As a result, there was more crowding in MOT and also more direction changes than in MIT. These differences in motion trajectories thus favored MIT over MOT. Nevertheless, MIT turned out to be more demanding than MOT.

Oksama and Hyönä interpret their results to point to two separate tracking systems: "position tracking in the MOT task is achieved by a covert parallel system, whereas identity tracking in the MIT task is achieved by an overt serial system" [20] (p. 407). The position tracking system yields only spatiotemporal information, which is sufficient for MOT but insufficient for MIT. In MIT target identity information needs to be bound with location information, which is a serial process requiring overt attention shifts between targets.

4.2. Wu and Wolfe (2018)

Wu and Wolfe [27] were critical of the notion of two parallel tracking mechanisms posited by Oksama and Hyönä [20]. They argue that since Oksama and Hyönä used different stimuli in MOT and MIT, their different results for MOT and MIT may rather reflect stimulus than task differences. In their study, Wu and Wolfe kept the stimuli the same between the two tasks. Both tasks were carried out for hidden stimuli; during the movement phase only the placeholders of objects were visually available. In MOT, participants were required to only keep track of target locations. After the movement stopped, one of the placeholders was probed and the participants responded whether or not the probed circle was a target. In MIT, participants were required to memorize the target identities prior to the movement phase. After the movement stopped, one of the targets was probed, after which a target identity was presented on the screen center and the participants were to respond whether it was the probed target. The targets presented prior to the movement phase (8 s) were 10 cartoon animals ($3 \times 3°$ in size), of which observers tracked 3, 4 or 5 that moved with a velocity of $6°/s$. MOT and MIT were presented in different blocks. Twelve observers took part in the experiment.

Tracking accuracy was 90% for MOT and 86% for MIT. For the eye movement analyses, four AoIs were delineated: target, distracter, centroid, and everywhere else. The AoI for target, distracter and centroid was $4°$ in size. Overall, observers spent about 35% of trial time fixating targets, about 35% the area outside targets, distracters and centroid, about 20% fixating the centroid, and about 8% fixating distracters (approximated on the basis of Figure 8 of Wu and Wolfe). Significantly more time was spent in MIT on target fixations and significantly less on centroid fixations than in MOT. Unlike Oksama and Hyönä [20], Wu and Wolfe [27] found no increase in either task in the number of fixations and target visits as a function of set-size (Oksama and Hyönä observed it for MIT but for MOT). The number of updated targets increased in both tasks, but unlike in the Oksama and Hyönä study the increase was not limited to MIT. Finally, a marginal increase was observed in pupil size as a function of set-size. In sum, the qualitative difference in the eye fixation patterns between MOT and MIT observed by Oksama and Hyönä was not replicated by Wu and Wolfe when tracking hidden targets. The main conclusion is that "a serial tracking process is not necessary in MIT since it is still possible to keep track of identities when those identities are hidden during tracking" [27] (p. 459).

4.3. Nummenmaa, Oksama, Glerean and Hyönä (2017)

Nummenmaa, Oksama, Glerean and Hyönä [35] studied the neural underpinnings of MOT and MIT in an fMRI investigation. In Experiment 1, they also registered observers' eye movements. Participants carried out MOT and MIT for 2 and 4 targets among 8 objects that moved for 14–18 s with a variable speed (average speed of $6.3°/s$). Tracking accuracy was at ceiling (an average accuracy of 94%) except for set-size 4 in MIT where the accuracy was about 75%. The number of fixations increased as a function of set-size in MIT but in MOT. No AoI analyses were conducted so it is not known where the fixations landed. Saccadic amplitudes were generally longer in MIT than MOT; in MIT they became longer with an increase in set-size but remained constant in MOT. Finally, pupil size increased more steeply during MIT and MOT, as the set-size increased from 2 to 4. In sum, the eye movement results are consistent with those of Oksama and Hyönä [20].

The results of brain activation pointed to a shared frontoparietal circuit between MOT and MIT and a unique resource for MIT in dorsolateral prefrontal cortex. The frontoparietal circuit is responsible for the control of attention and eye movements. Although shared between MOT and MIT, it was found to be more strongly activated during MIT than MOT. Dorsolateral prefrontal cortex has an important role in temporarily retaining information particularly in visuospatial working memory —in the case of MIT temporarily storing identity-location bindings.

4.4. Summary of the Comparison of MOT and MIT

When MOT and MIT were compared within the same experiment, the target-switching strategy was observed to prevalent in MIT, whereas in MOT participants stayed fixating a blank space or the centroid for longer time than the targets. Thus, these comparisons confirm the results obtained when MOT and MIT were studied separately. A linear increase in target visits was found in MIT as a function of target set-size, but not in MOT. The qualitative changes in eye behavior between MOT and MIT led Oksama and Hyönä [20] posit two distinct tracking mechanisms—a parallel mechanism for position tracking and a serial mechanism for identity tracking. Results on brain activation are compatible with the dual-mechanism view. Wu and Wolfe [27] challenge this view by demonstrating that MOT and MIT produce highly similar eye behavior when tracking hidden objects.

Interestingly, the hidden target tracking employed by Wu and Wolfe [27] is comparable to the None-Present presentation condition of Li et al. [26]. Yet, the results of the two studied differ markedly. Li et al. found that observers looked at the hidden targets 65% of the time, whereas in the Wu and Wolfe study the percentage was much smaller (~35%). A possible explanation for this is that in the Li et al. study hidden target tracking was performed as part of a task where most trials entailed visible objects. Thus, tracking of visibly moving targets may have carried over to hidden target tracking, which was not the case in the Wu and Wolfe study, where only hidden targets were tracked. Li et al. did not include MOT, so a direct comparison to Wu and Wolfe cannot be made. Yet, interestingly Li et al. found no difference in the prevalence of target fixations as a function of identity type. All in all, tracking by memory is an interesting visuospatial ability that deserves further study.

5. Future Directions

In the present review, we hope to have demonstrated that eye movements during MOT and MIT are not a mere epiphenomenon, but they play a functional role in tracking of multiple moving objects. This is especially the case in MIT, which more closely resembles tracking in real-life visual environments. In MIT, the default eye behavior is the target-switching strategy (moving the eyes between targets). Although MOT and MIT can be performed with reasonable success without eye movements, keeping the eyes centered on the screen decreases tracking accuracy, particularly for MIT (see e.g., [35]). In MIT, observers move their eyes between targets even when the target identities are occluded, presumably to facilitate refreshing and maintaining identity-location bindings for targets.

As argued below, future research should focus on MIT, as a more ecologically valid tracking task. Yet, we also think there are interesting issues to be solved with respect to MOT that can ideally be approached using the eye-tracking method. As reviewed above, crowding, possibility for target occlusion and abrupt form changes are likely to trigger a saccade toward "the problem area". A reasonable assumption is that such rescue saccades are executed so that foveal vision can be brought to bear on keeping track of the targets. Yet, there is evidence suggesting that their execution may actually hinder tracking accuracy. As crowding and occlusion are common phenomena in tracking (be it MOT or MIT), more research should be devoted to examine what role eye behavior plays in preventing disruption by crowding and occlusion. Such results will be relevant to both MOT and MIT.

It is curious that researchers of MOT or MIT motivate their eye-tracking studies by making reference to real-world visual environments (team sports, traffic, crowded areas, etc.) where dynamic tracking of moving objects is an integral part. Yet, practically all the reviewed studies do not make any efforts in mimicking tracking in the real world with three exceptions [10,11,34]. Landry et al. [34]

simulated a dynamic visual environment that air traffic controllers deal with. Zelinsky and his group [10,11] mimicked an underwater scene with moving sharks. Thus, it is fair to say that most studies lack ecological validity. Yet, MIT studies are ecologically more valid than MOT studies, as in virtually all real-world situations the to-be-tracked objects have distinct identities. Thus, in the future studies it is preferable to study MIT than MOT, which has not been the case to date.

The lack of ecological validity takes several forms. First, in real-world visual environments motion is quite seldom random, as it has been in the eye movement studies of MOT and MIT reviewed here. Consider, for example, a traffic scene where vehicle motion is heavily constrained by traffic rules. An example of such an approach can be found in the study of Huff, Papermeier and Zacks [36] where the motion of some football players was constrained by the ball motion. Second, in the conducted experiments objects move in a blank space; this is seldom the case in real life. Consider, for example football scene where players motion trajectories are constrained by their position on the field (e.g., near the goal vs. center field). Third, object identity heavily constrains the type of motion an object is capable of performing. For example, human beings are heavily tied to the ground, whereas birds are also equipped for vertical motion. Fourth, the studies have recruited observers inexperienced in tracking of moving objects. As dynamic tracking is a skill likely to improve by practice, it is important to also study expert behavior. Finally, and most importantly, in real-life dynamic environments tracking is not done for its own sake, but object tracking is carried out for the service of the performed task. For example, a football player tracks other players in order to make the decision of his/her next move. Such situations are heavily time-constrained, which means that the player needs to choose what players to track and what players to ignore. In other words, attentional priority is preferentially given to moving objects that are relevant to the performed task. Hence, in future studies MIT should be investigated in task environments where MIT is subsumed into the service of the primary task. This means, among other things, that target designation is not externally given to the observer but is instead determined by the observers themselves.

As reviewed above, eye movements play a functional role particularly in MIT; thus, future MIT studies, preferably along the lines suggested above, should include eye-tracking in their methodological arsenal. Why bother to do that? First, eye movements are closely coupled with overt attention shifts in visually and attentionally demanding tasks such as MIT. Thus, eye-tracking provides useful information about the allocation of attention as it fluctuates over time and space. It is then possible to reveal, for example, what targets are given attentional priority among all targets in a situation when there are multiple targets, of which some are more relevant to the performed task than others. Second, as eye movements are an integral part of MIT, they can be registered without introducing any secondary tasks to measure attentional allocation in MIT. Third, eye-tracking makes possible to study individual differences in the task performance MIT is subsumed into. Taking again an example from football, it would be possible to investigate how expert players track with overt attention other players when preparing to make the decision whom to pass the ball. Their eye behavior may then be compared to more novice players to determine the role of expertise in MIT. Here the role of MIT as a slave mechanism to the primary task is again stressed.

Is it feasible to register eye movements during multiple object tracking in situations mimicking real-world visual environments? Perhaps time is ripe to do so. With recent technological advancements, it has now become possible to register eye movements in virtual reality. Virtual reality itself opens possibilities to simulate and also manipulate visual scenes approximating real-life visual environments. Mobile eye-trackers, on the other hand, make it possible for the researchers to step out "to the wilderness". A downside of that is that experimentally controlled studies would not be feasible.

Author Contributions: Conceptualization, J.H., J.L., L.O.; writing—original draft preparation, J.H., J.L., L.O.

Funding: J.L. was funded by National Natural Science Foundation of China, grant number 31500887.

Conflicts of Interest: The authors declare no conflict of interest.

References

1. Endsley, M.R. Toward a Theory of Situation Awareness in Dynamic Systems. *Hum. Factors J. Hum. Factors Ergon. Soc.* **1995**, *37*, 32–64. [CrossRef]
2. Meyerhoff, H.S.; Papenmeier, F.; Huff, M. Studying visual attention using the multiple object tracking paradigm: A tutorial review. *Atten. Percept. Psychophys.* **2017**, *79*, 1255–1274. [CrossRef] [PubMed]
3. Oksama, L.; Hyönä, J. Is multiple object tracking carried out automatically by an early vision mechanism independent of higher-order cognition? An individual difference approach. *Vis. Cogn.* **2004**, *11*, 631–671. [CrossRef]
4. Pylyshyn, Z.W.; Storm, R.W. Tracking multiple independent targets: Evidence for a parallel tracking mechanism. *Spat. Vis.* **1988**, *3*, 179–197. [CrossRef] [PubMed]
5. Oksama, L.; Hyönä, J. Dynamic binding of identity and location information: A serial model of multiple identity tracking. *Cogn. Psychol.* **2008**, *56*, 237–283. [CrossRef] [PubMed]
6. Horowitz, T.S.; Klieger, S.B.; Fencsik, D.E.; Yang, K.K.; Alvarez, G.A.; Wolfe, J.M. Tracking unique objects. *Percept. Psychophys.* **2007**, *69*, 172–184. [CrossRef] [PubMed]
7. Hyönä, J.; Oksama, L.; Rantanen, E. Tracking the Identity of Moving Words: Stimulus Complexity and Familiarity Affects Tracking Accuracy. 2019; in press.
8. Fehd, H.M.; Seiffert, A.E. Eye movements during multiple object tracking: Where do participants look? *Cognition* **2008**, *108*, 201–209. [CrossRef] [PubMed]
9. Fehd, H.M.; Seiffert, A.E. Looking at the center of the targets helps multiple object tracking. *J. Vis.* **2010**, *10*, 19. [CrossRef]
10. Zelinsky, G.J.; Neider, M.B. An eye movement analysis of multiple object tracking in a realistic environment. *Vis. Cogn.* **2008**, *16*, 553–566. [CrossRef]
11. Zelinsky, G.J.; Todor, A. The role of "rescue saccades" in tracking objects through occlusions. *J. Vis.* **2010**, *10*, 29. [CrossRef]
12. Huff, M.; Papenmeier, F.; Jahn, G.; Hesse, F.W. Eye movements across viewpoint changes in multiple object tracking. *Vis. Cogn.* **2010**, *18*, 1368–1391. [CrossRef]
13. Vater, C.; Kredel, R.; Hossner, E.J. Detecting single-target changes in multiple object tracking: The case of peripheral vision. *Atten. Percept. Psychophys.* **2016**, *78*, 1004–1019. [CrossRef]
14. Vater, C.; Kredel, R.; Hossner, E.J. Detecting target changes in multiple object tracking with peripheral vision: More pronounced eccentricity effects for changes in form than in motion. *J. Exp. Psychol. Hum. Percept. Perform.* **2017**, *43*, 903–913. [CrossRef]
15. Vater, C.; Kredel, R.; Hossner, E.J. Disentangling vision and attention in multiple-object tracking: How crowding and collisions affect gaze anchoring and dual-task performance. *J. Vis.* **2017**, *17*, 21. [CrossRef]
16. Lukavský, J. Eye movements in repeated multiple object tracking. *J. Vis.* **2013**, *13*, 9. [CrossRef]
17. Děchtěrenko, F.; Lukavský, J. Models of Eye Movements in Multiple Object Tracking with Many Objects. In Proceedings of the 2014 5th European Workshop on Visual Information Processing (EUVIP), Paris, France, 10–12 December 2014.
18. Lukavský, J.; Děchtěrenko, F. Gaze position lagging behind scene content in multiple object tracking: Evidence from forward and backward presentations. *Atten. Percept. Psychophys.* **2016**, *78*, 2456–2468. [CrossRef]
19. Děchtěrenko, F.; Lukavský, J.; Holmqvist, K. Flipping the stimulus: Effects on scanpath coherence? *Behav. Res. Methods* **2017**, *49*, 382–393. [CrossRef]
20. Oksama, L.; Hyönä, J. Position tracking and identity tracking are separate systems: Evidence from eye movements. *Cognition* **2016**, *146*, 393–409. [CrossRef]
21. Yantis, S. Multielement visual tracking: Attention and perceptual organization. *Cogn. Psychol.* **1992**, *24*, 295–340. [CrossRef]
22. Papenmeier, F.; Huff, M. DynAOI: A tool for matching eye-movement data with dynamic areas of interest in animations and movies. *Behav. Res. Methods* **2010**, *42*, 179–187. [CrossRef]
23. Dorr, M.; Martinetz, T.; Gegenfurtner, K.R.; Barth, E. Variability of eye movements when viewing dynamic natural scenes. *J. Vis.* **2010**, *10*, 1–28. [CrossRef]
24. Doran, M.M.; Hoffman, J.E.; Scholl, B.J. The role of eye fixations in concentration and amplification effects during multiple object tracking. *Vis. Cogn.* **2009**, *17*, 574–597. [CrossRef]

25. Li, J.; Oksama, L.; Hyönä, J. Close coupling between eye movements and serial attentional refreshing during multiple-identity tracking. *J. Cogn. Psychol.* **2018**, *30*, 609–626. [CrossRef]
26. Li, J.; Oksama, L.; Hyönä, J. Model of Multiple Identity Tracking (MOMIT) 2.0: Resolving the serial vs. parallel controversy in tracking. *Cognition* **2019**, *182*, 260–274. [CrossRef]
27. Wu, C.; Wolfe, J.M. Comparing eye movements during position tracking and identity tracking: No evidence for separate systems. *Atten. Percept. Psychophys.* **2018**, *80*, 453–460. [CrossRef]
28. McConkie, G.W. Eye Movement Contingent Display Control: Personal Reflections and Comments. *Sci. Stud. Read.* **1997**, *1*, 303–316. [CrossRef]
29. Rayner, K. The perceptual span and peripheral cues in reading. *Cogn. Psychol.* **1975**, *7*, 65–81. [CrossRef]
30. Altmann, G.T. Language-mediated eye movements in the absence of a visual world: The 'blank screen paradigm'. *Cognition* **2004**, *93*, B79–B87. [CrossRef]
31. Ferreira, F.; Apel, J.; Henderson, J.M. Taking a new look at looking at nothing. *Trends Cogn. Sci.* **2008**, *12*, 405–410. [CrossRef]
32. Richardson, D.C.; Spivey, M.J. Representation, space and Hollywood Squares: Looking at things that aren't there anymore. *Cognition* **2000**, *76*, 269–295. [CrossRef]
33. Johansson, R.; Johansson, M. Look here, eye movements play a functional role in memory retrieval. *Psychol. Sci.* **2014**, *25*, 236–242. [CrossRef]
34. Landry, S.; Sheridan, T.; Yufik, Y. A methodology for studying cognitive groupings in a target-tracking task. *IEEE Trans. Intell. Transp. Syst.* **2001**, *2*, 92–100. [CrossRef]
35. Nummenmaa, L.; Oksama, L.; Glerean, E.; Hyönä, J. Cortical circuit for binding object identity and location during multiple-object tracking. *Cereb. Cortex* **2017**, *27*, 162–172. [CrossRef]
36. Huff, M.; Papenmeier, F.; Zacks, J.M. Visual target detection is impaired at event boundaries. *Vis. Cogn.* **2012**, *20*, 848–864. [CrossRef]

Review

The Changing Role of Phonology in Reading Development

Sara V. Milledge * and Hazel I. Blythe

Department of Psychology, University of Southampton, Southampton SO17 1BJ, UK; hib@soton.ac.uk
* Correspondence: s.milledge@soton.ac.uk

Received: 22 February 2019; Accepted: 28 May 2019; Published: 30 May 2019

Abstract: Processing of both a word's orthography (its printed form) and phonology (its associated speech sounds) are critical for lexical identification during reading, both in beginning and skilled readers. Theories of learning to read typically posit a developmental change, from early readers' reliance on phonology to more skilled readers' development of direct orthographic-semantic links. Specifically, in becoming a skilled reader, the extent to which an individual processes phonology during lexical identification is thought to decrease. Recent data from eye movement research suggests, however, that the developmental change in phonological processing is somewhat more nuanced than this. Such studies show that phonology influences lexical identification in beginning and skilled readers in both typically and atypically developing populations. These data indicate, therefore, that the developmental change might better be characterised as a transition from overt decoding to abstract, covert recoding. We do not stop processing phonology as we become more skilled at reading; rather, the nature of that processing changes.

Keywords: theories of learning to read; orthography; phonology; adults; children; eye-tracking

---

1. The Changing Role of Phonology in Reading Development

Learning to read is a vital process within modern societies given how much information is conveyed by the written word, ultimately affecting academic success, employability, and social and economic welfare. For example, it is estimated that the cost of illiteracy to the global economy is over $1 trillion each year, costing a developed nation 2% of its gross domestic product (GDP), an emerging economy 1.2% of its GDP, and a developing country 0.5% of its GDP [1]. Yet the acquisition of this skill, so pivotal to successful functioning within society, is a long, complicated and effortful process that can last for many years.

Reading is a process that requires the learning of associations between the visual forms of printed words (orthography) and their associated speech sounds (phonology) and meanings (semantics). The aim of reading is to construct meaning from text, i.e., for the reader to comprehend the written language. It is well-recognised, though, that making these links from orthography to semantics also involves phonological processing [2]. Oral language acquisition precedes written language acquisition, and so, a child's earliest cognitive representations of words include phonology and semantics; only later, as they learn to read, do those phonological and semantic representations map onto orthographic forms [3].

Within theoretical accounts of reading development, a broad consensus seems to be that as a child's reading skill increases, their lexical identification becomes increasingly based on direct orthographic–semantic links, and the contribution of phonology to lexical identification decreases, e.g., [4–10]. Consequently, skilled reading is often characterised as an individual's ability to access semantics directly from a word's printed form. This view has been supported by data from pen-and-paper tasks, such as hand-coding of a child's reading, spelling or pronunciation errors [11–13].

In recent years, though, eye movement research has indicated that children continue to process phonology during lexical identification as their reading skills increase [14]. These data indicate that developmental change in phonological processing is better characterised as a progression from early, overt decoding (the conscious, effortful sounding out of printed letters to identify a word) to more sophisticated, covert phonological recoding (the rapid, covert, pre-lexical processing of a printed word's phonology).

We begin by briefly reviewing the literature on theoretical models of children's reading development, which clearly documents a developmental change in phonological processing during lexical identification. We then review the literature on skilled adult readers' lexical identification which has examined, in considerable detail, the role of phonological processing. Subsequently, research within developmental populations, both typical and atypical, is discussed. Phonological processing in languages other than English is also briefly considered (given how theories of learning to read relate primarily to reading development within English, this paper's focus will predominantly be on research conducted in English). Finally, some models of word recognition are briefly outlined and then evaluated within the context of this paper. Taken together, we consider how these recent contributions to the experimental literature might contribute to both theoretical models of learning to read and models of word recognition.

2. Theories of Learning to Read

One prominent theory of how visual word recognition skills develop is Share's [15] self-teaching hypothesis. This hypothesis posits that phonology plays a central role in how readers acquire orthographic representations of words. Phonological decoding (to achieve a correct pronunciation) is assumed to be critical for the acquisition of orthographic representations, as it draws the child's attention to the order and identity of a word's constituent letters. As such, decoding provides children with the opportunity to set up direct connections between the spelling of a letter string and the phonology of the spoken word, which results in the growth and development of their lexicons. In this way, phonology serves as a powerful self-teaching device: the explicit learning of a few sets of grapheme to phoneme correspondences (GPCs) allows children to decode an increasing number of words, which, in turn, supports the growth of their lexicons.

A number of theories have been proposed in order to try to characterise the process that children go through as they progress from beginning to skilled reader, with many proposing that children progress through a series of phases as they become more experienced in dealing with written text, ultimately leading to fluent, skilled reading, e.g., [5–10,16,17]. It is assumed that whilst most children pass through these phases, they are not biologically determined [18]. These phases are described as representing the reader's dominant (but not sole) process for identifying words during reading at that point in the child's development. There are, of course, differences between the theories of reading development. For example, some theories suggest that there are three phases, e.g., [10], while others suggest four phases, e.g., [5–9]. Here, we focus upon the common aspects that are relevant to our interest in phonological processing. Broadly speaking, the earliest phase(s) of reading development is characterised by a child's attempts to learn associations between orthographic features of written text (although not complete word forms) and words that already exist in their oral vocabulary (e.g., recognising the word *camel* because it has two humps in the middle) [19]. Subsequently, children learn the alphabet and, consequently, learn grapheme to phoneme correspondences (e.g., learning that the word *cat* is pronounced /k/ /æ/ /t/), providing the capability to read words the child has not encountered before. Then, finally, a child progresses to the point where they are able to identify the majority of printed words that they encounter through whole word recognition, with the assumption that this process relies on direct orthographic–semantic links. At this point, a child does not engage in any observable, overt phonological decoding in order to identify words during reading (for a recent review, see [4]).

A major similarity between these theories of reading development is that they propose a developmental shift from beginning readers, who rely more on phonology to identify words, to more skilled readers, who form direct links between orthography and semantics, e.g., [5–10]. Inherent in this proposed trajectory is the decreased reliance on phonology, to the point where it no longer contributes to lexical identification for most words that a reader encounters. Such theories, though, were primarily formulated on the basis of findings from offline tasks, e.g., [12,20,21]. Whilst it is true that offline tasks, and isolated word recognition tasks (as discussed in the following section), have provided researchers with insight (albeit indirect) into the role that phonological processing plays in both skilled adult and beginning child readers and in the shift from effortful phonological decoding to fluent sight word reading, e.g., [8], it is eye movement research (discussed in Section 4) with skilled adult readers, and more recently with developmental populations, that has provided direct insight into how this proposed theoretical developmental shift may be more nuanced than these current theories account for.

3. The Role of Phonology: Isolated Word Recognition Tasks

This section outlines four key areas of evidence: (1) delineation of how isolated word recognition tasks have demonstrated the use of overt phonological decoding by beginner readers in order to achieve lexical access; (2) how this subsequently decreases based on reading skill; (3) how adults display covert phonological recoding; and (4) the display of this form of phonological processing by children.

A substantial body of evidence has documented how readers engage in overt phonological decoding in order to identify printed words, using a variety of experimental paradigms. For example, lexical decision tasks (LDTs), where participants are required to decide, as quickly as possible, whether a printed letter string is a real word or not; semantic categorisation tasks, which require the participant to decide whether or not each presented word is an exemplar of a particular semantic category; and naming tasks, which require participants to pronounce a written letter string, often at speed, have all been used.

First, such methods have documented overt phonological decoding in beginning readers. For example, Johnston and Thompson [22] found that 8-year-old English children were less accurate at rejecting pseudohomophones (e.g., *wotch-watch*) than ordinary nonwords (e.g., *cotch*) in a LDT (Experiment 1). It was noted that many of the children tended to sound the stimuli out loud prior to making the lexical decision. Sounding out is a clear indication of phonological decoding being undertaken by the children, and the children displayed reduced accuracy in rejecting the nonword pseudohomophones, indicating that lexical entries were being activated for their respective "real word" homophones. Phonological decoding was enabling the children to activate an existing lexical entry due to shared phonology, regardless of the status of the pseudohomophone as a nonword (with no possible lexical entry). This tendency for children to rely on phonological decoding seems to become particularly apparent when they encounter unfamiliar words. For example, Adams and Huggins [11] selected 50 exception words, such as *ocean*, *sword* and *yacht*, which were ordered by frequency (how often a word is typically encountered in text), so that easier words preceded harder words. The researchers found that children in Grades 2–5 typically read words accurately and without any overt decoding until they reached a point in the list where the words became unfamiliar (i.e., low frequency words). At this point, readers began sounding out and blending the words, which caused them to hesitate and often misread the words. Schmalz, Marinus, and Castles [23] found that children showed regularity effects (whereby a benefit is found for regular words, that is, words with pronunciations that conform to GPC rules, e.g., *spade*, over irregular words, with pronunciations that do not conform to GPC rules, e.g., *yacht*) for low frequency words (e.g., *desk* vs. *calm*) but not high frequency words (e.g., *mess* vs. *ghost*) in a LDT. The researchers argued that children were using phonological decoding for words that they encountered less frequently because the output for irregular words from phonological decoding conflicts with the correct entry in the mental lexicon. For high frequency words, however, the lack of regularity effects suggests that children as young as 8 years-old were relying predominantly on a direct route from orthography to semantics for high frequency words.

Second, the literature shows children's decreasing reliance on overt phonological decoding as their reading skill increases. It is posited that readers increasingly identify words by sight, with direct links from orthography to semantics, e.g., [8]. For example, Samuels, LaBerge, and Bremer [24] used a semantic categorisation task with children from Grades 2, 4 and 6 as well as college students. The words used in this task varied in length from three to six letters. Whilst second graders' response latencies increased as words grew longer, older students' latencies did not change as a function of word length. This suggests that the older participants were processing the words as wholes, whilst the second graders were processing component letters in order to read the words (although it is worth noting that this could be an orthographic effect rather than an effect of phonology). Nevertheless, other research has also demonstrated how phonological decoding decreases as reading skill increases. For example, Ehri and Wilce [20] measured the latencies of skilled and less skilled readers (from Grades 1–4) in a series of naming tasks using common words (e.g., *book*), number words (e.g., *four*), CVC nonwords (e.g., *jad*), and single digits (e.g., *6*). Skilled readers across the grades named words faster than nonwords and named words as quickly as digits, indicating that they were processing the words as wholes. In contrast, though, the less skilled readers only displayed this pattern of effects in Grade 4; only the oldest less skilled readers were equally as fast at naming words as digits. Overall, these data show that as children become increasingly skilled readers, decoding decreases. Researchers have often inferred from this an increasingly dominant process of direct access from orthography to semantics.

Third, a large body of evidence has documented phonological recoding in skilled adult readers. For example, Lesch and Pollatsek [25] had participants name target words (e.g., *nut*) after the presentation of a prime, either a semantic associate word (e.g., *beech*), a homophone of that associate (e.g., *beach*) or an orthographic control (e.g., *bench*). The researchers found that, at short prime durations, the target words were named faster following both the semantic associates and the homophone primes, in comparison to the orthographic controls. The researchers concluded, therefore, that phonological recoding contributed to readers' lexical access. Van Orden [26], in a semantic categorisation task, found that frequent errors were made to homophones of particular categories; for example, for the category 'flower' the word *rows* is homophonic to the category instance of *rose*, and participants frequently made false positive errors to *rows* relative to orthographic controls (e.g., *robs*). As such, phonology appears to play an important role in allowing adults to achieve lexical access through phonological recoding [25–29].

Fourth, children have also been shown to display phonological recoding, with this form of processing seeming to be pivotal in the development of visual word recognition skills. For example, Kyte and Johnson [30] had Grade 4 and 5 children make lexical decisions for monosyllabic words (e.g., *bean/meat*) and pseudowords (e.g., *meap/meep*) under two matched experimental conditions: one where items were named prior to lexical decision to promote phonological recoding (read aloud condition), and a condition presumed to limit phonological recoding (concurrent articulation condition; participants repeated a syllable (e.g., "LA") whilst completing the LDTs). Later, approximately 24 h after the LDTs, orthographic learning of the pseudowords was evaluated using orthographic choice, spelling and naming tasks. Target words learned with phonological recoding produced greater orthographic learning than those learned with concurrent articulation. This study provides some evidence for the importance of phonological processing in the development of visual word recognition skills and an orthographic lexicon, consistent with the self-teaching hypothesis [15]. However, it is important to note that this task requires overt phonological processing in order to name each stimulus aloud; such processing is not required in silent sentence reading. Error detection tasks have also been used to examine phonological recoding in children, where participants are required to decide whether an error is present in the context of a whole sentence. For example, Coltheart, Laxon, Rickard, and Elton [31] asked adults (Experiment 1) and children (Experiment 2) to judge whether printed sentences were correct or not. One of the unacceptable sentence conditions presented pseudohomophones (e.g., *Her bloo dress was new.*). The researchers argued that, in this condition, any observed effects of phonology must be pre-lexical because there are no lexical entries for nonwords (i.e., it is not possible for phonology to have a top-down influence, post-lexical access, as could be the case for known words). Pseudohomophone sentences resulted in significantly higher false positive rates for both adult and child readers, relative to control conditions. Thus, the authors argued that both the adults and the children were pre-lexically processing phonology (recoding). One possible caveat is that response times were not recorded, only accuracy. It is possible that readers were engaging in some form of subvocal phonological decoding in order to process the pseudohomophones.

Taken together, these studies provide strong evidence for phonological recoding in skilled adult readers, e.g., [25–29]. There is also clear evidence that beginning readers rely on phonological decoding and that this reduces over time as reading skill increases [20,24]. Finally, there is some evidence that once children are past the point in their reading development where they are engaging in effortful phonological decoding, they have made a transition to phonological recoding, e.g., [30,31]. Whilst these studies do suggest such a transition, they do not afford as direct insight into a reader's cognitive processing of text as eye movement research does, especially given the offline nature of some of the data, e.g., [31]. Consequently, seeking converging evidence from different approaches could prove useful.

4. The Role of Phonology: Eye Movement Research

Eye movement research provides a highly sensitive index of cognitive processing during reading, affording researchers an insight into the online, moment-to-moment operations involved in the reading process [32–34]. As such, researchers can gain insight into the cognitive processing of text using more naturalistic sentence reading, as opposed to isolated word recognition tasks or offline tasks. A body of literature has used eye movement recordings to examine the contribution of phonological processing to lexical identification during silent sentence reading.

Adults. Research has strongly indicated that adults continue to make use of phonology during reading. From the literature on skilled adult reading, two roles have been proposed for phonology during skilled reading: (1) phonology may play a pre-lexical role and aid the process of lexical access and word identification; or (2) phonological codes may be activated as a function of lexical access or after lexical access [2,3].

Rayner, Pollatsek, and Binder [35] provided evidence that phonological information is activated during silent reading. Participants read short passages that contained a correct target word, a homophone, or an orthographic control (e.g., *Murderers who kill many people according to a pattern are referred to as serial/cereal/verbal killers.*). Both the orthographic controls and the homophones were incongruent with the semantics of the sentence context, and, as such, longer reading times would be expected in both these conditions relative to the correct target word. Importantly, the orthographic controls and homophones were matched in terms of their orthographic overlap with the target word. Shorter reading times on the homophone relative to the orthographic control would, therefore, be attributable to the homophone's shared phonology with the correct target word. Strikingly, reading times on the homophone were not significantly different from reading times on the correct target word when it was orthographically similar to the target word (e.g., *heal-heel* vs. *right-write*). This suggests that readers' early activation of congruent phonological codes resulted in the reader not even noticing that the word they were fixating was an error word (that is, a word that was incorrect in the context of the sentence). Critically, across both orthographically similar and dissimilar conditions, participants displayed shorter reading times on homophones than on orthographic controls, and this effect was observed in early measures of processing (i.e., in first fixation duration—the duration of the first fixation on a word regardless of how many fixations it receives). It is worth noting that in the researchers' first experiment, a pseudohomophone condition (e.g., *brane-brain*) was also used, and the pattern of results was similar to that of the homophones. This provides further evidence for a pre-lexical role for phonology: pseudowords do not have lexical entries, so any characteristics of such words that facilitate lexical identification (i.e., shared phonology with real words) would have to be activated before lexical access is achieved [27]. Thus, phonological recoding was used by skilled adult readers in their initial fixation on a word, seemingly pre-lexically, facilitating lexical identification.

With respect to the pre- versus lexical/post-lexical phonology question, though, the strongest evidence comes from fast priming (Figure 1; [36]) and parafoveal pre-processing studies.

```
Stimulus during fixation on        The once popular| xxxxx now seems deserted.
pretarget word                                 *

Stimulus during initial 24 ms of   The once popular| beech now seems deserted.
first fixation on target word                          *

Stimulus after initial 24 ms of    The once popular| beach now seems deserted.
first fixation on target word                          *
```

Figure 1. An example of the fast priming technique. The asterisk underneath each sentence indicates the reader's fixation location. An invisible boundary is placed in a sentence in the space before a target word (the lines in the example above represent the location of the boundary, but this is not visible to participants). Before fixation, a string of xs is present where the target word should be. When the readers' eyes cross the invisible boundary and first fixate the target word location, a prime is presented for a very brief amount of time (e.g., 24 ms), before being replaced by the target word. This example shows a homophone prime (e.g., *beech*) for a target word (e.g., *beach*).

Rayner, Sereno, Lesch, and Pollatsek [37] used the fast priming technique to compare identity (e.g., *beach*), homophone (e.g., *beech*), orthographic control (e.g., *bench*), or dissimilar primes (e.g., *noise*). The critical comparison here was that of reading times on the target word when it was primed by a homophone relative to an orthographic control (i.e., looking for evidence of a phonological priming effect). Participants had shorter gaze durations on a target word when it was preceded by a homophone prime than when it was preceded by an orthographic control. Thus, it would appear that phonology can be coded quickly enough to facilitate lexical access and identification of the target word. Further evidence for this argument is provided by parafoveal pre-processing studies.

Parafoveal pre-processing refers to readers' extraction of information from the next word in a sentence (referred to as *n* + 1) before it is directly fixated (whilst processing is on-going for the currently fixated word- referred to as *n*). It is typically investigated using the boundary paradigm (Figure 2; [38]).

```
             Tom decided to take a long| brake for his lunch.
                           *

             Tom decided to take a long| brake for his lunch.
                                    *

             Tom decided to take a long| break for his lunch.
                                       *
```
(Sequence of fixations ↓)

Figure 2. An example of the boundary paradigm. The asterisk underneath each sentence indicates the reader's fixation location. An invisible boundary is placed in a sentence in the space immediately before a target word (the lines in the example above represent the location of the boundary, but this is not visible to participants). A preview letter string is available in the target word's location prior to the reader making a saccade that crosses this invisible boundary. After the reader's eyes cross the boundary, they move to directly fixate the target word. Then, a display change occurs wherein the preview letter string changes to the correct target word. By manipulating certain characteristics of the overlap (e.g., phonological similarity) between the preview string and the target word, parafoveal pre-processing can be studied. For example, phonologically consistent (e.g., *brake*) and inconsistent (e.g., *bread*) previews can be presented for a target (e.g., *break*) to examine the extent to which a reader is undergoing phonological pre-processing prior to direct fixation. If a reader does extract phonological information during parafoveal pre-processing, then reading times on the target word should be shorter following a consistent preview than an inconsistent preview. This decrease to reading times is referred to as preview benefit. If preview benefit is found, i.e., shorter reading times, on a word that was parafoveally available compared to when the parafoveal preview word was masked, this is strongly indicative of parafoveal pre-processing having occurred, as lexical identification has been facilitated. As such, parafoveal pre-processing and this paradigm enable researchers to investigate pre-lexical effects, as manipulations are conducted outside of direct fixation (i.e., lexical processing): if the manipulated characteristic of a given word in the parafovea confers preview benefit to the reader, the word must have been pre-lexically processed to some extent prior to it receiving a direct fixation.

Indeed, evidence from the use of the boundary paradigm has found that phonological recoding begins prior to direct fixation in skilled adult readers. For example, Pollatsek, Lesch, Morris, and Rayner [39] found that readers can pre-process phonological cues from an upcoming word. Previews were either homophones or orthographic controls for a target word that was presented after the reader's eyes had crossed the boundary. They found that reading times on the correct target word were shorter when the preview was a homophone than when it was an orthographic control. Such effects, indicating pre-lexical parafoveal processing of phonology, have now been shown in a number of studies looking at parafoveal pre-processing in skilled adult readers, e.g., [39–42], and the fast priming technique has provided similar findings [37]. This suggests that phonological recoding plays a key role in activating lexical entries during skilled adult reading; that is, a word's phonology plays a pre-lexical role rather than a lexical/post-lexical role.

Children. Far less research has been done with children using research methods that are sensitive to online cognitive processing during reading. To date, though, two studies have used eye movements to examine phonological processing during children's silent sentence reading, examining foveal reading processes. Blythe et al. [14] presented sentences containing correct target words, pseudohomophones, or orthographic controls, to both adults and children aged 7 to 9 years (e.g., *Today we had a huge water/worta/wecho fight in my back garden.*). Pseudohomophones were used due to this age group of children being limited in the number of homophone pairs known to them, especially with Age-of-Acquisition limited to earlier than 6 years to maximise the likelihood that all participants would be familiar with the target words. They found that children, similarly to adults, benefitted from the valid phonology of a pseudohomophone compared to an orthographic control. These data were argued to provide evidence for covert phonological recoding in children as young as 7 years old (contradictory to some isolated word recognition research; e.g., [22]). Two further points support this conclusion. First, all participants were reading silently, and no overt decoding was observed at any point. Clearly, these children were beyond the phase of reading development where overt decoding was their primary strategy for lexical identification. Second, and critically, when compared against reading times on the correct target word within a sentence, the cost associated with pseudohomophones was less than 200 ms, and reading times on the pseudohomophones were less than 600 ms in total in the children's data. These reading times are too short to plausibly incorporate the sounding out and then blending together of phonemes. These data are, therefore, most consistent with phonological recoding during lexical identification, suggesting that both adults and children are able to access the correct lexical representation on the basis of a letter string's phonology.

Moreover, Jared, Ashby, Agauas, and Levy (Experiment 3; [43]) provided further evidence that phonological representations are used in the initial activation of word meanings. The researchers monitored children's (10 to 11 year olds) eye movements as they read sentences silently, some of which contained a correct target word (e.g., *whether*), some a homophone (e.g., *weather*), and some an orthographic control (e.g., *winter*). Critically, the homophones were not as disruptive to the children's reading as the orthographic controls (i.e., the children displayed shorter reading times when a homophone was present than when an orthographic control was present). This observed homophone advantage reflects the contribution phonology made to activating the meanings of words for the child readers (regardless of word frequency). Phonology, therefore, seems to play a key role during children's lexical identification during silent sentence reading. Furthermore, similar to Blythe et al. [14], the mean reading times suggest that children were undertaking phonological recoding (as opposed to overt decoding).

This research [14,43] is consistent with the view that phonology continues to play a role in aiding lexical access, but in an increasingly covert manner as age and reading skill increase [4,44]. This argument is further supported by studies that have shown increased fixation times on long words (e.g., *medicine*) compared to short words (e.g., *salt*) in both children and adults [45,46]. There are two critical points to note with respect to these studies. First, Hyönä and Olson [45] used a reading aloud task with 8–12 year old children, and no overt decoding was observed for either the long or the short words. Second, the magnitude of the increase in reading times was between 22 ms per letter ([46]; silent reading in 7–11 year old children) and 58 ms per letter [45]. The magnitude of these increases to reading times are too small to conceivably argue that children were sounding out and blending phonemes together, either vocally or subvocally, in order to achieve lexical access (phonological decoding). Both of these points support the argument that children at this age have moved beyond overt phonological decoding during lexical identification.

It is widely recognised that adults continue to make use of phonology to aid lexical access and identification during reading, e.g., [2,3], but until recently, this issue has been somewhat neglected within the empirical literature on children's reading development. We contend that, while there is developmental change in phonological processing during reading, this is best characterised as a transition from phonological decoding to phonological recoding. Such a developmental transition is not currently accounted for in theoretical models of learning to read, which simply posit decreasing reliance on phonology as reading skill increases, e.g., [5–10].

It is worth noting that phonological processing in English, the focus of this paper, may differ from that in other languages, due to differences in orthographic depth (the consistency of a language's GPCs). For example, English has an opaque orthography, wherein GPCs are not very consistent (i.e., *ough* in *cough*, *through*, *though*, etc.), whilst other alphabetic languages, like Finnish and German, benefit from more transparent orthographies. One piece of research has investigated phonological pre-processing in German. Tiffin-Richards and Schroeder [47] found that German adults benefitted more from orthographic than phonological information in the parafovea. Whilst children also gained some preview benefit from orthographic information in the parafovea, this was only under certain conditions: when the target words only received a single fixation and when capitalisation of the word was present. In contrast, the children did show a clear preview benefit from pseudohomophones. This would suggest that, in German, for children, phonology plays a more important role in word identification than orthography, whilst, for adults, the opposite pattern seems to occur: orthography seems to play a more dominant role in facilitating lexical access than phonology. In Chinese, a morphosyllabic language [48], phonological information has been shown to be activated pre-lexically by children, whilst adults seem to use more direct access from orthography to semantics [49]. Within Chinese, the researchers argued, early, pre-lexical activation of phonology diminishes as readers become more skilled. It is worth noting though that this research focuses on parafoveal processing of orthographic and phonological information and so does not make claims that, for instance, children do not process orthographic information foveally in German. Overall though, this research on both adults' and children's parafoveal pre-processing in German and Chinese seems to be in contrast to the research looking at pre-processing of phonology in English adults, e.g., [39]. Indeed, concerns have been raised as to whether research conducted in English may have overestimated the importance of phonology, e.g., [50,51]. Consequently, the developmental transition from overt, effortful phonological decoding to covert, rapid phonological recoding that appears to occur in English, as outlined in this paper, may not be applicable to other languages. Whilst phonology does seem to play a role in reading development in other alphabetic languages besides English, it does seem to be modulated by orthographic depth [51]. Evidence suggests that readers of more transparent orthographies might make the transition from phonological decoding to phonological recoding at a faster rate than readers of English, with it suggested that the difficulty associated with progressing to phonological recoding is specific to English and its complex GPCs, e.g., [52]. Thus, the extent to which reading development within different languages is determined by phonological processing may differ.

Atypical development. Most recently, studies have begun to show evidence for pre-lexical phonological processing in populations with atypical reading development, specifically in individuals with permanent childhood hearing loss (PCHL; [53]) and individuals with developmental dyslexia [54]. Both of these participant populations are known to commonly experience substantial difficulties in learning to read, and one component of these difficulties is thought to be poor phonological processing skills, e.g., [55–57].

In the case of individuals with PCHL, their auditory perception since birth has been substantially impoverished, and it is likely that this results in underspecified cognitive representations of phonology. Indeed, on tasks that require overt awareness of, or conscious manipulation of, speech sounds, Blythe et al. [53] found that teenagers with PCHL scored significantly lower than both chronologically age-matched and reading-matched hearing peers. Despite their difficulties in overt phonological decoding and phonological awareness, these teenagers displayed a pseudohomophone advantage both during direct fixation and from parafoveal preview. In particular, the pseudohomophone advantage shown by teenagers with PCHL was very similar in terms of both time course and magnitude to the effect observed in their younger, reading-matched hearing peers. This strongly indicates that, despite their overall difficulties in learning to read, one particular aspect of lexical processing was maturing in a typical manner (albeit with a slight developmental delay)—the transition to phonological recoding.

In the case of developmental dyslexia, both overall reading difficulties and specific difficulties in phonological awareness and processing have been well-documented; indeed, poor phonological processing skills are largely accepted as the predominant cause of developmental dyslexia, e.g., [56,57]. Again, though, recent research has shown that teenagers with dyslexia still exhibit a pseudohomophone advantage during reading during both direct fixation and parafoveal preview [54]. Similar to the data from teenagers with PCHL, this pseudohomophone advantage during silent sentence reading was observed, in contrast to significantly poorer performance on overt tasks of phonological processing compared to their typically developing peers.

In sum, eye movement research in recent years has provided strong evidence for pre-lexical phonological recoding by adults, typically developing children, and even individuals with PCHL or dyslexia during silent sentence reading, e.g., [14,37,39–43,53,54]. These data challenge theoretical accounts of reading development which posit that phonological processing during lexical identification reduces with time and reading skill, e.g., [5–10]. Rather, these data are more consistent with the view that, as reading skill increases, there is a transition from phonological decoding to phonological recoding. This transition seems to occur relatively early and is remarkably robust across both typical and atypical reading development.

5. The Role of Phonology: Models of Word Recognition

A number of different models have been put forward by researchers in attempts to explain how printed word recognition occurs (e.g., the dual-route cascaded model—DRC; [58]; the multiple-route model; [59]; connectionist dual-process model—CDP+; [60]). It is noncontroversial that all of these models posit some role for phonology in visual word recognition, but they vary in terms of the importance that is ascribed to phonology (for a recent and comprehensive review, see [43]). Here, we briefly outline these models and how each of them incorporates phonological processing into printed word identification. Critically, we consider the degree to which these models can account for developmental change in this respect.

The DRC model. According to this model, processing is accomplished via two distinct but interactive routes: lexical and non-lexical, see [58] (Figure 7, p. 214). The lexical (direct) route relies on the activation of word-specific orthographic representations: the features of a word's letters activate the word's letter units (in parallel), and these letters then activate the word's entry in the orthographic lexicon. The non-lexical (indirect) route is based on the use of GPCs (operating serially from left to right); visual features and letter units are activated just as with the lexical route (as they are common to both routes). Processing along the direct lexical route gets faster each time a word is encountered, so the lexical representations of more common words are activated by the direct route before the slower, indirect, non-lexical route has finished processing the word, e.g., [11,23]. When tested, the DRC was 99% accurate in generating a pronunciation for the 7981 words in its orthographic lexicon. It can account for many phenomena that are observed in skilled adult reading, including frequency effects, regularity effects, the pseudohomophone advantage, and orthographic neighborhood effects. With respect to developmental change, however, the model has no learning mechanism, and " . . . has nothing to say about the actual process of learning to read" (p. 246). The authors argue that it does work well to characterise what a typically developing child reader has learned so far at any point during the process of learning to read, and that young readers (7 year olds) have reading systems similar to adults, albeit scaled-down versions. It is not clear, however, how the two routes would develop in a beginning reader or how the model would account for a developmental transition from decoding to recoding.

The multiple-route model. The multiple-route model, see [59] (Figure 2, p. 282) makes a distinction between the effortful phonological coding of beginning readers and the faster, more automatic use of phonology that develops with a reader's exposure to print. (Note that what Grainger et al. [59] refer to as "phonological recoding" is referred to within this paper as phonological decoding). The initial, overt coding process enables the development of parallel letter processing, involving an array of letter detectors that are location-specific (i.e., that encode the locations of letters within a printed word). Two orthographic codes are generated from this: a coarse-grained route that allows direct access to semantics and a fine-grained route that codes the precise ordering of letters within a string and then activates the corresponding phonemes as well as whole-word orthography. The model clearly predicts strong, phonologically-based effects (e.g., pseudohomophone effects) in younger children that reduce but do not disappear with increasing age as the reader transitions to phonological recoding. This model, therefore, seems to be entirely consistent with the experimental observations from the body of published literature reviewed within this paper.

The CDP+ model. The CDP+ model [60], similar to the DRC model, has two processes: a non-lexical one (sublexical route) that links orthography to phonology and learns GPCs very quickly and a direct lexical one that links orthography to phonology, where orthographic entries are linked to their phonological counterparts (an implementation of the DRC's lexical route). With respect to developmental change, Ziegler, Perry, and Zorzi [61] provided a computational simulation of the self-teaching hypothesis [15] within the framework of the CDP+ model. They examined the extent to which the model could learn to identify unknown words based on initial, explicit teaching of key GPC rules and its existing phonological lexicon, similar to what a child might experience. Ziegler et al. [61] argue that children receive phonics instruction early in their formal education, but they are not explicitly taught the correct pronunciation of every word that they encounter during reading. Rather, as they come across new printed word forms, they use their knowledge of phonics rules to generate a possible pronunciation and determine whether or not this matches a word that is already represented in their lexicon (through spoken language exposure). This learning loop is referred to as the phonological decoding self-teaching (PDST) hypothesis, and, indeed, the implementation of the PDST hypothesis worked in the context of a real computational model of learning to read (CDP+). Even starting with a small number of GPCs (as beginner child readers would do), the model was able to acquire word-specific orthographic representations for over 25,000 words and read novel words aloud. On the basis of these rudimentary GPCs (and decoding skills), the model could produce pronunciations for unfamiliar words. Despite the opaque orthography of English, the phonological decoding network was still able to learn up to 80% of the words. Overall, phonological decoding seems to serve as a powerful internal teaching device, as implemented in this model, allowing a basic set of GPCs to open up children's (and the model's) abilities to read novel words and gain orthographic knowledge. It is conceivable within the PDST hypothesis that there is a transition from beginner children's phonological decoding to skilled adult readers' phonological recoding, but this has not yet been operationalised.

In sum, all of these models propose that phonology plays a role in visual word recognition. To date, Grainger et al.'s [59] multiple-route model provides the clearest implementation that might account for the developmental transition from beginner child readers' effortful phonological decoding to skilled adult readers' unconscious, rapid phonological recoding.

6. Conclusions

Whilst it is widely recognised that children rely on phonological decoding in the early stages of learning to read, current theories do not fully account for skilled readers' pre-lexical processing of phonology, that is, phonological recoding [5–10], with only one recent model of word recognition seeming to account for this developmental transition (the multiple-route model; [59]). Eye movement research has shown pre-lexical processing of phonology in typically developing readers from the age of 7 years through to skilled adult readers, as well as in atypical developmental groups, despite the tasks used not requiring any overt phonological processing [14,35,39,43,53,54]. Thus, eye movement research provides compelling evidence for phonology having a continued and pervasive role in facilitating lexical identification during reading (consistent with the multiple-route model; [59]). As such, recent empirical findings from online research methods, such as eye movement recordings, need to be incorporated into theories of learning to read, and more consideration needs to be given to these findings in developmental models of word recognition. In order to accomplish this, further research is needed to understand the nature and time course of the transition from phonological decoding to recoding, by examining moment-to-moment cognitive processing during reading in beginning readers.

Funding: Sara Milledge was funded by the Economic and Social Research Council (grant number ES/J500161/1).

Conflicts of Interest: The authors declare no conflict of interest.

References

1. World Literacy Foundation. The Economic & Social Cost of Illiteracy: A Snapshot of Illiteracy in A Global Context. 2015. Available online: https://worldliteracyfoundation.org/wp-content/uploads/2015/02/WLF-FINAL-ECONOMIC-REPORT.pdf (accessed on 24 August 2015).
2. Leinenger, M. Phonological coding during reading. *Psychol. Bull.* **2014**, *140*, 1534–1555. [CrossRef]
3. Frost, R. Toward a strong phonological theory of visual word recognition: True issues and false trails. *Psychol. Bull.* **1998**, *123*, 71–99. [CrossRef]
4. Castles, A.; Rastle, K.; Nation, K. Ending the reading wars: Reading acquisition from novice to expert. *Psychol. Sci. Public Interest* **2018**, *19*, 5–51. [CrossRef] [PubMed]
5. Ehri, L.C. Phases of development in learning to read words by sight. *J. Res. Read.* **1995**, *18*, 116–125. [CrossRef]
6. Ehri, L.C. Word reading by sight and by analogy in beginning readers. In *Reading and Spelling: Development and Disorders*; Hulme, C., Joshi, R.M., Eds.; Lawrence Erlbaum Associates: Mahwah, NJ, USA, 1998; pp. 87–111.
7. Ehri, L.C. Phases of development in learning to read words. In *Reading Development and the Teaching of Reading: A Psychological Perspective*; Oakhill, J., Beard, R., Eds.; Blackwell: Oxford, UK, 1999; pp. 79–108.
8. Ehri, L.C. Learning to read words: Theory, findings, and issues. *Sci. Stud. Read.* **2005**, *9*, 167–188. [CrossRef]
9. Ehri, L.C. Development of sight word reading: Phases and findings. In *The Science of Reading: A handbook*; Snowling, M.J., Hulme, C., Eds.; Blackwell: Oxford, UK, 2007; pp. 135–154.
10. Frith, U. Beneath the surface of developmental dyslexia. In *Surface Dyslexia: Neuropsychological and Cognitive Studies of Phonological Reading*; Patterson, K.E., Marshall, J.C., Coltheart, M., Eds.; Erlbaum: Hillsdale, NJ, USA, 1985; pp. 301–330.
11. Adams, M.J.; Huggins, A.W.F. The growth of children's sight vocabulary: A quick test with educational and theoretical implications. *Read. Res. Q.* **1985**, *20*, 262–281. [CrossRef]

12. Doctor, E.A.; Coltheart, M. Children's use of phonological encoding when reading for meaning. *Mem. Cogn.* **1980**, *8*, 195–209. [CrossRef]

13. Greenberg, D.; Ehri, L.C.; Perin, D. Are word-reading processes the same or different in adult literacy students and third-fifth graders matched for reading level? *J. Educ. Psychol.* **1997**, *89*, 262–275. [CrossRef]

14. Blythe, H.I.; Pagán, A.; Dodd, M. Beyond decoding: Phonological processing during silent reading in beginning readers. *J. Exp. Psychol.* **2015**, *41*, 1244–1252. [CrossRef]

15. Share, D.L. Phonological recoding and self-teaching: Sine qua non of reading acquisition. *Cognition* **1995**, *55*, 151–218. [CrossRef]

16. Gough, P.B.; Hillinger, M.L. Learning to read: An unnatural act. *Bull. Orton Soc.* **1980**, *30*, 179–196. [CrossRef]

17. Stuart, M.; Coltheart, M. Does reading develop in a sequence of stages? *Cognition* **1988**, *30*, 139–181. [CrossRef]

18. Rayner, K.; Pollatsek, A.; Ashby, J.; Clifton, C., Jr. *Psychology of Reading*, 2nd ed.; Psychology Press: New York, NY, USA, 2012.

19. Gough, P.B.; Juel, C.; Griffith, P.L. Reading, spelling, and the orthographic cipher. In *Reading Acquisition*; Gough, P.B., Ehri, L.C., Treiman, R., Eds.; Lawrence Erlbaum Associates: Hillsdale, NJ, USA, 1992; pp. 35–48.

20. Ehri, L.C.; Wilce, L.S. Development of word identification speed in skilled and less skilled beginning readers. *J. Educ. Psychol.* **1983**, *75*, 3–18. [CrossRef]

21. Rosenthal, J.; Ehri, L.C. The mnemonic value of orthography for vocabulary learning. *J. Educ. Psychol.* **2008**, *100*, 175–191. [CrossRef]

22. Johnston, R.S.; Thompson, G.B. Is dependence on phonological information in children's reading a product of instructional approach? *J. Exp. Child. Psychol.* **1989**, *48*, 131–145. [CrossRef]

23. Schmalz, X.; Marinus, E.; Castles, A. Phonological decoding or direct access? Regularity effects in lexical decisions of Grade 3 and 4 children. *Q. J. Exp. Psychol.* **2013**, *66*, 338–346. [CrossRef] [PubMed]

24. Samuels, S.J.; LaBerge, D.; Bremer, C.D. Units of word recognition: Evidence for developmental changes. *J. Verbal Learn. Verbal Behav.* **1978**, *17*, 715–720. [CrossRef]

25. Lesch, M.F.; Pollatsek, A. Automatic access of semantic information by phonological codes in visual word recognition. *J. Exp. Psychol.* **1993**, *19*, 285–294. [CrossRef]

26. Van Orden, G.C. A rows is a rose: Spelling, sound, and reading. *Mem. Cogn.* **1987**, *15*, 181–198. [CrossRef]

27. Lukatela, G.; Turvey, M.T. Visual lexical access is initially phonological: I. Evidence from associative priming by words, homophones, and pseudohomophones. *J. Exp. Psychol.* **1994**, *123*, 107–128. [CrossRef]

28. Lukatela, G.; Turvey, M.T. Visual lexical access is initially phonological: 2. Evidence from phonological priming by homophones and pseudohomophones. *J. Exp. Psychol.* **1994**, *123*, 331–353. [CrossRef]

29. Rubenstein, H.; Lewis, S.S.; Rubenstein, M.A. Evidence for phonemic recoding in visual word recognition. *J. Verbal Learn. Verbal Behav.* **1971**, *10*, 645–657. [CrossRef]

30. Kyte, C.S.; Johnson, C.J. The role of phonological recoding in orthographic learning. *J. Exp. Child. Psychol.* **2006**, *93*, 166–185. [CrossRef] [PubMed]

31. Coltheart, V.; Laxon, V.; Rickard, M.; Elton, C. Phonological recoding in reading for meaning by adults and children. *J. Exp. Child. Psychol.* **1988**, *14*, 387–397. [CrossRef]

32. Rayner, K. Eye movements in reading and information processing: 20 years of research. *Psychol. Bull.* **1998**, *124*, 372–422. [CrossRef] [PubMed]

33. Rayner, K. Eye movements and attention in reading, scene perception, and visual search. *Q. J. Exp. Psychol.* **2009**, *62*, 1457–1506. [CrossRef] [PubMed]

34. Starr, M.S.; Rayner, K. Eye movements during reading: Some current controversies. *Trends Cogn. Sci.* **2001**, *5*, 156–163. [CrossRef]

35. Rayner, K.; Pollatsek, A.; Binder, K.S. Phonological codes and eye movements in reading. *J. Exp. Psychol.* **1998**, *24*, 476–497. [CrossRef]

36. Sereno, S.C.; Rayner, K. Fast priming during eye fixations in reading. *J. Exp. Psychol.* **1992**, *18*, 173–184. [CrossRef]

37. Rayner, K.; Sereno, S.C.; Lesch, M.F.; Pollatsek, A. Phonological codes are automatically activated during reading: Evidence from an eye movement priming paradigm. *Psychol. Sci.* **1995**, *6*, 26–32. [CrossRef]

38. Rayner, K. The perceptual span and peripheral cues during reading. *Cogn. Psychol.* **1975**, *7*, 65–81. [CrossRef]

39. Pollatsek, A.; Lesch, M.; Morris, R.K.; Rayner, K. Phonological codes are used in integrating information across saccades in word identification and reading. *J. Exp. Psychol.* **1992**, *18*, 148–162. [CrossRef]

40. Ashby, J.; Treiman, R.; Kessler, B.; Rayner, K. Vowel processing during silent reading: Evidence from eye movements. *J. Exp. Psychol.* **2006**, *32*, 416–424. [CrossRef]
41. Henderson, J.M.; Dixon, P.; Petersen, A.; Twilley, L.C.; Ferreira, F. Evidence for the use of phonological representations during transsaccadic word recognition. *J. Exp. Psychol.* **1995**, *21*, 82–97. [CrossRef]
42. Jouravlev, O.; Jared, D. Cross-script orthographic and phonological preview benefits. *Q. J. Exp. Psychol.* **2018**, *71*, 11–19. [CrossRef] [PubMed]
43. Jared, D.; Ashby, J.; Agauas, S.J.; Levy, B.A. Phonological activation of word meanings in Grade 5. *J. Exp. Psychol.* **2016**, *42*, 524–541. [CrossRef]
44. Ehri, L.C.; Wilce, L.S. Movement into reading: Is the first stage of printed word learning visual or phonetic? *Read. Res. Q.* **1985**, *20*, 163–179. [CrossRef]
45. Hyönä, J.; Olson, R.K. Eye fixation patterns among dyslexic and normal readers: Effects of word length and word frequency. *J. Exp. Psychol.* **1995**, *21*, 1430–1440. [CrossRef]
46. Joseph, H.S.S.L.; Liversedge, S.P.; Blythe, H.I.; White, S.J.; Rayner, K. Word length and landing position effects during reading in children and adults. *Vis. Res.* **2009**, *49*, 2078–2086. [CrossRef]
47. Tiffin-Richards, S.P.; Schroeder, S. Children's and adults' parafoveal processes in German: Phonological and orthographic effects. *J. Cogn. Psychol.* **2015**, *27*, 531–548. [CrossRef]
48. Perfetti, C.A.; Liu, Y.; Tan, L.H. The lexical constituency model: Some implications of research on Chinese for general theories of reading. *Psychol. Rev.* **2005**, *112*, 43–59. [CrossRef]
49. Zhou, W.; Shu, H.; Miller, K.; Yan, M. Reliance on orthography and phonology in reading of Chinese: A developmental study. *J. Res. Read.* **2018**, *41*, 370–391. [CrossRef]
50. Share, D.L. On the Anglocentricities of current reading research and practice: The perils of overreliance on an "outlier" orthography. *Psychol. Bull.* **2008**, *134*, 584–615. [CrossRef]
51. Ziegler, J.C.; Bertrand, D.; Tóth, D.; Csépe, V.; Reis, A.; Faísca, L.; Blomert, L. Orthographic depth and its impact on universal predictors of reading: A cross-language investigation. *Psychol. Sci.* **2010**, *21*, 551–559. [CrossRef]
52. Aro, M.; Wimmer, H. Learning to read: English in comparison to six more regular orthographies. *Appl. Psycholinguist.* **2003**, *24*, 621–635. [CrossRef]
53. Blythe, H.I.; Dickins, J.H.; Kennedy, C.R.; Liversedge, S.P. Phonological processing during silent reading in teenagers who are deaf/hard of hearing: An eye movement investigation. *Dev. Sci.* **2018**, *21*, e12643. [CrossRef]
54. Blythe, H.I.; Dickins, J.H.; Kennedy, C.R.; Liversedge, S.P. The role of phonology in lexical access in teenagers with dyslexia. 2019; in preparation.
55. Mayberry, R.I.; del Giudice, A.A.; Lieberman, A.M. Reading achievement in relation to phonological coding and awareness in deaf readers: A meta-analysis. *J. Deaf Stud. Deaf Educ.* **2011**, *16*, 164–188. [CrossRef]
56. Snowling, M.J. Phonemic deficits in developmental dyslexia. *Psychol. Res.* **1981**, *43*, 219–234. [CrossRef]
57. Vellutino, F.R.; Fletcher, J.M.; Snowling, M.J.; Scanlon, D.M. Specific reading disability (dyslexia): What have we learned in the past four decades? *J. Child. Psychol. Psychiatry* **2004**, *45*, 2–40. [CrossRef]
58. Coltheart, M.; Rastle, K.; Perry, C.; Langdon, R.; Ziegler, J. DRC: A dual route cascaded model of visual word recognition and reading aloud. *Psychol. Rev.* **2001**, *108*, 204–256. [CrossRef]
59. Grainger, J.; Lété, B.; Bertand, D.; Dufau, S.; Ziegler, J.C. Evidence for multiple routes in learning to read. *Cognition* **2012**, *123*, 280–292. [CrossRef]
60. Perry, C.; Ziegler, J.C.; Zorzi, M. Nested incremental modeling in the development of computational theories: The CDP+ model of reading aloud. *Psychol. Rev.* **2007**, *114*, 273–315. [CrossRef]
61. Ziegler, J.C.; Perry, C.; Zorzi, M. Modelling reading development through phonological decoding and self-teaching: Implications for dyslexia. *Philos. Trans. R. Soc. Lond. B Biol. Sci.* **2014**, *369*, 20120397. [CrossRef]

Review

Effects of Normative Aging on Eye Movements during Reading

Kevin B. Paterson [1,*], Victoria A. McGowan [1], Kayleigh L. Warrington [1], Lin Li [2], Sha Li [3], Fang Xie [2], Min Chang [2], Sainan Zhao [2], Ascensión Pagán [1], Sarah J. White [1] and Jingxin Wang [2]

[1] Department of Neuroscience, Psychology and Behaviour, University of Leicester, Leicester LE1 7RH, UK; vm88@leicester.ac.uk (V.A.M.); klw53@leicester.ac.uk (K.L.W.); appc1@leicester.ac.uk (A.P.); sjw92@leicester.ac.uk (S.J.W.)
[2] Faculty of Psychology, Tianjin Normal University, Tianjin 300387, China; linpsy@outlook.com (L.L.); xftjsf@126.com (F.X.); tjsfchangmin@126.com (M.C.); sainanpsy@163.com (S.Z.); wjxpsy@126.com (J.W.)
[3] School of Psychology, Fujian Normal University, Fuzhou 350117, China; llesq@outlook.com
* Correspondence: kbp3@le.ac.uk

Received: 7 April 2019; Accepted: 7 January 2020; Published: 14 January 2020

Abstract: Substantial progress has been made in understanding the mostly detrimental effects of normative aging on eye movements during reading. This article provides a review of research on aging effects on eye movements during reading for different writing systems (i.e., alphabetic systems like English compared to non-alphabetic systems like Chinese), focused on appraising the importance of visual and cognitive factors, considering key methodological issues, and identifying vital questions that need to be addressed and topics for further investigation.

Keywords: cognitive aging; eye movements during reading; alphabetic reading; Chinese reading

1. Introduction

When people read, their eyes make rapid movements (called saccades), separated by brief fixational pauses during which visual information is obtained from text (for reviews, [1–3]). These eye movements are closely coordinated with processes of perception, attention, language processing, and memory. This enables skilled readers to recognize words very rapidly as their eyes move across a text, at a rate of about 3 to 5 words per second, and for the meaning of these words to be integrated with the reader's evolving interpretation of the text. A substantial body of research over the past 40 years has focused on understanding how this feat is accomplished by skilled adult readers [2,3]. More recent research has also begun to investigate how this capability develops in beginner readers [4,5], as well as how it changes across the lifespan.

With the present review, we focus on the effects of normative, healthy aging. Being able to read well is important for individuals to function effectively in most modern societies, and to accomplish everyday tasks that enable them to live independently. However, changes in the visual and cognitive systems that occur naturally in healthy aging are likely to affect the ability of older adults to read effectively. Such changes begin in middle age but are especially marked in later adulthood (from about 65 years of age onwards; for reviews, see [6–10]). At the level of the eye, these include narrowing of the pupil (pupillary miosis) and increased opacity of the lens, reducing the passage of light through the eye, while a loss of elasticity of the lens limits the ability to focus on near objects (presbyopia). As a result, less light falls on the retina and the image it creates is blurred. These optical changes combine with retinal and neural changes, including loss of retinal photoreceptive cells (rods and cones), which limit visual sensitivity, and reductions in axons in the optic nerve and neurons in the visual cortex, which affect the transmission and processing of visual information.

Together, these changes decrease the efficiency of visual processing. Functionally, they are associated with lower acuity and reduced ability to perceive fine visual detail, especially in low light or when contrast is reduced [11–14]. This can contribute to the experience of visual blur [15], as well as problems reading fine-print [16] or in dim lighting conditions [17]. Reductions in visual sensitivity are also greater outside central vision [18] so that older adults see detail more poorly in peripheral vision, which may help explain why they appear to have a narrower field of effective vision [19,20]. Reductions in visual sensitivity are experienced even when vision is corrected optically. Consequently, even older adults who wear their habituation optical correction (glasses, contact lenses) will experience degradation in visual input and process visual information more slowly and with greater error. Age-related declines in attentional and motor control may also affect the efficiency of eye movement control. Non-reading studies consistently show that aging is associated with an increase in the time to initiate a saccade in response to a visual target [21–27], and the velocity of the resulting movement [28,29]. However, accuracy in saccade-targeting [23,25,30,31] and fixation stability [32] appear to be spared.

Finally, declines in memory capabilities may affect the retrieval of information from memory and short-term storage and manipulation of linguistic information in working memory [33–36]. It seems likely that effects on retrieval will impair processes of word recognition, while working memory deficits may impair comprehension [37,38]. However, while aging produces deterioration in elements of what often is described as fluid intelligence (e.g., memory, reasoning, problem-solving [9,39]), crystalized intelligence, which includes knowledge gained from a lifetime of experience of reading and exposure to language, appears stable across adulthood and only shows declines in very old age [8]. Crucially, crystalized knowledge may offset deficits in fluid intelligence and have a preservative effect on reading. For instance, even though older people can have difficulty bringing words to mind, they tend to outperform younger adults on vocabulary tests, due to superior knowledge of words gained from their greater reading experience [40,41]. Note, however, that an alternative account postulates that the slowdown in lexical access in older age is because of difficulty discriminating individual words in the mental lexicon as a result of a lifetime of accumulation of lexical items rather than cognitive decline [42,43]. Consequently, while research inevitably focuses on loss in abilities, it is important to remember that some losses may be offset by superior skills elsewhere, while other abilities may be highly resilient. Older adults may also use adaptive strategies to compensate for loss of abilities, including changing how they allocate processing resources [44–46].

With the present review, we focus on effects of normative aging on eye movements in reading. This is because measures of eye movements provide a highly naturalistic means of studying the process of reading, as participants read text presented normally at their own pace. A further advantage comes from the close yoking between eye movements and cognition, such that where a reader is looking is informative about what they currently are processing, while the length of time they look in this location is sensitive to any processing difficulty they are experiencing [2,3]. However, it is important to acknowledge valuable research that has used other methods to gain insight into aging effects on word recognition [47–51], sentence and discourse comprehension [52], and the neural correlates of these processes [53–58].

2. Aging Effects on the Perceptual Span

A first issue concerns whether aging limits the amount of linguistic information acquired on each fixational pause. Readers make eye movements because they cannot process an entire line of text on a single glance and so make multiple glances to obtain this information in discrete chunks. This is a consequence of limitations in retinal acuity, which is greatest within a narrow region of central vision (extending about 1° each side of fixation, corresponding roughly to the fovea) and declines sharply outside this region [59]. At a normal reading distance, about three or four letters will subtend about 1° of visual angle, so at most about six to eight letters can be seen clearly on each fixation. Saccadic eye movements, therefore, serve to shift the reader's gaze along each line of text so that successive chunks of linguistic information (i.e., words) are processed in high acuity.

Given that skilled reading essentially involves integrating these snapshots of linguistic information, it is important to understand how much text can be processed on a single fixation. The area of text from which useful information can be acquired on each fixation is called *the perceptual span*. This has been widely investigated using gaze-contingent moving window paradigms ([60,61]; for a review, see [62]). In these, text is shown normally within a narrow region (window) around gaze and text outside the window is masked (e.g., by replacing letters in words with 'x's). This window moves in synchrony with the reader's eye movements so that when the reader fixates a new location, text within the window at this new location is shown normally, and text at the previous fixation location is masked. This ensures only a small amount of text is seen normally on each fixation. These paradigms are implemented using high-speed computer programming and display screens with a fast refresh rate so that a participant's phenomenological experience is that the windows move in perfect synchrony with their eyes. Moreover, by systematically varying the size of the window across an experiment, it is possible to investigate which size produces normal reading rates, following the logic that this window must encompass the perceptual span. Studies taking this approach show that skilled young adult readers (of English) primarily obtain useful information from an asymmetrical region extending about 14–15 letters to the right of fixation and 3–4 letters to the left ([60,61,63], but see [64,65]). Roughly speaking, this means they acquire information about the word they are fixating and about two words to its right (including information about the boundaries between words). The perceptual span can vary, however, as a function of reading expertise and the script that is being read [66–76]. For instance, span size is smaller for beginning and dyslexic readers compared to skilled adult readers, and for readers of dense scripts like Chinese compared to English.

Surprisingly little aging research has been conducted on the perceptual span, however, although non-reading studies show older adults process information outside of central vision less effectively than young adults [19,20]. Consistent with this more limited visual processing, several studies by Keith Rayner and his colleagues suggest that older adults acquire less information on each fixation compared to young adults. The first study to examine these effects, by Rayner, Castelhano, and Yang (2009) [77], used a standard gaze-contingent paradigm in which letters in words outside a moving window were masked by replacing them with 'x's. As is typical for studies using this paradigm, young adults read fastest when the fixated word and two words to the right were visible within the moving window. By comparison, older adults showed no difference in performance when the fixated word and either one or two words to the right were visible, suggesting they obtained less information about upcoming words. Moreover, unlike the young adults, the older adults were disrupted when the word immediately to the left of the fixated word was masked. Rayner et al. took this to show that the perceptual span is smaller and more symmetrical for older compared to younger adults, the implication being that older readers make less use of information about upcoming words.

Support for this came from a subsequent study by Rayner, Castelhano, and Yang (2010) [78], using the boundary paradigm. In this paradigm, an invisible boundary is placed immediately in front of a specific target word in each sentence. This word is masked prior to the reader's gaze crossing the boundary, after which it is changed to normal. As with the moving window paradigm, these changes are made very fast, using high-speed computing and displays with fast refresh rates, and usually completed within the time it takes the reader to make a saccade. As a result, readers are generally unaware that a change has been made. Rayner et al. found that restricting the visibility of an upcoming word disrupted reading less for older than younger adults, again suggesting that older readers make less use of parafoveal (i.e., upcoming) information. Finally, Rayner, Yang Schuett, and Slattery (2014) [79] showed that older adults were significantly more impaired than younger adults when a moving window masked each fixated word. That is, as a reader moved their gaze along a line of the text, each word they looked at was masked while other text was shown normally. The greater difficulty this caused older adults was taken as further evidence that they use parafoveal information less effectively.

The overall picture from these studies is that the perceptual span is smaller for older than younger adult readers, quite possibly because of reduced parafoveal visual abilities or because higher attention demands associated with the processing of fixated words limit resources available for parafoveal processing. However, while these findings fit well with the notion that older readers process less information on each fixation, other studies are less supportive. First, Whitford and Titone (2016) [80] found no perceptual span differences for young and older adult bilinguals when both letters in words and the spaces between words were replaced by a dash (-). Second, Risse and Kliegl (2011) [81] provided evidence that older adults' parafoveal processing may actually be relatively well-preserved. They used an adaptation of the boundary paradigm, in which the availability of the second word following a target word (i.e., word N+2) was manipulated. This was shown normally or masked so that it was not identifiable until the reader made a saccade that crossed an invisible boundary immediately following the target word. The key consideration was whether masking this word would disrupt reading. Risse and Kliegl found that it did, but to the same extent for young and older adults. This suggested both age groups could use parafoveal information from up to two words ahead of a fixated word equally effectively, inconsistent with the view that older readers acquire less parafoveal information on each fixation.

Paterson, McGowan, and Jordan (2013) [82,83] took a different approach to this problem and speculated that older readers might obtain different qualities of visual information from outside central vision, based on evidence that their sensitivity to fine visual detail is reduced, especially in peripheral vision [11–14]. They used a moving window paradigm in which the spatial frequency content of text outside each window was filtered. This ensured only either low, medium, or high spatial frequency content was available. Low spatial frequencies convey coarse-grain blurry information that lacks detail but provides visual cues to the overall shape and location of words. By comparison, high spatial frequencies provide sharp edge-like features for letters without their normal density, while the quality of visual information provided by medium spatial frequencies is somewhere in-between. The findings showed that reading times were closer to normal for young adults for text containing only high rather than low spatial frequencies, but closer to normal for older adults when spatial frequencies were low rather than high. The indication, therefore, was that older readers require primarily coarse-scale information about the shape and location of words from outside central vision. A subsequent study by Jordan, McGowan and Paterson (2014) [84] used a version of this paradigm in which spatial frequencies were filtered within the moving window and text outside was shown normally. This produced a similar pattern of effects, suggesting that older readers have reduced sensitivity to visual detail even within central (i.e., foveal) vision.

Such findings raise broader questions concerning sensitivity to different types of parafoveal (and perhaps even foveal) information by different groups of readers. One issue relates to the use of regular patterned masks in many moving window experiments, where letters in words are replaced by 'x's or dashes. The regular pattern created by these masks is usually apparent to participants. This may, therefore, serve as a distraction that disrupts normal reading performance rather than a baseline condition in which access to parafoveal information is restricted. As considerable evidence suggests older adults suffer more from distraction than young adults [85–87], they may be more affected by these masks. Consequently, while research on adult age differences in perceptual span effects currently is inconclusive, a way ahead may be to consider more carefully potential age differences in the effects of different types of foveal and parafoveal masks.

3. Aging and Mechanisms of Eye Movement Control

The basic characteristics of eye movements during reading are relatively well understood, certainly for skilled young adult readers of alphabetic scripts like English (for reviews, see [2,3]). Most saccades are forward-directed and traverse a distance of about 7 to 9 character spaces, while a small proportion (typically 10%–15%) are backward eye movements (called regressions) made to re-read text in response to some difficulty. Forward eye movements tend to be shorter and regressions more frequent when text is more difficult. Skilled readers also fixate most words at least once, with each fixation lasting between 100 and 500 ms and averaging about 250 ms. Readers sometimes skip words, however, on about 10%–20% of saccades, by moving their gaze past a word without fixating it.

Research also shows that skilled readers target their saccades towards upcoming words extremely accurately and that the landing positions of these saccades (i.e., the location of the resulting fixation) is systematically related to the physical length of a word. In particular, saccades tend to land at a so-called preferred viewing location (PVL) between the beginning and middle letters of a word [88,89]. This has been interpreted as evidence that readers target their saccades towards the word center but often undershoot this location due to oculomotor error [90]. Such measures have generally been taken from only one eye in experimental studies (typically the right eye or dominant eye). However, studies that have examined the coordination of the movements of the two eyes show that the two eyes fixate the same letter in a word on about 50% of fixations ([91–93], for a review see [94]). On other fixations, the eyes are crossed (so that the left eye fixates to the right of the right eye) or uncrossed (so that right and left eye fixations are divergent), although the relative preponderance of crossed and uncrossed fixations varies across studies and may depend on lighting conditions. Moreover, while the fixations made by the two eyes on average are only about 2 characters apart, this disparity has a high variance so that sometimes the fixations can be much further apart and even on different words.

Studies comparing the eye movements of older adult readers (typically 60+ years) and young adult controls (18–30 years) show that older adults make more and, on average, longer fixations and more regressions [64,80,82–84,95–100]. These findings suggest the older adults read more slowly and experience greater reading difficulty. More controversially, findings from numerous studies suggest older readers make longer forward saccades (on average) and skip words more frequently compared to young adults. This has been attributed to older adults adopting a qualitatively different reading strategy to compensate for greater reading difficulty [97]. In particular, according to this account, older adults employ a "risky" reading strategy in which they are more likely, compared to young adults, to infer the identities of upcoming words based on contextual knowledge and only partial parafoveal word information, with the result that they also skip these words more frequently. It also is argued that this risky reading strategy results in the more frequent misidentification of words and so may contribute to an increased rate of regressions by older readers.

Increased skipping and longer forward saccades are not always observed in aging studies, however, leading some researchers to question the evidential support for risky reading [100,101]. For instance, both Choi et al. (2007) [101] and Whitford and Titone (2017) [100] found that older participants in their experiments did not skip words more frequently than young adults. However, one possibility, also considered by these critics, is that the reading strategy used by the older adults may be sensitive to task demands and so vary as a function of text difficulty. Wotschack and Kliegl (2013) [102] found that age differences in word-skipping can be modulated by the frequency and difficulty of the comprehension questions that follow sentence displays on at least a proportion of trials in most eye movement studies. In particular, whereas older adults showed higher skipping compared to young adults when questions were easy and asked infrequently, this difference was reduced when questions were harder and asked more often. This may explain the absence of an age difference in word-skipping effects in the Choi et al. study [101], as it included comprehension questions after every trial. Moreover, the paradigm employed by Choi et al. involved making surreptitious changes to upcoming words and used stimuli containing deliberate misspellings, which also may have encouraged participants to read carefully. The Whitford and Titone (2017) study [100] examined paragraph reading, and, having read each paradigm, participants had to answer multiple open-ended comprehension questions. As the authors note, this may have imposed attentional and working memory demands that also encouraged the older adults to read very carefully.

These studies are valuable in highlighting the potential that older readers may adopt different strategies depending on task demands. It will, therefore, be important to establish what factors cause older readers to employ a particular reading strategy (and for discussion of effects of reading strategy and goals on eye movements see [44,103–107]). It will also be important to establish whether risky reading is used to compensate for slower processing in older age as Rayner et al. (2006) [97] proposed. This should include considering the consequences of risky reading for processes of word recognition during reading. According to the original account, older readers are more likely to misidentify words based on their contextual expectations, lexical knowledge and parafoveal cues to the identity of the next word. As parafoveal visual cues are likely to be degraded [18], it seems likely that older readers might misidentify a word due to its visual similarity to another word that better fits the context. However, only one study to date, by Warrington et al. (2008) [108], has examined this possibility. This used stimuli from an earlier study [109] in which target words (e.g., "brunch") in sentences often had a lexical neighbor (i.e., a word sharing all but one of the same letters as the target, e.g., "branch" [110], that provided a similar or better fit to the sentence context. This neighbor word was also more commonly used in written language and so of higher lexical frequency. The original study showed that participants were more likely to make a regression to re-read the target word when it had a such a neighbor than when it did not, suggesting that readers often misidentify a word by temporarily mistaking it for its higher frequency neighbor (even though that neighbor does not appear in the text). The follow-up study by Warrington et al. [108] found young and older adults were both likely to misidentify a target word as its neighbor when this better fit the context, and so made more regressions to re-read the target. However, post hoc analyses (not reported in the paper) suggested this effect was greater for the older adults when the target and its neighbor were visually similar (e.g., "brunch" and "branch") rather than visually dissimilar (e.g., "story" and "stork"). Accordingly, a potentially fruitful approach to exploring the consequences of risky reading might involve investigating misreading errors by older adults.

Finally, in this section, we consider whether eye movement control and saccade-targeting are poorer for older readers and whether this might contribute to the difficulty they experience. In the Introduction, we reviewed evidence that eye movements might be slower in older age due to impaired motor and attentional control, but that accuracy in saccade-targeting was unlikely to be affected by age. Studies to date have not examined aging effects on eye movement dynamics in reading. Moreover, only a couple of studies have examined saccade-targeting. Consistent with evidence from non-reading tasks [23,25,30], this appears to be preserved. For instance, Rayner et al. (2006) [97] examined the location of initial fixations on words during reading and observed a very similar distribution for young and older adults, consistent with both age groups preferentially fixating near the PVL. A subsequent study by Paterson et al. (2013) [96] found that the pattern of landing positions for short (5-letter) and long (10-letter) words did not differ appreciably for young and older adults. This study additionally examined the coordination of binocular eye movements. This was motivated by evidence that reading difficulty is likely to produce larger fixation disparities [91,111], and other non-reading studies, suggesting older adults naturally produce larger disparities [112,113]. However, the results showed no age differences. Moreover, the magnitude of the disparity was similar to previous research, averaging about two letter spaces apart [93], with a similar predominance of crossed over uncrossed fixations. Consequently, while older adults are slower and experience greater reading difficulty, basic mechanisms of oculomotor control for reading appear to be preserved in older age.

In addition to these studies, two experiments have investigated the effects of removing or filling the spaces between words in sentences. Text in most alphabetic languages customarily includes spaces between words. By helping to demarcate word boundaries, these can aid the processing of words by reducing visual crowding and lateral masking (interference from flanking letters) of exterior letters in words [114]. Moreover, interword spaces provide useful coarse-scale (i.e., low spatial frequency) information about the location and physical extent of words and so can aid eye-guidance [115]. Removing or replacing these spaces disrupts normal reading (for languages that customarily include spaces [116–120]). Moreover, several studies show that this impairs word identification by producing larger than normal word frequency effects. An important question, therefore, is whether older readers suffer more when inter-word spaces are removed.

Rayner, Yang, Schuett, and Slattery (2013) [120] showed this is the case. Compared to young adults, the older adults in their study had much larger increases in sentence reading times when spaces were removed. They assessed effects on the lexical processing of words by examining the influence of this manipulation on the word frequency effect (i.e., the processing advantage for words that have a high frequency of written usage). The rationale for this approach was that if removing spacing disrupted word identification processes, this should impact more on low compared to high frequency words and so produce a larger word frequency effect. This is what was observed. However, the increased size of the word frequency effect was comparable for young and older adults, suggesting no age difference in the effects of spacing on word identification. A subsequent study by McGowan, Jordan and Paterson (2014) [64] confirmed these findings but also examined conditions in which spaces between words were filled using a closed square (■) or an open square (□). Disruption to normal reading was greater for older adults when an open rather than a closed square was used. McGowan et al. explained this in terms of the salience of these squares as cues to word boundaries. The closed squares provide a low spatial-frequency cue to word boundaries. By comparison, the open squares were composed of high spatial-frequency visual components (i.e., horizontal and vertical lines) similar to letter features. Older adults appeared, therefore, to have particular difficulty when word boundaries were delineated by a high-spatial frequency cue, and less difficulty when these were demarcated by low spatial-frequency cues, quite possibly due to reductions in sensitivity to higher spatial frequencies. Indeed, it seems likely that older readers had difficulty discriminating between the visual features of the open circles and those of adjacent letters in words, impairing their ability to delineate words in the text.

4. Aging and Eye Movements in Alphabetic Reading

A key research finding is that eye movements are under cognitive control, meaning that decisions about when and where the eyes move are governed by factors influencing how easily words can be identified. Research with alphabetic languages suggests that chief amongst these factors are the length of words, their familiarity to the reader (generally operationalized as the frequency of lexical usage), and their predictability from prior context (for reviews, see [2,3]). Consistent with this account, numerous studies show that words which are shorter, higher in lexical frequency, or more predictable have lower fixation probabilities (and so are more likely to be skipped) and shorter reading times compared to words that are longer, of lower frequency, or less predictable [88,95–97,101,121–130]. Such findings have been important in furthering our understanding of mechanisms of eye movement control in reading and led to the development of sophisticated computational models, such as E-Z Reader [131,132] and SWIFT [133].

Despite substantial evidence that older adults experience greater reading difficulty (see the previous section), few studies have investigated aging effects on the cognitive control of eye movements during reading. Both a corpus-based analysis [95] and an experimental study [96] suggest that aging and word length effects are additive so that longer words are not especially difficult for older readers to process. Other studies have examined age differences in the word frequency effect when other factors, including word length and predictability, are controlled. As noted above, higher frequency words receive shorter reading times and are more likely to be skipped. However, older adults typically produce larger word frequency effects than young adults by fixating for longer on words that are of lower rather than higher frequency [64,95,97,100,120]. This is consistent with older adults having greater difficulty identifying words due to slower lexical processing (due either to cognitive decline or as a consequence of increased lexical items in the mental lexical [42,43]).

These effects have been simulated computationally (in E-Z Reader and SWIFT) for alphabetic languages by limiting the area of text that is sampled on each fixation and slowing lexical processing [97,134]. These simulations capture the greater difficulty older readers experience when identifying lower frequency words. Simulations based on the E-Reader model additionally are in line with the "risky reading hypothesis" proposed by Rayner et al. (2006) [97]. This predicts that older readers skip words more often and make generally longer forward eye movements. Within the model, this is achieved by changing parameters associated with the role of contextual prediction in helping a reader to anticipate the next word in a sentence. This makes sense theoretically as it would be plausible for older readers to make greater use of context to compensate for slower lexical processing [97,135]. Such effects are also consistent with findings from research on spoken and written word recognition, suggesting differential use of context by adult age groups and, in particular, that older readers benefit more from supportive contexts when input is degraded [50,52,53,136–139].

The simulations predict that older readers are more likely than young adults to skip words that can be predicted from the prior context and also that reading times for these words will be shorter. However, evidence for an adult age difference in word predictability effects has proven elusive. A corpus study [95] showed that high word predictability resulted in faster reading by both young and older adults. However, this was realized differently for the two age groups, by increasing word-skipping by the young adults and decreasing the older adults' probability of making multiple fixations on words without affecting their word-skipping rates. Crucially, however, even the experiment that Rayner et al. (2006) [97] reported alongside their simulations provided little evidence of a larger word predictability effect for older readers. This study experimentally manipulated the cloze predictability of target words in sentences. This involved creating contexts in containing a target word that was either highly predictable or one that was less predictable, but where both words were plausible continuations. For instance, in a sentence like "Harriet sang while my brother played piano /flute for my birthday", "piano" is a more likely continuation than "flute", although both words make sense in the context. Rayner et al. established the greater predictability of one word over another using a task in which participants (who do not also take part in the experiment) provide a written continuation for the beginning fragment of a sentence (e.g., "Harriet sang while my brother played … ") and how frequently different words are used as completions is assessed. Sentences containing words that were highly predictable or less predictable continuations were then used as stimuli in the experiment. The results showed that higher word predictability increased skipping rates only marginally more for older compared to younger adults, and shortened reading times similarly for both age groups. There was therefore little evidence that older readers made greater use of predictability to skip words or speed the processing of words they fixated. However, as this study included a manipulation of font difficulty, participants may have had difficulty encoding text on some trials and so read more carefully than normal, which may explain why larger benefits of word predictability were not observed.

More propitiously, a recent experiment by Steen-Baker et al. (2017) [140] found a positive relationship between age and effects of the cloze predictability of a sentence-final word on reading times, such that the decrease in reading times produced by higher predictability was larger for older adults. However, this study was not informative about predictability effects on word-skipping as the target word was always the final word in a sentence (and so could not be skipped). Moreover, the reading time measure they used included time spent re-reading the sentence after encountering the target word and so will have captured sentence-wrap up effects as well as effects of predictability on the processing of the target word. Only one eye movement study to date, by Choi et al. (2017) [101], provides clear evidence for an adult age difference in predictability effects during the initial processing of words, although this was observed only in reading times for words without affecting word-skipping. The study showed that predictability speeded the recognition of words when these were fixated, with no evidence that it facilitated word-skipping. However, several other features of the study may have mitigated against observing such effects. In particular, the task used by Choi et al. included surreptitious changes to target words, deliberate misspellings, and included a comprehension question after each trial (see the previous section), which may have encouraged the older adults to read more carefully and skip words less often than normal.

The research reviewed above addressed key questions about aging effects on mechanisms of eye movement control during reading. This has developed our understanding of some influences of aging, especially in relation to effects of word length and word frequency, although even here relatively few studies have been conducted. Of greater concern is the paucity of studies focused on the claim that older readers make greater use of context to compensate for slower lexical processing. Only a few studies have addressed this question by investigating effects of word predictability and, while there is some indication of such an effect, further work is required to understand it more fully. It will be especially important to establish if, compared to younger adults, older adult readers make greater use of predictability to infer the identities of upcoming words and so skip them. However, more generally, it will be valuable to understand how older adults make use of contextual information and the extent to which their greater knowledge of words and experience of reading might lead them to use prediction or anticipation to a greater extent.

5. Aging and Eye Movement Control in Chinese Reading

Research on aging effects in reading have been conducted predominately in alphabetic languages in which words are created from letters, and words in sentences are clearly delineated using spaces. With this section, we review a relatively small number of studies conducted in Chinese, a non-alphabetic language with visual and linguistic characteristics very different to those of alphabetic languages like English and German. Investigating effects in this language is important for many reasons, including that Chinese is the most widely used writing system worldwide, and that China has the most rapidly aging population. Moreover, such investigations allow us to assess whether aging effects are similar cross-linguistically or reflect the specific visual and cognitive demands of the writing system.

Text in Chinese is composed of box-like logograms, called characters, which can differ considerably in their number of component strokes (lines, dashes), while always occupying the same square area of space [141,142]. For instance, simple characters may be created from a single stroke (e.g., 一 ["yi"], meaning "one") while more complex characters contain upwards of 20 strokes (such as 罐 ["guan"], meaning "pot"). Chinese characters can, therefore, vary substantially in visual complexity, and evidence from non-reading studies suggests such effects may be an important source of perceptual difficulty. For instance, assessments of character legibility show that higher acuity is required to recognize characters with more strokes [143,144]. Moreover, research investigating the visual span for Chinese characters (i.e., how many characters can be recognized on a single glance without moving the eyes) shows span size is smaller for more complex characters [145]. This effect appears to be even greater for older adults [146], suggesting they may recognize fewer characters on each fixational pause during reading, although published research to date has not investigated aging effects on the perceptual span in Chinese reading. Finally, while some characters can function as a word, most Chinese comprise two or more characters. Indeed, according to the Lexicon of Common Words in Contemporary Chinese (2008) [147], only 6% are one-character words, 72% two characters, 12% three characters, and the remainder mostly four characters. However, written Chinese does not include spaces or other cues to word boundaries. An important task when reading Chinese, therefore, is to segment this naturally unspaced text into words (for further discussion of the specific visual and cognitive demands of the Chinese writing system see [148,149]).

Despite these differences, research suggests the same fundamental variables are important in determining when and where the eyes move in Chinese reading. In particular, research with young adults shows that, as with alphabetic languages, reading times are faster and skipping rates higher for words that are shorter (and so composed of fewer characters), of higher frequency, or more predictable from context [150–158]. This suggests cross-linguistic similarity in basic mechanisms of eye movement control. These similarities have also enabled researchers to develop computational models of eye movement control in Chinese reading following the same principles used to develop models for alphabetic languages [159]. It has also led researchers to consider whether older readers produce patterns of age-related reading difficulty in Chinese reading similar to those observed with alphabetic languages. Relatively few such studies have been conducted to date. However, the effects show clearly that older readers experience considerably greater reading difficulty. Compared to young adults, they read much more slowly (often almost twice as slow) and make more and longer fixations and regressions [154,155,158,160,161]. However, by contrast with evidence from alphabetic languages, there is no indication that older Chinese readers use a more risky reading strategy to compensate for this slower reading. Compared to young adults, they appear to skip words infrequently, and their forward saccades are either shorter or of similar length, consistent with more careful reading. Therefore, while age-related reading difficulty is observed cross-linguistically, its consequences appear to differ depending on the writing system.

Studies that have examined aging effects on eye movements in Chinese reading also show that word frequency effects are larger for older than younger adults, suggesting that, as with alphabetic languages, older readers are slower to recognize words [154,155]. Others show that the visual complexity of Chinese characters has a larger effect on reading times by older adults than younger adults [159,162], indicating that this may be an important source of age-related reading difficulty. Only one study to date has examined adult age differences in word length effects [152], although this may be especially important in helping to understand the nature of the difficulties experienced by older Chinese readers. This study examined eye movements for two- and four-character words matched for frequency and embedded in identical sentence contexts. By contrast with additive effects of aging and word length in research with alphabetic languages [96], the effects of these variables produced an interaction in reading times, due to older adults having greater difficulty for long compared to short words. Why this interaction effect is observed became clearer once the landing positions of fixations was examined. Previous research on saccade landing positions in Chinese reading presents a complex picture in which the pattern differs for words receiving either only one fixation or multiple fixations during their initial reading [151,162]. Landing positions on words receiving only one fixation tended to be close to the center of short and long words. By comparison, landing positions on words receiving multiple fixations tended to land on the first character of words regardless of word length. This pattern has been interpreted in different ways. Yan et al. (2010) [162] proposed that readers select either the beginning or center of words as a saccade target depending on whether they can obtain parafoveal cues to word length. That is, if information about the length of the upcoming word is available, the reader will target their next saccade towards the center of that word. However, if this information is unavailable, the reader will employ a more cautious strategy that targets their next saccade towards the first character of the next word. By contrast with this account, X. Li et al. (2011) [151] argued that the effects are not due to use of parafoveal word length cues. They argued instead that the effects reflect the fact that words can be recognized more quickly, and so are less likely to be re-fixated, when a saccade just happens to land at an optimal intra-word location (i.e., word center). That is, readers are more likely to make only one fixation on a word when the initial saccade lands fortuitously at its center, and more likely to make multiple fixations on words when their initial saccade lands at a less optimal location. X. Li et al. proposed that parafoveal processing is character- rather than word-based in Chinese reading and that readers achieve processing efficiency by estimating how many upcoming characters they can identify on each fixation and targeting their next saccade to the right of these characters (see also [163,164]). Crucially, however, while the underlying mechanisms differ, both accounts highlight the importance of parafoveal processing for eye guidance in Chinese reading.

The young adults in the study by S. Li et al. (2018) [152] produced the same pattern of landing position effects as these previous studies. The older adults, however, made fewer single fixations on words, especially longer words, and their saccades were much more likely to land towards the beginning of the words. An exploratory analysis of subsequent fixations additionally showed that older adults were more likely to make a sequence of fixations that landed successively on each character in a word. This was taken as evidence that older readers employ a cautious strategy in which they inspect words character by character. This may explain why older Chinese readers exhibit a pattern of age-related reading difficulty different from that of older readers of alphabetic languages. It may also result from specific difficulties experienced by older Chinese readers when processing upcoming characters or segmenting unspaced text. However, research has yet to address these questions and so an important way forward will be to investigate age differences in parafoveal processing and word segmentation in Chinese reading. A final consideration that also has received little attention to date concerns whether older Chinese readers make greater use of word predictability to offset the difficulties they experience. Only one study has been reported to date [161]. This showed word predictability has a larger facilitatory influence on reading times for older adults but no corresponding effect on word-skipping. Strikingly, such effects may provide the strongest evidence to date (in alphabetic or non-alphabetic languages) for the greater use of contextual knowledge by older readers.

In sum, studies reviewed in this section show very clearly that investigations of aging effects in Chinese are important in demonstrating that age-related reading difficulty is experienced cross-linguistically but may reflect the specific visual and cognitive demands of the writing system. The review highlights that further work is necessary to fully understand the nature of these difficulties. In particular, the evidence that older readers have greater difficulty processing words due to the visual complexity of characters, and make greater use of context, requires more detailed investigation. Moreover, the indication that older readers have specific difficulties parafoveal processing and segmenting words in unspaced text has yet to be explored but likely to be central to our understanding of aging effects on Chinese reading.

6. Conclusions

It should be clear from this review that normative aging has pervasive effects on eye movements in reading. These are a consequence of changes in visual, attentional, and cognitive abilities that occur naturally in older adulthood. They also appear to depend on the characteristics of the writing system and differ subtly for readers of alphabetic languages compared to non-alphabetic languages like Chinese. For the most part, the age-related changes we observe are not catastrophic although clearly likely to affect the efficiency and ease of reading by older people, which will have practical implications and impact on their quality of life. However, it is notable that most research has been conducted with healthy, active, and well-educated older people, and it will be important to more fully understand effects of aging across a spectrum of older people, as well in relation to age-related visual impairment (e.g., macular degeneration) and neurodegenerative disease (e.g., Alzheimer's disease). Indeed, what is clearest from the review is that current understanding of aging effects is limited, and more work is required to fully understand the difficulties that older readers experience.

Author Contributions: All authors contributed to the preparation of this manuscript. All authors have read and agreed to the published version of the manuscript.

Funding: This research was funded by the British Academy, Leverhulme Trust, Ulverscroft Foundation, Economic and Social Research Council, grant number ES/R008957/1, National Science Foundation of China, grant number 81771823, China Scholarship Council scholarships to Lin Li, Sha Li, and Fang Xie, and a British Academy Mid-Career Research Fellowship, and a One Thousand Talents Visiting Professorship, to Kevin Paterson.

Conflicts of Interest: The authors declare no conflict of interest.

References

1. Liversedge, S.P.; Findlay, J.M. Saccadic eye movements and cognition. *Trends Cogn. Sci.* **2000**, *4*, 6–14. [CrossRef]
2. Rayner, K. Eye movements in reading and information processing: 20 years of research. *Psychol. Bull.* **1998**, *24*, 372–422. [CrossRef] [PubMed]
3. Rayner, K. The Thirty-fifth Sir Frederick Bartlett Lecture: Eye movements and attention in reading, scene perception, and visual search. *Q. J. Exp. Psychol.* **2009**, *62*, 1457–1506. [CrossRef] [PubMed]
4. Blythe, H.I. Developmental changes in eye movements and visual information encoding associated with learning to read. *Curr. Dir. Psychol. Sci.* **2014**, *23*, 201–207. [CrossRef]

5. Reichle, E.D.; Liversedge, S.P.; Drieghe, D.; Blythe, H.I.; Joseph, H.S.; White, S.J.; Rayner, K. Using E-Z Reader to examine the concurrent development of eye-movement control and reading skill. *Dev. Rev.* **2013**, *33*, 110–149. [CrossRef] [PubMed]

6. Fozard, J.L.; Gordon-Salant, S. Changes in vision and hearing with aging. In *Handbook of the Psychology of Aging*; Academic Press: San Diego, CA, USA, 2001; pp. 241–266.

7. Owsley, C. Aging and vision. *Vision Res.* **2011**, *1*, 1610–1622. [CrossRef] [PubMed]

8. Salthouse, T.A. When does age-related cognitive decline begin? *Neurobiol. Aging* **2009**, *30*, 507–514. [CrossRef] [PubMed]

9. Salthouse, T.A. Selective review of cognitive aging. *J. Int. Neuropsychol. Soc.* **2010**, *16*, 754–760. [CrossRef]

10. Schieber, F. Vision and Aging. In *Handbook of the Psychology of Aging*; Elsevier: Amsterdam, The Netherlands, 2006; pp. 129–161.

11. Elliott, D.B. Contrast sensitivity decline with ageing: A neural or optical phenomenon? *Ophthalmic Physiol. Opt.* **1987**, *7*, 415–419. [CrossRef]

12. Elliott, D.B.; Yang, K.C.; Whitaker, D. Visual acuity changes throughout adulthood in normal, healthy eyes: Seeing beyond 6/6. *Optom. Vis. Sci.* **1995**, *72*, 186–191. [CrossRef]

13. Derefeldt, G.; Lennerstrand, G.; Lundh, B. Age variations in normal human contrast sensitivity. *Acta Ophthalmol.* **1979**, *57*, 679–690. [CrossRef] [PubMed]

14. Owsley, C.; Sekuler, R.; Siemsen, D. Contrast sensitivity throughout adulthood. *Vis. Res.* **1983**, *23*, 689–699. [CrossRef]

15. Kline, D.W.; Buck, K.; Sell, Y.; Bolan, T.L.; Dewar, R.E. Older observers' tolerance of optical blur: Age differences in the identification of defocused text signs. *Hum. Factors* **1999**, *41*, 356–364. [CrossRef]

16. Akutsu, H.; Legge, G.E.; Ross, J.A.; Schuebel, K.J. Psychophysics of reading-X. Effects of age-related changes in vision. *J. Gerontol.* **1991**, *46*, 325–331. [CrossRef]

17. Kosnik, W.; Winslow, L.; Kline, D.; Rasinski, K.; Sekuler, R. Visual changes in daily life throughout adulthood. *J. Gerontol.* **1988**, *43*, 63–70. [CrossRef]

18. Crassini, B.; Brown, B.; Bowman, K. Age-related changes in contrast sensitivity in central and peripheral retina. *Perception* **1988**, *17*, 315–332. [CrossRef]

19. Ball, K.K.; Beard, B.L.; Roenker, D.L.; Miller, R.L.; Griggs, D.S. Age and visual search: Expanding the useful field of view. *J. Opt. Soc. Am. A* **1988**, *5*, 2210–2219. [CrossRef]

20. Sekuler, A.B.; Bennnett, P.J.; Mamelak, M. Effects of aging on the useful field of view. *Exp. Aging Res.* **2000**, *26*, 103–120.

21. Abel, L.A.; Troost, B.T.; Dell'Osso, L.F. The effects of age on normal saccadic characteristics and their variability. *Vis. Res.* **1983**, *23*, 33–37. [CrossRef]

22. Abrams, R.A.; Pratt, J.; Chasteen, A.L. Aging and movement: Variability of force pulses for saccadic eye movements. *Psychol. Aging* **1998**, *13*, 387–395. [CrossRef]

23. Warabi, T.; Kase, M.; Kato, T. Effect of aging on the accuracy of visually guided saccadic eye movement. *Ann. Neurol.* **1984**, *16*, 449–454. [CrossRef] [PubMed]

24. Huaman, A.G.; Sharpe, J.A. Vertical saccades in senescence. *Investig. Ophthalmol. Vis. Sci.* **1993**, *34*, 2588–2595.

25. Moschner, C.; Baloh, R.W. Age-related changes in visual tracking. *J. Gerontol.* **1994**, *49*, 235–238. [CrossRef] [PubMed]

26. Munoz, D.P.; Broughton, J.R.; Goldring, J.E.; Armstrong, I.T. Age-related performance of human subjects on saccadic eye movement tasks. *Exp. Brain Res.* **1998**, *121*, 391–400. [CrossRef] [PubMed]

27. Tedeschi, G.; Di, A.C.; Allocca, S.; Quattrone, A.; Casucci, G.; Russo, L.; Bonavita, V. Age-dependent changes in visually guided saccadic eye movements. *Funct. Neurol.* **1989**, *4*, 363–367.

28. Pitt, M.C.; Rawles, J.M. The effect of age on saccadic latency and velocity. *Neuro-Ophthalmology* **1988**, *8*, 123–129. [CrossRef]

29. Spooner, J.W.; Sakala, S.M.; Baloh, R.W. Effect of aging on eye tracking. *Arch. Neurol.* **1980**, *37*, 575–576. [CrossRef]

30. Hotson, J.R.; Steinke, G.W. Vertical and horizontal saccades in aging and dementia. Failure to inhibit anticipatory saccades. *Neuro-Ophthalmology* **1988**, *8*, 267–273. [CrossRef]

31. Sharpe, J.A.; Zackon, D.H. Senescent saccades. Effects of aging on their accuracy, latency and velocity. *Acta Otolaryngol.* **1987**, *104*, 422–428. [CrossRef]

32. Kosnik, W.; Kline, D.; Fikre, J.; Sekuler, R. Ocular fixation control as a function of age and exposure duration. *Psychol. Aging* **1987**, *2*, 302–305. [CrossRef]

33. Hasher, L.; Zacks, R.T. Working memory, comprehension, and aging: A review and new view. In *The Psychology of Learning and Motivation: Advances in Research and Theory*; Academic Press: New York, NY, USA, 1988; Volume 22, pp. 193–225.

34. Ryan, J.J.; Sattler, J.M.; Lopez, S.J. Age effects on Wechsler Adult Intelligence Scale-III subtests. *Arch. Clin. Neuropsychol.* **2000**, *15*, 311–317. [CrossRef] [PubMed]

35. Salthouse, T.A. The processing-speed theory of adult age differences in cognition. *Psychol. Rev.* **1996**, *103*, 403–428. [CrossRef] [PubMed]

36. Salthouse, T.A.; Meinz, E.J. Aging, inhibition, working memory, and speed. *J. Gerontol. B Psychol. Sci. Soc. Sci.* **1995**, *50*, 297–306. [CrossRef] [PubMed]

37. Almor, A.; Kempler, D.; MacDonald, M.C.; Andersen, E.S.; Tyler, L.K. Why do Alzheimer patients have difficulty with pronouns? Working memory, semantics, and reference in comprehension and production in Alzheimer's disease. *Brain Lang.* **1999**, *67*, 202–227. [CrossRef] [PubMed]

38. Kemper, S.; Crow, A.; Kemtes, K. Eye-fixation patterns of high- and low-span young and older adults: Down the garden path and back again. *Psychol. Aging* **2004**, *19*, 157–170. [CrossRef]

39. Horn, J.L.; Cattell, R.B. Refinement and test of the theory of fluid and crystallized general intelligences. *J. Educ. Psychol.* **1966**, *57*, 253–270. [CrossRef]

40. Ben-David, B.M.; Erel, H.; Goy, H.; Schneider, B.A. "Older is always better": Age-related differences in vocabulary scores across 16 years. *Psychol. Aging* **2015**, *30*, 856–862. [CrossRef]

41. Keuleers, E.; Stevens, M.; Mandera, P.; Brysbaert, M. Word knowledge in the crowd: Measuring vocabulary size and word prevalence in a massive online experiment. *Q. J. Exp. Psychol.* **2015**, *68*, 1665–1692. [CrossRef]

42. Ramscar, M.; Hendrix, P.; Shaoul, C.; Milin, P.; Baayen, H. The myth of cognitive decline: Non-linear dynamics of lifelong learning. *Top. Cogn. Sci.* **2014**, *6*, 5–42. [CrossRef]

43. Blanco, N.J.; Love, B.C.; Ramscar, M.; Otto, A.R.; Smayda, K.; Maddox, W.T. Exploratory decision-making as a function of lifelong experience, not cognitive decline. *J. Exp. Psychol. Gen.* **2016**, *145*, 284. [CrossRef]

44. Stine-Morrow, E.A.; Milinder, L.; Pullara, O.; Herman, B. Patterns of resource allocation are reliable among younger and older readers. *Psychol. Aging* **2001**, *16*, 69–84. [CrossRef] [PubMed]

45. Stine-Morrow, E.A.; Miller, L.M.; Hertzog, C. Aging and self-regulated language processing. *Psychol. Bull.* **2006**, *132*, 582–606. [CrossRef] [PubMed]

46. Stine-Morrow, E.A.; Soederberg Miller, L.M.; Gagne, D.D.; Hertzog, C. Self-regulated reading in adulthood. *Psychol. Aging* **2008**, *23*, 131–153. [CrossRef] [PubMed]

47. Allen, P.A.; Madden, D.J.; Crozier, L.C. Adult age differences in letter-level and word-level processing. *Psychol. Aging* **1991**, *6*, 261–271. [CrossRef]

48. Allen, P.A.; Madden, D.J.; Weber, T.A.; Groth, K.E. Influence of age and processing stage on visual word recognition. *Psychol. Aging* **1993**, *8*, 274–282. [CrossRef] [PubMed]

49. Cohen-Shikora, E.R.; Balota, D.A. Visual word recognition across the adult lifespan. *Psychol. Aging* **2016**, *31*, 488–502. [CrossRef]

50. Madden, D.J. Adult age differences in the effects of sentence context and stimulus degradation during visual word recognition. *Psychol. Aging* **1988**, *3*, 167–172. [CrossRef]

51. Tainturier, M.J.; Tremblay, M.; Lecours, A.R. Aging and the word frequency effect: A lexical decision investigation. *Neuropsychologia* **1989**, *27*, 1197–1203. [CrossRef]

52. Speranza, F.; Daneman, M.; Schneider, B.A. How aging affects the reading of words in noisy backgrounds. *Psychol. Aging* **2000**, *15*, 253–258. [CrossRef] [PubMed]

53. Dagerman, K.S.; Macdonald, M.C.; Harm, M.W. Aging and the use of context in ambiguity resolution: Complex changes from simple slowing. *Cogn. Sci.* **2006**, *4*, 311–345. [CrossRef] [PubMed]

54. Federmeier, K.D.; Kutas, M. Aging in context: Age-related changes in context use during language comprehension. *Psychophysiology* **2005**, *42*, 133–141. [CrossRef] [PubMed]

55. Hamberger, M.J.; Friedman, D.; Ritter, W.; Rosen, J. Event-related potential and behavioral correlates of semantic processing in Alzheimer's patients and normal controls. *Brain Lang* **1995**, *48*, 33–68. [CrossRef] [PubMed]

56. Wlotko, E.W.; Federmeier, K.D. So that's what you meant! Event-related potentials reveal multiple aspects of context use during construction of message-level meaning. *Neuroimage* **2012**, *62*, 356–366. [CrossRef] [PubMed]

57. Wlotko, E.W.; Federmeier, K.D.; Kutas, M. To predict or not to predict: Age-related differences in the use of sentential context. *Psychol. Aging* **2012**, *27*, 975–988. [CrossRef]

58. Payne, B.R.; Federmeier, K.D. Contextual constraints on lexico-semantic processing in aging: Evidence from single-word event-related brain potentials. *Brain Res.* **2018**, *15*, 117–128. [CrossRef]

59. Hilz, R.; Cavonius, C.R. Functional organization of the peripheral retina: Sensitivity to periodic stimuli. *Vis. Res.* **1974**, *14*, 1333–1337. [CrossRef]

60. McConkie, G.W.; Rayner, K. The span of the effective stimulus during a fixation in reading. *Percept. Psychophys.* **1975**, *17*, 578–586. [CrossRef]

61. McConkie, G.W.; Rayner, K. Asymmetry of the perceptual span in reading. *Bull. Psychon. Soc.* **1976**, *8*, 365–368. [CrossRef]

62. Rayner, K. The gaze-contingent moving window in reading: Development and review. *Vis. Cogn.* **2014**, *22*, 242–258. [CrossRef]

63. Rayner, K.; Well, A.D.; Pollatsek, A.; Bertera, J.H. The availability of useful information to the right of fixation in reading. *Percept. Psychophys.* **1982**, *31*, 537–550. [CrossRef]

64. McGowan, V.A.; White, S.J.; Jordan, T.R.; Paterson, K.B. Aging and the use of interword spaces during reading: Evidence from eye movements. *Psychon. Bull. Rev.* **2014**, *21*, 740–747. [CrossRef] [PubMed]

65. Jordan, T.R.; McGowan, V.A.; Kurtev, S.; Paterson, K.B. A further look at postview effects in reading: An eye-movements study of influences from the left of fixation. *J. Exp. Psychol. Learn. Mem. Cogn.* **2016**, *42*, 296–307. [CrossRef] [PubMed]

66. Bélanger, N.N.; Slattery, T.J.; Mayberry, R.I.; Rayner, K. Skilled deaf readers have an enhanced perceptual span in reading. *Psychol. Sci.* **2012**, *23*, 816–823. [CrossRef]

67. Häikiö, T.; Bertram, R.; Hyönä, J.; Niemi, P. Development of the letter identity span in reading: Evidence from the eye movement moving window paradigm. *J. Exp. Child Psychol.* **2009**, *102*, 167–181. [CrossRef] [PubMed]

68. Inhoff, A.W.; Liu, W. The perceptual span and oculomotor activity during the reading of Chinese sentences. *J. Exp. Psychol. Hum. Percept. Perform.* **1998**, *24*, 20–34. [CrossRef] [PubMed]

69. Ikeda, M.; Saida, S. Span of recognition in reading. *Vis. Res.* **1978**, *18*, 83–88. [CrossRef]

70. Jordan, T.R.; Almabruk, A.A.; Gadalla, E.A.; McGowan, V.A.; White, S.J.; Abedipour, L.; Paterson, K.B. Reading direction and the central perceptual span: Evidence from Arabic and English. *Psychon. Bull. Rev.* **2014**, *21*, 505–511. [CrossRef]

71. Paterson, K.B.; McGowan, V.A.; White, S.J.; Malik, S.; Abedipour, L.; Jordan, T.R. Reading direction and the central perceptual span in Urdu and English. *PLoS ONE* **2014**, *25*, e88358. [CrossRef]

72. Pollatsek, A.; Bolozky, S.; Well, A.D.; Rayner, K. Asymmetries in the perceptual span for Israeli readers. *Brain Lang.* **1981**, *14*, 174–180. [CrossRef]

73. Rayner, K. Eye movements and the perceptual span in beginning and skilled readers. *J. Exp. Child Psychol.* **1986**, *41*, 211–236. [CrossRef]

74. Rayner, K.; Murphy, L.A.; Henderson, J.M.; Pollatsek, A. Selective attentional dyslexia. *Cogn. NeuroPsychol.* **1989**, *6*, 357–378.

75. Rayner, K.; Slattery, T.J.; Bélanger, N.N. Eye movements, the perceptual span, and reading speed. *Psychon. Bull. Rev.* **2010**, *17*, 834–839. [CrossRef] [PubMed]

76. Sperlich, A.; Meixner, J.; Laubrock, J. Development of the perceptual span in reading: A longitudinal study. *J. Exp. Child Psychol.* **2016**, *146*, 181–201. [CrossRef] [PubMed]

77. Rayner, K.; Castelhano, M.S.; Yang, J. Eye movements and the perceptual span in older and younger readers. *Psychol. Aging* **2009**, *24*, 755–760. [CrossRef] [PubMed]

78. Rayner, K.; Castelhano, M.S.; Yang, J. Preview benefit during eye fixations in reading for older and younger readers. *Psychol. Aging* **2010**, *25*, 714–718. [CrossRef]

79. Rayner, K.; Yang, J.; Schuett, S.; Slattery, T.J. The effect of foveal and parafoveal masks on the eye movements of older and younger readers. *Psychol. Aging* **2014**, *29*, 205–212. [CrossRef]

80. Whitford, V.; Titone, D. Eye movements and the perceptual span during first- and second-language sentence reading in bilingual older adults. *Psychol. Aging* **2016**, *31*, 58–70. [CrossRef]

81. Risse, S.; Kliegl, R. Adult age differences in the perceptual span during reading. *Psychol. Aging* **2011**, *26*, 451–460. [CrossRef]
82. Paterson, K.B.; McGowan, V.A.; Jordan, T.R. Filtered text reveals adult age differences in reading: Evidence from eye movements. *Psychol. Aging* **2013**, *28*, 352–364. [CrossRef]
83. Paterson, K.B.; McGowan, V.A.; Jordan, T.R. Effects of adult aging on reading filtered text: Evidence from eye movements. *PeerJ* **2013**, e63. [CrossRef]
84. Jordan, T.R.; McGowan, V.A.; Paterson, K.B. Reading with filtered fixations: Adult age differences in the effectiveness of low-level properties of text within central vision. *Psychol. Aging* **2014**, *29*, 229–235. [CrossRef] [PubMed]
85. Connelly, S.L.; Hasher, L.; Zacks, R.T. Age and reading: The impact of distraction. *Psychol. Aging* **1991**, *6*, 533–541. [CrossRef] [PubMed]
86. Rozek, E.; Kemper, S.; McDowd, J. Learning to ignore distracters. *Psychol. Aging* **2012**, *27*, 61–66. [CrossRef] [PubMed]
87. Healey, M.K.; Campbell, K.L.; Hasher, L. Cognitive aging and increased distractibility: Costs and potential benefits. *Prog. Brain Res.* **2008**, *169*, 353–363. [CrossRef] [PubMed]
88. Rayner, K. Eye guidance in reading: Fixation locations within words. *Perception* **1979**, *8*, 21–30. [CrossRef]
89. Plummer, P.; Rayner, K. Effects of parafoveal word length and orthographic features on initial fixation landing positions in reading. *Atten. Percept. Psychophys.* **2012**, *74*, 950–963. [CrossRef]
90. McConkie, G.W.; Kerr, P.W.; Reddix, M.D.; Zola, D. Eye movement control during reading: I. The location of initial eye fixations on words. *Vis. Res.* **1988**, *28*, 1107–1118. [CrossRef]
91. Heller, D.; Radach, R. Are two eyes better than one? In *Current Oculomotor Research*; Plenum Press: New York, NY, USA, 1999; pp. 341–348.
92. Kliegl, R.; Nuthmann, A.; Engbert, R. Tracking the mind during reading: The influence of past, present, and future words on fixation durations. *J. Exp. Psychol. Gen.* **2006**, *135*, 12–35. [CrossRef]
93. Liversedge, S.P.; Rayner, K.; White, S.J.; Findlay, J.M.; McSorley, E. Binocular coordination of the eyes during reading. *Curr. Biol.* **2006**, *16*, 1726–1729. [CrossRef]
94. Kirkby, J.A.; Webster, L.A.D.; Blythe, H.I.; Liversedge, S.P. Binocular coordination during reading and non-reading tasks. *Psychol. Bull.* **2008**, *134*, 742–763. [CrossRef]
95. Kliegl, R.; Grabner, E.; Rolfs, M.; Engbert, R. Length, frequency, and predictability effects of words on eye movements in reading. *Eur. J. Cogn. Psychol.* **2004**, *16*, 262–284. [CrossRef]
96. Paterson, K.B.; McGowan, V.A.; Jordan, T.R. Aging and the control of binocular fixations during reading. *Psychol. Aging* **2013**, *28*, 789–795. [CrossRef] [PubMed]
97. Rayner, K.; Reichle, E.D.; Stroud, M.J.; Williams, C.C.; Pollatsek, A. The effect of word frequency, word predictability, and font difficulty on the eye movements of young and older readers. *Psychol. Aging* **2006**, *21*, 448–465. [CrossRef]
98. Stine-Morrow, E.A.; Shake, M.C.; Miles, J.R.; Lee, K.; Gao, X.; McConkie, G. Pay now or pay later: Aging and the role of boundary salience in self-regulation of conceptual integration in sentence processing. *Psychol. Aging* **2010**, *25*, 168–176. [CrossRef] [PubMed]
99. Warrington, K.L.; McGowan, V.A.; Paterson, K.B.; White, S.J. Effects of aging, word frequency, and text stimulus quality on reading across the adult lifespan: Evidence from eye movements. *J. Exp. Psychol. Learn. Mem. Cogn.* **2018**, *44*, 1714–1729. [CrossRef] [PubMed]
100. Whitford, V.; Titone, D. The effects of word frequency and word predictability during first- and second-language paragraph reading in bilingual older and younger adults. *Psychol. Aging* **2017**, *32*, 158–177. [CrossRef]
101. Choi, W.; Lowder, M.W.; Ferreira, F.; Swaab, T.Y.; Henderson, J.M. Effects of word predictability and preview lexicality on eye movements during reading: A comparison between young and older adults. *Psychol. Aging* **2017**, *32*, 232–242. [CrossRef]
102. Wotschack, C.; Kliegl, R. Reading strategy modulates parafoveal-on-foveal effects in sentence reading. *Q. J. Exp. Psychol.* **2013**, *66*, 548–562. [CrossRef]
103. Kaakinen, J.K.; Hyönä, J. Task effects on eye movements during reading. *J. Exp. Psychol. Learn. Mem. Cogn.* **2010**, *36*, 1561–1566. [CrossRef]
104. O'Regan, J.K. Eye movements and reading. In *Eye Movements and Their Role in Visual and Cognitive Processes*; Elsevier: Amsterdam, The Netherlands, 1990; pp. 395–453.

105. O'Regan, J.K. Optimal viewing position in words and the strategy-tactics theory of eye movements in reading. In *Eye Movements and Visual Cognition: Scene Perception and Reading*; Springer: New York, NY, USA, 1992; pp. 333–354.

106. Reichle, E.D.; Reineberg, A.E.; Schooler, J.W. Eye movements during mindless reading. *Psychol. Sci.* **2010**, *21*, 300–310. [CrossRef]

107. White, S.J.; Warrington, K.L.; McGowan, V.A.; Paterson, K.B. Eye movements during reading and topic scanning: Effects of word frequency. *J. Exp. Psychol. Hum. Percept. Perform.* **2015**, *41*, 233–248. [CrossRef] [PubMed]

108. Warrington, K.L.; White, S.J.; Paterson, K.B. Ageing and the misperception of words: Evidence from eye movements during reading. *Q. J. Exp. Psychol.* **2018**, *71*, 75–80. [CrossRef] [PubMed]

109. Slattery, T.J. Word misperception, the neighbor frequency effect, and the role of sentence context: Evidence from eye movements. *J. Exp. Psychol. Hum. Percept. Perform.* **2009**, *35*, 1969–1975. [CrossRef]

110. Coltheart, M.; Davelaar, E.; Jonasson, J.T.; Besner, D. Access to the internal lexicon. In *Attention and Performance VI*; Erlbaum: Hillsdale, NJ, USA, 1977; pp. 535–555.

111. Pickwell, D.; Jenkins, T.; Yekta, A.A. The effect on fixation disparity and associated heterophoria of reading at an abnormally close distance. *Ophthalmic Physiol. Opt.* **1987**, *7*, 345–347. [CrossRef]

112. Yekta, A.A.; Pickwell, D.; Jenkins, T. Binocular vision, age and symptoms. *Ophthalmic Physiol. Opt.* **1989**, *8*, 115–120. [CrossRef] [PubMed]

113. Zaroff, C.M.; Knutelska, M.; Frumkes, T.E. Variation in stereoacuity: Normative description, fixation disparity, and the roles of aging and gender. *Investig. Ophthalmol. Vis. Sci.* **2003**, *44*, 891–900. [CrossRef] [PubMed]

114. Bouma, H. Interaction effects in parafoveal letter recognition. *Nature* **1970**, *226*, 177–178. [CrossRef]

115. Pollatsek, A.; Rayner, K. Eye movement control in reading: The role of word boundaries. *J. Exp. Psychol. Hum. Percept. Perform.* **1982**, *8*, 817–833. [CrossRef]

116. Malt, B.C.; Seamon, J.G. Peripheral and cognitive components of eye guidance in filled-space reading. *Percept. Psychophys.* **1978**, *23*, 399–402. [CrossRef]

117. Morris, R.K.; Rayner, K.; Pollatsek, A. Eye movement guidance in reading: The role of parafoveal letter and space information. *J. Exp. Psychol. Hum. Percept. Perform.* **1990**, *16*, 268–281. [CrossRef]

118. Perea, M.; Acha, J. Space information is important for reading. *Vis. Res* **2009**, *49*, 1994–2000. [CrossRef] [PubMed]

119. Rayner, K.; Fischer, M.H.; Pollatsek, A. Unspaced text interferes with both word identification and eye movement control. *Vis. Res.* **1998**, *38*, 1129–1144. [CrossRef]

120. Rayner, K.; Yang, J.; Schuett, S.; Slattery, T.J. Eye movements of older and younger readers when reading unspaced text. *Exp. Psychol.* **2013**, *60*, 354–361. [CrossRef] [PubMed]

121. Altarriba, J.; Kroll, J.F.; Sholl, A.; Rayner, K. The influence of lexical and conceptual constraints on reading mixed-language sentences: Evidence from eye fixations and naming times. *Mem. Cogn.* **1996**, *24*, 477–492. [CrossRef] [PubMed]

122. Brysbaert, M.; Drieghe, D.; Vitu, F. Word skipping: Implications for theories of eye movement control in reading. In *Cognitive Processes in Eye Guidance*; Oxford University Press: Oxford, UK, 2005; pp. 53–77.

123. Frisson, S.; Harvey, D.R.; Staub, A. No prediction error cost in reading: Evidence from eye movements. *J. Mem. Lang.* **2017**, *95*, 200–214. [CrossRef]

124. Hand, C.J.; Miellet, S.; O'Donnell, P.J.; Sereno, S.C. The frequency-predictability interaction in reading: It depends where you're coming from. *J. Exp. Psychol. Hum. Percept. Perform.* **2010**, *36*, 1294–1313. [CrossRef]

125. Inhoff, A.W.; Rayner, K. Parafoveal word processing during eye fixations in reading: Effects of word frequency. *Percept. Psychophys.* **1986**, *40*, 431–439. [CrossRef]

126. Joseph, H.S.; Liversedge, S.P.; Blythe, H.I.; White, S.J.; Rayner, K. Word length and landing position effects during reading in children and adults. *Vis. Res.* **2009**, *49*, 2078–2086. [CrossRef]

127. Rayner, K.; Ashby, J.; Pollatsek, A.; Reichle, E.D. The effects of frequency and predictability on eye fixations in reading: Implications for the E-Z Reader model. *J. Exp. Psychol. Hum. Percept. Perform.* **2004**, *30*, 720–732. [CrossRef]

128. Rayner, K.; Sereno, S.C.; Raney, G.E. Eye movement control in reading: A comparison of two types of models. *J. Exp. Psychol. Hum. Percept. Perform.* **1996**, *22*, 1188–1200. [CrossRef]

129. Rayner, K.; Slattery, T.J.; Drieghe, D.; Liversedge, S.P. Eye movements and word skipping during reading: Effects of word length and predictability. *J. Exp. Psychol. Hum. Percept. Perform.* **2011**, *37*, 514–528. [CrossRef] [PubMed]

130. Staub, A.; Benatar, A. Individual differences in fixation duration distributions in reading. *Psychon. Bull. Rev.* **2013**, *20*, 1304–1311. [CrossRef] [PubMed]

131. Reichle, E.D.; Pollatsek, A.; Fisher, D.L.; Rayner, K. Toward a model of eye movement control in reading. *Psychol. Rev.* **1998**, *105*, 125–157. [CrossRef]

132. Reichle, E.D.; Rayner, K.; Pollatsek, A. The E-Z reader model of eye-movement control in reading: Comparisons to other models. *Behav. Brain Sci.* **2003**, *26*, 445–476. [CrossRef] [PubMed]

133. Engbert, R.; Nuthmann, A.; Richter, E.M.; Kliegl, R. SWIFT: A dynamical model of saccade generation during reading. *Psychol. Rev.* **2005**, *112*, 777–813. [CrossRef] [PubMed]

134. Laubrock, J.; Kliegl, R.; Engbert, R. SWIFT explorations of age differences in eye movements during reading. *Neurosci. Biobehav. Rev.* **2006**, *30*, 872–884. [CrossRef]

135. McGowan, V.A.; Reichle, E.D. The "risky" reading strategy revisited: New simulations using E-Z Reader. *Q. J. Exp. Psychol.* **2017**, *71*, 179–189. [CrossRef]

136. Federmeier, K.D.; McLennan, D.B.; Ochoa, E.; Kutas, M. The impact of semantic memory organization and sentence context information on spoken language processing by younger and older adults: An ERP study. *Psychophysiology* **2002**, *39*, 133–146. [CrossRef]

137. Lash, A.; Rogers, C.S.; Zoller, A.; Wingfield, A. Expectation and entropy in spoken word recognition: Effects of age and hearing acuity. *Exp. Aging Res.* **2013**, *39*, 235–253. [CrossRef]

138. Pichora-Fuller, K. Use of supportive context by younger and older adult listeners: Balancing bottom-up and top-down information processing. *Int. J. Audiol.* **2008**, *47*, S72–S82. [CrossRef]

139. Stine, E.A.L.; Wingfield, A. Older adults can inhibit high-probability competitors in speech recognition. *Aging Neuropsychol. Cogn.* **1994**, *1*, 152–157. [CrossRef]

140. Steen-Baker, A.A.; Ng, S.; Payne, B.R.; Anderson, C.J.; Federmeier, K.D.; Stine-Morrow, E.A.L. The effects of context on processing words during sentence reading among adults varying in age and literacy skill. *Psychol. Aging* **2017**, *32*, 460–472. [CrossRef] [PubMed]

141. Hoosain, R. Psycholinguistic implications for linguistic relativity: A case study of Chinese. *J. Neurolinguistics* **1991**, *8*, 157–161.

142. Hoosain, R. Psychological reality of the word in Chinese. *Adv. Psychol.* **1992**, *90*, 111–130.

143. Zhang, J.Y.; Zhang, T.; Xue, F.; Liu, L.; Yu, C. Legibility variations of Chinese characters and implications for visual acuity measurement in Chinese reading population. *Investig. Ophthalmol. Vis. Sci.* **2007**, *48*, 2383–2390. [CrossRef]

144. Zhang, J.Y.; Zhang, T.; Xue, F.; Liu, L.; Yu, C. Legibility of Chinese characters in peripheral vision and the top-down influences on crowding. *Vis. Res.* **2009**, *49*, 44–53. [CrossRef]

145. Wang, H.; He, X.; Legge, G.E. Effect of pattern complexity on the visual span for Chinese and alphabet characters. *J. Vis.* **2014**, *14*, 6. [CrossRef]

146. Xie, F.; Li, L.; Zhao, S.; Wang, J.; Paterson, K.B.; White, S.J.; Warrington, K.W. Aging and Pattern Complexity Effects on the Visual Span: Evidence from Chinese Character Recognition. *Vision* **2019**, *3*, 11. [CrossRef]

147. Lexicon of Common Words in Contemporary Chinese Research Team. *Lexicon of Common Words in Contemporary Chinese*; The Commercial Press: Beijing, China, 2008.

148. Li, X.; Zang, C.; Liversedge, S.P.; Pollatsek, A. The role of words in Chinese reading. In *The Oxford Handbook of Reading*; Oxford University Press: New York, NY, USA, 2015; pp. 232–244.

149. Zang, C.; Liversedge, S.P.; Bai, X.; Yan, G. Eye movements during Chinese reading. In *The Oxford Handbook of Eye Movements*; Oxford University Press: Oxford, UK, 2011; pp. 961–978.

150. Li, X.; Bicknell, K.; Liu, P.; Wei, W.; Rayner, K. Reading is fundamentally similar across disparate writing systems: A systematic characterization of how words and characters influence eye movements in Chinese reading. *J. Exp. Psychol. Gen.* **2014**, *143*, 895–913. [CrossRef]

151. Li, X.; Liu, P.; Rayner, K. Eye movement guidance in Chinese reading: Is there a preferred viewing location? *Vis. Res.* **2011**, *51*, 1146–1156. [CrossRef]

152. Li, S.; Li, L.; Wang, J.; McGowan, V.A.; Paterson, K.B. Effects of word length on eye guidance differ for young and older Chinese readers. *Psychol. Aging* **2018**, *33*, 685–692. [CrossRef] [PubMed]

153. Liversedge, S.P.; Zang, C.; Zhang, M.; Bai, X.; Yan, G.; Drieghe, D. The effect of visual complexity and word frequency on eye movements during Chinese reading. *Vis. Cogn.* **2018**, *22*, 441–457. [CrossRef]

154. Wang, J.; Li, L.; Li, S.; Xie, F.; Chang, M.; Paterson, K.B.; White, S.J.; McGowan, V.A. Adult age differences in eye movements during reading: The evidence from Chinese. *J. Gerontol. Ser. B Psychol. Sci. Soc. Sci.* **2018**, *73*, 584–593. [CrossRef] [PubMed]

155. Wang, J.; Li, L.; Li, S.; Xie, F.; Liversedge, S.P.; Paterson, K.B. Effects of aging and text-stimulus quality on the word-frequency effect during Chinese reading. *Psychol. Aging* **2018**, *33*, 693–712. [CrossRef] [PubMed]

156. Yan, G.; Tian, H.; Bai, X.; Rayner, K. The effect of word and character frequency on the eye movements of Chinese readers. *Br. J. Psychol.* **2006**, *97*, 259–268. [CrossRef] [PubMed]

157. Zang, C.; Fu, Y.; Bai, X.; Yan, G.; Liversedge, S.P. Investigating word length effects in Chinese reading. *J. Exp. Psychol. Hum. Percept. Perform.* **2018**, *44*, 1831–1841. [CrossRef]

158. Zang, C.; Zhang, M.; Bai, X.; Yan, G.; Paterson, K.B.; Liversedge, S.P. Effects of word frequency and visual complexity on eye movements of young and older Chinese readers. *Q. J. Exp. Psychol.* **2016**, *69*, 1409–1425. [CrossRef]

159. Rayner, K.; Li, X.; Pollatsek, A. Extending the E-Z reader model of eye movement control to Chinese readers. *Cogn. Sci.* **2007**, *31*, 1021–1033. [CrossRef]

160. Li, L.; Li, S.; Xie, F.; Chang, M.; McGowan, V.A.; Wang, J.; Paterson, K.B. Establishing a role for the visual complexity of linguistic stimuli in age-related reading difficulty: Evidence from eye movements during Chinese reading. *Atten. Percept. Psychophys.* **2019**. [CrossRef]

161. Zhao, S.; Li, L.; Chang, M.; Xu, Q.; Zhang, K.; Wang, J.; Paterson, K.B. Older adults make greater use of word predictability in Chinese reading. *Psychol. Aging* **2019**, *34*, 780–790. [CrossRef]

162. Yan, M.; Kliegl, R.; Richter, E.M.; Nuthmann, A.; Shu, H. Flexible saccade-target selection in Chinese reading. *Q. J. Exp. Psychol.* **2010**, *63*, 705–725. [CrossRef] [PubMed]

163. Liu, Y.; Reichle, E.D.; Li, X. Parafoveal processing affects outgoing saccade length during the reading of Chinese. *J. Exp. Psychol. Learn Mem. Cogn.* **2015**, *41*, 1229–1236. [CrossRef] [PubMed]

164. Wei, W.; Li, X.; Pollatsek, A. Word properties of a fixated region affect outgoing saccade length in Chinese reading. *Vis. Res.* **2013**, *80*, 1–6. [CrossRef] [PubMed]

Review

New Perspectives on Serialism and Parallelism in Oculomotor Control During Reading: The Multi-Constituent Unit Hypothesis

Chuanli Zang [1,2]

1   School of Psychology, University of Central Lancashire, Preston, PR1 2HE, UK; czang@uclan.ac.uk
2   Academy of Psychology and Behaviour, Faculty of Psychology, Tianjin Normal University, Tianjin 300074, China

Received: 3 September 2019; Accepted: 9 September 2019; Published: 25 September 2019

Abstract: Currently there are several computational models of eye movement control that provide a good account of oculomotor behavior during reading of English and other alphabetic languages. I will provide an overview of two dominant models: E-Z Reader and SWIFT, as well as a recently proposed model: OB1-Reader. I will evaluate a critical issue of controversy among models, namely, whether words are lexically processed serially or in parallel. I will then consider reading in Chinese, a character-based, unspaced language with ambiguous word boundaries. Finally, I will evaluate the concepts of serialism and parallelism of process central to these models, and how these models might function in relation to lexical processing that is operationalized over parafoveal multi-constituent units.

Keywords: serialism; parallelism; oculomotor control; reading

1. Introduction

Reading is a visually mediated psychological process. Humans process visual information via the eyes, and during reading, text is visually encoded and an abstract orthographic representation is formed. This abstract orthographic representation is then used to undertake lexical processing whereby a word's syntactic category and meaning are accessed. The physiological basis of the human retina is important in relation to reading. At (approximately) the middle of the retina, the fovea, a small circular region (roughly two degrees across) where visual acuity is the highest, delivers detailed visual information about the environment [1]. Thus, readers visually perceive a small detailed region of text immediately around the point at which they are directly fixating, while beyond this, in the parafovea (extending to five degrees on each side of fixation), the text is visually degraded. In order to obtain clear information for comprehension, readers have to make a series of saccades and move their eyes frequently to place the point of fixation on the upcoming text. When a reader fixates a word, they not only process that word itself, but also the upcoming words from parafovea [2,3]. Readers spend less time fixating a word when it is parafoveally available compared to when it is masked or removed [4]. This advantage is referred to as the *preview benefit* [2] and indicates that partial information about parafoveal words is available prior to their direct fixation. It is well known that during reading the oculomotor control system makes two moment-to-moment decisions: one temporal—when to move the eyes, and one spatial—where to move next. The "when" decision refers to how long it takes to fixate a word, captured by fixation duration measures. The "where" decision (in relation to progressive saccades) refers to which word is selected as the upcoming saccadic target, and the specific position where the eyes actually land on a target word, captured by fixation probability and fixation location measures. A number of researchers have endeavored to explain how the two decisions occur by developing various types of computational models of eye movement control during reading.

It is widely accepted that eye movements are under cognitive control during reading [5–8], and are both central to the process of reading, and constrain the rate at which orthographic information is encoded and processed by the written comprehension system [9]. Currently, there are several cognitive computational models of eye movement control that provide a good account of oculomotor behavior during reading of English and other alphabetic languages. I will begin by briefly providing an overview of two dominant cognitive models that are currently used to guide reading research: E-Z Reader [10–12] and SWIFT [13–15], as well as a model that has been recently proposed: OB1-Reader [16]. I will then evaluate a critical issue of controversy among these models regarding whether words are lexically processed serially or in parallel, and consider reading in Chinese, a character-based, unspaced language with ambiguous word boundaries. Finally, a hypothesis concerning processing of multi-constituent units will be proposed as a potential solution to the current serialism/parallelism impasse.

2. An Overview of Models of Oculomotor Control during Reading: Serial or Parallel?

2.1. E-Z Reader Model

E-Z Reader [10–12] is the most elaborate of sequential attention shift (SAS) models during reading which share the two basic assumptions: (1) attention acts like a spotlight and focuses on only one word at a time; (2) attentional shifts occur sequentially from one word to the next, in order to keep serial word order for comprehension. In this model lexical identification is the engine moving the eyes forward during reading. Specifically, lexical identification is a two-stage process: the early stage of processing, L_1, corresponds to a *familiarity check*, an assessment of the familiarity of the upcoming word based on its frequency of occurrence and predictability from the preceding context in a sentence. L_1 is modulated by visual acuity (determined by the mean absolute distance between the current fixation location and each letter in the word being processed [17]) such that long words and/or words located further away from the center of the fovea are processed less efficiently and are thus fixated for longer than short words and/or words located closer to the center of the fovea. E-Z Reader posits that the completion of L_1 causes the oculomotor system to start programming a saccade to the next word. The later stage of processing, L_2, corresponds to *lexical access*. The completion of L_2 causes attention to shift from the currently fixated word (the word that has been lexically identified) to the next parafoveal word, after which parafoveal processing of that word can occur (though note that the eyes remain on the currently fixated, foveal word, as the attentional shift is fast and usually happens before the eyes actually move to directly fixate the parafoveal word [18]). To reiterate, lexical processing of the next parafoveal word only starts after lexical processing of the currently fixated word has been completed.

With respect to saccade targeting, E-Z Reader presumes that the oculomotor system uses low-spatial frequency information such as word boundaries (indicated by interword spaces in English and most other alphabetic languages) to select the next, unidentified, parafoveal word (by default) as a saccadic target. Furthermore, saccades are often targeted towards the center of a word—the best place to fixate a word in order to recognize it most efficiently ([19,20], though see also references [21,22]). Due to several factors including systematic bias and random motor errors, saccades do not land precisely at the word center but actually at a position slightly to the left of the center of a word, the preferred viewing location (PVL, [23]). If an upcoming word is not selected as a target, then it is skipped. E-Z Reader assumes that if the L_1 stage of lexical processing of the parafoveal word is completed very rapidly, then prior to the completion of programming of the saccade from the foveal word to the parafoveal word, a new saccade to the word beyond the adjacent parafoveal word will be generated and the parafoveal word will be skipped.

According to E-Z Reader, lexical processing drives the eyes to move from one word to the next and saccade targeting is made on a word by word basis, with upcoming words being selected as the next saccadic target. E-Z Reader provides an account for a wide range of findings including effects of word length, frequency, predictability, parafoveal preview, foveal load with parafoveal preview, spillover, skipping cost (fixations are longer prior to skipping words than fixating words), post lexical processing and so on [10]. However, critically, in the framework of E-Z Reader, both foveal and parafoveal processing occur in a serial manner mediated by a process of sequential attention allocation, and they do not take place simultaneously. In other words, more than one word cannot be processed lexically at a time. On the basis of these assumptions, according to this view, lexical or sub-lexical properties of the parafoveal word do not have a direct influence on processing of the currently fixated, foveal word. Thus no so-called lexical parafoveal-on-foveal effects are expected (e.g., reference [24]).

2.2. SWIFT Model

The SWIFT model [13–15] is the most developed of the parallel processing gradient (PG) models during reading. It assumes that attention is spatially distributed across an attentional gradient spreading over multiple words to support parallel lexical processing. PG models are based on the dynamic field theory of movement planning such that a field of activation is spatially distributed over several potential movement targets and the spatially-distributed pattern of activation determines the probability of selecting a saccade target in the activation field. In the SWIFT model, each word of a sentence indicated by spaces is the unit of the activation field that changes over time due to word identification. As soon as words fall within the activation field, they start to accumulate activation. Activation is built up in the first stage of preprocessing, decreases in a second stage of lexical completion, and tends to zero after a word is completely identified. The relative activation associated with lexical identification determines the probability of which word is selected as a saccade target, therefore saccade target selection is a competitive process among all of the activated words within the attention gradient (the span of effective vision), and the one with highest activation is most likely to be selected as the next target.

Note, in a recent version of the SWIFT model [25], a dynamically modulated processing span was incorporated such that attentional deployment might vary in size from a sharp, narrow focus to a widely broad area (i.e., a zoom lens of attention) and this is modulated by processing difficulty of the fixated word. If the foveal word is difficult (e.g., a low frequency word) and its activation is increased, this causes the processing span to be narrow and probably only the fixated word to be processed during a fixation. However, if the foveal word is easy (e.g., a high frequency word) and its activation is reduced, this causes the processing span to be dynamically increased and extend over a number of words. In terms of when to move the eyes, SWIFT assumes that the decision to program a new saccade is generated by a random timer, but an inhibitory control process, foveal inhibition, modulates the progress of the random timer and the processing time, via the difficulty of the currently fixated word. Lexical processing rate is constrained by visual acuity, with increased processing speed for words closer to the point of foveal region. Therefore, it is likely that foveal words are identified more rapidly than parafoveal words. However, the lexical activation level of a word is related to its processing difficulty constrained by its frequency. Thus, when the eyes are fixating the foveal word n, if word $n + 1$ is a high frequency word but word $n + 2$ is a low frequency word, then word $n + 2$ might likely have a higher level of lexical activation compared to word $n + 1$. As a consequence, word $n + 2$ will become the next saccade target. Similarly, if word $n - 1$ has been previously fixated or skipped but has not yet been fully recognized, then it might have a higher level of lexical activation, resulting in a regression to word $n - 1$.

SWIFT captures many patterns of eye movements and provides accounts for reading related phenomena including effects of word length, frequency, predictability, skipping cost and benefit (fixations are shorter prior to skipping short and/or high frequency words) and regressions. Critically, SWIFT assumes that multiple words within the perceptual span [26] can be lexically identified in parallel. This claim of parallel processing is supported by observations of (1) parafoveal-on-foveal effects mentioned earlier whereby processing of the parafoveal word $n + 1$ influences fixation durations on the currently fixated foveal word n; and (2) word $n + 2$ effects whereby there is an observable influence of the lexical properties of words two to the right of the currently fixated word n. However, there have been questions as to whether both types of effect are reliable. Parafoveal-on-foveal effects are relatively weak and the observation of them is mainly restricted to corpus studies as opposed to carefully controlled experiments that require reading for comprehension [24,27,28]. And word $n + 2$ effects are subtle and most often reported when word $n + 1$ is a short or high frequency word [17,29,30]. Furthermore, the SWIFT model has received criticism in relation to how the attention gradient mechanism might account for effects associated with comprehension difficulty should words be identified out of order, and how readers might maintain word order to support incremental interpretation when this occurs [31].

2.3. OB1- Reader

Recently, Snell et al. [16] sought to integrate ideas associated with models of visual word recognition and those with eye movement control in reading. They proposed a computational model of reading called the OB1-Reader. In line with the SWIFT model, OB1 adopts the approach of parallel graded attention (PG) supporting the position that multiple words can be identified in parallel within an attentional visual input window (comprising five words—the fixated word n, along with words $n - 2$, $n - 1, n + 1, n + 2$). Furthermore, OB1 adopts the approach of relative letter position coding for word recognition to support parallel processing at the letter level. OB1 assumes that visual input activates nodes that represent the relative position of letter pairs of a word (open bigram nodes). For example, the visual input *word* can activate nodes for *wo, wr, wd, or, od,* or *rd*, these nodes (e.g., *wo*) in turn activate all related lexical representations (e.g., *word, work, world, wonder,* etc.). The activation of letters in the visual input is constrained by visual acuity, attention, and crowding, with stronger activation of letters that are closer to the fixation and spatial attention, and weaker activation of letters that are crowded by other letters (central letters receiving more benefit from acuity, but outer letters receiving more benefit from reduced crowding, [32]). The open-bigram nodes subsequently activate word nodes via bigram-to-word excitation (activating bigram nodes that are part of the word node) and word-to-word inhibition (inhibiting word nodes that share the same bigrams). The activation of a word node is determined by its length, frequency, and predictability from the preceding context in a sentence. When its activation reaches a recognition threshold, it is identified. Also, if there is orthographic overlap between parafoveal and foveal words, then parafoveal information has a facilitatory influence on the word representation associated with the foveal word, resulting in an orthographic parafoveal-on-foveal effect [33].

Recall earlier, in the context of the parallel processing framework such as SWIFT, if words are identified out of order, the question has been raised as to how readers are able to keep track of the correct sequential order of the words, arguably, necessary for a veridical interpretation of sentential meaning [31]. In order to handle this issue, OB1 posits that all activated words are mapped onto a spatial location in a spatiotopic sentence-level representation, based on expectations about the number of to-be-recognized words, as well as the approximate length of each word indicated by interword spaces in the visual input. If the activation of a word reaches its recognition threshold, the length of the activated lexical representation must match with the length of an expected word at a location in the sentence-level representation in order for identification to occur. Note that these expectations can only be generated when spaces between words demarcate word boundaries (and therefore word lengths). The sentence-level representation is also constrained by syntactic structure. If word n is an article, then there may be an increased expectation for word $n + 1$ to be a noun or an adjective [33]. When a word is identified, attention moves forward to the next most salient location. The visual salience of each word in the visual input is modulated by visual acuity with which its constituent letters are processed relative to the point of fixation. Words that are closer to the point of fixation are more salient, and more likely to be selected as saccade targets. Therefore, in terms of the saccade targeting, OB1 operates similarly to the SWIFT model.

OB1 can account for a range of effects such as word length, frequency, predictability, neighborhood size and orthographic parafoveal-on-foveal effects. However, it cannot explain lexical parafoveal-on-foveal effects even though it adopts the framework of parallel processing in reading.

According to OB1, a high frequency parafoveal word $n + 1$ increases rather than decreases fixation durations on word n, as the activation of nodes associated with a high frequency word $n + 1$ exerts increased inhibition of nodes associated with word n. This is a prediction opposite to that of the SWIFT model. Furthermore, one of the theoretical advances of OB1 is considered to be the use of spatiotopic representations guided by word length. However, it is not clear how readers might generate differential expectations for words based on length if, for example, a sentence is comprised of words of equal length (e.g., *That tall girl must want some food*, see references [21,22]). Though, recently, Mirault, Snell and Grainger examined reading without word spaces, and argued that readers might need to identify the currently fixated word and use its word length information to estimate the left boundary of the next word, then make a saccade beyond that boundary. In this case, readers might engage in a more serial word-by-word identification strategy rather than the parallel process that occurs during spaced text reading. The question arises regarding how such word identification processes might operate in the absence of word boundaries in reading unspaced text [34]. Finally, when word length information is not immediately obvious in reading unspaced languages such as Chinese, a language in which it has been shown that the length of a word influences word identification and saccadic targeting [35], it is an interesting challenge for OB1 to explain how readers segment the text into words in order to use word length information to keep track of word order and maintain a sentence-level representation.

3. The Current Challenge: The Concept of a Word and Its Role in Chinese Reading

A critical issue of controversy and conflict among the three models concerns whether words are lexically processed serially or in parallel during sentence reading. According to E-Z Reader, lexical processing occurs in a serial manner whereby words are lexically processed sequentially one at a time. In contrast, SWIFT proposes that multiple words around the point of fixation (within the perceptual span) are lexically processed in parallel, and therefore, potentially out of sequential order. OB1 adopts the approach of the SWIFT model stipulating parallel processing of multiple words, and also integrates the approach of relative letter position coding to support parallel letter identification in multiple letter strings. In order for the parallel processing system to identify multiple words at a time without misperceiving word order in reading, OB1 constructs a spatiotopic sentence-level representation and uses visual word length information to generate expectations regarding which activated word representation belongs to which spatial location in working memory. In these accounts fixations and saccades are considered to fundamentally constrain the delivery of visual information to the brain for linguistic processing, and therefore, are formative with respect to the nature of such processing. Thus, in the context of these competing models, these theoretical issues (serialism vs. parallelism; sequential vs. non-sequential lexical processing; formation of spatial mappings etc.) are extremely important in relation to eye movement control during reading, because it is widely accepted that word identification is a primary determinant of when a reader makes the decision to move their eyes from the current word to the next point of fixation.

It should be clear that the current debate among these competing models has been mainly limited to the reading of alphabetic languages like English, German, and French. Written word identification in these languages seems relatively straightforward, as it involves words that are comprised of adjacent letters, have pronunciation and meaning, and are visually separated from other words in a sentence by spaces [31]. However, this is not the case in a number of alphabetic scripts. For example, there are no spaces to define words in Thai [36], and there is often ambiguity as word segmentation in Thai relies heavily on sentential context [37]. Even in a spaced language like Finnish, a highly agglutinative language, a word might be comprised of multiple constituent sub-words that appear together without spacing. For example, "lumi" is the basic form of "snow", "lumipallo" means "snowball", "lumipallosota" means "snowball fight", and "lumipallosotatantere" means "snow ball fight field" [9,38]. It is likely that Finnish readers segment and encode the orthography of such words in chunks or units smaller than the entire word, though it remains unclear as to exactly how they determine the units of orthography to encode during a fixation [38–40]. Also in English, the same compound can be written as a single word such as "lifestyle", a hyphenated word that may be processed as a single word or two words such as "life-style", or two separate words such as "life style" [41]. It should be clear that it is challenging to consider what actual form a lexical unit takes in reading of alphabetic languages, and whether the concept of a word in the process of word identification is constant across different language scripts.

The issue becomes more complex when taking into account a non-alphabetic language with completely different orthography. Chinese is logographic and character based, being formed of closely packed box like characters that comprise sentences. Characters are formed from strokes with different visual complexity but occupying the same area of space. Words are comprised of one or more characters. However, word units in Chinese are not clearly demarcated by spaces at their beginning and end. There are no visual cues or inflectional indicators (e.g., lexical categories, number demarcations, tense demarcations, etc.) to mark words' syntactic properties. In fact, the Chinese did not have a term for "word" until the concept was imported from the West at the beginning of the 20th century [42]. It is, perhaps, not so surprising therefore, that Chinese readers sometimes do not have a clear concept of what a word is in Chinese, and often do not discriminate words from other linguistic units like phrases [43–47]. In word segmentation tasks where participants are required to put a "/" between the words of a sentence, different native Chinese readers often segment the same sentence into different word units, and they frequently demarcate strings of characters comprising several words as a single word [46,48]. Despite the ambiguity regarding word boundaries in Chinese, there is considerable evidence suggesting that the word has psychological reality during Chinese reading (note, in this line of research the word unit is most often defined according to the dictionary definition, also any strings with ambiguous word status are precluded from experiments following prescreening procedures, see reference [45] for a review). For example, characters that belong to a word are processed efficiently as a whole unit (e.g., reference [49]). If spaces are inserted between the constituent characters of a word, reading is slowed [43]. By contrast, if spaces are artificially introduced between words, reading is facilitated for children [50,51] and learners of Chinese as a second language [52]. Furthermore, it has been demonstrated that Chinese readers use statistical lexicality cues (i.e., some characters are more likely to be a single character word—single character word likelihood [53]; and some characters are more likely to appear at the beginning or end of a word—within word character positional frequency [54–56]) to facilitate word segmentation processes during reading.

To this extent, word segmentation and the nature of lexical processing are tightly intertwined in Chinese reading. In order to lexically identify a written word in a Chinese sentence, it is necessary to make decisions about which characters are constituents of the word, and which are not. That is, it is essential to segment the characters that comprise the word in relation to those in the sentence around it. Note again, there are no spaces to demarcate words in Chinese text. Currently, not only do the E-Z Reader, SWIFT, and OB1 models offer conflicting accounts of the nature of lexical processing that occurs across fixations during reading, but they also have no mechanism for word segmentation within them. This represents a significant limitation with respect to the generality of these models to unspaced languages.

There has been some effort to explore how words may be segmented during Chinese reading. Li, Rayner, and Cave [57] proposed an account of word segmentation on the basis of the interactive activation framework [58]. They assumed that Chinese characters within the perceptual span are processed in parallel, with processing of each character being constrained by visual acuity and attention. Activation of representations at the character-level passes to corresponding word unit representations at the word-level. Subsequently, activation from the word units passes back to the constituent character representations, and then forward once more through the system. Characters that are part of a word are activated faster and to a greater degree compared to the other characters. The word-level representations compete with each other until the most activated word unit wins the competition, at which point the word is recognized and it is automatically segmented from the character string within which it is embedded. Thus, in Li et al.'s account, word segmentation and lexical identification are part of a single unified process. However, this model currently only accounts for lexical identification of four-character units, and offers no account of lexical identification in relation to saccadic targeting during normal sentence reading. Until now, no model that specifically focuses on eye movement control during Chinese reading has been put forward, though Rayner, Li, and Pollatsek [59] examined whether E-Z Reader could be extended to Chinese reading. Their work did suggest that the model may offer a potential account, though the centrality of the word unit and its identification to oculomotor decisions in this framework meant that the issue of word segmentation remained a critical (unexplained) aspect of processing. It seems fair to suggest that at present it is far from clear how a sequence of characters are segmented into words and how attention is allocated and processing operationalized across those characters, or words, or phrases, in unspaced languages like Chinese.

4. Beyond Serialism and Parallelism: A Multi-Constituent Unit Hypothesis

The characteristics of the Chinese writing system provide challenges for the debate over serialism versus parallelism in oculomotor control during reading. As mentioned earlier, whilst a significant amount of eye movement research has suggested that words are processed serially [31], it is also the case that an increasing number of studies have shown effects that indicate, under certain circumstances, words may be processed in parallel [28]. For example, when word $n + 1$ is a short or very high frequency word, or when word $n + 1$ is grouped with its spatially adjacent word(s) to form a single unit (as will be discussed below), readers are able to obtain useful linguistic information from word $n + 2$ while fixating word n and parafoveal processing can extend across more than two spatially adjacent words. Also semantic information about the parafoveal word $n + 1$, under some circumstances (e.g., for synonyms), can be acquired sometimes while fixating the foveal word [17]. Such effects occur particularly in Chinese reading [60,61]. If one adopts a traditional word-based processing approach, then such evidence would be favorable for parallel models and problematic for serial models. Note though, as mentioned earlier, there has been some effort with SWIFT to show that the attentional window can be dynamically modulated by the properties of the fixated word (the zoom lens model of attention), and on this basis the model can be made to behave in a more serial-like manner if foveal load is high, but a more parallel-like manner if foveal load is low. This seems to be a comparatively flexible position with regards to a stipulation of serialism or parallelism of lexical processing. However, further research is needed to investigate the extent that foveal load has an influence on the depth and/or spatial extent of parafoveal processing [62]. Currently, it is reasonable to say that we have reached something of an impasse in the serial/parallel debate, and this situation necessitates some rethinking in relation to the central concepts of the debate (i.e., serialism and parallelism) and how models based on these ideas might function. Specifically, I will consider how lexical processing may be operationalized over a more flexible unit of text than the word.

In an attempt to move the issue forward, a theoretical hypothesis, the Multi- Constituent Unit (MCU) Hypothesis has been developed that may, at least in certain circumstances, offer a way of reconciling parallel processing of words during reading with serial accounts. The account was put forward in a keynote address to the European Conference on Eye Movements by Liversedge [63] and has been outlined briefly in published work [64,65]. The idea behind the MCU Hypothesis is that some linguistic units that occur in the language, are comprised of multiple words, such as spaced compound words, binomial word pairs, idioms, and other common phrases, and these may be represented lexically as single representations. To be clear, parallel processing of multiple individual words (that are not MCUs) remains entirely consistent with a parallel account (e.g., SWIFT), and fundamentally problematic for advocates of the serial processing account (e.g., E-Z Reader). However, if words that are processed in parallel comprise a MCU, and MCUs are represented and stored in the lexicon as a single lexical entry, then in fact any demonstration of parallelism over the constituent elements of a MCU need not violate serial processing assumptions. This is the central theoretical claim of the MCU Hypothesis.

Some empirical evidence has shown that spaced compounds operate as MCUs in English [64], and that foveal processing can occur across MCUs in Chinese [65]. Cutter et al. [64] constructed sentences containing spaced compounds, that is, MCUs comprised of two frequently co-occurring words that refer to a single concept (e.g., *teddy bear*). They employed the boundary paradigm [2] and positioned the invisible boundary before the first constituent of a spaced compound (e.g., *teddy* in the sentence *"The small child gently cuddled his fluffy teddy bear while trying to get to sleep"*). In this paradigm, prior to the eyes crossing the boundary and directly fixating the target word, a parafoveal preview is presented to the reader. When readers' eyes cross the boundary, the preview is changed to the target word and reading times on the target provide an index of the degree to which the preview is processed before it is actually fixated. In Cutter et al.'s experiment, they orthogonally manipulated the preview of each constituent to be either an identity preview, or a nonword, to examine whether preview benefit could be observed for each constituent (e.g., word $n + 1$ 'teddy' and $n + 2$ 'bear') while readers were currently fixating the preceding word n (e.g., 'fluffy'). The whole of the spaced compound was displayed correctly when a saccade crossed the boundary. They found an interaction between the previews of each word, such that there was an $n + 2$ preview benefit only when there was an identical preview of word $n + 1$. In other words, processing of the second constituent of the spaced compound occurred only if this was "licensed" by the presence of the first constituent (note, as mentioned earlier, that $n + 2$ preview effect has only been reported when word $n + 1$ is short and highly frequent, and that E-Z Reader normally is not ready to explain $n + 2$ effects when word $n + 1$ is relatively long and of relatively low frequency making it unlikely to be skipped as per the Cutter et al. experiment). Cutter et al. argued that these results are due to the two constituent words being represented as a single unified lexical entry, a MCU, and that lexical identification of the whole MCU occurred directly. They also argued that the results are consistent with the view that MCUs are psychologically real and are represented lexically.

The Cutter et al. study has important implications for computational models of eye movement control in reading. As per E-Z Reader, processing operates serially with lexical processing being operationalized sequentially over each adjacent lexical unit within a sentence. Most often such units are individual words, but of course, some lexical units are MCUs that have a single lexical representation (corresponding to the entire MCU). Note also that lexical identification operates in the standard way according to E-Z Reader based on the parafoveal familiarity of the MCU. Thus, within the E-Z Reader framework, it is possible to maintain the central tenet of seriality that this model relies upon, whilst also readily explaining how "parallel" processing of some words (MCUs) might occur. In this way, the MCU Hypothesis offers a step change in theoretical thinking, moving us beyond the current serialism/parallelism impasse.

Next, let us consider Chinese reading. To date, there has been some preliminary research investigating whether the lexical status of multi-character strings (i.e., whether they are processed as a single word or two separate words) exerted an influence on how they were processed. For example, Cui et al. [66] used the boundary paradigm and manipulated the preview of the second constituent (an identity character, or a nonsense character) of a monomorphemic word, a compound word, or a phrase. They found increased fixation durations on the first constituent when the preview of the second constituent was a nonsense character rather than when it was the character itself—a reliable parafoveal-on-foveal effect, and this effect only occurred when the first constituent was part of a monomorphemic word, but not when it was part of a compound word or a phrase (for similar findings in English reading, see [67]). Also, preview benefit on the second constituent was numerically larger for monomorphemic words than compound words and phrases, but comparable between the latter two. Apparently, the linguistic category of the character string modulates parafoveal processing in Chinese reading. Presumably, due to morphological or probabilistic characteristics of monomorphemic words, the first constituent licenses preprocessing of the second constituent since the two together form a meaningful unit, and thus, they were processed in parallel. In contrast, when the characters formed a compound word or a phrase, no such licensing occurred and the two constituents were processed serially. This argument was confirmed in an off-line assessment of judgements regarding the lexical status of the target character strings for phrases and compounds in the sentential contexts. 74% of compounds were rated as words (MCUs), whereas only 45% of phrases were rated as words (MCUs). The ambiguity with regards to the concept of a word or a phrase and the very low frequency of the compound word (3 per million) in this experiment might result in the two constituents of the compound word being processed serially.

In a follow-up study, Drieghe et al. [68] examined parafoveal processing of the second constituent of an adjective-noun or a noun-noun Chinese compound to investigate whether the morphological structure of Chinese character strings affected how they were processed. They observed a larger preview effect for the adjective-noun compound than for the noun-noun compound (for similar findings in English reading, see reference [69]). They argued that for the noun-noun compound, each constituent contributes similarly to lexical identification, whereas for the adjective-noun compound, the adjective modifies the noun, and the noun plays a greater role in lexical identification. It is clear that the lexical characteristics of compounds influenced parafoveal processing in Chinese reading. That is to say, how the compound was represented lexically directly influenced how processing associated with the identification of the compound was operationalized over its constituents when they were in the parafovea. This is a significant point in relation to the MCU Hypothesis.

Beyond the investigation of two-character compounds and phrases, Yu et al. [65] examined whether there are linguistically meaningful lexical representations associated with idioms that are comprised of three characters (a single-character verb and a two-character noun). Yu et al. obtained evidence to suggest that the idioms were processed foveally as a single lexical unit. More recently, with more careful and systematic stimulus generation, we obtained strong evidence that some idioms with a modifier-noun structure are processed both parafoveally and foveally as MCUs [70]. Overall, these studies provide evidence that the unit over which parafoveal and foveal processing is operationalized in reading should be considered more flexibly. Indeed, the fact that Chinese is dense, character based, unspaced and extensively ambiguous in relation to word boundaries, provides an optimal written language situation in which to study how real time word identification and oculomotor processing occur over stimuli with indefinite lexical status during natural reading.

It is important to note that an existing body of work on theories of alphabetic language use and processing is directly relevant to the MCU hypothesis, and the current proposal grows from this research. Earlier studies have provided explanations of why our mental lexicon might store multiple word units. For example, the usage-based theory [71] posits that the representations associated with a language are based on experiences with it. When a particular word sequence is encountered frequently, it gradually comes to be processed as a single unit, since frequent exposure strengthens its representation in memory, making it easier to be accessed as a whole. Similarly, the exemplar-based theory, often referred to as data-oriented parsing [72], contends that there is "universal representation" (rather than universal grammar) in language cognition, and assignment of the representation during language acquisition and processing relies solely on statistics. That is, the probability of a word sequence being represented as a certain constituent is computed entirely from linguistic experience. Therefore, both theories presume that linguistic units including words, as well as phrases, are represented and processed similarly, and thus are comparably affected by the frequency of occurrence. To be clear, frequently occurring multiple word units can be lexicalized alongside individual words as individual units in the lexicon. Even in the Words-and-Rules theory wherein there is a distinction between the lexicon and the grammar [73], it has been proposed that some memorized chunks (sometimes called listemes) that are larger than a word such as idioms and collocations, and that cannot be generated according to rules but should be lexicalized and processed as wholes. In line with these models, an increasing amount of evidence has indicated that our mental lexicon not only contains representations of individual words, but also frequently occurring multi-word units (so-called formulaic sequences, see references [74–76]) including collocations, idioms, binomials and lexical bundles, because formulaic sequences are processed more quickly and easily than matched non-formulaic phrases [74–78]. Some analyses show that at least 30%-50% of the written and spoken discourse is comprised of formulaic sequences [79,80]. The widespread existence of MCUs in language provides an excellent opportunity to establish what units can, and are, lexicalized, and what the criteria are for lexicalization during reading. The lack of distinct boundaries within multi-word sequences suggests that a flexible (perhaps more probabilistic rather than categorical) view of the processing unit should be incorporated in models of eye movement control during reading (for more discussions regarding the existing probabilistic models of morphological processing and sentence comprehension that may indirectly implicate multiword sequences, see reference [77]). Of course, much more empirical work is necessary to evaluate the MCU Hypothesis in natural reading of alphabetic languages like English, as well as non-alphabetic languages like Chinese.

5. Conclusions

In summary, the current models of eye movement control have provided a substantial understanding of the cognitive and oculomotor processes involved in reading of most alphabetic languages such as English, German, and French, and they have motivated a large amount of experimental research on eye movements and reading. These models have different assumptions with respect to the nature of attention allocation, lexical processing and saccade targeting. Regardless of their differing stipulations as to whether words are processed serially or in parallel, they state that word identification fundamentally determines when to move the eyes and can also affect where to move next. In other words, both these decisions in relation to oculomotor commitments in reading are considered to be word based, and the concept of a word has been considered to be rather constant, static and fixed across scripts. However, as suggested here, this may not always be the case, particularly when we consider reading of non-alphabetic, unspaced languages.

Funding: This research was funded by ESRC Grant (ES/R003386/1), Natural Science Foundation of China Grant (31571122), the Special Program of Talents Development for Excellent Youth Scholars in Tianjin and a scholarship from the China Scholarship Council.

Acknowledgments: The author would like to thank Simon Liversedge for helpful comments and discussions on this paper. The ideas presented here formed the basis of ESRC Grant (ES/R003386/1). The author also would like to thank Denis Drieghe and two anonymous reviewers for comments on an earlier version of this paper.

Conflicts of Interest: The author declares no conflict of interest.

References

1. Balota, D.A.; Rayner, K. Word recognition processes in foveal and parafoveal vision. In *Basic Processes in Reading: Visual Word Recognition*; Besner, D., Humphreys, G., Eds.; Lawrence Erlbaum Associates: Hillsdale, NJ, USA, 1991; pp. 198–232.
2. Rayner, K. The perceptual span and peripheral cues during reading. *Cogn. Psychol.* **1975**, *7*, 65–81. [CrossRef]
3. Rayner, K.; Liversedge, S.P. Linguistic and cognitive influences on eye movements during reading. In *The Oxford Handbook on Eye Movements*; Liversedge, S.P., Gilchrist, I.D., Everling, S., Eds.; Oxford University Press: New York, NY, USA, 2011; pp. 751–766.
4. Rayner, K.; Liversedge, S.P.; White, S.J. Eye movements when reading disappearing text: The importance of the word to the right of fixation. *Vis. Res.* **2006**, *46*, 310–323. [CrossRef] [PubMed]
5. Liversedge, S.P.; Findlay, J.M. Saccadic eye movements and cognition. *Trends Cogn. Sci.* **2000**, *4*, 6–14. [CrossRef]
6. Rayner, K. Eye movements in reading and information processing: 20 years of research. *Psychol. Bull.* **1998**, *124*, 372–422. [CrossRef] [PubMed]
7. Rayner, K. The thirty-fifth Sir Frederick Bartlett Lecture: Eye movements and attention in reading, scene perception, and visual search. *Q. J. Exp. Psychol.* **2009**, *62*, 1457–1506. [CrossRef] [PubMed]
8. Rayner, K.; Liversedge, S.P.; White, S.J.; Vergilino-Perez, D. Reading disappearing text: Cognitive control of eye movements. *Psychol. Sci.* **2003**, *14*, 385–389. [CrossRef]
9. Liversedge, S.P.; Drieghe, D.; Li, X.; Yan, G.; Bai, X.; Hyönä, J. Universality in eye movements and reading: A trilingual investigation. *Cognition* **2016**, *147*, 1–20. [CrossRef] [PubMed]
10. Reichle, E.D. Serial-Attention Models of Reading. In *The Oxford Handbook of Eye Movements*; Liversedge, S.P., Gilchrist, I.D., Everling, S., Eds.; Oxford University Press: New York, NY, USA, 2011; pp. 767–786.

11. Reichle, E.D.; Pollatsek, A.; Fisher, D.L.; Rayner, K. Toward a model of eye movement control in reading. *Psychol. Rev.* **1998**, *105*, 125–157. [CrossRef]

12. Reichle, E.D.; Rayner, K.; Pollatsek, A. The E-Z Reader model of eye movement control in reading: Comparisons to other models. *Behav. Brain Sci.* **2003**, *26*, 445–476. [CrossRef]

13. Engbert, R.; Longtin, A.; Kliegl, R. A dynamical model of saccade generation in reading based on spatially distributed lexical processing. *Vis. Res.* **2002**, *42*, 621–636. [CrossRef]

14. Engbert, R.; Nuthmann, A.; Richter, E.M.; Kliegl, R. SWIFT: A dynamical model of saccade generation during reading. *Psychol. Rev.* **2005**, *112*, 777–813. [CrossRef] [PubMed]

15. Engbert, R.; Kliegl, R. Parallel graded attention models of reading. In *The Oxford Handbook of Eye Movements*; Liversedge, S.P., Gilchrist, I.D., Everling, S., Eds.; Oxford University Press: New York, NY, USA, 2011; pp. 787–800.

16. Snell, J.; van Leipsig, S.; Grainger, J.; Meeter, M. OB1-Reader: A model of word recognition and eye movements in text reading. *Psychol. Rev.* **2018**, *125*, 969–984. [CrossRef] [PubMed]

17. Schotter, E.R.; Reichle, E.D.; Rayner, K. Rethinking parafoveal processing in reading: Serial-attention models can explain semantic preview benefit and N + 2 preview effects. *Vis. Cogn.* **2014**, *22*, 309–333. [CrossRef]

18. Reichle, E.D. Computational models of reading: A primer. *Lang. Linguist. Compass* **2015**, *9*, 271–284. [CrossRef]

19. McConkie, G.W.; Kerr, P.W.; Reddix, M.D.; Zola, D. Eye movement control during reading: I. The location of initial fixations in words. *Vis. Res.* **1988**, *28*, 1107–1118. [CrossRef]

20. McConkie, G.W.; Kerr, P.W.; Reddix, M.D.; Zola, D.; Jacobs, A.M. Eye movement control in reading: II. Frequency of refixating a word. *Percept. Psychophys.* **1989**, *46*, 245–253. [CrossRef] [PubMed]

21. Cutter, M.G.; Drieghe, D.; Liversedge, S. Reading sentences of uniform word length: Evidence for the adaptation of the preferred saccade length during reading. *J. Exp. Psychol. Hum. Percept. Perform.* **2017**, *43*, 1895–1911. [CrossRef]

22. Cutter, M.; Drieghe, D.; Liversedge, S.P. Reading sentences of uniform word length II: Very rapid adaptation of the preferred saccade length. *Psychon. Bull. Rev.* **2018**, *25*, 1435–1440. [CrossRef]

23. Rayner, K. Eye guidance in reading: Fixation locations within words. *Perception* **1979**, *8*, 21–30. [CrossRef]

24. Brothers, T.; Hoversten, L.J.; Traxler, M.J. Looking back on reading ahead: No evidence for lexical parafoveal-on-foveal effects. *J. Mem. Lang.* **2017**, *96*, 9–22. [CrossRef]

25. Schad, D.J.; Engbert, R. The zoom lens of attention: Simulating shuffled versus normal text reading using the SWIFT model. *Vis. Cogn.* **2012**, *20*, 391–421. [CrossRef] [PubMed]

26. McConkie, G.W.; Rayner, K. The span of the effective stimulus during a fixation in reading. *Percept. Psychophys.* **1975**, *17*, 578–586. [CrossRef]

27. Clifton, C.; Ferreira, F.; Henderson, J.M.; Inhoff, A.W.; Liversedge, S.P.; Reichle, E.D.; Schotter, E.R. Eye movements in reading and information processing: Keith Rayner's 40 year legacy. *J. Mem. Lang.* **2016**, *86*, 1–19. [CrossRef]

28. Drieghe, D. Parafoveal-on-foveal effects on eye movements during reading. In *The Oxford Handbook of Eye Movements*; Liversedge, S.P., Gilchrist, I.D., Everling, S., Eds.; Oxford University Press: New York, NY, USA, 2011; pp. 839–856.

29. Kliegl, R.; Risse, S.; Laubrock, J. Preview benefit and parafovealon-foveal effects from words n + 2. *J. Exp. Psychol. Hum. Percept. Perform.* **2007**, *33*, 1250–1255. [CrossRef] [PubMed]

30. Radach, R.; Inhoff, A.W.; Glover, L.; Vorstius, C. Contextual constraints and N + 2 preview effects in reading. *Q. J. Exp. Psychol.* **2013**, *66*, 619–633. [CrossRef] [PubMed]

31. Reichle, E.D.; Liversedge, S.P.; Pollatsek, A.; Rayner, K. Encoding multiple words simultaneously in reading is implausible. *Trends in Cogn. Sci.* **2009**, *13*, 115–119. [CrossRef] [PubMed]

32. Grainger, J.; Dufau, S.; Ziegler, J.C. A vision of reading. *Trends Cogn. Sci.* **2016**, *20*, 171–179. [CrossRef]

33. Snell, J.; Meeter, M.; Grainger, J. Evidence for simultaneous syntactic processing of multiple words during reading. *PLoS ONE* **2017**, *12*, e0173720. [CrossRef]

34. Mirault, J.; Snell, J.; Grainger, J. Reading without spaces revisited: The role of word identification and sentence-level constraints. *Acta Psychol.* **2019**, *195*, 22–29. [CrossRef]

35. Zang, C.; Fu, Y.; Bai, X.; Yan, G.; Liversedge, S.P. Investigating word length effects in Chinese reading. *J. Exp. Psychol. Hum. Percept. Perform.* **2018**, *44*, 1831–1841. [CrossRef]

36. Kasisopa, B.; Reilly, R.G.; Luksaneeyanawin, S.; Burnham, D. Eye movements while reading an unspaced writing system: The case of Thai. *Vis. Res.* **2013**, *86*, 71–80. [CrossRef] [PubMed]

37. McBride-Chang, C.; Chen, H.C.; Kasisopa, B.; Burnham, D.; Reilly, R.; Leppanen, P. What and where is the word. *Behav. Brain Sci.* **2012**, *35*, 295–296. [CrossRef] [PubMed]

38. Pollatsek, A.; Hyönä, J.; Bertram, R. The role of morphological constituents in reading Finnish compound words. *J. Exp. Psychol. Hum. Percept. Perform.* **2000**, *26*, 820–833. [CrossRef] [PubMed]

39. Bertram, R.; Hyönä, J. The length of a complex word modifies the role of morphological structure: Evidence from eye movements when reading short and long Finnish compounds. *J. Mem. Lang.* **2003**, *48*, 615–634. [CrossRef]

40. Hyönä, J.; Bertram, R.; Pollatsek, A. Are long compound words identified serially via their constituents? Evidence from an eye-movement-contingent display change study. *Mem. Cogn.* **2004**, *32*, 523–532. [CrossRef] [PubMed]

41. Siyanova, A. On-Line Processing of Multi-Word Sequences in a First and Second Language: Evidence from Eye-Tracking and ERP. Ph.D. Thesis, University of Nottingham, Nottingham, UK, 2010.

42. Packard, J.L. Introduction. In *New Approaches to Chinese Word Formation: Morphology, Phonology and the Lexicon in Modern and Ancient Chinese*; Packard, J.L., Ed.; Mouton de Gruyter: Berlin, Germany; New York, NY, USA, 1998; pp. 1–34.

43. Bai, X.; Yan, G.; Liversedge, S.P.; Zang, C.; Rayner, K. Reading spaced and unspaced Chinese text: Evidence from eye movements. *J. Exp. Psychol. Hum. Percept. Perform.* **2008**, *34*, 1277–1287. [CrossRef] [PubMed]

44. Hoosain, R. Psychological reality of the word in Chinese. In *Language Processing in Chinese*; Chen, H.C., Tzeng, O.J.L., Eds.; North-Holland: Amsterdam, The Netherlands, 1992; pp. 111–130.

45. Li, X.; Zang, C.; Liversedge, S.P.; Pollatsek, A. The role of words in Chinese reading. In *The Oxford Handbook of Reading*; Pollatsek, A., Treiman, R., Eds.; Oxford University Press: New York, NY, USA, 2015; pp. 232–244.

46. Liu, P.; Li, W.; Lin, N.; Li, X. Do Chinese readers follow the national standard rules for word segmentation during reading? *PLoS ONE* **2013**, *8*, e55440. [CrossRef] [PubMed]

47. Zang, C.; Liversedge, S.P.; Bai, X.; Yan, G. Eye movements during Chinese reading. In *The Oxford Handbook on Eye Movements*; Liversedge, S.P., Gilchrist, I.D., Everling, S., Eds.; Oxford University Press: New York, NY, USA, 2011; pp. 961–978.

48. He, L.; Song, Z.; Chang, M.; Zang, C.; Yan, G.; Liversedge, S.P. Contrasting off-line word segmentation with on-line word segmentation during reading. *J. Exp. Psychol. Learn. Mem. Cogn.* **2019**. Invited resubmission.

49. Li, X.; Gu, J.; Liu, P.; Rayner, K. The advantage of word-based processing in Chinese reading: Evidence from eye movements. *J. Exp. Psychol. Learn. Mem. Cogn.* **2013**, *39*, 879–889. [CrossRef]

50. Blythe, H.I.; Liang, F.; Zang, C.; Wang, J.; Yan, G.; Bai, X.; Liversedge, S.P. Inserting spaces into Chinese text helps readers to learn new words: An eye movement study. *J. Mem. Lang.* **2012**, *67*, 241–254. [CrossRef]

51. Zang, C.; Liang, F.; Bai, X.; Yan, G.; Liversedge, S.P. Interword spacing and landing position effects during Chinese reading in children and adults. *J. Exp. Psychol. Hum. Percept. Perform.* **2013**, *39*, 720–734. [CrossRef] [PubMed]

52. Shen, D.; Liversedge, S.P.; Tian, J.; Zang, C.; Cui, L.; Bai, X.; Yan, G.; Liversedge, S.P. Eye movements of second language learners when reading spaced and unspaced Chinese text. *J. Exp. Psychol. Appl.* **2012**, *18*, 192–202. [CrossRef] [PubMed]

53. Zang, C.; Wang, Y.; Bai, X.; Yan, G.; Drieghe, D.; Liversedge, S.P. The use of probabilistic lexicality cues for word segmentation in Chinese reading. *Q. J. Exp. Psychol.* **2016**, *69*, 548–560. [CrossRef] [PubMed]

54. Liang, F.; Blythe, H.I.; Bai, X.; Yan, G.; Li, X.; Zang, C.; Liversedge, S.P. The role of character positional frequency on Chinese word learning during natural reading. *PLoS ONE* **2017**, *12*, e0187656. [CrossRef] [PubMed]

55. Liang, F.; Blythe, H.I.; Zang, C.; Bai, X.; Yan, G.; Liversedge, S.P. Positional character frequency and word spacing facilitate the acquisition of novel words during Chinese children's reading. *J. Cogn. Psychol.* **2015**, *27*, 594–608. [CrossRef]

56. Yen, M.H.; Radach, R.; Tzeng, O.J.L.; Tsai, J.L. Usage of statistical cues for word boundary in reading Chinese sentences. *Read. Writ.* **2012**, *25*, 1007–1029. [CrossRef]

57. Li, X.; Rayner, K.; Cave, K.R. On the segmentation of Chinese words during reading. *Cogn. Psychol.* **2009**, *58*, 525–552. [CrossRef] [PubMed]

58. McClelland, J.L.; Rumelhart, D.E. An interactive activation model of context effects in letter perception: Part 1. An account of basic findings. *Psychol. Rev.* **1981**, *88*, 375–407. [CrossRef]

59. Rayner, K.; Li, X.; Pollatsek, A. Extending the E-Z Reader model of eye movement control to Chinese readers. *Cogn. Sci.* **2007**, *31*, 1021–1033. [CrossRef]

60. Yan, M.; Kliegl, R.; Shu, H.; Pan, J.; Zhou, X. Parafoveal load of word n + 1 modulates preprocessing effectiveness of word n + 2 in Chinese reading. *J. Exp. Psychol. Hum. Percept. Perform.* **2010**, *36*, 1669–1676. [CrossRef]

61. Yang, J.; Wang, S.; Xu, Y.; Rayner, K. Do Chinese readers obtain preview benefit from word n + 2? Evidence from eye movements. *J. Exp. Psychol. Hum. Percept. Perform.* **2009**, *35*, 1192–1204. [CrossRef] [PubMed]

62. Zhang, M.; Liversedge, S.P.; Bai, X.; Yan, G.; Zang, C. The influence of foveal lexical processing load on parafoveal preview and saccadic targeting during Chinese reading. *J. Exp. Psychol. Hum. Percept. Perform.* **2019**, *45*, 812–825. [CrossRef] [PubMed]

63. Liversedge, S.P. Rethinking theoretical frameworks: Studies of eye movements during non-alphabetic reading and reading development. In Proceedings of the 17th European Conference on Eye Movements, Lund, Sweden, 11–16 August 2013.

64. Cutter, M.G.; Drieghe, D.; Liversedge, S.P. Preview benefit in English spaced compounds. *J. Exp. Psychol. Learn. Mem. Cogn.* **2014**, *40*, 1778–1786. [CrossRef] [PubMed]

65. Yu, L.; Cutter, M.G.; Yan, G.; Bai, X.; Fu, Y.; Drieghe, D.; Liversedge, S.P. Word n + 2 preview effects in three-character Chinese idioms and phrases. *Lang. Cogn. Neurosci.* **2016**, *31*, 1130–1149. [CrossRef]

66. Cui, L.; Drieghe, D.; Yan, G.; Bai, X.; Chi, H.; Liversedge, S.P. Parafoveal processing across different lexical constituents in Chinese reading. *Q. J. Exp. Psychol.* **2013**, *66*, 403–416. [CrossRef] [PubMed]

67. Drieghe, D.; Pollatsek, A.; Juhasz, B.J.; Rayner, K. Parafoveal processing during reading is reduced across a morphological boundary. *Cognition* **2010**, *116*, 136–142. [CrossRef] [PubMed]

68. Drieghe, D.; Cui, L.; Yan, G.; Bai, X.; Chi, H.; Liversedge, S.P. The morphosyntactic structure of compound words influences parafoveal processing in Chinese reading. *Q. J. Exp. Psychol.* **2018**, *71*, 190–197. [CrossRef] [PubMed]

69. Juhasz, B.J.; Inhoff, A.W.; Rayner, K. The role of interword spaces in the processing of English compound words. *Lang. Cogn. Process.* **2005**, *20*, 291–316. [CrossRef]

70. Zang, C.; Fu, Y.; Bai, X.; Yan, G.; Liversedge, S.P. Preview effects in three-character Chinese idioms and phrases with a modifier-noun structure. **2019**, in preparation.

71. Bybee, J. From usage to grammar: The mind's response to repetition. *Language* **2006**, *82*, 711–733. [CrossRef]

72. Bod, R. Exemplar-based syntax: How to get productivity from examples. *Linguist. Rev.* **2006**, *23*, 291–320. [CrossRef]

73. Pinker, S. *Words and Rules: The Ingredients of Language*; Harper-Collins: New York, NY, USA, 1999.

74. Conklin, K.; Schmitt, N. Formulaic sequences: Are they processed more quickly than nonformulaic language by native and nonnative speakers? *Appl. Linguist.* **2008**, *29*, 72–89. [CrossRef]

75. Conklin, K.; Schmitt, N. The processing of formulaic language. *Annu. Rev. Appl. Linguist.* **2012**, *32*, 45–61. [CrossRef]

76. Wulff, S.; Titone, D. Introduction to the special issue, bridging the methodological divide: Linguistic and psycholinguistic approaches to formulaic language. *Ment. Lex.* **2014**, *9*, 371–376.

77. Shaoul, C.; Webtbury, C.F. Formulaic sequences: Do they exist and do they matter? *Ment. Lex.* **2011**, *6*, 171–196.

78. Siyanova-Chanturia, A.; Conklin, K.; Van Heuven, W.J. Seeing a phrase time and again matters: The role of phrasal frequency in the processing of multiword sequences. *J. Exp. Psychol. Learn. Mem. Cogn.* **2011**, *37*, 776–784. [CrossRef]

79. Erman, B.; Warren, B. The idiom principle and the open choice principle. *Text Talk* **2000**, *20*, 29–62. [CrossRef]

80. Foster, P. Rules and routines: A consideration of their role in the task-based language production of native and non-native speakers. In *Researching Pedagogic Tasks: Second Language Learning, Teaching and Testing*; Bygate, M., Skehan, P., Swain, M., Eds.; Longman: Harlow, UK, 2001; pp. 75–93.

Article

What Can Eye Movements Tell Us about Higher Level Comprehension?

Anne E. Cook * and Wei Wei

Department of Educational Psychology, University of Utah, Salt Lake City, UT 84112, USA
* Correspondence: anne.cook@utah.edu

Received: 28 February 2019; Accepted: 1 September 2019; Published: 6 September 2019

Abstract: The majority of eye tracking studies in reading are on issues dealing with word level or sentence level comprehension. By comparison, relatively few eye tracking studies of reading examine questions related to higher level comprehension in processing of longer texts. We present data from an eye tracking study of anaphor resolution in order to examine specific issues related to this discourse phenomenon and to raise more general methodological and theoretical issues in eye tracking studies of discourse processing. This includes matters related to the design of materials as well as the interpretation of measures with regard to underlying comprehension processes. In addition, we provide several examples from eye tracking studies of discourse to demonstrate the kinds of questions that may be addressed with this methodology, particularly with respect to the temporality of processing in higher level comprehension and how such questions correspond to recent theoretical arguments in the field.

Keywords: eye tracking; reading; anaphor; discourse comprehension

---

1. Introduction

The use of eye tracking technology to study reading has roots in early work on physical movements of the eyes during reading, e.g., [1–5], as well as on the properties of the perceptual span, which refers to the area of visual acuity within a single fixation, e.g., [6,7]. Subsequent work focused on how studying individuals' eye movements during reading can provide information about underlying processes, such as those involved in word recognition, semantic access, syntactic parsing, and higher level comprehension. Although there have been a large number of studies that have used eye movements to examine word level and sentence level processing during reading (for reviews of that work, see, e.g., [8–10]), there have been far fewer studies that have applied eye tracking technology to the study of higher level comprehension. This latter issue is the topic of this article.

One primary reason that higher level comprehension has received less attention in eye tracking research on reading is that the texts necessary to study processing at this level are longer and, thus, more complex to construct than the sentence-level texts typically used for studies on word- and sentence-level processing. Most eye tracking studies of reading examine eye movement measures for a targeted word or short phrase that is embedded within a sentence; processing time on these regions is dependent upon properties of the word/phrase itself and/or properties of the immediately preceding information; see [11]. This is not meant to imply that the word and sentence level studies are limited to single-sentence stimuli; several studies have employed paragraph-level texts to study these processes, e.g., [12–14]. However, the great majority of stimuli used to study word- and sentence-level processes are composed of one to a few sentences.

Since higher level comprehension entails integrating and validating information across multiple sentences, paragraphs, or even texts, the materials that are used to study processes that contribute to higher level comprehension are necessarily different in nature. In addition to being longer, the nature of what is being studied is different; researchers may be interested in issues such as whether paragraph contexts support activation of particular inferences, or how and when readers process inconsistencies in information from various points of a text. Thus, it is a more complex task to design paragraph-level stimuli such that the evidence for the type of higher level processing under investigation can be isolated to a particular word or phrase. For example, in self-paced reading studies that employ the inconsistency paradigm to investigate reactivation of information from memory during reading, materials are constructed so that a target sentence is either consistent or inconsistent with respect to previous information presented in the passage [15]. For example, in [15], the target sentence "Mary ordered a cheeseburger and fries" can be consistent with the prior context about her being a junk food lover, but inconsistent with the context about her being a strict vegetarian. Participants in these studies advance through each passage line by line, and reading time for the entire target sentence is the dependent measure, which is predicted to vary as a function of the preceding context. In eye tracking studies, however, it is important to be able to isolate the critical region of text to a single word or short phrase. With longer target regions of text, the researcher runs the risk that processing difficulty will be distributed across multiple words, thus diluting the ability to determine where processing difficulty begins and ends and how those effects play out over time. Eye tracking researchers interested in discourse processing must be careful to select and design their materials carefully; it is not advisable to just assume that materials created for line-by-line studies can be readily used in eye tracking studies. We will argue that careful construction of stimulus materials is essential to the investigation of critical issues that are at the crux of theoretical debates in discourse processing.

Another pragmatic issue in using eye tracking to study higher level comprehension concerns the use and interpretation of the specific measures that are derived from the eye movement record. For example, studies of word-level processing are often focused on issues that impact processing very early on during reading, thus the measures that provide the most information are often those that reflect the earliest stages of processing a word (e.g., probability of skipping, single fixation duration, first fixation duration). The probability of skipping a word refers to the probability that the reader does not fixate on a word/region when moving from left to right across a text. Single fixation duration is the duration of a fixation if only one fixation is made on a word before the reader moves past it, and first fixation duration is the duration of the first fixation made on a word. Single fixation duration and first fixation duration are highly correlated; they only differ when a reader makes multiple fixations on a word before moving past it in the text. Researchers may also report other measures that reflect ongoing processing of the word and subsequent integration of it with the surrounding sentence/passage context (i.e., first-pass duration, go-past duration). First-pass duration is defined as the amount of time from when a reader first fixates on a region to when they first leave that region, whereas go-past duration (also sometimes referred to as regression path duration) is the amount of time from when a reader first fixates on a region to when they first leave that region to the right. First pass and go-past duration differ when readers leave a region to regress back to and reread previous content; go-past includes the time spent rereading whereas first-pass duration does not. For a more detailed discussion of these measures, see the overview provided by Cook and Wei [11]. Higher level comprehension reflects processes that occur somewhat downstream in the time course from word recognition and lexical access, so the most useful measures for studies in this area also tend to reflect processing that occurs later in the time course of reading (i.e., first-pass duration, go-past duration) as well as attempts to review/reread previous material in an attempt to resolve processing difficulty.

Across all levels of processing, it is also important to examine the extent to which readers reread previously presented portions of text in attempts to resolve comprehension difficulty. Researchers often report information about regressions, or eye movements made to previously encountered material; regressions are typically reported in terms of the probability of regressing into a region of text or the probability of regressing out of a region of text. Another commonly reported measure is rereading, or second-pass duration, which includes all refixations on a region of text after the eye has already moved past that region in the text. Many researchers hesitate to report second-pass duration, though, because not every reader rereads every item in every experimental condition, so there are many empty cells in the resulting data matrix. Instead, there is a recent trend in the literature to report total duration, which is the sum of initial processing of a target region (i.e., first-pass duration) and any subsequent rereading of that region (i.e., second-pass duration). The problem with reporting and interpreting total duration as a measure of delayed processing is that it includes initial processing time as well. We feel a better alternative is to report second pass reading time, but then include convergent measures of rereading behavior, such as probability of regressions into and/or out of the target region [8].

How individual eye tracking measures are interpreted is also influenced by the type of texts that are used. In single sentence stimuli, the entirety of the text is available on a single line on the screen, and likely, in the reader's working memory when the target region is encountered. In these cases, readers often regress out of problematic content to reread preceding information. Thus, measures of rereading reflect not only probability of regressing out of the target region, but also regressions to earlier content, and rereading of both earlier content and the target region. In contrast, in studies of discourse comprehension where stimuli consist of multiple lines/sentences of text, the information needed to resolve processing difficulty may be several sentences back and, thus, no longer available in working memory and well outside the range of the reader's perceptual span. This may make planning regressions several lines back to specific content presented earlier in the text much more difficult (and much less likely) than if the entire text were available on a single line. However, just how far back in a text readers will regress to reread content in attempts to resolve difficulty when processing extended paragraph level texts has not been studied extensively in the discourse processing literature. The distance that readers will regress, and the amount of previous content that they will reread, is likely to differ from sentence-level to paragraph-level studies, meaning that the interpretation of measures that involve rereading should also differ between these types of studies. For example, eye tracking researchers focused on discourse-level phenomena tend to focus their analyses on probability of regressions out of a target region and back to the target region, as well as rereading of the target region, rather than on regressions to or rereading of specific information presented earlier in a text.

We have just established that the use of longer passages of text in research on reading is important for the understanding of eye movements during higher level comprehension, especially when texts require readers to establish connections between incoming content and previously encountered information that may no longer be available in the reader's working memory. One particular phenomenon in which such connections are essential to comprehension is the case of anaphoric references. An anaphor is a word or phrase that refers to previously encountered content (i.e., an antecedent). Much of the research on anaphoric references has focused on the processes through which the antecedent is reactivated after the reader encodes the anaphoric reference. Studies on this topic have primarily utilized self-paced line-by-line reading paradigms, paired with probe response methodologies (e.g., [16–18]). Although early work in this area focused on questions concerning the processes governing reactivation of the antecedent, more recent work has focused on questions about what happens after the reactivation process. That is, what happens in anaphor processing after an antecedent has been reactivated? How are the two concepts integrated with and validated against one another, and how does this play out over time?

Since anaphoric references are typically single words or short phrases, the study of anaphoric processing is ideally suited to eye tracking. However, relatively few eye tracking studies in this area have been conducted. O'Brien and colleagues [19,20] used eye tracking to demonstrate that ease of processing anaphoric references depends on the strength of the connection between the anaphor and its referent, as well as on the nature of the anaphoric phrase itself. However, the goal of those studies was to explore the antecedent reactivation process, not necessarily the time course of processing the anaphor. However, Duffy and Rayner [21] and Ehrlich and Rayner [22] found that processing difficulty on the text immediately following the anaphor was a function of the relation between the anaphor and its antecedent. This means that processing of the anaphor was not complete even when readers' eyes had moved past it in the text. Thus, although researchers have long assumed that establishing antecedents for anaphors is "necessary" for comprehension [23], it may be that anaphoric processing is not as straightforward as originally assumed. This raises a critical question for anaphoric processing: is full reactivation of an antecedent required for successful comprehension of an anaphor, or is initial processing of the anaphor based on the goodness of fit of reactivated content with the anaphor?

Although there has been research on incomplete processing, or shallow, or "good enough" processing in other domains [24,25], there has been little done in the realm of anaphoric processing, and to our knowledge, none with eye tracking. Most work in the area of shallow processing has been conducted with materials in which anomalous information replaced correct content in sentences or short paragraphs, and participants were explicitly asked to detect the anomalies. Researchers have consistently found that participants are less likely to detect anomalies when they are highly-related to the correct content than when they are low-related [24,25]. Putting this in the context of anaphor processing, Cook [26] argued that if processing of anaphors is not complete before readers move on in the text [21,22], then highly-related, but incorrect anaphors may be less likely to cause processing difficulty than low-related incorrect anaphors. If, on the other hand, full activation and resolution of anaphors is required for comprehension, the semantic relation between the anaphor and the antecedent should not matter if an anaphor is an incorrect referent for the antecedent. Cook [26] tested these arguments with a self-paced line-by-line reading paradigm in which she asked participants to read passages in which an anaphor (e.g., cello) was either correct with respect to an antecedent presented several sentences earlier in the passage (e.g., cello), incorrect but highly-related to the antecedent (e.g., violin), or incorrect and lowly-related to the antecedent (e.g., oboe). Across multiple experiments, Cook found that reading times on the target line containing the anaphor were a function of the semantic overlap between the anaphor and the antecedent, and that this processing difficulty played out across multiple sentences. Participants' reading times on the target sentence were faster in the correct condition than in the incorrect conditions, and they were faster in the incorrect-high overlap condition than in the incorrect-low overlap condition. By the time participants reached the next sentence, the difference between the two incorrect conditions was no longer significant, although reading times in both conditions were still slower than in the correct condition. Cook suggested that initial reading times on the anaphor may have been based on their goodness of fit with reactivated information about the antecedent, thereby supporting an incomplete processing account of anaphor resolution. Additionally, consistent with the argument that processing of anaphors continues even after the eyes move past it in the text [21,22], incorrect anaphors influenced processing of information in the text after readers had moved past the line containing the anaphor. However, as discussed earlier, in the line-by-line reading paradigm, the unit of analysis is time to read an entire line; thus, it is not clear whether the processing difficulty on the target line in the incorrect conditions occurred on the anaphor itself, or after a delay. In addition, when reading line-by-line, readers are not able to regress back to previously encountered content to resolve comprehension difficulties.

In a follow-up to Cook [26], Rayner and colleagues [27] varied whether anaphors were consistent or inconsistent with respect to their antecedents, as well as the distance between the anaphor and the antecedent. They used eye tracking to measure processing on the anaphor and found that when the anaphor was near the antecedent in the text (i.e., in adjacent sentences), readers spent more time processing incorrect anaphors and were more likely to regress back to the antecedent. When the anaphor was more distant (i.e., several sentences after the antecedent), however, there was no reliable effect of inconsistency on either time spent processing the anaphor or probability of regressing back to the antecedent. This suggests that comprehension of anaphors depends more on what information may be available in memory when the anaphor is encountered than on what the reader physically has access to in the text. The goal of the present study was to provide an additional test of the incomplete processing account in anaphor resolution by conducting an extension of Cook's [26] and Rayner et al.'s [27] work. We used eye tracking to examine incomplete processing of anaphors, the timing of anaphoric processing (i.e., immediate or delayed), and the nature of information used to resolve difficulties in anaphoric processing.

With respect to the first question about whether anaphoric processing is incomplete, fixation times should replicate the pattern of times observed by Cook [26] and Rayner et al. [27]; reading times should be faster in the correct condition than in the incorrect conditions, and they should be faster in the incorrect-high overlap condition than in the incorrect-low overlap condition. This would also mean that processing difficulty due to incorrect anaphors may be most likely to occur after the reader moved past the anaphor [21,22]; differences as a function of condition would be observed only in measures that reflect delayed processing of the anaphor, such as rereading (i.e., second pass) of the anaphor and probability of regressing into the anaphor.

With respect to the question about what information readers utilize in resolving comprehension difficulty due to incorrect anaphors, we examined regressions back to previous content from the target line. Although Rayner and colleagues [27] did measure the probability of regressing out of the anaphor and back to the antecedent, they only found significant effects of anaphor inconsistency in the near condition. It may be that readers regressed out of the anaphor when it was more distant from the antecedent, but their regressions never reached the antecedent itself. This suggests that anaphor resolution depends on reactivated information about the antecedent in working memory. Providing a more detailed analysis of readers' regression behaviors during comprehension of passage-level texts will provide information about whether readers actually revisit the explicit mention of the antecedent in order to resolve comprehension difficulty, or whether they mostly utilize content that has been reactivated in working memory in response to the anaphor. There is considerable evidence in the research literature that readers do consult previously read information when processing difficult text. However, much of this comes from work on expository texts, or in looking at individual differences and/or reader strategies [28–30]. The present study examines the extent to which readers reread previously encountered content under "normal" reading demands during narrative comprehension, when there is no specific task or strategy imposed on them other than reading for understanding.

2. Materials and Methods

Participants. Twenty-four members of a large University community in the Northeastern United States participated in exchange for either money or course credit.

Apparatus. Eye movements were recorded by a Fourward Technologies Inc. (San Marcos, TX, USA) Dual Purkinje Eye tracker that has a resolution of 10 min of arc. The eye tracker was interfaced with a computer that ran the experiment. Viewing was binocular, with eye location recorded from the right eye. The position of the participant's eye was sampled every millisecond by the computer and averaged over four consecutive samples. The averaged horizontal and vertical positions of the eye were compared with those of the previous sample to determine whether the eye was fixated or moving.

Passages were presented in their entirety on an NEC (Minato, Tokyo, Japan) 4FG monitor with up to 60 character spaces per line. During the experiment, the participant was seated 62 cm from the monitor, where four characters of text equaled 1° of visual angle. Luminance on the monitor was adjusted to a comfortable brightness for the participant, then held constant. The room was dark except for an indirect light source that enabled the experimenter to keep notes.

Materials. The materials used were modified versions of the 24 passages from Cook [26]. An example appears in Table 1. Passages consisted of a brief introductory section, a context section that described an antecedent with one explicit and two implicit mentions, a transition sentence, and then a target sentence that contained an anaphoric reference to the antecedent. This anaphor was either a correct referent for the antecedent, incorrect but had high semantic overlap with the antecedent, or incorrect and had low semantic overlap with the antecedent. Note that the target sentence was exactly the same across all three conditions; the information about the antecedent was the only content that varied across conditions. Target sentences were positioned within the text such that the anaphor appeared in the middle of the sentence and did not appear at the beginning of a line of text, and several words followed the anaphor prior to the end of the sentence/line. The target region consisted of the anaphor; target regions ranged from five to nine characters, and were, on average, 6.58

characters in length (SD = 1.18). Passages ended with a brief closing sentence. Mean lengths of the passages for the correct, incorrect-high, and incorrect-low antecedent conditions were 96.88, 97.5, and 97.54 words, respectively.

Table 1. Sample passage from Cook [26], modified for the eye tracking study.

Correct Antecedent Condition
Terry and her friend Jill drove to a music shop. As they entered the store, Terry saw a beautifulcello. The large instrument was almost bigger than she was. Terry decided she would teach herself how to play it. She imagined herself sitting down to play the heavy instrument. After thinking for a few minutes, she decided to buy it. Just then, Jill walked over to where Terry was standing. Terry showed Jill the *cello* she had bought at the store that day. She even tried to play a few notes.
Incorrect-High Overlap Condition
Terry and her friend Jill drove to a music shop. As they entered the store, Terry saw a beautiful violin. The small instrument fit perfectly between her chin and shoulder. Terry decided she would teach herself how to play it. She imagined herself dancing as she playedthe lightweight instrument. After thinking for a few minutes, she decided to buy it. Just then, Jill walked over to where Terry was standing. Terry showed Jill the *cello* she had bought at the store that day. She even tried to play a few notes.
Incorrect-Low Overlap Condition
Terry and her friend Jill drove to a music shop. As they entered the store, Terry saw a beautifuloboe. The keys were bright and shiny, and the case was lined in black velvet. Terry decided she would teach herself how to play it. She imagined herself fingering the keys tocreate perfect notes. After thinking for a few minutes, she decided to buy it. Just then, Jill walked over to where Terry was standing. Terry showed Jill the *cello* she had bought at the store that day. She even tried to play a few notes.

Three materials sets were constructed, such that each set contained eight passages that appeared in each of the three conditions. Across the three materials sets, each passage appeared once in each of the three conditions. Each set of 24 experimental passages always appeared intermixed with a set of 48 additional filler passages that were designed to mask the purpose of the experiment; of the 48 filler passages, 12 contained incorrect information (although not in anaphoric references), while the remaining 36 did not contain any incorrect content. Thus, across all experimental and filler passages, 28 items contained incorrect information, and 54 items did not contain any incorrect content.

Procedure. All participants gave their informed consent for inclusion before they participated in the study. The study was conducted in accordance with the Declaration of Helsinki, and the protocol was approved by the University's Internal Review Board Committee (Protocol 13440). Each individual participated in a session that lasted approximately 60 min. For each participant, a clay bite bar was prepared to eliminate head movements, and the eye tracker was calibrated. The initial calibration procedure took approximately five minutes. Prior to reading each passage, calibration of the eye tracking system was checked to ensure that accurate records were obtained. Each participant read three practice passages followed by the set of 24 experimental and 48 filler passages. Participants were told that they would be reading a series of paragraphs displayed on a computer monitor. They were told to read for comprehension so that they would be able to answer an occasional "yes/no" oral comprehension question; comprehension questions focused on content from the passage other than the anaphor and appeared after one fourth of the passages. The comprehension question for the sample passage in Table 1 was "Was Terry with her friend Jill?" At the beginning of each trial, five boxes appeared across the top of the screen, one box appeared in the middle, and five boxes appeared at the bottom of the screen. Each participant was instructed to look at the middle box until the experimenter said, "Ready," and then to look at the left-most box. Once the experimenter had determined that the participant was fixating on the left-most box, the entire passage was presented on the screen. When the participant was finished reading the passage, he or she was instructed to press a button that would end the trial. Participants were given a brief break approximately halfway through the experiment.

3. Results

Across all analyses, F_1 and t_1 indicate analyses based on participants variability and F_2 and t_2 indicate analyses based on items variability. All contrasts were significant at the $p < 0.05$ level, unless otherwise indicated.

Overall, comprehension question accuracy was high, with a mean of 85%; there was no difference in accuracy across the three experimental conditions, Fs < 1. In addition, there was no change in the size of the effect of antecedent condition from the beginning to the end of the experiment, Fs < 1. This was also true for all additional analyses reported below, Fs < 1; the size of the effect did not change over the course of the experiment, suggesting that readers' reactions to or strategies for processing incorrect content did not change with multiple exposures.

Mean first-pass reading time, go-past reading time, and second-pass reading times for the anaphor are reported in Table 2, as well as the mean probability of regressions into the anaphor from subsequent text. Consistent with the argument that individual fixations below 100 ms and above 1000 ms are uncommon and more likely to reflect measurement error [8], any fixations outside this range were excluded from the analysis. Any other outliers more than three standard deviations beyond the cell mean were excluded from analysis; this resulted in the elimination of less than 2% of the data.

Table 2. Mean (and standard deviations) for first-pass duration, go-past duration, and second-pass duration (in milliseconds), with probability of regressions into and out of the anaphor, as a function of antecedent condition.

Measure	Antecedent Condition		
	Correct	Incorrect-High Overlap	Incorrect-Low Overlap
First-pass duration	274 (48.98)	279 (52.91)	283 (46.53)
Go-past Duration	367 (201.79)	341 (68.75)	378 (111.45)
Second-pass duration	15.75 (35.42)	45.58 (50.45)	69.42 (78.75)
Probability of Regression out of the Anaphor	0.17 (0.19)	0.17 (0.2)	0.20 (0.18)
Probability of Regression into Anaphor	0.02 *(0.05)*	0.11 *(0.13)*	0.11 *(0.15)*

Antecedent condition had no impact on measures that reflect initial processing on the anaphor (i.e., first pass), or before the eyes moved past it (i.e., go-past), all Fs < 1. In addition, there was no effect of the antecedent condition on the probability of regression out of the antecedent, all Fs < 1. Thus, it must be the case that Cook's [26] results were due to processing that took place after readers had moved past the anaphor in the target sentence. Consistent with this, the main effect of antecedent condition on second pass reading times was significant for the anaphor, $F_1(2, 46) = 5.43$, MSe = 3193, $p = 0.008$, partial $\eta^2 = 0.19$; $F_2(2, 46) = 6.67$, MSe = 2325, $p = 0.003$, partial $\eta^2 = 0.23$. Second-pass reading times on the anaphor were faster in the correct condition than in both the incorrect-high overlap condition, $F_1(1, 23) = 5.95$, MSe = 3589, d = 0.68; $F_2(1, 23) = 7.71$, MSe = 2381, d = 0.83, and the incorrect-low overlap condition, $F_1(1, 23) = 8.12$, MSe = 8517, d = 0.88; $F_2(1, 23) = 16.3$, MSe = 3799, d = 1.08, but the difference between the two incorrect-overlap conditions was not significant, $F_1(1, 23) = 1.93$, $p = 0.18$, MSe = 7056; $F_2(1, 23) = 1.65$, $p = 0.21$, MSe = 7772.

In addition, the main effect of antecedent condition was significant for probability of regressions to the anaphor, $F_1(2, 46) = 4.82$, MSe = 124, $p = 0.01$, partial $\eta^2 = 0.17$; $F_2(2, 46) = 3.68$, MSE = 145.98, $p = 0.03$, partial $\eta^2 = 0.14$. Readers regressed to the anaphor less often in the correct antecedent condition than in either the incorrect-high overlap condition, $F_1(1, 23) = 9.91$, MSe = 178, d = 0.86; $F_2(1, 23) = 6.86$, MSe = 221, d = 0.79 or the incorrect-low overlap condition, $F_1(1, 23) = 6.81$, MSe = 267, d = 0.78; $F_2(1, 23) = 7.22$, MSe = 235, d = 0.74. The difference between the two incorrect-overlap conditions was not significant, both Fs < 1.

Recall that Cook [26] found that initial reading times on the target sentence containing an anaphor were a function of the semantic overlap between the anaphor and the antecedent, and that this supported an account in which readers do not fully resolve anaphors before they move on in the text.

She argued that it took additional time for information about the antecedent to be reactivated, integrated, and validated against incoming information about the anaphor, such that readers did not experience comparable difficulty in both incorrect conditions until downstream in the time course of processing, when readers had moved on to a subsequent sentence. Since her target sentences were not designed for the kind of fine-grained analyses used in eye tracking studies, it is possible that her effects were distributed across the entire sentence instead of isolated to a single word or short phrase. The results presented here indicate that processing difficulty did not occur on the anaphor itself, but downstream, after readers had already moved on to subsequent text. Effects of the antecedent condition were observed only in measures that reflected delayed processing (i.e., second-pass, probability of regressions into the anaphor). This highlights the importance for inclusion of measures that reflect different points in the time course of processing. Moreover, the usage of a self-paced line by line paradigm did not allow participants to regress to previously encountered content from the passage. The question remains, though, whether they would have done so if the entire text had been available. That is, if participants do regress back to earlier information in a text to resolve encountered inconsistencies, how far back in the text do they go?

In order to answer that question, we next turned to an exploratory analysis of the overall patterns of regressions in the text out of the target region. As mentioned previously, there have been several studies examining overall reading patterns in extended discourse [28–30], but those studies either investigated general reading strategies and/or used expository texts. The present experimental context is different in that narrative texts were used to measure processing in a normal comprehension task, and each text contained a one-word anaphoric reference in the target line that was specifically designed to evoke processing difficulty. Thus, we can use the anaphor as a starting point to gain information about how far back in the text readers will go to reread. To our knowledge, this exploratory analysis of general regression patterns has not been presented in the research literature.

When considering regressions made from the target line across all participants and items, readers made more regressions between words than across lines; on average, participants regressed 1.88 words (SD = 3.86), and 0.08 lines (SD = 0.42) back in the text. A frequency analysis of participants' regression behaviors revealed that readers made regressions from content in the target line approximately 84.4% of the time. However, 78.3% of all regressions were between only one and four words back in the text. Consistent with this, when examining regressions within versus across lines, 94.1% of regressions were to material within the same line rather than to text on preceding lines. The mean number of words and lines regressed as a function of passage condition appear in Table 3. Note that these exploratory analyses are based only on regressions launched from the target line. Since not all participants made regressions from the target line in each item or condition, our analyses are tested only against error terms based on participants' variability.

Table 3. Mean number of words and lines regressed (and standard deviations) from the target region as a function of antecedent condition.

	Antecedent Condition		
Measure	**Correct**	**Incorrect-High Overlap**	**Incorrect-Low Overlap**
Number of Words Regressed from Target	1.21 (0.68)	1.48 (1.11)	2.57 (2.56)
Number of Lines Regressed from Target	0.01 (0.04)	0.04 (0.09)	0.13 (0.24)

Number of Words Regressed from Target. There was a significant main effect of condition on the number of words regressed from the target region, F(2,52) = 5.15, MSE = 2.73, p = 0.01, partial η^2 = 0.17. Planned comparisons demonstrated that the difference between the correct and incorrect-high overlap conditions was not significant, t(26) = −1.28, p = 0.21. However, participants regressed back fewer words from the target in the correct condition than in the incorrect-low overlap condition, t(26) = −2.57, p = 0.02, d = 0.73, and they regressed back fewer words in the incorrect-high overlap condition than in the incorrect-low overlap condition, t(26) = −2.07, p = 0.049, d = 0.55.

Number of Lines Regressed from Target. The same pattern appeared when the number of lines regressed from the target was analyzed, F(2, 52) = 4.29, MSE = 0.02, p = 0.02, partial η^2 = 0.14. The difference between the correct and incorrect-high overlap conditions, although in the right direction, was not significant, p > 0.1. However, the contrast for the correct and incorrect-low overlap conditions was significant, t(26) = −2.48, p = 0.02, d = 0.7, but the difference in number of lines regressed from the target in the incorrect-high and incorrect-low overlap conditions did not reach criteria for significance, t(26) = 1.73, p = 0.096, d = 0.5.

This experiment demonstrated that, at least in some discourse processing studies, effects of interest may not appear immediately upon encoding a target region; instead, effects may be observed across a wider time course of processing. In the case of anaphoric processing, effects did not appear until after the reader had already moved past the anaphor in the target sentence, supporting the argument that processing of anaphors may be incomplete, even when readers move on in the text [21,22]. Effects of antecedent condition appeared on regressions into and rereading of the anaphor, as well as in regressions back to previously read content. The purpose of the more exploratory analysis was to examine, when readers do have the opportunity to regress back to previously encountered content in the text, just how far back will they go? The answer is: not very far. When the entire text was available for rereading, readers rarely regressed back more than a line or two–meaning they rarely regressed back to the antecedent itself in order to resolve difficulty caused by an incorrect anaphor. Thus, resolution of difficulty due to incorrect anaphors had to be based on information reactivated from memory either when the anaphor was encountered, or soon thereafter. These eye tracking results, in combination with previous work on anaphor processing [16–18], and the previous findings of Cook [25] and Rayner and colleagues [27], support a view in which higher level comprehension of text results from a continuous process of integrating incoming information with and validating it against information that has been reactivated from memory [31–33].

4. Discussion

Given our observation that readers resolve processing difficulty during discourse comprehension without extensive rereading of earlier portions of the text, it must be that reading times on the target sentence reflect difficulty in integrating incoming information and evaluating it against information that has been reactivated from memory. This is the same argument made by researchers who use line-by-line self-paced reading paradigms in which it is not possible for readers to regress back to previous portions of the text (e.g., [15,31]). In anaphor resolution in particular, this allows readers to connect anaphors with distant antecedents without engaging in extensive rereading or experiencing large coherence breaks. The downside of this, however, is that comprehension relies on algorithmic processes that are not perfect in nature. In the present case and in earlier work by Cook [26] and by Rayner and colleagues [27], time to process incorrect anaphors was a function of their overlap with the antecedent, instead of based on whether the anaphor was a correct referent or not. Indeed, Klin and colleagues [34–36] argued that in some cases, readers may never fully activate the specific lexical item for an antecedent and instead rely on a partially activated set of conceptual features about the antecedent during the initial stages of anaphor resolution. If the reactivated content is "good enough" [24,25,37], comprehension proceeds. However, as demonstrated here and in Cook's [26] original study, additional information may become available and lead to processing difficulty downstream of the anaphor. Since Klin and colleagues' [34–36] studies used a single response probe paradigm, they were not able to observe the continuum of processing that the use of eye tracking allowed for here and in Rayner's [27] study. In general, the results reported here add to a growing body of literature that supports a view in which information is continually being reactivated from memory, integrated with incoming content, and validated with respect to the information in active memory [31–33].

The benefit of eye tracking studies beyond line-by-line self-paced reading paradigms, then, is not in the kinds of phenomena that can be studied but in the level of analysis that can be obtained. Eye tracking allows for a more fine-grained measure of where in the time course of reading processing difficulty occurs and what information readers may utilize to resolve that difficulty. This is particularly important when examining processing of particular words or phrases that play out over time. The study reported here focused on the time course of anaphor resolution, demonstrating that resolution is based on the "goodness of fit" between the anaphor and reactivated information, that it continues after readers have moved past the anaphor in the text, and that resolution depends on reactivated information rather than direct access in the text to the previously encountered content.

This ability to examine processing over time in higher level of comprehension is important, because reliance on measures that fail to examine the full time course of processing may paint a misleading picture of the comprehension process. In another study, Cook and colleagues [38] used eye tracking with Moses Illusion items in which highly related, but incorrect target concepts were embedded in general knowledge statements (e.g., "It was Moses who took two animals of each kind on the Ark). Consistent with previous eye tracking studies on semantic anomalies [39–41], Cook and colleagues [38] demonstrated that readers incorrectly responded "true" to illusion statements and did not have any differences in initial reading times (i.e., first fixation duration, first-pass duration) between correct and incorrect content. However, different from previous studies, Cook and colleagues [38] also found that relatively late measures of reading on the target (i.e., regressions and second-pass duration) showed that readers spent more time reprocessing incorrect targets than correct ones, even if they had initially failed to detect the incorrect information and responded "true" to the item. This suggests that participants' explicit responses to incorrect content in text may not be reflective of the extent to which that information is actually processed.

The previous paragraphs presented examples of how eye tracking can be used to examine critical issues in higher level comprehension—particularly processing of inconsistent or difficult content in text. In the course of this discussion, we want to revisit our earlier discussion of the importance of careful stimulus design in eye tracking studies of discourse processing. As illustrated in the analysis presented in this article, it is useful to construct text-level stimuli such that comprehension of a very specific region of text (i.e., a target word or phrase) is dependent upon previous portions of text. This allows researchers to understand how comprehension of information may change as a function of the preceding content, even if that content appeared several sentences or lines back in the text. For example, even though studies of lexically ambiguous words (e.g., "bank") may be focused on lexical access, which is a lower, word-level process, researchers have studied how access of word meaning is influenced by discourse level variables. Wiley and Rayner [42] investigated how processing time on ambiguous words embedded in paragraph length passages was influenced by passage titles as well as passage context. Similarly, Colbert-Getz and Cook [13] examined: (1) whether elaboration of passage context that supported the subordinate meaning of a lexically ambiguous word would influence word processing time; and (2) whether a prior encounter of an ambiguous word in its subordinate sense would influence subsequent processing of the same word in its dominant sense. In both studies just described, the target regions consisted of a lexically ambiguous word and a disambiguating word or phrase; processing of these regions depended upon the preceding passage context. Although there have been fewer studies in which researchers studied the influence of discourse context on sentence processing, researchers use the same general stimulus design strategy. Processing of a particular word or phrase depends on how it is parsed, and parsing may be influenced by the global passage content [43]. Across these studies, though, measures are typically limited to data taken from the target line itself. The results from the present study illustrate why: readers do not appear to reread distant portions of the text to resolve comprehension difficulty, even when those earlier portions of the text are still present on the screen and, thus, available to the reader. For a more detailed discussion of stimulus design issues for eye tracking studies in reading research, see Cook and Wei [11].

Of greatest relevance to the present discussion, though, are studies in which processing of information in a target sentence is dependent upon readers making connections between the target information and previously encountered information. Although the general design of the target region may be the same, the types of questions that are asked may be different. Researchers interested in discourse processing are generally focused on how a developing representation of a text in memory influences processing over time. For example, do readers activate inferences based on preceding contextual information, and do they instantiate those inferences into the evolving discourse model in long-term memory? O'Brien and colleagues [19,20] found that processing times on words that were only implied in a text were just as fast as when those same words had been explicitly mentioned, indicating that the implied concepts had been inferred during reading (i.e., activated; see [23,44,45]) and instantiated into the representation of the text in memory.

As illustrated in the study in this article, eye tracking studies of discourse can also reveal how processing plays out over time. Although researchers interested in word- and sentence-level processing also examine the time course of processing, what researchers mean by "early" and "later" processing differs across levels of processing. For example, word-level researchers may examine early recognition processes related to orthography and phonology, followed by later processes of semantic access or integration with sentence context. In contrast, discourse processing researchers are generally focused on the time course of encoding new information, linking it with the current contents of active memory, and verifying it against information in long-term memory, e.g., [31–33]. Since passage-level texts contain more information that can be held in the reader's working memory, readers must rely on information that is reactivated from long-term memory. Although this may include information previously presented within the text, as in the examples described in the preceding paragraph, it may also include information that is reactivated from the reader's general world knowledge, or semantic memory. Thus, in eye tracking studies of higher level comprehension, "early" processing may reflect influences of content that is active when a target word is encoded, whereas "later" processing may reflect influences of information that is not activated and incorporated into the ongoing discourse representation until the reader has already moved past the target in the text.

This interpretation of "early" and "late" influences in comprehension has been applied to studies of the time course of influences of previously encountered contextual content versus information from general world knowledge on processing of incoming information. Using eye tracking technology, Garrod and Terras [46] examined whether readers' processing of role fillers was initially influenced by either information from general world knowledge or the previous discourse context. They had participants read short texts in which the target region indicated either an appropriate or inappropriate role filler (based on general world knowledge) for an action presented in a previous sentence. For example, the target phrase "the pen dropped" is an appropriate role filler for the preceding sentence "The teacher was busy writing a letter of complaint to a parent" but is inappropriate if the preceding sentence was "The teacher was busy writing an exercise on the blackboard." Eye tracking measures revealed no initial effect of appropriateness on processing the noun, "pen." However, times in the region of the verb, "dropped," and regressions from it back to the noun "pen" indicated delayed processing difficulty when the pen was an inappropriate role filler for the preceding action. Garrod and Terras argued that early processing of the role filler represented low-level associative bonding of the role filler (pen) to the preceding action (writing), but this link was subsequently resolved against the broader discourse context (writing a letter vs. writing on the chalkboard). Thus, processing difficulty due to a mismatch between the role filler and the context was not observed until relatively late in the time course.

Cook and Myers [47] extended this work by creating scripted narrative texts (e.g., a rock band context) in which the initial encounter with a role filler was either appropriate (a song was played by a guitarist) or inappropriate (a song was played by the manager) with respect to general world knowledge. Consistent with Garrod and Terras' [46] findings, processing times were a function of the appropriateness of the role filler for the action described. The passage continued, however, and a second encounter with the role filler was also either appropriate or inappropriate with respect to general script-based knowledge. More important, though, the second encounter either matched or mismatched the first encounter. Cook and Myers found that when the second encounter matched the first encounter, regardless of whether it was appropriate or not, initial processing of this encounter was facilitated. Subsequent processing on the second encounter, though, showed a delayed effect of appropriateness of the role filler; readers had increased regressions and longer second pass reading times for the inappropriate role fillers. Cook and Myers argued that the early effects of appropriateness on the first encounter, but the delayed effects of appropriateness on the second encounter, suggested that either general world knowledge or context has the potential to be reactivated and influence initial processing of incoming information. However, as additional information continues to be reactivated, it has the potential to influence processing downstream in the time course, even if the reader has moved on in the text. Although one source of knowledge may dominate early processing of target content, the fine-grained nature of eye tracking measures allow researchers to examine the extent to which the other sources of knowledge come into play downstream.

The argument that initial processing is influenced by the winner of a "race" for activation between contextual information and general world knowledge is consistent with assumptions of the RI-Val model of discourse comprehension proposed by Cook and O'Brien [31–33]. They argued that comprehension can be explained in terms of three parallel asynchronous stages of processing that each operate on the output of the preceding stage. In the first stage (R), information is reactivated from long-term memory in response to incoming content via a passive retrieval mechanism, e.g., [48,49], and this includes both previously read content as well as information from general world knowledge. As soon as information becomes available, it is linked to, or integrated (I), with the contents of working memory on the basis of goodness of fit in the second stage. The third stage involves validating (Val) linkages against the contents of active memory via a feature-based partial matching mechanism [50–52]. These stages are assumed to be passive in nature and, thus, run to completion; they are also continuously operating. Thus, new information may be reactivated even as the validation stage is starting. This is true regardless of whether readers have reached their coherence threshold, the point in time at which attention shifts to new information in the text. This means that new information may still be coming available in working memory even after the reader has moved on in the text. Since processing operates on either side of the coherence threshold, it is possible to observe processing difficulty either immediately upon encountering the problematic content, or after a delay.

Cook [26] used the RI-Val model to explain her finding that early processing of anaphors was based on goodness of fit; as contextual information about the antecedent continued to become available in memory, however, that content influenced processing downstream from the anaphor. Although Cook's results were based on line-by-line self-paced reading data, the same general pattern of results was found with the eye tracking data reported here; incorrect anaphors resulted in processing difficulty, but only in measures that reflected processing relatively late in the time course (i.e., regressions, second-pass duration). The present findings also show, though, that readers did not utilize the entirety of the text to resolve that processing difficulty; most regressions were within the same line, and there were relatively few regressions more than one or two lines back in the text—not far enough to reread the portion of the text containing the explicit mention of the antecedent. As suggested previously, this means that processing difficulty was resolved based on the information that had been reactivated in memory. Given the continuous nature of processing assumed by the RI-Val model, information about the antecedent becomes available in working memory over time, meaning that early processing of an anaphor may be based on incomplete content. Resolution continues as more information becomes available, and this may continue occur even after the reader has moved on in the text.

In another discourse processing study, Creer, Cook, and O'Brien [53] examined how narrator perspective (i.e., first-person, third-person) influenced processing of spatial inconsistencies embedded in texts. Across multiple self-paced line-by-line experiments, they found that under normal reading conditions, readers were disrupted by spatial inconsistencies involving the protagonist when texts were written in the first-person perspective, but not when they were written in the third-person perspective. Creer et al. argued that the disruption was due to readers having difficulty validating incoming content against information reactivated from the discourse representation in long-term memory. Consistent with the view that validation occurs relatively late in the time course of processing, an eye tracking experiment isolated the inconsistency effects to measures that reflected processing that occurred after participants had initially encountered the inconsistent content (e.g., go-past duration, second-pass duration).

Although the present study demonstrated that readers do not typically regress very far in the text to reread information that may help in resolving inconsistencies, it may be possible to push them to do so by increasing their coherence threshold, within the assumptions of the RI-Val model [32,33]. Recent studies have demonstrated that subtle changes to the study procedure can result in large shifts in the reader's coherence threshold. For example, Williams and colleagues [54] argued that changing the number of comprehension questions asked at the end of each passage may shift the coherence threshold, such that readers will either wait more or less time for validation processes to complete before they move on in the text. When comprehension questions were increased, the coherence threshold was high, meaning that the validation process had more time to complete before readers moved on to subsequent text (see also [53]). When comprehension questions were decreased, the coherence threshold was low, and readers waited very little time for validation to complete before moving on to subsequent information. Within the present study context, it is possible that shifting the coherence threshold with similar manipulations would alter the extent to which readers experience difficulty validating the incorrect anaphor. By this logic, within the present study context, a higher coherence threshold would result in more efforts to validate the incorrect anaphor before readers move on to subsequent information, possibly leading to more regressions back to previous text, including the antecedent. Additionally, a lower coherence threshold may reduce the extent to which readers attempt to validate the anaphor before moving on in the text, possibly reducing difficulty due to incorrect anaphors altogether.

The distribution of processing effects over time is a growing area of interest in discourse comprehension research. This area of research is uniquely suited to paradigms and measures that allow for observation of the time course of processing effects—such as eye tracking. Even before the positing of theoretical models of discourse comprehension in which the timing of effects is critical (e.g., RI-Val), we have long argued for the importance of using measures that allow more than a single window into processing. This is now more important than ever. As tests of theoretical assumptions in discourse comprehension research hinge on which sources of information influence processing and when, it is essential that researchers utilize measures that provide a wider view of the time course of processing. Although several studies have accomplished this with careful development and presentation of stimuli in line-by-line self-paced reading paradigms, the use of eye tracking technology can complement that work by providing finer-grained analyses that allow researchers to isolate effects to critical words or phrases and to determine how they are processed over time. However, we want to end with a note of caution–researchers should be careful not to equate specific measures with specific processes. As Cook and Wei [11] argued, the considerable overlap among measures makes mapping specific measures onto specific cognitive processes a complex and unwise task. Instead, we recommend the approach long recommended by Rayner [8], in which researchers use a variety of convergent measures that cover a range of points on the temporal continuum of processing.

Author Contributions: Both authors contributed equally to the conception, design, and analysis of the information presented in this article.

Funding: This research received no external funding.

Acknowledgments: We would like to thank Ed O'Brien for previous conversations about the issues discussed in this article.

Conflicts of Interest: The authors declare no conflict of interest.

References

1. Dodge, R. Visual perception during eye movement. *Psychol. Rev.* **1900**, *7*, 454–465. [CrossRef]
2. Holt, E.B. Eye-movement and central anaesthesia. *Psychol. Rev.* **1903**, *4*, 1–45.
3. Tinker, M.A. Recent studies of eye movements in reading. *Psychol. Bull.* **1958**, *55*, 215–231. [CrossRef] [PubMed]
4. Tinker, M.A. *Legibility of Print*; Iowa State University Press: Ames, IA, USA, 1963.
5. Tinker, M. *Bases for Effective Reading*; University of Minnesota Press: Minneapolis, MN, USA, 1965.
6. McConkie, G.W.; Rayner, K. The span of the effective stimulus during a fixation in reading. *Percept. Psychophys.* **1975**, *17*, 578–586. [CrossRef]
7. Poulton, E.C. Peripheral vision, refractoriness and eye movements in fast oral reading. *Br. J. Psychol.* **1962**, *53*, 409–419. [CrossRef] [PubMed]
8. Rayner, K. Eye movements in reading and information processing: Twenty years of research. *Psychol. Bull.* **1998**, *124*, 372–422. [CrossRef] [PubMed]
9. Rayner, K. The Thirty-Fifth Sir Frederick Bartlett Lecture: Eye movements and attention during reading, scene perception, and visual search. *Quart. J. Exp. Psychol.* **2009**, *62*, 1457–1506. [CrossRef]
10. Rayner, K.; Pollatsek, A.; Ashby, J.; Clifton, C. *Psychology of Reading*, 2nd ed.; Psychology Press: New York, NY, USA, 2012.
11. Cook, A.; Wei, W. Using eye movements to study reading processes: Methodological considerations. In *Eye Tracking Technology Applications in Educational Research*; Was, C.A., Sansoit, F.J., Morris, B.J., Eds.; IGI Global: Hershey, PA, USA, 2017; pp. 27–47.
12. Rayner, K.; Binder, K.S.; Duffy, S.A. Contextual strength and the subordinate bias effect: Comment on Martin, Vu, Kellas, and Metcalf. *Quart. J. Exp. Psychol. Sec. A* **1999**, *52*, 841–852. [CrossRef]
13. Colbert-Getz, J.; Cook, A.E. Revisiting effects of contextual strength on the subordinate bias effect: Evidence from eye movements. *Mem. Cognit.* **2013**, *41*, 1172–1184. [CrossRef]
14. Rayner, K.; Pacht, J.M.; Duffy, S.A. Effects of prior encounter and global discourse bias on the processing of lexically ambiguous words: Evidence from eye movements. *J. Mem. Lang.* **1994**, *33*, 527–544. [CrossRef]
15. Albrecht, J.E.; O'Brien, E.J. Updating a mental model. *J. Exp. Psychol. Learn Mem. Cognit.* **1993**, *19*, 1061–1070. [CrossRef]
16. O'Brien, E.J. Antecedent search processes and the structure of text. *J. Exp. Psychol. Learn Mem. Cognit.* **1987**, *13*, 278–290. [CrossRef]
17. OBrien, E.J.; Albrecht, J.E. The role of context in accessing antecedents in text. *J. Exp. Psychol. Learn Mem. Cognit.* **1991**, *17*, 94–102. [CrossRef]
18. O'Brien, E.J.; Plewes, P.; Albrecht, J.E. Antecedent retrieval processes. *J. Exp. Psychol. Learn Mem. Cognit.* **1990**, *16*, 241–249. [CrossRef]
19. O'Brien, E.J.; Shank, D.M.; Myers, J.L.; Rayner, K. Elaborative inferences during reading: Do they occur on-line? *J. Exp. Psychol. Learn Mem. Cognit.* **1988**, *14*, 410–420. [CrossRef]
20. Garrod, S.; O'Brien, E.J.; Morris, R.K.; Rayner, K. Elaborative inferencing as an active or passive process. *J. Exp. Psychol. Learn Mem. Cognit.* **1990**, *16*, 250–257. [CrossRef]
21. Duffy, S.A.; Rayner, K. Eye movements and anaphor resolution. Effects of antecedent typicality and distance. *Lang. Speech* **1990**, *33*, 103–119. [CrossRef]
22. Ehrlich, S.F.; Rayner, K. Pronoun assignment and semantic integration during reading: Eye movements and immediacy of processing. *J. Verbal Learn Verbal Behav.* **1981**, *22*, 75–87. [CrossRef]
23. Cook, A.E.; O'Brien, E.J. Passive activation and instantiation of inferences during reading. In *Inferences during Reading*; O'Brien, E.J., Cook, A.E., Lorch, R.F., Eds.; Cambridge University Press: New York, NY, USA, 2015; pp. 19–41.
24. Sanford, A.J.; Graesser, A.C. Shallow processing and underspecification. *Discourse Process.* **2006**, *42*, 99–108. [CrossRef]
25. Ferreira, F.; Ferraro, V.; Bailey, K.G.D. Good-enough representations in language comprehension. *Curr. Dir. Psychol. Sci.* **2002**, *11*, 11–15. [CrossRef]
26. Cook, A.E. Processing anomalous anaphors. *Mem. Cognit.* **2014**, *42*, 1171–1185. [CrossRef] [PubMed]
27. Rayner, K.; Chace, K.H.; Slattery, T.J.; Ashby, J. Eye movements as reflections of comprehension processes in reading. *Sci. Stud. Read.* **2006**, *10*, 241–255. [CrossRef]

28. Hyönä, J.; Lorch, R.F.; Kaakinen, J.K. Individual differences in reading to summarize expository text: Evidence from eye fixation patterns. *J. Educ. Psychol.* **2002**, *94*, 44–55. [CrossRef]

29. Hyönä, J.; Lorch, R.F.; Rinck, M. Eye movement measures to study global text processing. In *The Mind's Eye*; Hyonä, J., Radach, R., Deubel, H., Eds.; Elsevier Science: Amsterdam, The Netherlands, 2003; pp. 313–334.

30. Hyönä, J.; Nurminen, A.M. Do adult readers know how they read? Evidence from eye movement patterns and verbal reports. *Br. J. Psychol.* **2006**, *97*, 31–50. [CrossRef] [PubMed]

31. Cook, A.E.; O'Brien, E.J. Knowledge activation, integration, and validation during narrative text comprehension. *Discourse Process.* **2014**, *51*, 26–49. [CrossRef]

32. O'Brien, E.J.; Cook, A.E. Coherence threshold and the continuity of processing: The RI-Val model of comprehension. *Discourse Process.* **2016**, *53*, 326–338. [CrossRef]

33. O'Brien, E.J.; Cook, A.E. Separating the activation, integration, and validation components of reading. In *The Psychology of Learning and Motivation*; Ross, B.H., Ed.; Academic Press: Cambridge, MA, USA, 2016; Volume 65, pp. 249–276.

34. Klin, C.M.; Guzmán, A.E.; Weingartner, K.M.; Ralano, A.S. When anaphor resolution fails: Partial encoding of anaphoric inferences. *J. Mem. Lang.* **2006**, *54*, 131–143. [CrossRef]

35. Klin, C.M.; Weingartner, K.M.; Guzmán, A.E.; Levine, W.H. Readers' sensitivity to linguistic cues in narratives: How salience influences anaphor resolution. *Mem. Cognit.* **2004**, *32*, 511–522. [CrossRef]

36. Levine, W.H.; Guzmán, A.E.; Klin, C.M. When anaphor resolution fails. *J. Mem. Lang.* **2000**, *43*, 594–617. [CrossRef]

37. Ferreira, F.; Patson, N.D. The 'good enough' approach to language comprehension. *Lang. Linguist. Compass* **2007**, *1*, 71–83. [CrossRef]

38. Cook, A.E.; Walsh, E.; Bills, M.A.A.; Kircher, J.C.; O'Brien, E.J. Validation of semantic illusions independent of anomaly detection: Evidence from eye movements. *Quart. J. Exp. Psychol.* **2018**, *7*, 113–121. [CrossRef] [PubMed]

39. Bohan, J.; Sanford, A. Semantic anomalies at the borderline of consciousness: An eye-tracking investigation. *Quart. J. Exp. Psychol.* **2008**, *61*, 232–239. [CrossRef] [PubMed]

40. Daneman, D.; Lennertz, T.; Hannon, B. Shallow semantic processing of text: Evidence from eye movements. *Lang. Cognit. Process.* **2007**, *22*, 83–105. [CrossRef]

41. Sanford, A.J.; Leuthold, H.; Bohan, J.; Sanford, A.J. Anomalies at the borderline of awareness: An ERP study. *J. Cognit. Neurosci.* **2011**, *23*, 514–523. [CrossRef] [PubMed]

42. Wiley, J.; Rayner, K. Effects of titles on the processing of text and lexically ambiguous words: Evidence from eye movements. *Mem. Cognit.* **2000**, *28*, 1011–1021. [CrossRef]

43. Camblin, C.C.; Gordon, P.C.; Swaab, T.Y. The interplay of discourse congruence and lexical association during sentence processing: Evidence from ERPs and eye tracking. *J. Mem. Lang.* **2007**, *56*, 103–128. [CrossRef] [PubMed]

44. Cook, A.E.; O'Brien, E.J. Fundamentals of inferencing during reading. *Lang. Linguist. Compass* **2017**. [CrossRef]

45. O'Brien, E.J.; Cook, A.E. Models of discourse comprehension. In *Handbook on Reading*; Pollatsek, A., Treiman, R., Eds.; Oxford University Press: New York, NY, USA, 2015; pp. 217–231.

46. Garrod, S.; Terras, M. The contribution of lexical and situational knowledge to resolving discourse roles: Bonding and resolution. *J. Mem. Lang.* **2000**, *42*, 526–544. [CrossRef]

47. Cook, A.E.; Myers, J.L. Processing discourse roles in scripted narratives: The influences of context and world knowledge. *J. Mem. Lang.* **2004**, *50*, 268–288. [CrossRef]

48. Myers, J.L.; O'Brien, E.J. Accessing the discourse representation during reading. *Discourse Process.* **1998**, *26*, 131–157. [CrossRef]

49. O'Brien, E.J.; Myers, J.L. Text comprehension: A view from the bottom up. In *Narrative Comprehension, Causality, and Coherence: Essays in Honor of Tom Trabasso*; Goldman, S.R., Graesser, A.C., van den Broek, P., Eds.; Lawrence Erlbaum Associates: Mahwah, NJ, USA, 1999; pp. 35–53.

50. Kamas, E.N.; Reder, L.M. The role of familiarity in cognitive processing. In *Sources of Coherence in Reading*; Lorch, R.F., O'Brien, E.J., Eds.; Erlbaum: Hillsdale, NJ, USA, 1995; pp. 177–202.

51. Kamas, E.N.; Reder, L.M.; Ayers, M.S. Partial matching in the Moses illusion: Response bias not sensitivity. *Mem. Cognit.* **1996**, *24*, 687–699. [CrossRef] [PubMed]

52. Reder, L.M.; Kusbit, G.W. Locus of the Moses illusion: Imperfect encoding, retrieval, or match? *J. Mem. Lang.* **1991**, *30*, 385–406. [CrossRef]
53. Creer, S.D.; Cook, A.E.; O'Brien, E.J. Taking the perspective of the narrator. *Quart. J. Exp. Psychol.* **2018**, *72*, 1055–1067. [CrossRef] [PubMed]
54. Williams, C.; Cook, A.E.; O'Brien, E.J. Validating semantic illusions during narrative comprehension. *J. Exp. Psychol. Learn Mem. Cognit.* **2018**, *44*, 1414–1429. [CrossRef] [PubMed]

Review

Regressions during Reading

Albrecht W. Inhoff [1,*], Andrew Kim [1] and Ralph Radach [2]

1   Department of Psychology, Binghamton University, Binghamton, NY 13902, USA
2   Department of Psychology, Bergische Universitaet, 42103 Wuppertal, Germany
*   Correspondence: inhoff@binghamton.edu; Tel.: +1-607-777-3958

Received: 28 February 2019; Accepted: 24 June 2019; Published: 9 July 2019

Abstract: Readers occasionally move their eyes to prior text. We distinguish two types of these movements (regressions). One type consists of relatively large regressions that seek to re-process prior text and to revise represented linguistic content to improve comprehension. The other consists of relatively small regressions that seek to correct inaccurate or premature oculomotor programming to improve visual word recognition. Large regressions are guided by spatial and linguistic knowledge, while small regressions appear to be exclusively guided by knowledge of spatial location. There are substantial individual differences in the use of regressions, and college-level readers often do not regress even when this would improve sentence comprehension.

Keywords: reading; eye movements; regressions; individual differences

1. Introduction

Visual text consists of symbols that are spatially ordered along horizontal rows or vertical columns. Typically, a large number of symbols is visible concurrently, and they are visible for an extended period until a screen is changed or a page is turned. Speech, by contrast, consists of a temporally ordered sequence of acoustic symbols, and only a very limited amount of linguistic information is available at each point in time. The extraction of linguistic information during reading thus requires modality-specific skills. With reading, these skills include the programming of eye movements that position the eyes at—or near—individual words, as high acuity vision is confined to a relatively small retinal area: The fovea and adjoining parafovea. The spatial targeting of eye movement programming needs to be coordinated with linguistic processing, so that high acuity vision is moved to words when their identification becomes relevant for text comprehension. Most eye movements (saccades) progress with word order, from left-to-right for Roman and modern Chinese script, right-to-left for Hebrew and Arabic script, and also from top to bottom with traditional Chinese script.

A distinct subset of saccades, 5–20%, however, moves the eyes in a direction that is opposite to word order [1–4]. Kolers [5] noted that these reversals of saccade direction do not interfere with reading comprehension, and Rayner [4] suggested that they are responses to reading difficulties. The current review extends prior overviews in several aspects: We argue that there are two distinct types of regressions, that they serve distinct functions, and that their targeting is controlled by somewhat different types of representations. We also review individual differences in the use of regressions and consider potential implications for the teaching of reading.

2. The Spatial Targeting of Regressions

Corpus analyses show that regressions differ substantially in size [2,3]. They can reposition the eyes closer to the beginning of the fixated word, move the eyes to an adjacent prior word, or move across several prior words. When multi-line text is read, regressions can even cross one or more lines. The implications of these substantial differences in the size of regressions have remained unexplored.

2.1. Large Regressions

The literature has primarily focused on "large" regressions that traverse across more than one prior word [4]. Though words that are the target of a large regression are viewed out of order, Kolers [5] suggested that their viewing does not interfere with text comprehension because grammatical word order is determined by words' spatial location. Kennedy [6] (see also [7,8]) elaborated on this view. In his conception, identified words are represented in conjunction with a spatial tag that indexes their location on a line, or their relation to a spatial reference frame. This spatial indexing is assumed to be part and parcel of visual word recognition and to occur automatically. When processing difficulties can be linked to a prior word or a prior text segment, the indexes of corresponding text are retrieved and used for regression targeting.

In this scheme, spatial indexes assume two useful functions: They guide a regression to a previously read text segment, the hypothesized source of a processing difficulty, and they correct the ensuing mismatch between the temporal order with which words are viewed and grammatical word order. If, for instance, a reader executes a regression from word location seven to word location three on a line of text, then the spatial index of the regressed-to word informs the reader that word three will be re-inspected.

Kennedy's experimental work [8,9] appeared to provide compelling support for this hypothesis. Since the occurrence and targeting of large regressions cannot be controlled under normal reading conditions, a probe classification task was devised that resulted in the likely execution of a regression with controlled starting and ending points. Specifically, sentences were constructed that contained a target word at a specific location. Participants were asked to read the sentence, and then to view a probe word that was shown to the right of the sentence. The task was to determine whether the probe had appeared in the sentence (see sample sentences 1a and 1b). In the original study [9], the probe was either identical to the target word in the sentence (1a) or a semantically related word (1b).

1a. The man was looking for a spade in the shed next to the barn. Spade
1b. The man was looking for a spade in the shed next to the barn. Shovel

Under these conditions, regressions toward the target occurred on approximately a quarter of trials. These regressions were remarkably accurate, as they positioned the eyes consistently near the center of the target word, *spade* in the example, irrespective of probe-target distance. Consistent with the spatial coding hypothesis, readers were assumed to use the probe to find the related spatial index in the sentence, and to use it for the targeting of the regression.

Shortcomings of the Spatial Coding Hypothesis

By design, the spatial location of words on a line of text is confounded with their grammatical order. When this relationship is removed, by presenting the words of a sentence at arbitrary screen locations, knowledge of a word's spatial location is poor and short-lived [10]. This finding is difficult to reconcile with the spatial coding hypothesis. The hypothesis was also challenged by additional experiments that examined the accuracy of regressions whose start and end points are experimentally controlled. Although the size of these regressions increased with the distance of the regression target, the reaching of these targets was error prone, in particular, when the target was far. In this case, regressions often missed a designated word location by several words [11–13].

In Weger and Inhoff [11], the regression error for text that extended across two lines increased not only with spatial (horizontal and vertical) distance but also with the number of words that intervened between the starting and end points of a regression. Furthermore, readers executed additional, search-like saccades that moved the eyes onto the target, when the initial regression to the designated word location was inaccurate. Linguistic information thus influenced the accuracy of regression targeting. Related work by Guerard and collaborators showed larger regression errors, in particular for far targets, when the reading task was accompanied by an articulatory suppression task that involved the repeated articulation of the consonant sequence "ABCD" at a pace of two second per letter [14,15]. These results suggest that phonological working memory may be involved in the specification of far locations, as articulatory suppression was likely to interfere with the representation of phonological forms in working memory.

Rather than retrieving the spatial index of a designated regression target, readers may use the working memory representation of previously recognized words (and linguistic knowledge) to estimate how far back a prior word had been read, and this could be used to be used to estimate a target's spatial coordinates [15]. The difficulty of this estimate may increase with the number of words that intervene between the fixated word and with the target's spatial distance.

2.2. Small Regressions

The vast majority of regressions reposition the eyes within a word or moves the eyes to the spatially adjacent prior word. Several findings indicate that the programming of these small regressions differs from the programming of large regressions. They often follow an oculomotor targeting error [2,3], and regressions up to 10 letter spaces move the eyes with a remarkable degree of accuracy to the center of the preceding word [16], that is, they differ from similar size forward-directed saccades in that they are not subject to systematic targeting errors. Presumably, this occurs because the spatial properties of the regression target are still represented in working memory [17]. The results of a recently completed study [18] are consistent with this view. Readers with upper quartile spatial working memory scores executed more accurate regressions than readers with lower quartile scores when the regression target was near but not when it was far.

3. The Function of Regressions

Text remains generally available after it has been read, and it can thus be used like an external storage device. If a processing impasse can be related to a previously read text segment, a reader can re-view it with a flick of the eyes and re-read the segment. Re-processing may reveal whether and why represented linguistic content differed from the actual text. This can be used to validate or update the representation of corresponding linguistic content. In the case of small regressions, two reasons for making movements against the direction of reading can be specified. First, there are cases when the eyes land on a word that was not the intended recipient of a saccade, especially when a word was accidentally "skipped". Second, there can be a need to return to a preceding word when it was not fully processed during fixation [19]. These possibilities will be explored later, but for now the focus is on long range regression in the service of comprehension.

3.1. Regressions for Text Comprehension

The "re-viewing for reprocessing" hypothesis [5,20] is the prevailing account for the execution of large regressions. It maintains that a previously read word or sentence segment become the target of a regression when their represented meaning (or grammatical role) disagrees with subsequently read linguistic content. For instance, the heterophonic homograph "bass" in the sentence "The fisherman was looking for the bass" should become the target of a regression when the representation of "bass" as a type of fish disagrees with the remainder of the sentence, "that the guitarist had dropped in the lake" [21], which implies that "bass" refers to a type of musical instrument. Similarly, syntactic parsing may go wrong in sentences with garden path constructions. In the sentence, "Because Ed drinks vodka is everywhere in the house", comprehension is likely to be break down when the words "is everywhere" are identified. This occurs because "vodka" is preferentially parsed as the object of "Ed drinks" rather than as the subject of "is everywhere". Readers may regress to reprocess prior text in order to correct the parsing error.

One alternative to the reprocessing hypothesis is that large regressions are not aimed at a particular text location [22]. Instead, they benefit processing primarily because they provide additional processing time. Consistent with this view, detailed analyses of the starting and end locations of regressions in response to syntactic garden pathing showed that regressions did not consistently move the eyes to the source of the processing difficulty (the misparsed sentence segment). In other experiments, readers occasionally regressed toward a previously read sentence segment even when it was no longer visible on a screen for reviewing and reprocessing [8,23].

These eye movements to blank screen locations are also consistent with another alternative, according to which the primary function of regressions is to enhance retrieval from working memory. Laeng and Teodorescu [24] showed that the sequence of saccades during the generation of a mental image of a previously viewed visual pattern was similar to the sequence of saccades when the visual pattern was originally viewed, as if moving the eyes to previously viewed locations supported content-specific imagery. Moreover, memory for a previously viewed visual pattern was more accurate when participants could move their eyes during the memory task than when their eyes had to remain fixated. Presumably, this occurred because saccades to the location of previously depicted objects supported the retrieval of objects that were viewed at those locations.

To distinguish between these accounts, Booth and Weger [25] constructed sentences with endings that required knowledge of a critical target word that had occurred earlier in the sentence, the assumption being that readers would regress to the target to validate it or revise its representation. In Experiment 3, the target was either unchanged or it was replaced with another word when the eyes regressed to the target location (e.g., the word "driver" was replaced with the word "dancer"). Sentence reading was followed by a multiple choice task with three alternatives, one that referred to the meaning of the originally presented target word, another that referred to the regression-contingent substitute, and yet another to a word that did not appear in the sentence (the target and replacement word were not presented in the multiple choice task). In the absence of regressions, or when regressions did not land on the target word location, the target-related meaning was selected on 70% of the trials. Conversely, a meaning related to the substituted word was selected on 68% of the trials when a regression to the target location had resulted in the "re-viewing" of the replacement word at that location. This indicates that regressions were used to re-process the word. and that the meaning of the original word was replaced with the meaning of the re-processed substitute.

Similar results emerged from the key condition of a follow-up study [26] in which a target word, e.g., "house" was inconsistent with the meaning of a subsequent verb, e.g., "ridden". A regression to the target either left it unchanged or replaced it with a congruent word, "horse" in the example. Substantially less re-viewing time was spent on the target when it was changed to a congruent word, presumably because the replaced target offered a better contextual fit. Consistent with the reprocessing hypothesis, these findings demonstrate that readers re-processed the word that was reached with a regression, and that the outcome of re-processing was used to update represented sentence content.

Other studies also show that large regressions can improve comprehension. Using eye-movement-contingent display changes to manipulate the visibility of text to the left of a fixation, so that it was either masked or re-readable after a regression, Schotter et al. [27] showed that sentences with garden path constructions were understood more successfully when readers re-read text after a regression. Readers also appear to use regressions to correct word identification errors [28], and to fill in "missed" sentence and story parts that were viewed during episodes of mind wandering and attentional lapses [29].

3.1.1. The Frequency of Regression Usage

Processing difficulties increase the frequency of regressions. When readers read sentences with contextually incongruent and congruent words [26], regressions out of the end of the sentence occurred on 60% of the trials in which a previously read target word was incongruent with a subsequently read verb, and on 20% of the trials in which the two words were congruent. Surprisingly, the frequencies of outgoing regressions differed little when the mismatching verb itself, "ridden" in the prior example, was fixated, which had an outgoing regression rate of 25%, when it was incongruent with the target [house -> ridden], and 18%, when it was congruent [horse -> ridden].

Using a similar approach, we examined the rate of regressions that moved the eyes onto the source of a presumed processing difficulty [28]. The materials were constructed so that controlled target words would be the source of a processing difficulty in an incongruent condition but not in a congruent condition. For instance, in the sentence "The midwife thought that the birch was successful despite the mother's concerns", the target "birch" is incongruent with the subsequent sentence context. A visually similar word, "birth", was the target for the same sentence frame in the congruent condition (with a different sentence frame "birch" was congruent and "birth" incongruent). With this approach, regressions to incongruent targets were approximately twice as common as regressions to congruent targets. However, regressions to incongruent targets were not routinely executed. Across experiments, the highest rate of regressions onto incongruent targets was 32%.

Similarly, a detailed examination of regression usage during the reading of sentences with garden-path constructions [22] (see also [30]) showed that parsing difficulties were not routinely responded to with a regression. In the study, readers regressed out of the breakdown region of garden path sentences on approximately 25% of trials, when the region was on the same line as the misparsed phrase, and on approximately 17% of trials, when the breakdown region and the misparsed phrase were on different lines. Moreover, regressions out of the breakdown region were not preferentially directed at the misparsed phrase, and, as noted earlier, regressions out of verbs ("ridden") were approximately as common when they were congruent with a prior target as when they were incongruent.

Why were comprehension difficulties not routinely and/or not immediately responded to with a regression? According to Folk and Morris [31] (see also [32]) this may occur because readers prefer to use the represented forms of previously read words (in phonological working memory) for the resolution of processing difficulties.

In these experiments, prior context did not constrain the meaning of lexically ambiguous target words with biased meanings. Subsequent text was, however, constructed so that it implied the subordinate meaning of a target, that is, it disagreed with the likely selection of the preferred target meaning. In two pertinent conditions, targets were either homographic homophones (e.g., "bank") or homographic heterophones ("tear"). Homophones and their controls differed in that readers spent more time viewing post-homophone context. With heterophones, by contrast, readers increased the rate of regressions from post-target context to targets (see also [21]). According to Folk and Morris [31], homophone representations could be corrected with a retrieval of working memory (WM) forms, as represented phonological target forms could be used to retrieve a context-consistent subordinate meaning. Regressions were executed when the retrieval of phonological forms did not provide access to an alternative meaning for the resolution of the integration difficulty. To obtain a different phonological form, homographic heterophones had to be reviewed with a regression.

3.1.2. The Timeline of Regression Programming

The timeline of eye movement programming appears to be similar when large regressions are executed for the re-viewing and reprocessing of text and when forward-directed saccades are executed for the identification of upcoming words. A comparison of regressive and progressive saccades out of the 'breakdown region' of sentences with garden path constructions [30] showed that the mean fixation duration prior to the execution of a saccade was approximately 235 ms, irrespective of whether the outgoing saccade regressed or progressed in the sentence when several words were visible to the right and left of the fixated breakdown region. Regressions and progressions differed, however, in size, with a mean of 18 character spaces for regressions and of eight character spaces for progressions.

The programming of large regressions and of forward-directed saccades could have similar timelines because their programming occurs in response to information that is extracted from the fixated segment of text. If the extracted information is incongruent with represented content (and when use of working memory did not resolve the processing difficulty), readers may decide to regress to reprocess prior text. If the extracted information is congruent, readers may move their eyes forward to identify upcoming words.

3.2. Small Regressions

Corpus analyses show that regressions are, on average, smaller than forward-directed saccades [2,11,33], extending typically across one to three letter spaces. The viewing duration preceding these regressions is typically short [34], as if these regressions corrected a poor viewing location [34,35] that was not well suited for word recognition [36].

Small regressions often move the eyes from the ending of a word toward the center or the beginning to improve its recognition. Other small regressions move the eyes from a fixated word to the immediately preceding word, and these inter-word regressions are particularly common when the prior word has not been fixated (i.e., when it was skipped) [32]. Similar to regressions within a word, post-skip regressions could be used to obtain a better viewing location for the identification of skipped word.

Although word skipping can occur when an upcoming word is recognized before a saccade to it is committed to execution [33,37], corpus analyses not only indicate that readers generally direct the eyes at the center of an upcoming word [34,37,38], but also that this location is often missed. These presumed targeting errors are attributed to a range error that biases the size of executed saccades toward a default (mean) saccade size, and an additional random error. These findings imply that some word skipping is due to oculomotor targeting errors, and computational simulations (e.g., [39]) are consistent with that. An erroneous skipping of a word is likely to impair its recognition, and this could be corrected with a small regression to the skipped word.

Readers may also execute small regressions to correct oculomotor timing errors. That is, the eyes may leave a fixated word before its processing has been completed. Akin to a spatial targeting error, premature saccades will position the eyes off a to-be-processed word, and this divergence can be corrected with a small regression [35,40].

3.2.1. Frequency of Small Corrective Regressions

It appears plausible to assume that oculomotor targeting errors would be corrected routinely when they impair word recognition, and this appears to be the default assumption in current conceptions of eye movement control during reading [35,41]. However, corpus analyses cannot examine the link between word skipping, recognition failures, and corrective regressions, as the success of visual word recognition is not self-evident. Even when the recognition of a skipped word failed, a reader could move the eyes forward after skipping and perhaps use upcoming text to infer the identity of the skipped word. To illuminate the link between skipping, word recognition failure, and corrective regressions, we examined the occurrence of regressions after the skipping of words that could not be recognized prior to skipping.

In the experiment (Experiment 1) [42], the display of three-letter target words (e.g., "tax") was manipulated so that they were either visible (pre-viewable) throughout sentence reading or masked with a length-matched string of visually dissimilar random letters ("gfj"). The mask occupied the target location until the eyes reached a display-change boundary, the blank space preceding the target location. A saccade to the right of the boundary replaced the random letter mask with the target word, and the intact target was always shown when the eyes landed on the target location or to the right of it. Since targets were relatively short words, they were liable to skipping, either because pre-viewable targets were recognized before a target-reaching saccade was committed to execution, or due to erroneous oculomotor overshoot (irrespective of whether the target had been previewed or masked).

Examination of boundary-crossing saccades showed that the skipping of the target area was relatively common (31%; 687 of 2167 trials). This rate is similar to the skipping rate of three-letter words in other studies (e.g., [43]). Slightly more than half of the target skips (58%; 395 of 687) occurred in the target visible condition; on the remaining skipping trials, a random letter mask, such as "gfj", was skipped, presumably due to oculomotor targeting error. When this occurred, the erroneous fixation of the post-target word should have impeded target word recognition, and regressions from post-target words to target words should have occurred routinely.

The relative frequency of saccades from the post-target word to the target after the skipping of previously masked and visible target is shown as a function of the direction of the outgoing saccade in Figure 1. It is evident that the skipping of a masked target was not routinely corrected with a regression. To the contrary, forward-directed saccades out of the post-target word were approximately three times as common as regressions back to the target after masked target skipping. Moreover, regressions to skipped masked targets were as common as regressions to skipped visible targets. These findings imply that erroneous oculomotor targeting is not routinely corrected with a regression.

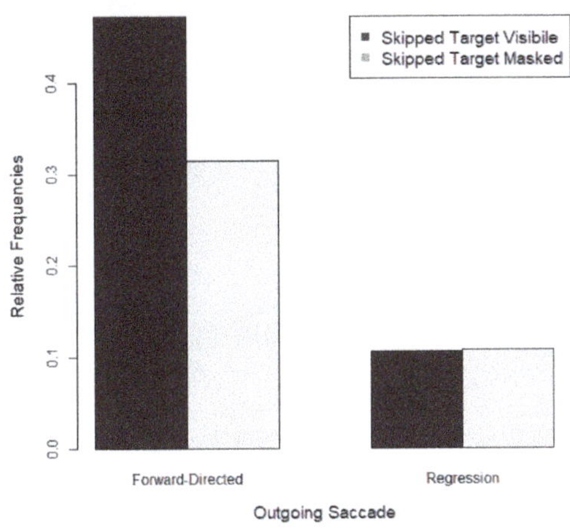

Figure 1. The relative frequencies of saccades out of post-target words after the preceding three-letter target area had been skipped. Relative frequencies are shown as a function of the visibility of the target word prior to the skipping of the target area and as a function of the direction of the outgoing saccade (relative frequencies add up to 1).

An analysis of the viewing time (gaze) for post-target words after target skipping (Figure 2) showed that the skipping of a target resulted in relatively long post-target viewing durations (approximately half a second) when readers moved out of the post-target word with a forward-directed saccade, and this occurred, in particular, when a masked target had been skipped. The relatively long post-target viewing duration when readers exited with a forward-directed saccade suggests that readers sought to identify the skipped target while the post-target word was fixated. This was particularly difficult when the skipped target word had been masked.

Together, these analyses indicate that oculomotor targeting errors are relatively common, and that they increase the difficulty of word recognition. Critically, they also show that the increase in word recognition difficulty is not routinely responded to with a regression.

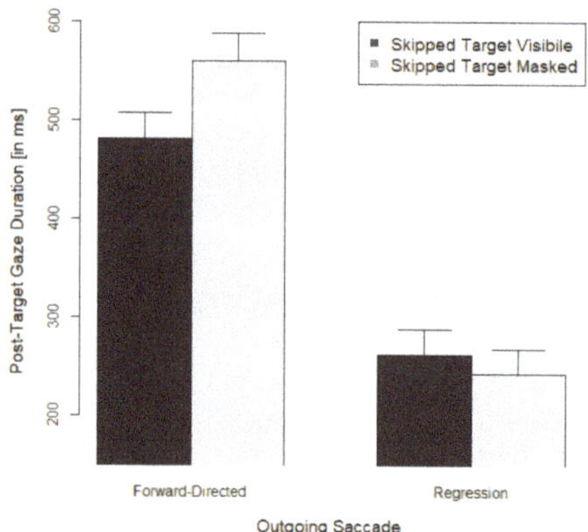

Figure 2. Mean gaze durations and their standard errors for post-target words when the preceding target word had been skipped. Means are shown as a function of the visibility of the target word prior to skipping and of the direction of the saccade out of the fixated post-target word.

3.2.2. The Time Course for the Programming of Corrective Regressions

Figure 2 also shows relatively short post-target viewing durations (less than a quarter of a second) when a regression to the skipped target was executed. The large size of the saccade direction effect suggests that the extraction of post-target information was short-circuited prior to the execution of a corrective regression to a skipped target. To determine whether useful information was extracted from a post-target word prior the execution of a regression to a skipped target, we analyzed post-target viewing durations as a function of a skipped target's visibility prior to boundary crossing and of the information that could be extracted at the onset of post-target viewing (Experiment 2) [42]. As in Experiment 1, the visibility of a target prior to boundary crossing was manipulated so that the three-letter target was either visible or masked before the eyes crossed the boundary. Again, the intact target was shown when the eyes landed on it or skipped it. In addition, Experiment 2 manipulated the visibility of the post-target word after target skipping. The post-target word was now either visible (in upper case) at the onset of the post-skip fixation or its presentation was delayed by 50 ms (a string of random letters occupied the post-target position for 50 ms in this case). In both viewing conditions, the post-target word was visible in lower case 50 ms after the onset of its fixation. If readers sought to obtain linguistic information from the fixated post-target word prior to regression programming, then fixation durations preceding the regression should be longer when a post-target word's onset was delayed.

Examination of eye movements showed that target skipping was once more relatively common (951 skips on 2657 trials). Again, only slightly more than half of the target skipping ($n = 518$) occurred when the target had been visible prior to skipping; the remaining skipping occurred when the target was masked. After skipping, there were 248 regressions where the eyes moved back to the target. Figure 3 shows the post-target viewing durations for these cases, that is, when target skipping was followed by a regression back to the target. These post-target viewing durations were examined as a function of the visibility of the target prior to skipping (visible vs. masked) and also as a function of post-target onset, that is, whether the fixated post-target word was visible immediately or whether the onset of useful information was briefly delayed.

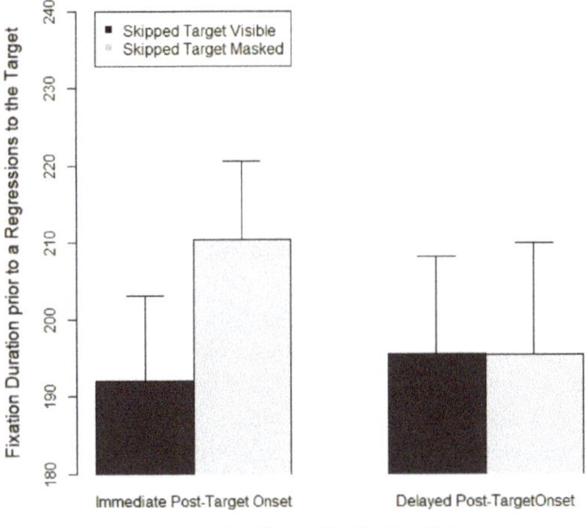

Figure 3. Fixation durations on post-target words prior to a regression to a skipped target word. Fixation durations and standard errors are shown as a function of the visibility of the skipped target and the timeline of post-target onset.

A linear mixed model yielded a relatively small estimated effect of post-target onset on pre-regression fixation duration, less than 4 ms (b = 3.41, SE = 14.20, t = 0.24, p > 0.7) and no effect to target visibility (p > 0.5). Although Figure 3 appears to suggest a moderating influence of the visibility of the skipped target in the immediate post-target onset condition, the corresponding interaction did not approach significance (p = 0.36). The main finding, a null effect for post-target onset effect, suggests that readers did not seek to extract linguistic information from the post-target word prior to the programming of a regression to a skipped target. Since the data set was small and null effects are difficult to interpret, this account needs to be considered tentative.

4. Individual Differences

A recent examination of basic oculomotor indexes (fixation duration and saccade size) in four perceptual tasks revealed stable individual differences across tasks and over time [44]. Since eye movements are part and parcel of fluent reading, individuals may also develop characteristic regression strategies.

Cluster analyses applied to indexes of eye movements during the reading of structurally different story segments [45] are consistent with this view. Clustering of readers yielded three approximately equal-size groups of fluent readers with distinct eye movement preferences. The hallmark of one group was the frequent use of regressions. These readers were assumed to resolve comprehension difficulties through the re-reading of structurally important parts of the story. Members of the two other groups preferred to move the eyes forward in the text, and they may have attempted to resolve difficulties through the reading of upcoming text.

Similarly, path analyses, that examine the directionality of associated effects [46], yielded a relatively weak link between oculomotor variables that control the encoding of text (skipping rates, the size of forward directed saccades, and fixation duration) and successful reading comprehension. Reprocessing (use of regressions and re-reading time), by contrast, was strongly associated with successful comprehension. In the most effective model, encoding and reprocessing were also strongly linked, suggesting that regressions are an oculomotor tool that can be used to achieve a more accurate representation of linguistic content when the initial encoding of text was poor (i.e., when skipping was common and when viewing durations were short). Cluster analysis (with a three group solution) yielded one group with low comprehension scores. This group consisted of slow readers who regressed rarely. The two other groups achieved a relatively high reading comprehension score: One group consisted of fast readers who regressed rarely, and the other of slower readers who regressed relatively often. With slower readers, it was thus the use of regressions that distinguished readers with good and poor text comprehension.

These analyses suggest that some readers learned to use regressions to improve their text comprehension. Other readers may "under-use" regressions because they have not learned to regress or because the benefit of rereading is not transparent. To gain some insight into the "under-use" of regressions, we examined whether readers' would regress more often after the benefits of re-reading have become transparent.

For this, we re-analyzed the results of two experiments [28] in which the identity of a target word was manipulated in one of the experimental conditions so that the execution of a regression to it resolved a processing difficulty when a target was incongruent with prior context. As noted earlier, targets were followed by different sentence context: One in which the target and its subsequent context were congruent, one in which they were incongruent, and one in which their relationship was "neutral". For instance, the target word birch in "The midwife thought that the birch … " was followed by "was the most beautiful tree in the yard" (congruent), or "was successful despite the mother's concerns" (incongruent), or "was a strange sight first thing in the morning" (neutral). By design, all targets had an orthographic neighbor, "birth" in the example, that agreed with incongruent context. In the incongruent condition, regression-contingent display changes were implemented that replaced the target with a much better fitting neighbor when readers regressed to the target location. The execution of a regression to an incongruent target was thus rewarded with the viewing of a word, a congruent orthographic neighbor, that resolved the processing difficulty.

Regressions to targets during the first half of the trials were compared with regressions to targets during the second half, the assumption being that the usefulness of regressions in the incongruent condition would increase their occurrence over the course of the experiment. No change in regression frequency, or a lesser change, was expected for the congruent and the neutral conditions. Figure 4 shows regression rates to targets as a function of trial sequence (beginning vs. ending half) and sentence context (74 participants who participated in Experiments 1 and 2 [28]). As can be seen, regression rates did not increase over the course of the experiment in the incongruent condition, and the interaction of trial sequence with context was negligible ($p > 0.3$).

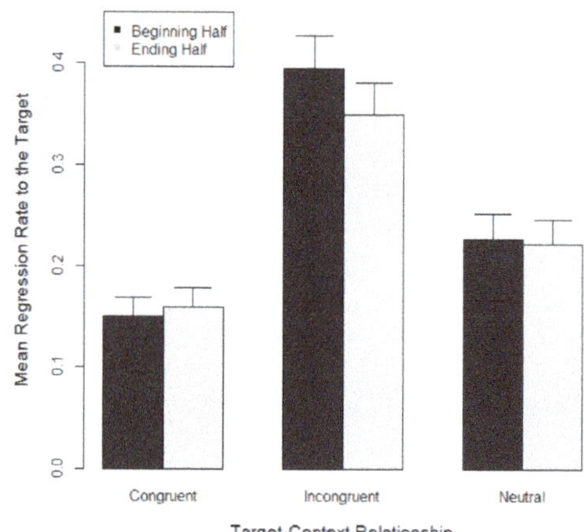

Figure 4. Mean regression rate to target words and standard errors as a function of targets' consistency with subsequent sentence content. Incongruent target words were replaced with congruent targets when readers regressed back to the target.

These results indicate that readers under-use regressions even when the benefits of rereading are relatively transparent.

5. Regressions and the Teaching of Reading

Why did readers' use of regressions not increase over the course of the experiments in the incongruent condition? Some readers may have under-used large regressions because they were content with a somewhat odd—but "good-enough"—representation of sentence meaning [28]. Other readers may not have learned to use regressions strategically. In a large scale ($n = 632$) study of children's eye movements during reading [47], increases in reading skill from grades one to five were associated with the execution of fewer and larger saccades and the use of shorter fixation durations. However, the proportion of inter-word regressions increased only after grade four. From then on, more skillful readers "may go back to words that might have caused processing difficulty" [47] (p. 477). Even older readers may not acquire this strategic skill without explicit instruction.

Many readers may also prefer to resolve difficulties through a reprocessing of working memory content [31]. For instance, they could "normalize" the representation of sentence meaning by assuming that a represented word form was misperceived, or, as mentioned before, they could be content with a somewhat odd sentence meaning [48–50]. Current reading instruction emphasizes forward-directed reading, as this is necessary for successful text comprehension. Regressions could even be discouraged, as they interrupt the speech-like forward-directed flow of information uptake. This is particularly likely in the case of reading aloud, which is the predominant form of reading in lower grades, as regressions for meaning will likely compromise eye-voice-coordination [48]. Oculomotor studies of reading development indicate that there is a transition from a more sequential and sub-lexical beginning stage towards lexical reading, which appears common at about grade four [44]. More generally, large-scale psychometric analyses of individual differences in reading development indicate a systematic increase in the variance accounted for by higher order comprehension at the expense of decoding [49]. As this development progresses, it may be beneficial to systematically teach the use of re-inspection strategies when comprehension is suboptimal. Indeed, successful developing readers routinely engage in the monitoring of comprehension and use regressions to resolve inconsistencies [50]. As we have argued above, this skill may not develop automatically and should therefore become a more explicit focus of reading instruction.

6. Conclusions

Reading differs from spoken language processing in that linguistic symbols have spatial properties that are relatively stable. The forms of depicted words thus remain accessible after they have been identified, and some readers discover that regressions to prior text can help resolve word recognition and text comprehension difficulties during silent reading. Somewhat surprisingly, supplementary analyses of pertinent data suggest that even skilled readers do not take full advantage of this opportunity.

Funding: This research received no external funding.

Conflicts of Interest: The authors declare no conflict of interest.

References

1. Henderson, J.M.; Pollatsek, A.; Rayner, K. Covert visual attention and extrafoveal information use during object identification. *Percept. Psychophys.* **1989**, *45*, 196–208. [CrossRef] [PubMed]
2. Vitu, F.; McConkie, G.W. *Reading as a Perceptual Process*; Kennedy, A., Radach, R., Heller, D., Pynte, J., Eds.; North-Holland/Elsevier Science Publishers: Amsterdam, The Netherlands, 2000; pp. 301–326, ISBN 978-0-08-043642-5.

3. Vitu, F. About the global effect and the critical role of retinal eccentricity: Implications for eye movements in reading. *J. Eye Mov. Res.* **2008**, *2*, 1–18.

4. Rayner, K. Eye movements and attention in reading, scene perception, and visual search. *Q. J. Exp. Psychol.* **2009**, *62*, 1457–1506. [CrossRef] [PubMed]

5. Kolers, P.A. Introduction. In *The Psychology and Pedagogy of Reading*; Huey, E.B., Ed.; MIT Press: Boston, MA, USA, 1968; pp. XIII–XXXIX, ISBN 9780262080293.

6. Rayner, K.; Kennedy, A. Eye movements and visual cognition: Scene Perception and reading. In *Eye Movements and Visual Cognition: Scene Perception and Reading*; Springer-Verlag: New York, NY, USA, 1992; pp. 379–396, ISBN 978-1-4612-2852-3.

7. Kennedy, A.; Baccino, T. The Effects of Screen Refresh Rate on Editing Operations Using a Computer Mouse Pointing Device. *Q. J. Exp. Psychol. Sect. A* **1995**, *48*, 55–71. [CrossRef]

8. Radach, R.; Kennedy, A.; Rayner, K. Eye Movements and Information Processing during Reading: Preface. *Eur. J. Cogn. Psychol.* **2004**, *16*, 1–2.

9. Holmes, V.M.; Kennedy, A.; Murray, W.S. Syntactic Structure and the Garden Path. *Q. J. Exp. Psychol. Sect. A* **1987**, *39*, 277–293. [CrossRef]

10. Fischer, M.H. Memory for Word Locations in Reading. *Memory* **1999**, *7*, 79–116. [CrossRef] [PubMed]

11. Weger, U.W.; Inhoff, A.W.; Weger, U.W. Long-range regressions to previously read words are guided by spatial and verbal memory. *Mem. Cogn.* **2007**, *35*, 1293–1306. [CrossRef]

12. Weger, U.W.; Meier, B.P.; Robinson, M.D.; Inhoff, A.W.; Weger, U.W. Things are sounding up: Affective influences on auditory tone perception. *Psychon. Bull. Rev.* **2007**, *14*, 517–521. [CrossRef]

13. Inhoff, A.W.; Greenberg, S.N.; Solomon, M.; Wang, C.-A. Word integration and regression programming during reading: A test of the E-Z reader 10 model. *J. Exp. Psychol. Hum. Percept. Perform.* **2009**, *35*, 1571–1584. [CrossRef]

14. Guérard, K.; Saint-Aubin, J.; Maltais, M. The role of verbal memory in regressions during reading. *Mem. Cognit.* **2013**, *41*, 122–136. [CrossRef] [PubMed]

15. Guérard, K.; Saint-Aubin, J.; Maltais, M.; Lavoie, H. The role of verbal memory in regressions during reading is modulated by the target word's recency in memory. *Mem. Cognit.* **2014**, *42*, 1155–1170. [CrossRef] [PubMed]

16. Radach, R.; McConkie, G.W. Determinants of Fixation Positions in Words During Reading. *Eye Guid. Read. Scene Percept.* **1998**, 77–100. [CrossRef]

17. Inhoff, A.W.; Weger, U.W.; Radach, R. Sources of Information for the Programming of Short-and Long-Range Regressions during Reading. In *Cognitive Processes in Eye Guidance*; Underwood, G., Ed.; Oxford University Press: Oxford, UK, 2005; pp. 33–52.

18. Friede, A.; Inhoff, A.W.; Vorstius, C.; Radach, R. Visuomotor strategies and the role of spatial memory for regressive saccades in reading. Under reviews.

19. Binder, K.S.; Pollatsek, A.; Rayner, K. Extraction of information to the left of the fixated word in reading. *J. Exp. Psychol. Hum. Percept. Perform.* **1999**, *25*, 1162–1172. [CrossRef] [PubMed]

20. Rayner, K. Eye Movements in Reading and Information Processing: 20 Years of Research. *Psychol. Bull.* **1998**, *124*, 372–422. [CrossRef] [PubMed]

21. Carpenter, P.A.; Daneman, M. Lexical retrieval and error recovery in reading: A model based on eye fixations. *J. Verbal Learn. Verbal Behav.* **1981**, *20*, 137–160. [CrossRef]

22. Mitchell, D.C.; Shen, X.; Green, M.J.; Hodgson, T.L. Accounting for regressive eye-movements in models of sentence processing: A reappraisal of the Selective Reanalysis hypothesis. *J. Mem. Lang.* **2008**, *59*, 266–293. [CrossRef]

23. Inhoff, A.W.; Weger, U.W. Memory for word location during reading: Eye movements to previously read words are spatially selective but not precise. *Mem. Cognit.* **2005**, *33*, 447–461. [CrossRef] [PubMed]

24. Laeng, B.; Teodorescu, D.S. Eye scanpaths during visual imagery reenact those of perception of the same visual scene. *Cogn. Sci.* **2002**, *26*, 207–231. [CrossRef]

25. Booth, R.W.; Weger, U.W. The function of regressions in reading: Backward eye movements allow rereading. *Mem. Cogn.* **2013**, *41*, 82–97. [CrossRef] [PubMed]

26. Sturt, P.; Kwon, N. Processing Information during Regressions: An Application of the Reverse Boundary-Change Paradigm. *Front. Psychol.* **2018**, *9*, 1630. [CrossRef] [PubMed]

27. Schotter, E.R.; Tran, R.; Rayner, K. Don't believe what you read (Only Once): Comprehension is supported by regressions during reading. *Psychol. Sci.* **2014**, *25*, 1218–1226. [CrossRef] [PubMed]

28. Gregg, J.; Inhoff, A.W. Misperception of Orthographic Neighbors during Silent and Oral Reading. *J. Exp. Psychol. Hum. Percept. Perform.* **2016**, *42*, 799–820. [CrossRef] [PubMed]

29. Reichle, E.D.; Reineberg, A.E.; Schooler, J.W. Eye movements during mindless reading. *Psychol. Sci.* **2010**, *21*, 1300–1310. [CrossRef] [PubMed]

30. Apel, J.K.; Henderson, J.M.; Ferreira, F. Targeting regressions: Do readers pay attention to the left? *Psychon. Bull. Rev.* **2012**, *19*, 1108–1113. [CrossRef] [PubMed]

31. Folk, J. Multiple Lexical Codes in Reading: Evidence from Eye Movements, Naming Time, and Oral Reading The operation of the lexical and sublexical systems during spelling View project. *J. Exp. Psychol. Learn. Mem. Cogn.* **1995**, *21*, 1412–1429. [CrossRef]

32. Folk, J. Phonological Codes Are Used to Access the Lexicon during Silent Reading The operation of the lexical and sublexical systems during spelling View project. *Artic. J. Exp. Psychol. Learn. Mem. Cogn.* **1999**, *25*, 892–906. [CrossRef]

33. Vitu, F.; McConkie, G.W.; Zola, D. About Regressive Saccades in Reading and Their Relation to Word Identification. *Eye Guid. Read. Scene Percept.* **1998**, 101–124. [CrossRef]

34. Vitu, F.; McConkie, G.W.; Kerr, P.; O'Regan, J.K. Fixation location effects on fixation durations during reading: An inverted optimal viewing position effect. *Vis. Res.* **2001**, *41*, 3513–3533. [CrossRef]

35. Reichle, E.D.; Rayner, K.; Pollatsek, A. The E–Z reader model of eye-movement control in reading: Comparisons to other models. *Behav. Brain Sci.* **2003**, *26*, 445–476. [CrossRef] [PubMed]

36. O'Regan, J.K.; Jacobs, A.M. Optimal Viewing Position Effect in Word Recognition: A Challenge to Current Theory. *J. Exp. Psychol. Hum. Percept. Perform.* **1992**, *18*, 185–197. [CrossRef]

37. Mcconkie, G.W.; Kerr, P.W.; Reddix, M.D.; Zola, D.; Jacobs, A.M. Eye movement control during reading: II. Frequency of refixating a word. *Percept. Psychophys.* **1989**, *46*, 245–253. [CrossRef] [PubMed]

38. McConkie, G.W.; Kerr, P.W.; Reddix, M.D.; Zola, D. Eye movement control during reading: I. The location of initial eye fixations on words. *Vis. Res.* **1988**, *28*, 1107–1118. [CrossRef]

39. Nuthmann, A.; Engbert, R.; Kliegl, R. Mislocated fixations during reading and the inverted optimal viewing position effect. *Vis. Res.* **2005**, *45*, 2201–2217. [CrossRef] [PubMed]

40. Reichle, E.D.; Pollatsek, A.; Rayner, K. E–Z Reader: A cognitive-control, serial-attention model of eye-movement behavior during reading. *Cogn. Syst. Res.* **2006**, *7*, 4–22. [CrossRef]

41. Engbert, R.; Nuthmann, A.; Richter, E.M.; Kliegl, R. Swift: A dynamical model of saccade generation during reading. *Psychol. Rev.* **2005**, *112*, 777–813. [CrossRef]

42. Wang, C.-A.; Inhoff, A.W. Extraction of Linguistic Information from Successive Words during Reading: Evidence for Spatially Distributed Lexical Processing. *J. Exp. Psychol. Hum. Percept. Perform.* **2013**, *39*, 662–677. [CrossRef]

43. Brysbaert, M.; Drieghe, D.; Vitu, F. Word skipping: Implications for theories of eye movement control in reading. In *Cognitive Processes in Eye Guidance*; Oxford University Press: Oxford, UK, 2012; ISBN 9780191693618.

44. Henderson, J.M.; Luke, S.G. Stable individual differences in saccadic eye movements during reading, pseudoreading, scene viewing, and scene search. *J. Exp. Psychol. Hum. Percept. Perform.* **2014**, *40*, 1390–1400. [CrossRef]

45. Hyönä, J.; Lorch, R.F.; Kaakinen, J.K. Individual differences in reading to summarize expository text: Evidence from eye fixation patterns. *J. Educ. Psychol.* **2002**, *94*, 44–55. [CrossRef]

46. Inhoff, A.W.; Gregg, J.; Radach, R. Eye movement programming and reading accuracy. *Q. J. Exp. Psychol.* **2018**. [CrossRef] [PubMed]

47. Vorstius, C.; Radach, R.; Lonigan, C.J. Eye movements in developing readers: A comparison of silent and oral sentence reading. *Vis. cogn.* **2014**, *22*, 458–485. [CrossRef]

48. Christianson, K.; Hollingworth, A.; Halliwell, J.F.; Ferreira, F. Thematic roles assigned along the garden path linger. *Cogn. Psychol.* **2001**, *42*, 368–407. [CrossRef] [PubMed]

49. Ferreira, F.; Bailey, K.G.D.; Ferraro, V. Good-enough representations in language comprehension. *Curr. Dir. Psychol. Sci.* **2002**, *11*. [CrossRef]
50. Ferreira, F.; Christianson, K.; Hollingworth, A. Misinterpretations of garden-path sentences: Implications for models of sentence processing and re analysis. *J. Psycholinguist. Res.* **2001**, *30*, 3–20. [CrossRef] [PubMed]

Review

Eye Movements and Fixation-Related Potentials in Reading: A Review

Federica Degno * and Simon P. Liversedge

School of Psychology, University of Central Lancashire, Marsh Ln, Preston PR1 2HE, UK;
SPLiversedge@uclan.ac.uk
* Correspondence: fdegno@uclan.ac.uk

Received: 31 March 2019; Accepted: 28 January 2020; Published: 3 February 2020

Abstract: The present review is addressed to researchers in the field of reading and psycholinguistics who are both familiar with and new to co-registration research of eye movements (EMs) and fixation related-potentials (FRPs) in reading. At the outset, we consider a conundrum relating to timing discrepancies between EM and event related potential (ERP) effects. We then consider the extent to which the co-registration approach might allow us to overcome this and thereby discriminate between formal theoretical and computational accounts of reading. We then describe three phases of co-registration research before evaluating the existing body of such research in reading. The current, ongoing phase of co-registration research is presented in comprehensive tables which provide a detailed summary of the existing findings. The thorough appraisal of the published studies allows us to engage with issues such as the reliability of FRP components as correlates of cognitive processing in reading and the advantages of analysing both data streams (i.e., EMs and FRPs) simultaneously relative to each alone, as well as the current, and limited, understanding of the relationship between EM and FRP measures. Finally, we consider future directions and in particular the potential of analytical methods involving deconvolution and the potential of measurement of brain oscillatory activity.

Keywords: reading; eye movements; event-related potentials; fixation-related potentials

---

1. The Timing Conundrum of Eye Movements and Event-Related Potentials

Decades of research recording eye movements (EMs) have revealed much about the processes that underlie written language comprehension and their temporal course [1,2]. One of the most important findings in the EM literature is that reading processing is fast and highly incremental [3]. Readers construct an incremental interpretation of the sentence, roughly on a word-by-word basis, as successive fixations are made along a sentence (e.g., [4,5]). Fixations are short periods of time, which on average last approximately 250 ms, during which information associated with the currently fixated word in the fovea, and to some extent with the upcoming word in the parafovea, is extracted and processed [1,2,6]. It is widely accepted that fixations reflect online cognitive processing [7,8], as the duration, and to a certain degree the location, of each fixation is determined by a number of cognitive factors (e.g., word frequency [9,10]; word predictability [11,12]). By implication, a sequence of processes occurs during each eye fixation. The sequence must include, as a minimum, transmission of the signal associated with the written word from the retina to the visual cortex, visual encoding, initiation of word identification, and programming of the next eye movement [13].

The large amount of robust evidence reflecting the rapid time course of reading from the EM literature however is in contrast with the likewise robust and compelling evidence of effects with a later time course reported in the ERP literature [3]. ERPs are EEG signals recorded at the scalp (with no measurable conduction delay between scalp potential and underlying source activity [14]) and time-locked to specific events [15]. They reflect postsynaptic potentials generated by populations of neurons active in synchrony, spatially aligned, and with the same direction of current flow [16,17]. A great number of studies recording ERPs during written language comprehension have consistently shown electrical signatures of linguistic processing in late time windows, associated with the N400 (between 300–500 ms after stimulus onset [18,19], although onsets are observed also from 200–250 ms after stimulus onset, e.g., [20]) and P600 (between 500–800 ms [21]) components.

ERP measures have been traditionally recorded using the rapid serial visual presentation paradigm (RSVP; e.g., [22]) in experiments focusing on the identification of individual words within or without a sentence context. In these traditional ERP experiments, one word at a time is displayed in the center of the screen, and blank screens are presented in between words. Each word is typically displayed for between 400–1000 ms, that is, for a period much longer than the average fixation duration in natural reading. Under these circumstances, ERP waveforms can be considered as a single stream of data corresponding to the cognitive processing associated with a single word during the entirety of the exposure period. Thus, finding that linguistic manipulations modulate late ERP components, and therefore, that observable effects associated with higher levels of cognitive processing occur at time points beyond the duration of an eye fixation, might seem unremarkable. However, we know from the EM literature that during natural reading, late time windows associated with these components are periods of time when the eyes have already moved to the next word, and identification of that word may remain underway, or indeed, may have been completed. Processing of a printed word in context, as reflected in EM measures, is determined by processing of both the individual word and integration of that word with the syntactic structure and the semantic representation of the sentence context constructed up to that point. Thus, observing modulation of ERP components that occur in relatively late time windows might reflect processing associated with fixations on words downstream in the sentence from the word in relation to which it was initiated. The important point to note here is that making a single long fixation on a word, or multiple refixations on a word, or even making multiple fixations after having left that word, might reasonably reflect qualitatively different aspects of cognitive processing [23].

Combining the on-line recording of EMs under natural reading conditions, and the real-time ms-by-ms recording of ERPs has great potential to unravel the nature and time course of the processes underlying reading [24,25]. Investigating neural correlates of foveal processing when both foveal and parafoveal information are available to the reader could lead to at least three potential scenarios. If effects associated with a linguistic manipulation are observed in late time windows, this would support the more traditional results that exist in the ERP literature and provide some evidence that a certain amount of time needs to pass for a linguistic manipulation to show a measurable effect in electrical brain activity. That is, cortical processing associated with word processing might outlast the fixational pause and behavioural response [26,27]. In contrast, if effects are observed exclusively in early time windows, this would bring into question the validity of the traditional ERP effects and raise the possibility that the nature of the paradigm used might affect the nature of the differences. For example, it might be possible that the foveal effects observed in previous traditional ERP studies are the result of a delay in processing due to the unavailability of parafoveal information, or due to differences in the deployment of attention under these experimental conditions relative to natural reading. Finally, if effects are observed in both early and late time windows, this would provide evidence that modulation of early ERP components might reflect cognitive processing associated with the identification of a fixated word, while later ERP components might also reflect similar processes as well as cognitive processing related to the integration of that word with its sentence context (e.g., semantic and syntactic processing).

2. A Tool to Discriminate between Theoretical Accounts

A large body of evidence from EM studies has demonstrated that readers not only process the word they are fixating, but also the upcoming word in the parafovea, that is to the right of fixation in alphabetic languages such as English (see [1,2] for reviews). Pre-processing of parafoveal information facilitates subsequent foveal processing of that word, and this contributes significantly to the rapid rate at which we read [6]. While in the EM literature parafoveal processing has been investigated for over 40 years (e.g., using gaze contingent paradigms [28,29]), parafoveal processing could not be investigated with ERPs until very recently, due to the nature of the paradigms being used. To reduce contamination of the EEG signal by EM artefacts and component overlap, words were presented one-by-one in the middle of the screen (e.g., [30–32]), or in the periphery away from central vision (e.g., [33–35]), such that normal parafoveal processing in reading could not occur. Recently, new paradigms such as the RSVP-with-flanker-word presentation method have been developed to address this issue (e.g., [36–39]). According to this paradigm, sentences or lists of words are presented word-by-word in the centre of the screen, with the preceding and following word(s) of the sentence, or of the word list, displayed laterally. The lateral presentation of the preceding and following words allows for parafoveal processing of the upcoming word, as in natural reading. However, in this situation, participants are required to keep their gaze on the centrally presented word, and not to make any eye movements. It is well documented that the allocation of attention and saccadic eye movements are most often tightly yoked (e.g., [7]), and for this reason, when participants are not required to plan and execute a saccade to the right, there are strong a priori grounds to anticipate that attentional allocation will not proceed in the same way as during natural reading. Thus, although ERPs have the potential to offer insights into the fine-grained timeline of parafoveal processing, and of foveal processing when parafoveal information is available, the experimental paradigms being used might not permit ready investigation of these issues. In this regard, co-registration of eye movements and brain potentials offers a methodological advance, and the possibility of investigating important theoretical questions that could not be addressed through the use of one of the two techniques alone.

Parafoveal processing is at the heart of the historical debate between serial versus parallel models of saccadic control in reading. The extent to which lexical processing of a parafoveal word is carried out during processing of the currently fixated word, and the temporal course of the lexical processing of the parafoveal word, are important issues in the reading literature. According to the serial accounts (e.g., E-Z Reader model [40–42]), words are fully lexically identified one word at a time. Extraction of information from the word in the parafovea initially occurs when attention is shifted to the parafoveal word but whilst the eyes remain fixating the word in the fovea. From a serial perspective, parafoveal pre-processing is largely limited to visual, orthographic and phonological properties of the word in the parafovea. Lexical processing of the upcoming word is initiated whilst it still lies in the parafovea. The initiation of parafoveal lexical processing occurs only after the reader is assured of the familiarity of the currently fixated word [43,44]. In contrast, advocates of parallel models of reading (e.g., SWIFT model [45,46]; see also OB1-reader, [47]) argue that more than one word is lexically processed at a time, with the degree to which parafoveal words are processed being determined by a number of factors including the word's frequency and where it lies within the graded attentional window. According to parallel models, extensive (lexical) pre-processing of the word in the parafovea is expected during processing of the currently fixated word (e.g., [45,46]).

Co-registration of eye movements and brain potentials might allow for discrimination between theoretical accounts, as manipulation of characteristics of the word in the parafovea should have a different temporal influence according to the different models. Indeed, by time-locking the ERPs to the fixation onset of the word in foveal vision, it is possible to examine whether and which characteristics of the parafoveal word might be extracted and processed during processing of the foveal word. If processing of the parafoveal word is initiated only after the foveal word is identified, then we should observe an effect in time windows that follow latencies associated with foveal lexical processing. Alternatively, if visuospatial attention is distributed across multiple words but lexical access proceeds serially, manipulation of visual, orthographic and phonological properties of the word in the parafovea might elicit an effect in the time windows associated with foveal lexical processing, while manipulation of higher level linguistic properties of the parafoveal word might produce an effect in later time windows. Both scenarios would provide evidence in support of serial models of saccadic control in reading. In contrast, if lexical processing proceeds in parallel across multiple words, manipulation of the higher-level linguistic characteristics of the parafoveal word might show an effect in the same time windows associated with lexical processing of the currently fixated (foveal) word. Furthermore, whether word position coding is flexible and expectation driven (as in the OB1-reader model; Snell et al., 2018 [47]) might also be tested with co-registration.

It is possible that serial and parallel models are two extreme accounts, and new theoretical frameworks might be able to better explain the existing empirical data (e.g., see the Multi-Constituent Unit account advocated by Zang [48] in the current Special Issue). Despite this, however, given the more fine-grained nature of the FRPs, as recorded with experimental paradigms that allow participants to freely read and make saccadic eye movements, and time-locked to specific oculomotor events, co-registration can be adopted to potentially shed light on this debate.

3. Phases of Co-Registration Research

3.1. Pioneering Co-Registration Studies

The idea of using a single technique to record eye movements and brain potentials has been developed over years of pioneering research. As early as in the 1950s, researchers investigated the existence of brain responses associated with EMs (see [49]). These studies revealed the existence of a sequence of components associated with saccadic EMs [50,51]. First, a presaccadic slow negative waveform has been reported, starting up to 1s before onset of a voluntary saccade over posterior frontal areas, and then extending over parietal areas, being maximal over the vertex [52–55], which some hypothesised to be similar to the 'readiness potential' [55]. Following this, a presaccadic slow positivity, also known as the antecedent potential, was observed between approximately 30–300 ms prior to saccade onset. This effect occurred primarily over occipito-parietal areas, but also over frontal areas of the scalp (e.g., [53,54,56,57]). This slow positive wave was found to be associated with processes that precede saccade execution, such as saccade planning and shifting of attention towards the next saccade target (e.g., [54,56,58,59]). Next, a biphasic wave shape (first negative and then positive), also called the presaccadic spike potential, was observed at saccade onset (with a sharp positive potential approximately 10–40 ms prior to saccade onset [54,55]) caused by the contraction of extra-ocular muscles associated with saccade execution. This potential, positive over centro-parietal areas of the scalp contralateral to the direction of the next saccade, and negative ipsilateral to the direction of the next saccadic EM, was present regardless of light or dark visual conditions [60,61], and modulated by saccade size and direction [62–64]. In addition, a positive response was also observed, originating in the visual cortex of awake individuals about 80–100 ms after fixation onset [51,65,66] in response to changes in the retinal image that accompanied the saccadic EMs [52,67]. This visually evoked response, labelled 'lambda wave', was observed when saccadic EMs were required [49,65], appeared to be modulated by physical properties of the stimulus (e.g., luminance and spatial frequency [68–70]), was not detected in darkness [49], and was considered to be associated with uptake of visual information [61].

However, it was in the 1980s that the first attempts to concurrently record EMs and ERPs were made in order to understand the cognitive processes that occur during reading. Marton and colleagues [33–35,71] conducted a series of ground-breaking experiments time-locking ERPs to fixation onsets (labelled 'saccade-related potentials', SRPs, due to the focus being on saccade offsets). In their experiments, participants were asked to move their eyes to a word presented in the periphery of the visual field, which was located about 20 degrees to the left or to the right of either the midline point of the screen [33], or the margin of a new row of text [34,35]. This approach was adopted to study reading under conditions that approximated natural reading, while keeping saccade amplitude constant and while controlling for the direction of the saccades, with the assumption being that ocular artifacts associated with left and right saccades cancel out EOG contamination of the waveform during averaging [54,72]. Marton and Szirtes found that execution of EMs produced an advantage both in the peak latencies of the SRPs (compared to the visual evoked potentials, VEPs) and in the mean reaction times [33], suggesting that differences can be observed depending on the naturalness of the task. Furthermore, for the first time they presented entire sentences on the screen (over multiple rows of text) and manipulated the final word of the sentence such that it could be congruous or incongruous with the previous context [34,35]. In addition to the presence of a more pronounced negative deflection for incongruous final words relative to congruous words (the established N400 component [30,31,73]), their findings also revealed significant differences between the two conditions at approximately 110–160 ms from fixation onsets, with SRPs being more negative for incongruous compared to congruous final words over frontal and central regions of the scalp [35]. These differences were observed simultaneously with the appearance of the lambda wave (peaking at approximately 130 ms after saccade offset over the occipital region of the scalp). Because the lambda wave was thought to reflect analysis of the physical features of the word, and the N400 component was thought to be associated with lexical access, the authors concluded that their findings represented evidence for an effect of sentence context before lexical access.

The theoretical questions investigated and the paradigm used (i.e., free reading of natural sentences) make these studies of relevance to the present discussion. However, a severe limitation was the very low spatial accuracy in determining eye fixations. EMs were measured via electro-oculogram (EOG) channels placed around the eyes, which is not an optimal method for measuring eye gaze. Thus, although subsequent efforts were made to improve this initial approach (e.g., [74]), research moved towards experimentation recording EMs with high-precision eye tracking devices which were being developed in the meantime (although see [75,76] for recent examples of reading studies using EOG channels).

3.2. Separate Recording of Eye Movements and Event-Related Potentials

In the 1990s, the scientific community became increasingly aware that comparing results from studies conducted with EMs and ERPs investigating the same theoretical question could provide a more complete understanding of the processes underlying reading [77]. However, the idea of concurrently recording EMs and ERPs was still considered to be out of reach. Two main issues were considered problematic: the disruption of the EEG signal caused by the ocular artifacts (saccades, blinks, etc.), and the component overlap across successive fixations [77,78]. Thus, a second phase of co-registration research was characterised by the comparison of EM and ERP data obtained from two separate experimental sessions [77–80]. Typically (except for [79]), in an EM experiment, sentences were presented normally while EMs of one group of participants were recorded. In a separate ERP experiment, testing different participants, target words, either within a contextual frame, or presented in isolation, were presented word-by-word according to the RSVP paradigm while the EEG signal was recorded. Although this approach was limited in a number of respects, the work raised a number of important theoretical questions.

Raney and Rayner [77] first compared converging results from two different existing studies ([81] for the EM data; [82] for the ERP data) that investigated the nature of changes in processing associated with rereading (i.e., when text was read for the second time). In these experiments, participants were required to read and remember as much text as possible for later recall. The EM data showed that multiple aspects of reading behaviour were facilitated during rereading (e.g., duration and number of forward fixations was reduced, amplitude of forward saccades increased, number of regressive fixations fell), and that the facilitation associated with the second reading affected high and low frequency target words similarly. The ERP data showed that changes in rereading behaviour were likely due, on one side, to decreased perceptual and comprehension demands (denoted by increased N1-P2 amplitudes), and on the other side, to increased memory demands (as suggested by decreased P300 amplitudes). Thus, converging evidence from EM and ERP data supported the theoretical view of the existence of different stages of processing during reading. However, Raney and Rayner did also report instances of diverging EM and ERP results. Different findings were observed in two experiments investigating processing of a target word that was related or unrelated to the preceding sentence context ([83] for the EM data; [84] for the ERP data). The EM data showed that facilitation occurred only when both the subject noun and verb were related to the target word. In contrast, ERP data showed facilitation when both or only one of either the subject noun or the verb were related to the target word. Although Raney and Rayner did not discuss in detail the diverging results, the paper raised an important issue concerning the relationship between EM and ERP measures. This relationship was later investigated by Dambacher and Kliegl [78]. These authors compared EM and ERP results from two separate experiments in which two different groups of participants were required to read the same sentences for comprehension ([85] for the EM data; [86] for the ERP data). The aim of the study was to investigate whether both EM and ERP measures (i.e., fixation durations and N400 amplitudes respectively) were modulated by the same word properties, and by implication, by common mechanisms associated with word recognition. They found that longer single fixation durations were correlated with more negative N400 amplitudes on the corresponding word, a relationship accounted for by both word frequency and word predictability, as well as by the predictability of the upcoming word. In addition, more negative N400 amplitudes were associated with longer single fixation durations on the following word, and this relationship was accounted for by word frequency. Thus, their findings confirmed that both EM and ERP measures were similarly modulated by word frequency (considered a bottom-up variable) and word predictability (considered a top-down variable), and, as a consequence, by at least one common stage of processing.

Taking advantage of these complementary methods, Ashby and Martin [79] also compared EM and ERP results. Aiming to investigate the nature of prelexical phonological processing, they used a boundary-change lexical decision task for the EM experiment, and a masked priming semantic judgment task for the ERP study. In both experiments, isolated target words were presented, preceded by a partial word parafoveal preview (in the EM experiment) or prime (in the ERP experiment). Both were comprised of a syllable congruent or incongruent with the initial syllable of the target word. Shorter fixation times and more positive ERP amplitudes between 250–350 ms were observed for the syllable congruent condition compared to the syllable incongruent condition. These converging results provided support for the view that activation of prelexical phonological representations includes activation of prosodic (i.e., suprasegmental) information, such as syllable information, that is not encoded in the writing system. In addition, they suggested a role for memory in the activation of suprasegmental information, since preserving congruent syllable information in memory during a saccade, or during a backward mask, facilitated word recognition.

Although the nature of the relationship between EM and ERP measures remains unclear and it has not yet been determined whether different results might be explained by differences in the specific paradigm used, the three studies [77–79] showed that considering converging, as well as diverging, EM and ERP results offers potential benefit for our understanding of reading, and of human cognitive processing more generally.

In this second phase of co-registration research, another important issue was investigated, namely, sensitivity of early ERP components to lexical manipulations. Sereno, Rayner, and Posner [80] investigated timing discrepancies for effects thought to index lexical access (e.g., word frequency [87]) between methodologies, that is, as reported in independent EM and ERP studies. They argued that it was not self-evident why experimenters should investigate relatively late ERP components, such as the N400, by default when investigating aspects of lexical processing. An alternative, arguably more plausible, approach might be to attempt to account for lexical effects within a more typical EM time-line (i.e., considering lexical processing effects within 250 ms from fixation onset). By recording EM and ERP measures in response to the same target words (within a sentence in the EM experiment and in isolation in the ERP experiment) their study showed effects of lexicality as early as 100 ms from stimulus onset (on the P1 component), effects of word frequency starting at 132 ms from stimulus onset (on the N1 component), and effects of word regularity (in terms of spelling sound correspondence) as early as 164 ms from stimulus onset (on the P2 component). Thus, as they argued, their study did show lexical effects on early ERP components, although the two groups of participants were engaged in quite different tasks in the two experimental sessions: silent reading for comprehension in the EM experiment, and lexical decision task in the ERP experiment. This study emphasised the need for more research on the early ERP components when investigating the cognitive processes underlying reading, which had been largely ignored up to that point.

It was during this period that the idea of a single technique that would simultaneously combine EMs and ERPs was formalized [24,25], thereby initiating the modern conception of co-registration research. It was evident that the existence of electrophysiological signatures associated with specific cognitive processes within an eye fixation had the potential to reveal a more precise timing of the sequence of processing and to offer insight into the nature of pre-lexical, lexical, or post-lexical processes underlying reading.

3.3. Simultaneous Recording of Eye Movements and Fixation-Related Potentials

Despite the existing challenges associated with simultaneously recording EMs and ERPs (see [26, 88–91] for a discussion), joint efforts from the international research community, alongside technological advances (see [92–95] for a discussion) and advances in computational algorithms used for the correction of ocular artifacts [26,96–99], have allowed for a third, ongoing, phase of co-registration research to become possible. These studies have involved simultaneously recording EM and ERP data from the same group of participants, reading the same stimuli, and performing the same task. This has ensured an exact correspondence between EM and ERP data under identical experimental testing conditions in the same individual. In this approach ERPs are referred to as fixation-related potentials (FRPs; or EFRPs, e.g., [100]), as the time-locked events of interest are fixation onsets on particular words in a trial. Note, though, that ERPs can also be time-locked to saccade onsets, in which case we speak about saccade-related potentials (SRPs; see for example [91] for a discussion). A variety of different paradigms have been used under such testing circumstances, for example, free reading of pairs of words, lists of words, sentences or even paragraphs, during which participants make saccadic EMs. Via these paradigms, it is possible to shed light on the neural correlates of not only foveal processing, but also parafoveal processing, an important aspect of reading that was not possible to investigate in experiments conducted in the first two phases of co-registration research.

A complete list of studies involving simultaneous recording of EMs and FRPs is presented in Table 1. Note that we include in Table 1 only co-registration experiments that have investigated reading through the analysis of EMs and FRP components, in which participants were allowed to make at least forward eye movements in each trial, and in which eye movements were recorded with an eye tracking system. Studies using co-registration but analysing oscillatory brain activity time-locked to fixation onsets on particular target words [101–105] will be discussed only in relation to future directions.

Below we present three tables (Tables 2–4) in which we summarize every experiment that has been conducted to investigate parafoveal and foveal processing, presenting, and in some cases manipulating, information in the parafovea under co-registration conditions (see Figure 1 for a visualization of the investigated effects).

Table 1. Co-Registration Studies Presenting Pairs of Words (+), Lists of Words (++), Sentences (*) or Paragraphs (**).

Study	Language	Participants	Paradigm	Task	Variables	Investigated Effects
BM2005 [+] [106]	French	Age range: 22–38 Average age: 27.0 Typical readers	Priming	Semantic association judgment	Parafoveal preview, Semantic relatedness	Parafoveal processing (PoF)
Hetal2007 [++] [107]	German	Age range not reported Average age: 24.6 Typical readers	Free reading RSVP	Recognition	Reading modality, Repetition	Foveal processing
KBSS2009 * [108]	German	Age range: 19–31 Average age: 23.7 Typical readers	Free reading	Reading for comprehension	Target word predictability, Semantic relatedness	Foveal processing, Parafoveal processing (PoF)
SHL2009 [+] [109]	Swedish	Age range not reported Average age: 27.5 Typical readers	Priming	Semantic association judgment	Parafoveal preview, Visual field of presentation	Parafoveal processing (PoF)
DSHJK2011 * [26]	German	Age range: 17–37 Average age: 23.0 Typical readers	Free reading	Reading for comprehension	Target word predictability	Foveal processing
DKS2012 [++] [110]	German	Age range: 19–36 Average age: 24.4 Typical readers	Boundary, Free reading	Semantic decision	Parafoveal preview, Semantic relatedness, Repetition	Foveal processing, Parafoveal processing (PoF, Preview benefit)
HLSR2013 ** [111]	English	Age not reported Typical readers	Free reading	Reading	Text type	Foveal processing
Hetal2013 [++] [112]	German	Age range not reported Average age: 24.0 Typical readers	Boundary, RSVP fixed-pace, RSVP self-pace	Recognition	Parafoveal preview, Reading modality, Repetition	Foveal processing, Parafoveal processing (Preview baseline)
KSS2015 * [113]	English	Age range: 18–29 Average age: 20.3 Typical readers	Free reading	Reading for comprehension	Target word frequency, Target word predictability	Foveal processing, Parafoveal processing (PoF)
MvdMVR2015 * [104]	German	Age range: 19–34 Average age: 25 Typical readers	Free reading, RSVP	Reading for comprehension	Target word predictability	Foveal processing, Parafoveal processing (PoF)

Table 1. *Cont.*

Study	Language	Participants	Paradigm	Task	Variables	Investigated Effects
KNSD2016 [++] [114]	German	Age range: 18–34 Average age: 24.8 Typical readers	Boundary, RSVP-with-flankers	Semantic decision	Parafoveal preview, Reading modality, Pretarget word frequency	Foveal processing, Parafoveal processing (Preview benefit)
LPDHCB2016 [+] [115]	Spanish	Age range: 18–29 Average age: 20.0 Typical readers	Priming, Boundary	Semantic association judgment	Preview semantic relatedness, Target semantic relatedness	Parafoveal processing (PoF, Preview benefit)
MvdMVR2016 * [116]	German	Age range not reported Average age: 25.0 Typical readers	Free reading, RSVP	Sentence well-formedness judgment	Syntactic/semantic violations, Violation position, Reading modality	Foveal processing
ND2016 [++] [117]	German	Age range: 18–34 Average age: 26.1 Typical readers	Boundary, RSVP-with-flankers	Semantic decision	Parafoveal preview, Foveal load, Target word frequency	Foveal processing, Parafoveal processing (PoF, Preview benefit)
WKV2016 * [118]	Hungarian	Age range: 20–26 Average age: 22.3 Typical readers Fast and slow readers	Free reading	Reading for comprehension	Inter-letter spacing, Reading ability	Foveal processing
LHHL2018 * [119]	Finnish	Age range: 12–13.5 Average age not reported Typical readers	Free reading	Sentence plausibility judgm0nt	Semantic violations	Foveal processing, Parafoveal processing (PoF for EM data only)
DLZZDL2019 * [120]	English	Age range: 18–26 Average age: 19.3 Typical readers	Boundary	Reading for comprehension	Parafoveal preview, Target word frequency	Foveal processing, Parafoveal processing (PoF, Preview benefit)
DLZZDL2019 * [121]	English	Age range: 18–26 Average age: 19.3 Typical readers	Boundary	Reading for comprehension	Inter-word spacing, Parafoveal preview	Foveal processing, Parafoveal processing (Preview benefit)
LHHL2019 * [122]	Finnish	Age range: 12–13.5 Average age not reported Slow and typical readers	Free reading	Sentence plausibility judgment	Word length, Fixation order, Reading ability	Foveal processing

Note: The only other co-registration study to date that has used a reading task is [123]. However, the study focused on issues related to problem solving rather than aspects of linguistic processing. For this reason, this study is not discussed in the present review.

Table 2. Summary of the Findings Reported in Co-Registration Studies Investigating Parafoveal-on-Foveal (PoF) Effects. In these studies, the parafoveal word was manipulated, and the eye movements (EM) and fixation related-potentials (FRP) data were time-locked to the fixation onset on the pretarget word. Thus, these results are associated with effects derived from parafoveal manipulations measured during fixation on the pretarget word.

Investigated Effect	Study	FRP Data				EM Data		
		Significant	Time-Window	Electrode Sites	Direction of the Effect	Significant	EM Measure	Direction of the Effect
Word form	BM2005 [106]	yes	peak 119 ms (N1)	LO	related words > letter-string	yes	TFD	words > letter-string
			from 100 ms, peak 140 ms (P)	RC, RF	unrelated words > letter-string			
	SHL2009 [109]	yes	200–280 ms (P2)	O	RVF words > RVF letter-string	yes	FFD	RVF unrelated words > letter-string
							TRT	words > letter-string
	KNSD2016 [114]	no	200–280 ms (N1)	OT		not analysed		
	DLZZDL2019 [120]	yes	70–120 ms (P1)	RO, RP, MO, MP	X-string > identity	yes	(FFD)	X-string > identity
				RO, RP, MO, MP	X-string > letter-string		SFD	X-string > identity
				LO, LP, T, F	letter-string > X-string		GD	X-string > identity and letter-string
				C	identity > letter-string			
			120–300 ms (N1)	RO, RP, MO, MP	identity > X-string			
				RO, RP, MO, MP	letter-string > X-string			
				LO, LP, T, F	X-string > letter-string			
				C	letter-string > identity			
				F, C, T	X-string > identity			
			300–500 ms (N400)	RO, RP, MO, MP	identity > X-string			
				RO, RP, MO, MP	letter-string > X-string			
				LO, LP, T, F	X-string > letter-string			
				C	letter-string > identity			

Table 2. *Cont.*

Investigated Effect	Study	FRP Data				EM Data		
		Significant	Time-Window	Electrode Sites	Direction of the Effect	Significant	EM Measure	Direction of the Effect
				F, C, T	X-string > identity			
Word repetition	DKS012 [110]	no	from 0–40 ms to 560–600 ms	all	X	yes	FFD	repeated < different
							SFD	repeated < different
							GD	repeated < different
Preview frequency	KSS2015 [113]	no†	from 150–200 to 350–400 ms (N400)	CP, M	X	no	LFD	X
	ND2016 [117]	yes	130–140 ms (P)	RF	LF > HF	yes	GD	LF > HF
			630–640 ms (P)	LP	HF > LF			
	DLZZDL2019 [120]	no	70–120 ms (P1), 120–300 ms (N1), 300–500 ms (N400)	all	X	no	FFD, SFD, GD	X
Semantic relatedness	BM2005 [106]	yes	from 160 ms, peak 215 ms (P2)	C, F	related words > unrelated words	yes	TFD	unrelated words > related words
	SHL2009 [109]	no	90–140 ms (P1), 140–200 ms (N1), 200–280 ms (P2)	O	X	no	FFD, GD, TRT	X
			70–120 ms (N1), 140–280 ms (P2)	FP	X			
	KBS2009 [108]	yes	250–400 ms (N400)	P, C	unpredicted unrelated > predicted antonym	no	LFD	X
				P, C	unpredicted unrelated > unpredicted related			
	DKS2012 [110]	no	from 0–40 ms to 560–600 ms	all	X	no	FFD, SFD, GD	X

Table 2. *Cont.*

Investigated Effect	Study	FRP Data				EM Data		
		Significant	Time-Window	Electrode Sites	Direction of the Effect	Significant	EM Measure	Direction of the Effect
	LDHB2016 [115] *	yes	400–550 ms (N400)	all	unrelated > related	no	FFD, GD	X
Preview predictability	KBS2009 [108]	no	250–400 ms (N400)	P, C	X	no	LFD	X
	DSHJK2011 [26]	no	300–500 ms (N400)	X	X	not analysed	X	X
	KSS2015 [113]	†	from 150–200 to 350–400 ms (N400)	CP, M	X	no	LFD	X
	MvdMVR2015 [104]	+	334–826 ms, peak 608 ms (N) peak 658 ms (P)	CP F	incongruent > congruent incongruent > congruent	no	FFD, GD, RP	X

Note: * ERPs were time-locked to the onset presentation of the prime-preview pair. †Some short time-windows did show significant effects, but the authors disregarded those effects as meaningless. + The authors observed significant differences but pointed out that they could actually reflect an effect in response to the target word. RVF = right visual field; C = central; CP = centro-parietal; F = frontal; LO = left occipital; LP = left parietal; M = midline; MO = midline occipital; MP = midline parietal; O = occipital; P = parietal; FP = fronto-parietal; RC = right central; RF = right frontal; RO = right occipital; RP = right parietal; T = temporal; FFD = first fixation duration; GD = gaze duration; LFD = last fixation duration; RP = regression probability; SFD = single fixation duration; TFD = total fixation duration; TRT = total reading time; HF = high frequency word; LF = low frequency words. ">" = amplitudes associated with the left-hand side conditions were greater than amplitudes associated with conditions on the right-hand side of the symbol (i.e., more negative if a negative (N) component was observed in that time-window, more positive if a positive (P) component was present in that time-window). Fixation durations associated with the left-hand side conditions were longer (or there was an increased regression probability) than the fixation durations associated with conditions on the right-hand side of the symbol. "<" = amplitudes associated with the left-hand side conditions were lower than amplitudes associated with conditions on the right-hand side of the symbol (i.e., less negative if a negative (N) component was observed in that time-window, less positive if a positive (P) component was present in that time-window). Fixation durations associated with the left-hand side conditions were shorter (or there was a reduced regression probability) than the fixation durations associated with conditions on the right-hand side of the symbol.

Table 3. Summary of the Findings Reported in Co-Registration Studies Investigating Preview Effects. In these studies, the effects associated with parafoveal manipulation were measured when both EMs and FRPs were time-locked to the initial fixation onset on the target word, to examine the influence that the pre-processing of an upcoming word in the parafovea exerts on the processing of that word when currently fixated.

Investigated Effect	Study	FRP Data				EM Data		
		Signicant	Time-Window	Electrode Sites	Direction of the Effect	Signicant	EM Measure	Direction of the Effect
Identity parafoveal preview	DKS012 [110]	yes	200–240 ms (N1)	OT	invalid > identity	yes	FFD	invalid > valid previews
			240–280 ms (N1)	OT	invalid > identity		SFD	invalid > valid previews
			360–400 ms (N400)	CP	invalid > identity		GD	invalid > valid previews
	KNSD2016 [114]	yes	200–280 ms (N1)	OT	X-string > 1 letter	yes	FFD	X-string > 3 letters
					X-string > 2 letters			X-string > full preview
					X-string > 3 letters			
					X-string > full preview			
			400–500 ms (N400)	CP	invalid > identity			
	ND2016 [117]	yes	140–200 ms (N1)	OT	invalid > identity	yes	FFD	invalid > valid previews
			200–300 ms (N1)	OT	invalid > identity		SFD	invalid > valid previews
							GD	invalid > valid previews
	LDHB2016 [115]	yes	300–500 ms (N400)	not reported	invalid > identity	not analysed	X	X
			500–800 ms (P600)	C	invalid > identity			

Table 3. *Cont.*

Investigated Effect	Study	FRP Data				EM Data		
		Signicant	Time-Window	Electrode Sites	Direction of the Effect	Signicant	EM Measure	Direction of the Effect
	DLZZDL2019 [120]	yes	0–70 ms (N)	RO, MO, RP, MP	identity > X-string	yes	FFD	invalid > valid previews
				RO, MO, RP, MP	letter-string > X-string		SFD	invalid > valid previews
				RO, T, P	identity > letter-string		GD	invalid > valid previews
				C	letter-string > X-string			
				C	letter-string > identity			
				T, F	X-string > letter-string			
			70–120 ms (P1)	RO, RP, MO, MP	X-string > identity			
				RO, RP, MO, MP	X-string > letter-string			
				RO, T, P	letter-string > identity			
				C	X-string > letter-string			
				C	identity > letter-string			
				RT, RF	letter-string > X-string			
			120–300 ms (N1)	O, P (120–200 ms)	identity > X-string			
				O, T, P (200–300 ms)	X-string > identity			
				O, T, P	X-string > letter-string			
				O, T, P	identity > letter-string			
				T, C, F (120–200 ms)	X-string > identity			
				M, C	letter-string > X-string			

Table 3. Cont.

Investigated Effect	Study	FRP Data				EM Data		
		Signicant	Time-Window	Electrode Sites	Direction of the Effect	Signicant	EM Measure	Direction of the Effect
				M, C (200–300 ms)	identity > X-string			
				C	letter-string > identity			
			300–500 ms (N400)	O, T, P	X-string > identity			
				O, T, P	X-string > letter-string			
				O, P	identity > letter-string			
				M, C	letter-string > X-string			
				M, C	identity > X-string			
	DLZZDL2019 [121]	yes	0–70 ms	X		yes	FFD	invalid > valid previews
			70–120 ms (P1)	O, P	string > identity		SFD	invalid > valid previews
				C, F	identity > string			
			120–300 ms (N1)	O, RT, LT, RP, LP (120–180 ms)	identity > string		GD	invalid > valid previews
				C, F (120–180 ms)	identity < string			
				O, RT, LT, RP, LP (185–300 ms)	identity < string			
				MP, C, LF, MF (185–300 ms)	identity > string			
			300–500 ms (N400)	O, LP, LC, MC	identity > string			
				RT, RC, MC, F	identity < string			

Table 3. *Cont.*

Investigated Effect	Study	FRP Data				EM Data		
		Signicant	Time-Window	Electrode Sites	Direction of the Effect	Signicant	EM Measure	Direction of the Effect
Semantic relatedness	DKS2012 [110]	no	from 0–40 ms to 560–600 ms	all	X	no	FFD, SFD, GD	
	LDHB2016 [115]	yes	0–200 ms (N)	O, P, C	unrelated > related	no	FFD,	
			300–500 ms (N400)	all	unrelated > related		GD	
			500–750 ms (P600)	all	unrelated > related			

Note: OT = occipito-temporal; RT = right temporal. See Table 2 for a legend of the other abbreviations.

Table 4. Summary of the Findings Reported in Co-Registration Studies Investigating Foveal Processing from Fixation Onset on the Target Word. In these studies, both EMs and FRPs were time-locked to the initial fixation onset on the target word, to examine variables that affect processing of a word from (at least) its initial fixation onward and their time course.

Investigated Effect	Study	FRP Data				EM Data		
		Significant	Time-Window	Electrode Sites	Direction of the Effect	Significant	EM Measure	Direction of the Effect
Text type	HLSR2013 [111]	yes	75–125 ms (P1)	O, T	text > pseudotext	yes	FFD	pseudotext > text
			125–210 ms (N1)	O, T	text > pseudotext			
Inter-word spacing X	DLZZDL2019 [121]	yes	0–70 ms^ (N)	LP, LO	unspaced > spaced	yes	FFD	unspaced > spaced
			70–120 ms (P1)	O, P	spaced > unspaced		SFD	unspaced > spaced
				F, LC, MC	unspaced > spaced		GD	unspaced > spaced

Table 4. *Cont.*

Investigated Effect	Study	FRP Data				EM Data		
		Significant	Time-Window	Electrode Sites	Direction of the Effect	Significant	EM Measure	Direction of the Effect
			120–300 ms (N1)	O, P (135215– ms)	spaced > unspaced			
				F, C, T (145205– ms)	spaced < unspaced			
				O, P (215300– ms)	spaced < unspaced			
				RC (145300– ms)	spaced < unspaced			
				F, C, MP (220300– ms)	spaced > unspaced			
			300–500 ms (N400)	X	X			
Inter-letter spacing	WKV2016 [118]	yes	120–175 ms (N)	OT, P	normal spacing >reduced and double spacing	yes	FD *	reduced > normal spacing
			155–220 ms (N)	ROT, RP	reduced > normal > double spacing			normal > double spacing
			230–265 ms (P)	ROT, P	normal spacing >reduced and double spacing		SA *	double > normal spacing
			345–380 ms (N)	LOT	normal spacing >reduced and double spacing			normal > reduced spacing
Word length	LHHL2019 [122]	yes	130–300 ms (P)	F	TP: long > short words for additional fixation	yes	FFD	long > short words
			130–300 ms (N)	O	TP: long > short words for additional fixation		GD	long > short words
			170–280 ms (N)	RO	SR: long > short words for additional fixation		REFIX	long > short words

Table 4. *Cont.*

Investigated Effect	Study	FRP Data				EM Data		
		Significant	Time-Window	Electrode Sites	Direction of the Effect	Significant	EM Measure	Direction of the Effect
Word frequency	KSS2015 [113]	no	from 150–200 to 650–700 ms (N400)	CP, M	X	yes	FFD	LF > HF
							GD	LF > HF
							skipping	LF < HF
	ND2016 [117]	yes	140–200 ms (N1)	OT	LF > HF	yes	FFD	LF > HF
			200–300 ms (N1)	OT	LF > HF		SFD	LF > HF
							GD	LF > HF
	DLZZDL2019 [120]	no	0–70 ms ˆ, 70–120 ms (P1),	all	X	yes	FFD	LF > HF
			120–300 ms (N1), 300–500 ms (N400)				SFD	LF > HF
							GD	LF > HF
Repetition (old/new)	Hetal2007 [107]	yes	250–600 ms (P)	P, C, F +	old > new	not analysed		
	DKS2012 [110]	yes	80–480 ms (N400)	CP	new > old	yes	FFD	new > old
							SFD	new > old
							GD	new > old
	Hetal2013 [112]	yes	176–390 ms (P)	P, C, F +	old > new	not analysed		
Semantic Relatedness	KBS2009 [108]	yes	450–740 ms (P600)	P	unpredicted unrelated > unpredicted related	no	FFD	X
				P, C	unpredicted unrelated > predicted antonyms			

Table 4. *Cont.*

Investigated Effect	Study	FRP Data				EM Data		
		Significant	Time-Window	Electrode Sites	Direction of the Effect	Significant	EM Measure	Direction of the Effect
				LC	unpredicted related > predicted antonyms			
	DKS2012 [110]	yes	160–480 ms (N400)	CP	unrelated > related	yes	FFD	unrelated > related
							SFD	unrelated > related
							GD	unrelated > related
	LDHB2016 [115]	yes	300–500 ms (N400)	all	unrelated > related	no	FFD, GD	X
			500–750 ms (P600)	all	unrelated > related			
Word predictability	KBS2009 [108]	yes	250–400 ms (N400)	RP	unpredicted related > predicted antonym	yes	FFD	unpredicted > predicted
				P	unpredicted unrelated > predicted antonym			
	DSHJK2011 [26]	yes	248–500 ms (N400)	CP	low predictable > high predictable	yes	FFD	LP > HP
							GD	LP > HP
							TSR	LP > HP
	KSS2015 [113]	yes	150–250 ms (P200)	CP, M	predictable > unpredictable	yes	FFD	LP > HP
			250–650 ms (N400)	CP, M	unpredictable > predictable		GD	LP > HP
							regressions	LP > HP
							skipping	LP < HP

Table 4. *Cont.*

Investigated Effect	Study	FRP Data				EM Data		
		Significant	Time-Window	Electrode Sites	Direction of the Effect	Significant	EM Measure	Direction of the Effect
	MvddMVR2015 [104]	yes	222–514 ms, peak 378 ms (N400)	ROT	incongruent > congruent	yes	FFD	incongruent > congruent
			318–626 ms, peak 476 ms (P)	F, FC	incongruent > congruent		GD	incongruent > congruent
			692–1400 ms, peak 1382 ms(P)	CP	incongruent > congruent		RP	incongruent > congruent
Syntactic & semantic violations	MvdMVR2016 [116]	yes	290–1000 ms (P600)	X	Mid-sentence syntactic violations regression trials > control	yes	FFD	violations > control
			540–1000 ms (P600)	X	Mid-sentence semantic violations regression trials > control		GD	violations > control
			24–378 ms (N400)	CP	Sentence-final syntactic violations regression trials > control		RP	violations > control
			244–1000 ms (P600)	CP	Sentence-final syntactic violations regression trials > control			
			98–392 ms (N400)	OT	Sentence-final semantic violations regression trials > control			
			412–1000 ms (P600)	CP	Sentence-final semantic violations regression trials > control			

Table 4. *Cont.*

Investigated Effect	Study	FRP Data				EM Data		
		Significant	Time-Window	Electrode Sites	Direction of the Effect	Significant	EM Measure	Direction of the Effect
			310–1000 ms (N)	CP	Sentence-final syntactic violations no-regression trials > control			
			336–646 ms (N)	CP	Sentence-final semantic violations no-regression trials > control			
			652–774 ms (N)	CP	Sentence-final semantic violations no-regression trials > control			
	LHHL2018 [119]	yes	167–547 ms (N)	RFEF	anomalous word neighbour > plausible	yes	FFD	anomalous word neighbour > plausible
			238–738 ms (N)	RFEF	unrelated anomalous > plausible			unrelated anomalous > plausible
			254–445 ms (N400)	LP	unrelated anomalous > plausible		GD	anomalous word neighbour > plausible
			263–447 ms (N400)	CP	unrelated anomalous > plausible			unrelated anomalous > plausible
			309–535 ms (N400)	LP	anomalous word neighbour > plausible		REFIX	anomalous word neighbour > plausible
			484–683 ms (P600)	LP	unrelated anomalous > anomalous word neighbour			unrelated anomalous > plausible
			558–899 ms (P600)	LP	unrelated anomalous > plausible			
			564–709 ms (P600)	RP	unrelated anomalous > plausible			

Table 4. *Cont.*

Investigated Effect	Study	FRP Data				EM Data		
		Significant	Time-Window	Electrode Sites	Direction of the Effect	Significant	EM Measure	Direction of the Effect
			648–739 ms (P600)	CP	anomalous word neighbour > plausible			
			710–899 ms (P600)	LP	anomalous word neighbour > plausible			
			739–813 ms (P600)	RP	unrelated anomalous > plausible			
			792–869 ms (P600)	CP	unrelated anomalous > plausible			
Foveal load	KNSD2016 [114]	yes	200–280 ms (N1)	OT	HF > LF	yes	FFD	LF > HF
Reading ability	LHHL2019 [122]	yes	140–250 ms (P)	C	Slow > Typical readers	yes	FFD	Slow > Typical readers
			250–300 ms (P)	O	Slow > Typical readers		GD	Slow > Typical readers
							REFIX	Slow > Typical readers

Note: * Fixation duration (FD) and saccade amplitude (SA) were calculated based on median values. + Although the effect was more pronounced on the right central and frontal scalp sites. ^ Note that effects between 0–70 m after fixation onset are, in fact, parafoveally triggered. That is, those effects are related to the processing of the stimulus that was in parafovea immediately preceding the fixation on the target word. X We have classified this effect as foveal, but strictly speaking, this manipulation involved both foveal and parafoveal change, in that the word final space was masked until the eyes moved onto the next word. FC = fronto-central; LC = left central; LOT = left occipito-temporal; OT = occipito-temporal; RFEF = right frontal eye field; ROT = right occipito-temporal; TSR = total sentence reading; REFIX = refixation probability; LP = low predictable words; HP = high predictable words. See Table 2 for a legend of the other abbreviations.

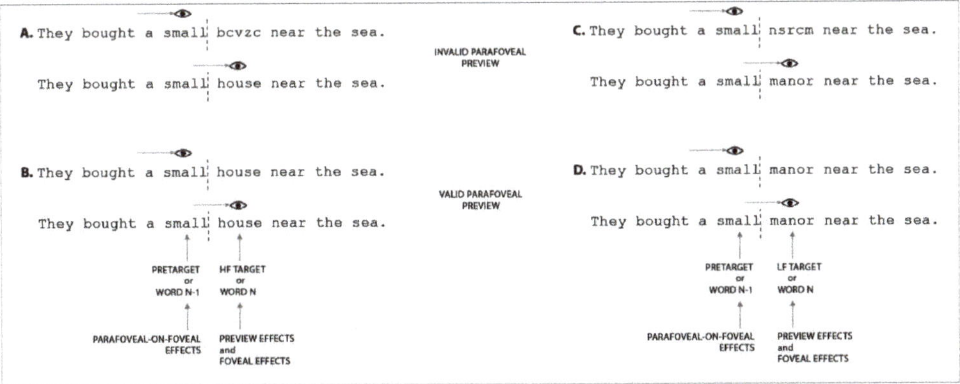

Figure 1. Illustration of example stimuli and investigated effects. The image shows example sentences presented according to the boundary paradigm [28]. An invisible boundary is placed at the end of the pretarget word (i.e., 'small'). When the eyes fixate the pretarget word, a preview is displayed in the parafovea (i.e., 'bcvzc', 'house', 'nsrcm', 'manor'). When the eyes cross the invisible boundary, the target word is displayed (i.e., 'house', 'house', 'manor', 'manor'). Panels A and B differ from panels C and B in the frequency with which the target words 'house' and 'manor' occur in the English language, such that 'house' is a high frequency (HF) word, 'manor' is a low frequency (LF) word. In addition, panels A and C differ from panels B and D as the preview stimulus is an invalid preview in the first two panels, but a valid preview for the other two panels. That is, in panels A and C, a string of random letters is presented in the parafovea, and this string does not share many features with the target word (i.e., bcvzc'–'house', 'nsrcm'–'manor'). Instead, in panels B and D, preview and target words are identical (i.e., 'house'–'house', 'manor'–'manor'). Parafoveal-on-Foveal (PoF) effects are examined by time-locking EM and FRP data to the onset of the first fixation on the pretarget word. Thus, researchers examining PoF effects, compare the effect that the different parafoveal previews (i.e., 'bcvzc', 'house', 'nsrcm', 'manor') have on the processing of the pretarget word (i.e., 'small') that is currently being fixated. Preview effects are studied by time-locking EM and FRP data to the onset of the first fixation on the target word. Here, researchers compare the effect that the different parafoveal previews (i.e., 'bcvzc', 'house', 'nsrcm', 'manor') have on the processing of the target word (i.e., 'house' or 'manor') when it is subsequently fixated. Foveal effects are investigated by time-locking EM and FRP data to the onset of the first fixation on the target word and comparing how the characteristics of the stimulus in fovea (i.e., a HF word 'house' vs. A LF word 'manor') affect processing of that word.

Table 2 provides a summary of the results associated with effects derived from parafoveal manipulations measured during fixation of the pretarget word. In these studies, the parafoveal word was manipulated, and the EM and FRP data were time-locked to the fixation onset of the pretarget word, thus allowing for investigation of parafoveal-on-foveal (PoF) effects. As discussed in Section 2 of this review, the time course of PoF effects is crucial in the context of the debate between serial and parallel models. Whether lexical characteristics of the parafoveal word are extracted after or during lexical processing of the foveal word pertains directly to serial or parallel accounts, and whether visuospatial attention and processing operates in a focused or a distributed manner. Note, though, that whilst early effects might be clearly associated with processing of the parafoveal word during fixation on the pretarget word, FRPs associated with later time windows might be contaminated by activation associated with foveal information processed during subsequent fixations on the following (target) word, or refixations on the same (pretarget) word, implying that effects in these later time windows require very careful analysis and consideration.

Table 3 provides a summary of experiments demonstrating effects at the target word when it was manipulated in some way whilst presented in the parafovea. The effects associated with this manipulation were measured when both EMs and FRPs were time-locked to the initial fixation onset on the target word. These studies aimed to examine the influence that the pre-processing of an upcoming word in the parafovea exerts on the processing of that word when currently fixated. In the EM and reading literature this effect is generally called the preview benefit effect and it is usually investigated using the boundary paradigm [28]. In the boundary paradigm, the target word is replaced by a preview string and an invisible boundary is placed immediately prior to the target word. In this way, when the eyes are fixating the pretarget word, parafoveal processing of the preview occurs. However, when the eyes cross the boundary, the preview is replaced by the target. If efficient parafoveal processing of the preview occurs, and the preview is related in some way to the target, then if the characteristic that is shared between the target and the preview is processed parafoveally, there will be a processing benefit at the target (though see [124–126] for discussion of parafoveal preview cost relative to parafoveal preview benefit). Thus, by systematically manipulating the characteristics of the preview in relation to the target word, it is possible to make inferences regarding which properties of the preview were parafoveally pre-processed prior to direct fixation. For example, if a parafoveal preview and target word are semantically associated, then if the preview is lexically processed, semantic facilitation of the target word should be observed upon fixation. Alternatively, if parafoveal pre-processing is limited to visual and orthographic (but not lexical) properties of a preview, then no facilitation at the target should occur. Again, investigations of this kind seek to discriminate between serial or parallel models of reading.

Table 4 provides a summary of the existing findings observed for foveally triggered effects as measured during fixation of the target word. That is, in these studies variables that affect processing of a word from (at least) its initial fixation onward and their time course are examined. Thus, the effects observed in these co-registration experiments should closely match with the results observed in traditional ERP studies. However, knowing that the time course of word processing is affected by the availability of parafoveal information, this research might investigate fairly directly the timing conundrum that exists in relation to EM and ERP effects (see Section 1).

The information that is provided in the tables is comprehensive. Space limitations preclude an extended discussion of all the aspects of the studies that are reviewed in these tables, however, evaluating all the information together leads us to form two conclusions.

First, studies have consistently observed two neural correlates of identity preview benefit. The first neural correlate is such that between 200–280 [110,114], between 140–200 and 200–300 [117], and between 120–300 [120,121], N1 amplitudes are more positive for identical previews compared to invalid previews, largely over occipito-parietal and temporal areas of the scalp. The only study which failed to find such an effect [115] involved a methodological difference such that linked mastoids were used as an offline reference (see [39] for discussion). The second neural correlate of identity preview benefit is observed between 360–400 ms, when valid previews elicit more positive amplitudes than invalid previews over central sites of the scalp [102,110,115], and between 300–500 ms, when valid previews elicit more negative amplitudes than invalid previews over occipital areas of the scalp [121]. This late preview effect on the N400 component was not observed in Degno et al. [120], indicating that this might be related to naturalness of the reading task or to baseline choices. The eye movement results also mapped onto the FRP data. Overall, and to date, this identity preview effect represents the most robust and well documented effect in the co-registration literature, having been investigated and demonstrated in word list reading experiments, prime-target pair experiments, as well as in natural reading experiments. Current understanding of the neural correlates of the identity preview benefit suggests that the latency range associated with the N1 component might reflect the period of time when efficient orthographic and/or phonological processing occurs, because orthographic and/or phonological characteristics of a word are correctly activated based on parafoveal information (see [110,121]). The effect observed on the N400 component might likely be a consequence of the effect shown on the N1 component, suggesting that when orthographic and/or phonological processing is disrupted, later cognitive processing (likely lexical or semantic processing) is also slowed down and disrupted.

The second conclusion is that the literature is limited, the findings are mixed and sometimes inconsistent. This leads us to the view that much more experimental work using these techniques is necessary to allow us to identify those effects that occur with reliability and that are robust across studies. Fairly consistent results have been observed between EM and FRP measures for PoF effects of parafoveal preview [106,109,120], preview effects of predictability [26,108,113], frequency [113,117,120], and type of preview [110,114,117,120], foveal effects of text type [111], inter-letter spacing [118], repetition [107,110,112], word predictability [26,104,108,113], syntactic and semantic violations [116,119] and foveal load [114]. However, inconsistencies in EM and FRP results have been observed for PoF effects of semantic relatedness [106,108–110,115], preview effects of semantic relatedness [110,115], foveal effects of semantic relatedness [108,110,115] and word frequency [113,120]. It remains the case, though, that co-registration investigations of aspects of reading are in their formative stages (with some effects being investigated in single studies only) and a greater body of experimental data is a necessity before firm conclusions may be formed as to the kinds of experimental manipulations that regularly and consistently produce FRP effects of specific kinds. Given the current limitations with respect to the restricted empirical investigations employing the co-registration method alongside the relatively small number of data sets, to us, it feels appropriate to place emphasis on only those effects that appear strongest across studies at this point.

4. Are FRP Components Reliable Correlates of Cognitive Processes in Reading?

As discussed in the previous section, existing co-registration studies demonstrate a series of prominent FRP components (e.g., P1, N1, N400, etc.) that are elicited during reading tasks (see Figure 2). Understanding of what these FRP components might represent in relation to cognitive processing, at this stage, is currently developing. In this section, we briefly describe each component and consider the aspects of processing associated with reading with which each may be associated.

The earliest effect that we observe is a positive visually evoked response that originates from the occipital regions of the scalp with a peak at approximately 100 ms after fixation onset on a written word. This visually evoked response, once labelled 'lambda wave', is now considered the equivalent of the P1 component [26,110]. Indeed, both responses appear to be associated with the uptake of visual information [61], and to have a common neural generator in the visual cortex [127]. The P1 component seems to be modulated by the nature of parafoveal preview [120,121], as well as the visual form of the text [111].

The P1 component is followed by the N1 component, a negative potential which is largest over left occipital, parietal and temporal areas of the scalp, with a peak approximately 200 ms after fixation onset. Existing co-registration studies support the view that the N1 component is a time-window during which orthographic and/or phonological representations of a written word become activated (see effects of parafoveal preview on the N1 component [110,114,117,120,121]), and that the negativity in this latency range is increased when less effective orthographic and/or phonological pre-processing of parafoveal information occurs [114,120,121]. In addition, there is some evidence that effects of word frequency [117] and foveal load [114] also modulate the N1 latency. At a first glance, these timing effects might fit well with previous ERP studies that have shown word frequency effects on early components (e.g., [80,128,129]), and with the time-line constraints derived from EM effects [25]. However, currently, given the sparse evidence for such effects, and the failure to find these effects in sentence reading experiments [113,120], further investigation is needed.

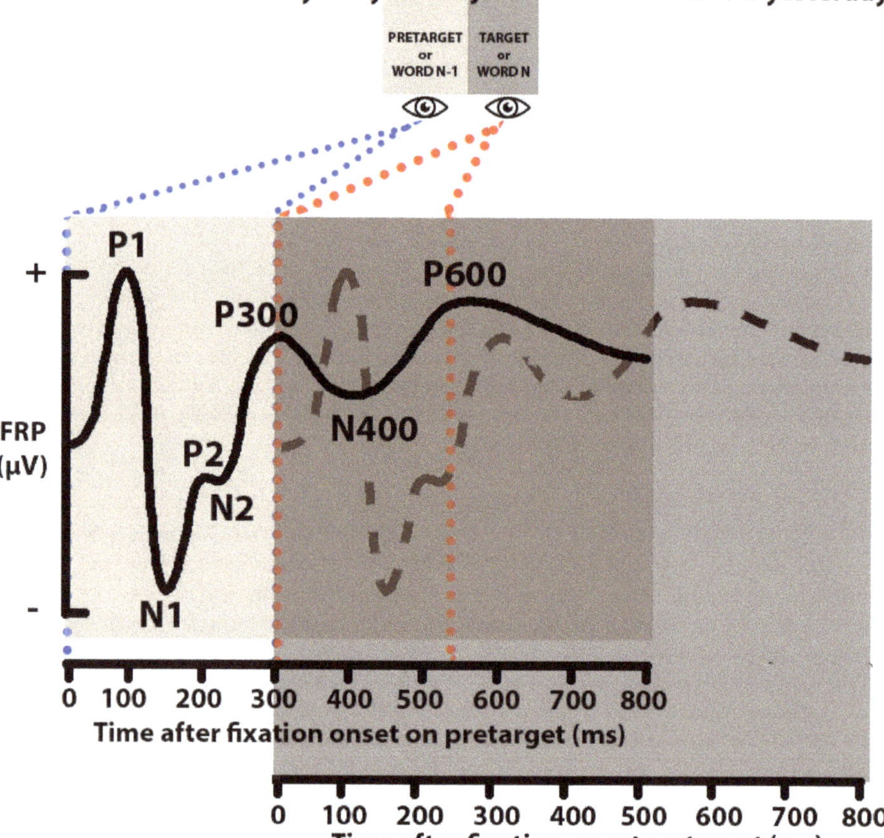

Figure 2. Here we offer a stylized characterisation of deconcolved waveforms that might (ideally) be revealed if deconvolution processes were applied successfully to an average FRP data stream recorded across two successive fixations made on the word "carrot" and then on the word "juice" as the sentence "John made a very tasty carrot juice with fresh carrots yesterday" was read. The solid black line represents the waveform and components (i.e., P1, N1, P2, N2, P300, N400, P600) that result from processes associated with the fixation on the pretarget word, and these have been separated from the waveform and components (dashed line) associated with the fixation on the target word. Note that the actual waveform that would be recorded during the experimental trial would be a convolved signal comprised of multiple waveforms deriving from fixations on the word(s) prior to the pretarget word, the target word (and potentially) the posttarget word ('with' in the current example). For simplicity, we have not included the convolved waveform in this figure, but the overlapping portion of the two panels (time-locked to fixation onset on pretarget and target words) shows where that convolved signal would occur (for only pretarget and target words) if it had been illustrated. See Ehinger and Dimigen (2019) [130]) for a discussion of deconvolution and its mathematical properties.

Following the N1 component, a P2 component has also been observed in several co-registration studies [106,109,113]. This positive potential was observed first over anterior and central areas of the scalp between approximately 140–280 ms after fixation onset, being modulated by parafoveal semantic relatedness [106] and word predictability [113], and then later between approximately 200–280 ms over occipital scalp sites being modulated by parafoveal word form [109]. However, again, given the small number of studies to have found modulation of this component, the functional description of the P2 component currently remains unclear.

Another prominent FRP component is the well-established N400, a negativity observed over centro-parietal areas of the scalp. This FRP component appears to be modulated by both parafoveal and foveal manipulations. The parafoveal N400 component time-locked to the fixation onset on pretarget words is modulated by parafoveal preview type [120] and parafoveal semantic relatedness [108,115]. The N400 component time-locked to fixation onset on the target word seems to be modulated by parafoveal preview type [110,114,115,121], semantic relatedness [110,115], word predictability [26,104,108,113], as well as syntactic and semantic violations [116,119]. Observing modulation of this component during natural reading is noteworthy. In one respect, it suggests that there is a stage of processing in this time latency that is affected by linguistic manipulations regardless of the paradigm used, and therefore that the component might reflect aspects of processing that continue beyond a single eye fixation [27]. In another respect, it raises the interesting possibility that the same cognitive mechanism might underlie effects observed in both early and late components (e.g., the parafoveal preview effect observed on both the N1 and N400 components).

The N400 component is followed by the P600 component, a positive response with a widespread distribution (largest over centro-parietal sites), with a peak at approximately 600 ms from fixation onset. The P600 component time-locked to the fixation onset on target words appears to be modulated by parafoveal preview type [115], semantic relatedness [108,115], word predictability [104], as well as syntactic and semantic violations [116,119].

The existing findings suggest that FRP components that are observed in existing co-registration studies are quite comparable to components observed in more standard ERP investigations, and these appear to be elicited by similar experimental manipulations. Given this, it seems appropriate to argue that FRP components are as reliable electrophysiological correlates of cognitive processing underlying reading as those observed in more traditional (non-co-registration) ERP data sets. In addition, the ecological validity advantage of the FRPs over the traditional ERP methods allows for a more comprehensive investigation of the neural correlates of reading.

5. Do Co-Registration Studies add Value to Our Understanding of Reading?

In answering this question, we consider that there are at least two relevant perspectives, in that co-registration adds value (and here we mean in the sense of providing increased scientific insight) both to studies in which EMs alone are recorded, as well as to studies in which ERPs are solely recorded.

Let us first consider whether co-registration studies, potentially at least, offer greater insight than studies in which only EM data are recorded. In relation to this issue, note that EM studies investigating reading very often produce patterns of effects that are statistically robust with clear and predicted numerical effects. Furthermore, as our discussion of the existing co-registration literature above should elucidate, such clear EM effects are in stark contrast to patterns of effects in FRP studies that can frequently be mixed and much more difficult to interpret. Despite this, based on our assessment of this body of work, it is our view that co-registration studies do offer added value. The study by Degno et al. (2019) [120] might serve as an example to substantiate our claim. In their experiment, Degno et al. (2019) [120] obtained no significant difference in the EM data between X-string and letter string parafoveal preview conditions at the target word. The early first-pass reading EM results associated with both X-string and letter string parafoveal previews showed similar disruption to reading compared to the identity preview conditions (X-string vs. identity preview condition: difference of 41 ms in first fixation duration, 63 ms in single fixation duration; letter string vs. identity preview condition: difference of 41ms in first fixation duration, 57 ms in single fixation duration). Thus, on the basis of these data, there was no observable processing difference between the X-string and letter-string preview conditions. However, significant differences between these conditions were observed in the FRP data. Our interpretation of this result is that during the fixations (of numerically very similar duration), the nature of the processing that occurred in the brain produced an amount of disruption under each condition that was comparable, therefore increasing reading behaviour similarly. Yet, even though this was the case, it seems entirely reasonable to us to suggest that the disruption that did occur during these fixations may well have been qualitatively different in nature. If this suggestion is correct, then these results exemplify one way in which FRP data may add value by way of scientific insight, in that they offer the potential to discern between qualitatively different types of processing that each cause quantitatively comparable disruption to fixation durations.

Next, let us consider how co-registration as an empirical method may offer increased potential insight relative to studies recording EEG data alone. From an ERP perspective, it might initially be difficult to appreciate the potential of co-registration under free reading conditions, in that experiments of this kind add disruption to the EEG signal via ocular artifacts that arise due to saccadic EMs intrinsic to reading. However, let us briefly consider the studies that have directly compared results obtained under active (i.e., saccadic reading) and passive (i.e., RSVP or RSVP-with-flankers paradigm) reading conditions ([107,112,114,116,117]). These studies have shown several effects that have similar scalp distributions in both the passive ERP and active FRP settings (although see Niefind and Dimigen (2016) [117], where the parafoveal preview effect and preview frequency effect were observed only in the FRPs, and Metzner et al. (2016) [116], where sentence-final semantic violations elicited different voltages in the N400 latency range for the RSVP condition, and in both the N400 and P600 latency ranges in the FRP reading condition). However, and critically in relation to our argument here, in all these studies those effects were larger in size, were longer lasting and had an earlier onset in the FRP active reading conditions compared to the passive ERP reading conditions. For example, Kornrumpf and colleagues (2016)[114] observed a parafoveal preview effect between 230–250 ms for passive reading ERPs, but between 160–300 ms in FRPs under active reading conditions. We consider that although this difference in the onset of the effect is small, it is likely very important in that it might reflect a more advanced timeline associated with natural reading due to the rapid dynamics of attentional deployment in such circumstances. Clearly, it is the case that the time of processing has been an issue of central investigation in reading research, and given this, to us at least, any such differences are potentially informative regarding the precise nature of processing.

In addition, with co-registration it is possible to investigate neural correlates of specific and different aspects of oculomotor behaviour that might occur under different experimental conditions during reading ([116,122]). For example, Metzner et al. (2016) [116] separated trials in which participants did, or did not, make a regressive saccade in order to re-read previous parts of the sentence, and revealed different FRP effects associated with re-reading of syntactic and semantic violations. That is, whilst trials with regressions elicited effects on the N400 (for sentence-final syntactic and semantic violations) and P600 component (for sentence-medial and sentence-final syntactic and semantic violations), in trials without regressions effects associated with sentence-final syntactic and semantic violations elicited a sustained negativity only. In this way, again, FRP data offer potential for added value.

Finally, an area of reading research that has received little attention, but which may be fruitful for co-registration research, concerns the investigation of oculomotor planning per se. For example, we question whether there may be differential FRP signatures associated with alternative patterns of oculomotor activity. For instance, whether there is a different FRP "signature" associated with a fixation prior to a progressive saccade relative to that for a fixation prior to a regressive saccade.

6. What is the Nature of the Relationship between Eye Movements and Fixation-Related Potentials?

Let us state at the outset that the currently limited number of studies mean that a completely clear perspective is not, at present, possible. Moreover, this is an extremely difficult question to tackle. Nonetheless, we do feel that it is at least necessary to consider aspects of those results that do currently exist that might inform an answer. For example, we consider it important, as we have already noted, that there are relationships and consistencies in EM and FRP data sets that are suggestive of common (potentially causative) links at the level of neural and cognitive processing. If the suggestion that such commonality across data streams is reflective of common aspects of psychological function, then such results are extremely encouraging in respect of the future value of the co-registration approach as a tool to further future understanding. At this stage, from our perspective, we see enough such effects in existing data sets to persuade us that this venture is worthwhile. All of this said, it remains the case that there are a number of situations when there are inconsistencies in effects across the two data streams. Furthermore, such inconsistencies may take the form of a particular effect in one data stream seemingly corresponding to multiple different effects in the other data stream. Alternatively, inconsistencies might even occur such that whilst an effect occurs in one data stream, there is no evidence of any corresponding effect in the other stream. If both EM and FRP data streams do reflect common causative neural and cognitive aspects of processing, then it is quite unclear why such inconsistency should arise. Again, at this stage, we feel it is prudent not to speculate to any great degree as to potential explanations for such data patterns. Instead, it seems likely that over time as results from co-registration studies proliferate, the patterns of consistency, as well as the patterns of inconsistency will develop and become substantiated, and at such a point downstream, we should then be better equipped to provide an answer to the question that we have posed.

Even though we are cautious in our interpretation of existing co-registration data set, based on our assessment of current data patterns, it appears that during natural reading, when a clear expectation is not met, as for example with predictability and semantic and syntactic violation effects, both behavioural and neural systems exhibit pronounced disruption. The disruption can be observed clearly and comparably in both the EM and FRP data streams, with counterpart "signatures" in both. In contrast, when subtler, less disruptive manipulations are adopted, such as word frequency and semantic relatedness manipulations, where processing may proceed relatively unhindered (i.e., without pronounced disruption), then it appears that less consistent counterpart effects appear in the two data streams. Thus, potentially, the degree of disruption (that is, both the magnitude of cognitive disruption, as well as the extent to which effects are statistically robust) that a manipulation causes to ongoing cognitive processing in reading may be the mediator of the degree to which comparable counterpart EM and FRP effects are observed. It is for future research to refine explanations of why EM and FRP recordings sometimes offer consistency in effects, and on other occasions inconsistency. The application of linked mixed models [131] to co-registration data might be a useful analytical tool to further our understanding of the relationship between EM and FRP measures.

A final point that we will consider concerns the contrasting manner in which measures of EM and FRP are used in the literature. Within the EM research community, different EM measures (e.g., first fixation duration, single fixation duration, gaze duration, etc.) are used to show the temporal course of an effect, and it is well documented that different EM measures can be influenced by the same experimental variable (e.g., [132]), and that the same EM measure can provide insights into different aspects of processing associated with reading [23,77]. Thus, the reason why different, and often numerous, EM measures are analysed and reported is because the entire time course of an effect is being considered such that claims can be made as to the point in time at which an experimental manipulation first exerted an observable influence on processing, as well as the duration (and arguably, nature) of its influence. There is a crucial and fundamental difference between the approach to measures in the EM and ERP literatures. Broadly speaking, EEG researchers favour an approach in which effects are directly investigated through the examination of specific components known to be modulated by a particular experimental variable (although see for example [128,133] for a 'time course processing' approach). Of course, current understanding has moved on from the idea that individual components within an EEG data set index particular cognitive functions, almost on a "one-to-one" basis (e.g., N400 as a language measure of semantic processing). Instead, it is now widely considered that the EEG data stream provides an electrophysiological index of activity in the cortical mechanisms underlying a mental operation (e.g., N400 as an index of processing in semantic memory of a range of meaningful stimuli across different modalities; [18]). However, despite this, it remains the case that many ERP studies focus on particular ERP components, and seek to demonstrate an influence of one or more experimental variables on that component. In contrast to the approach in EM studies where the continuity of processing over time is investigated, in ERP studies, often, "snapshots" in time are offered, and in this way, insight into the entire time course of the influence of an experimental variable may be missed. In our view, co-registration research offers an opportunity to bring the "time course of processing" approach to analysis that is often utilised in EM research to the analysis of EEG data sets. For example, existing co-registration studies have analysed a series of FRP components that are initiated and developed from fixation onset, and such studies do adopt more of a time course approach to their analyses. In addition, the adoption of raster plots that illustrate data acquired from all scalp sites within relatively extended time windows of analysis, also seems to offer insight into the nature of change in FRP data over time.

7. Future Directions

An important issue that faces co-registration researchers concerns how we might most effectively separate components that overlap temporally (i.e., from one fixation to the next) within the EEG data (e.g., [26,77,78,88,111]). Figure 2 presents a visual example of the issue. As can be seen, each fixation has associated with it a waveform containing a series of components that overlap with the waveform and components associated with the preceding and following fixations. In the figure, the composite waveform has been separated (on the assumption of the successful application of deconvolution) to illustrate how components associated with each successive fixation might be isolated and identified. It should be clear that without deconvolution, the composite waveform would have a form represented by a combination of the underlying waveforms (and this is not represented in Figure 2).

Progress has been made in developing approaches to deal with this issue (e.g., [130,134–139]), but the majority of the existing, published, co-registration experiments investigating reading do not report analyses that separate temporally overlapping components deriving from different fixations, and therefore, have not yet directly addressed this challenge (although see [119] for an assessment of spatial overlap of components and [122] for temporal overlap). Limiting the analysis to time windows in which the next fixation does not yet overlap (e.g., selecting only fixations with a minimum fixation duration and analysing the FRP components within those intervals; see Nikolaev et al., 2016 [91] for a discussion) does directly address this issue. However, this approach leads to a series of other important considerations (e.g., fixations that could reflect meaningful cognitive processes might likely be systematically excluded from the analyses), which make the deconvolution approach a much better solution to date.

New advances in techniques to deconvolve the EEG signal whilst at the same time controlling for covariation due to other effects have been developed (e.g., [130]). Loberg et al. [122] have taken the first step to try to adopt this approach to investigate school-aged children with slow or typical reading speed as they performed a natural reading task where a target word in each sentence was manipulated for length. Deconvolved FRPs were analysed, and saccade amplitudes were used as covariates. Loberg et al. did not provide comparative analyses of convolved vs. deconcolved data sets, however, Ehinger and Dimigen [130] did undertake such analyses (though their task involved face processing, not reading). These analyses indicated that whilst the main components and effects are observable in both types of analysis, the analyses of the deconvolved data appeared more tightly and definitively indicative of specific processes associated with experimental manipulations. We anticipate that researchers adopting co-registration to investigate reading will increasingly engage with these techniques into the future.

A second area that we see as offering future promise is the investigation of brain oscillatory activity. Increases and decreases in power in specific frequency bands are considered to reflect synchronization and desynchronization, respectively, of the underlying neural networks. To date, a very limited number of co-registration studies has investigated brain responses in the time-frequency domain during reading [102–105]. Based on published studies to date, two considerations emerge that may be of particular importance. First, Kornrumpf, Dimigen, and Sommer [102] have shown that analysis of these data might best suit investigation of more covert aspects of cognitive processing in reading, such as distribution/deployment of attention in relation to foveal and parafoveal processing. In this sense, analysis of brain oscillation might become a crucial tool to use in the context of distinguishing between different models of reading that specify how and when attention is allocated across words over fixations. At a more general level though, brain oscillations might also offer an opportunity to investigate induced (rather than evoked) meta-cognitive aspects of reading, such as task engagement, attentivity, dual-task costs, distraction effects, discourse coherence and text comprehension, as well as tasks demands (see e.g., [103]). Again, we must reiterate that the investigation of brain oscillatory activity is currently limited in silent reading research (see [140]), and to an even greater extent in co-registration research. It is for future research to demonstrate the true value of the approach.

8. Conclusions

The present review has provided an overview of the history and literature on co-registration of EMs and FRPs in reading, and discussed the potential of this methodology for providing novel scientific insight into the nature of processes underlying reading. The appraisal of the existing literature has allowed us to raise questions for consideration that we hope will challenge current understanding and stimulate debate concerning the neural correlates of cognitive processing in reading. We consider the theoretical assumptions that underlie the co-registration approach to be plausible, and that consideration of both data streams simultaneously provides additional, complementary value to the insights we can obtain by considering either data stream alone. However, current understanding of the relationship between oculomotor events and neural correlates of those events remains unclear. It appears crucial to conduct a significantly larger number of experiments with this methodology to develop our understanding of how neural and cognitive processes associated with oculomotor events relate to their FRP correlates in reading. The timing and nature of these correlates time-locked to specific words will be instrumental in the development of better specified and more comprehensive models of reading.

Funding: This research was funded by the Economic and Social Research Council, grant number ES/R003386/1.

Acknowledgments: We thank Olaf Dimigen, Raymond Klein and two anonymous reviewers for their insightful comments on earlier versions of the manuscript.

Conflicts of Interest: The authors declare no conflict of interest. The funder had no role in the design or writing of the manuscript.

References

1. Rayner, K. Eye movements in reading and information processing: 20 Years of research. *Psychol. Bulletin* **1998**, *124*, 372–422. [CrossRef]
2. Rayner, K. Eye movements and attention in reading, scene perception, and visual search. *Q. J. Exp. Psychol.* **2009**, *62*, 1457–1506. [CrossRef]
3. Rayner, K.; Clifton, C., Jr. Language processing in reading and speech perception is fast and incremental: Implications for event-related potential research. *Biol. Psychol.* **2009**, *80*, 49. [CrossRef] [PubMed]
4. Frazier, L.; Rayner, K. Making and correcting errors during sentence comprehension: Eye movements in the analysis of structurally ambiguous sentences. *Cogn. Psychol.* **1982**, *14*, 178–210. [CrossRef]
5. Just, M.A.; Carpenter, P.A. A theory of reading: From eye fixations to comprehension. *Psychol. Rev.* **1980**, *87*, 329–354. [CrossRef] [PubMed]
6. Schotter, E.R.; Angele, B.; Rayner, K. Parafoveal processing in reading. *Atten. Percept. Psychophys.* **2012**, *74*, 5–35. [CrossRef]
7. Liversedge, S.P.; Findlay, J.M. Saccadic eye movements and cognition. *Trends Cogn. Sci.* **2000**, *4*, 6–14. [CrossRef]
8. Rayner, K. Eye movements in reading and information processing. *Psychol. Bull.* **1978**, *85*, 618–660. [CrossRef]
9. Inhoff, A.W.; Rayner, K. Parafoveal word processing during eye fixations in reading: Effects of word frequency. *Percept. Psychophys.* **1986**, *40*, 431–439. [CrossRef]

10. Rayner, K.; Duffy, S.A. Lexical complexity and fixation times in reading: Effects of word frequency, verb complexity, and lexical ambiguity. *Mem. Cogn.* **1986**, *14*, 191–201. [CrossRef]

11. Ehrlich, S.F.; Rayner, K. Contextual effects on word perception and eye movements during reading. *J. Verbal Learn. Verbal Behav.* **1981**, *20*, 641–655. [CrossRef]

12. Rayner, K.; Well, A.D. Effects of contextual constraint on eye movements in reading: A further examination. *Psychon. Bull. Rev.* **1996**, *3*, 504–509. [CrossRef] [PubMed]

13. Reichle, E.D.; Reingold, E.M. Neurophysiological constraints on the eye-mind link. *Front. Hum. Neurosci.* **2013**, *7*, 1–6. [CrossRef] [PubMed]

14. Nunez, P.L.; Srinivasan, R. *Electric Fields of the Brain: The Neurophysics of EEG*; Oxford University Press: New York, NY, USA, 2006.

15. Callaway, E.; Tueting, P.; Koslow, S.H. *Event-Related Potentials in Man*; Academic Press: New York, NY, USA, 1978.

16. Luck, S.J. *An Introduction to the Event-Related Potential Technique*; MIT Press: Cambridge, MA, USA, 2014.

17. Luck, S.J.; Kappenman, E.S. *The Oxford Handbook of Event-Related Potential Components*; Oxford University Press: New York, NY, USA, 2011.

18. Kutas, M.; Federmeier, K.D. Thirty years and counting: Finding meaning in the N400 component of the event-related brain potential (ERP). *Annu. Rev. Psychol.* **2011**, *62*, 621–647. [CrossRef]

19. Kutas, M.; Hillyard, S.A. Reading senseless sentences: Brain potentials reflect semantic incongruity. *Science* **1980**, *207*, 203–205. [CrossRef]

20. Kutas, M.; Lindamood, T.E.; Hillyard, S.A. Word expectancy and event-related brain potentials during sentence processing. In *Preparatory States and Processes*; Kornblum, S., Requin, J., Eds.; Erlbaum: Hillsdale, NJ, USA, 1984; pp. 217–237.

21. Osterhout, L.; Holcomb, P.J. Event-related brain potentials elicited by syntactic anomaly. *J. Mem. Lang.* **1992**, *31*, 785–806. [CrossRef]

22. DeLong, K.A.; Urbach, T.P.; Kutas, M. Probabilistic word pre-activation during language comprehension inferred from electrical brain activity. *Nat. Neurosci.* **2005**, *8*, 1117–1121. [CrossRef]

23. Liversedge, S.P.; Paterson, K.B.; Pickering, M.J. Eye movements and measures of reading time. In *Eye Guidance in Reading and Scene Perception*; Underwood, G., Ed.; Elsevier Science Ltd.: Oxford, UK, 1998; pp. 55–75.

24. Sereno, S.C.; Rayner, K. The when and where of reading in the brain. *Brain Cogn.* **2000**, *42*, 78–81. [CrossRef]

25. Sereno, S.C.; Rayner, K. Measuring word recognition in reading: Eye movements and event-related potentials. *Trends Cogn. Sci.* **2003**, *7*, 489–493. [CrossRef]

26. Dimigen, O.; Sommer, W.; Hohlfeld, A.; Jacobs, A.M.; Kliegl, R. Coregistration of eye movements and EEG in natural reading: Analyses and review. *J. Exp. Psychol.* **2011**, *140*, 552–572. [CrossRef]

27. Kliegl, R.; Dambacher, M.; Dimigen, O.; Jacobs, A.M.; Sommer, W. Eye movements and brain electric potentials during reading. *Psychol. Res.* **2012**, *76*, 145–158. [CrossRef]

28. Rayner, K. The perceptual span and peripheral cues in reading. *Cogn. Psychol.* **1975**, *7*, 65–81. [CrossRef]

29. McConkie, G.W.; Rayner, K. The span of the effective stimulus during a fixation in reading. *Percept. Psychophys.* **1975**, *17*, 578–586. [CrossRef]

30. Kutas, M.; Hillyard, S.A. Event-related potentials to semantically inappropriate and surprisingly large words. *Biol. Psychol.* **1980**, *11*, 99–116. [CrossRef]

31. Kutas, M.; Hillyard, S.A. Reading between the lines: Event-related brain potentials during natural sentence processing. *Brain Lang.* **1980**, *11*, 354–373. [CrossRef]

32. Kutas, M.; Hillyard, S.A. Event-related brain potentials to grammatical errors and semantic anomalies. *Mem. Cogn.* **1983**, *11*, 539–550. [CrossRef]

33. Marton, M.; Szirtes, J.; Breuer, P. Electrocortical signs of word categorization in saccade-related brain potentials and visual evoked potentials. *Int. J. Psychophysiol.* **1985**, *3*, 131–144. [CrossRef]

34. Marton, M.; Szirtes, J. Context effects on saccade-related brain potentials to words during reading. *Neuropsychologia* **1988**, *26*, 453–463. [CrossRef]

35. Marton, M.; Szirtes, J. Saccade-related brain potentials during reading correct and incorrect versions of proverbs. *Int. J. Psychophysiol.* **1988**, *6*, 273–280. [CrossRef]

36. Barber, H.A.; Ben-Zvi, S.; Bentin, S.; Kutas, M. Parafoveal perception during sentence reading? An ERP paradigm using rapid serial visual presentation (RSVP) with flankers. *Psychophysiology* **2011**, *48*, 523–531. [CrossRef]

37. Barber, H.A.; Doñamayor, N.; Kutas, M.; Münte, T. Parafoveal N400 effect during sentence reading. *Neurosci. Lett.* **2010**, *479*, 152–156. [CrossRef]

38. Barber, H.A.; van der Meij, M.; Kutas, M. An electrophysiological analysis of contextual and temporal constraints on parafoveal word processing. *Psychophysiology* **2013**, *50*, 48–59. [CrossRef] [PubMed]

39. Li, N.; Niefind, F.; Wang, S.; Sommer, W.; Dimigen, O. Parafoveal processing in reading Chinese sentences: Evidence from event-related brain potentials. *Psychophysiology* **2015**, *52*, 1361–1374. [CrossRef] [PubMed]

40. Reichle, E.D.; Pollatsek, A.; Fisher, D.L.; Rayner, K. Toward a model of eye movement control in reading. *Psychol. Rev.* **1998**, *105*, 125. [CrossRef] [PubMed]

41. Reichle, E.D.; Warren, T.; McConnell, K. Using EZ Reader to model the effects of higher level language processing on eye movements during reading. *Psychon. Bull. Rev.* **2009**, *16*, 1–21. [CrossRef]

42. Reichle, E.D.; Liversedge, S.P.; Pollatsek, A.; Rayner, K. Encoding multiple words simultaneously in reading is implausible. *Trends Cogn. Sci.* **2009**, *13*, 115–119. [CrossRef]

43. Schotter, E.R. Reading ahead by hedging our bets on seeing the future: Eye tracking and electrophysiology evidence for parafoveal lexical processing and saccadic control by partial word recognition. *Psychol. Learn. Motiv.* **2018**, *68*, 263–298.

44. Schotter, E.R.; Reichle, E.D.; Rayner, K. Rethinking parafoveal processing in reading: Serial-attention models can explain semantic preview benefit and N+ 2 preview effects. *Vis. Cogn.* **2014**, *22*, 309–333. [CrossRef]

45. Engbert, R.; Nuthmann, A.; Richter, E.M.; Kliegl, R. SWIFT: A dynamical model of saccade generation during reading. *Psychol. Rev.* **2005**, *112*, 777–813. [CrossRef]

46. Schad, D.J.; Engbert, R. The zoom lens of attention: Simulating shuffled versus normal text reading using the SWIFT model. *Vis. Cogn.* **2012**, *20*, 391–421. [CrossRef]

47. Snell, J.; van Leipsig, S.; Grainger, J.; Meeter, M. OB1-reader: A model of word recognition and eye movements in text reading. *Psychol. Rev.* **2018**, *125*, 969–984. [CrossRef]

48. Zang, C. New Perspectives on Serialism and Parallelism in Oculomotor Control During Reading: The Multi-Constituent Unit Hypothesis. *Vision* **2019**, *3*, 50. [CrossRef]

49. Evans, C.C. Spontaneous excitation of the visual cortex and association areas; lambda waves. *Electroencephalogr. Clin. Neurophysiol.* **1953**, *5*, 69–74. [CrossRef]

50. Kurtzberg, D.; Vaughan, H.G., Jr. Electrophysiological observations on the visuomotor system and visual neurosensorium. In *Visual Evoked Potentials in Man: New Developments*; Desmedt, J.E., Ed.; Clarendon Press: Oxford, UK, 1977.

51. Yagi, A. Saccade size and lambda complex in man. *Physiol. Psychol.* **1979**, *7*, 370–376. [CrossRef]

52. Barlow, J.S.; Cigánek, L. Lambda responses in relation to visual evoked responses in man. *Electroencephalogr. Clin. Neurophysiol.* **1969**, *26*, 183–192. [CrossRef]

53. Becker, W.; Hoehne, O.; Iwase, K.; Kornhuber, H.H. Bereitschaftspotential, prfimotorische Positivierung und andere Hirnpotentiale bei sakkadischen Augenbewegungen. *Vis. Res.* **1972**, *12*, 421–436. [CrossRef]

54. Kurtzberg, D.; Vaughan, H.G., Jr. Topographic analysis of human cortical potentials preceding self-initiated and visually triggered saccades. *Brain Res.* **1982**, *243*, 1–9. [CrossRef]

55. Thickbroom, G.W.; Mastaglia, F.L. Presaccadic 'spike' potential: Investigation of topography and source. *Brain Res.* **1985**, *339*, 271–280. [CrossRef]

56. Csibra, G.; Johnson, M.H.; Tucker, L.A. Attention and oculomotor control: A high-density ERP study of the gap effect. *Neuropsychologia* **1997**, *35*, 855–865. [CrossRef]

57. Richards, J.E. Cortical sources of event-related potentials in the prosaccade and antisaccade task. *Psychophysiology* **2003**, *40*, 878–894. [CrossRef]

58. Everling, S.; Krappmann, P.; Flohr, H. Cortical potentials preceding pro-and antisaccades in man. *Electroencephalogr. Clin. Neurophysiol.* **1997**, *102*, 356–362. [CrossRef]

59. Wauschkuhn, B.; Verleger, R.; Wascher, E.; Klostermann, W.; Burk, M.; Heide, W.; Kömpf, D. Lateralized human cortical activity for shifting visuospatial attention and initiating saccades. *J. Neurophysiol.* **1998**, *80*, 2900–2910. [CrossRef]

60. Riggs, L.A.; Merton, P.A.; Morton, H.B. Suppression of visual phosphenes during saccadic eye-movements. *Vis. Res.* **1974**, *14*, 997–1011. [CrossRef]

61. Marton, M.; Szirtes, J. Averaged lambda potential and visual information processing. *Studia Psychol.* **1982**, *24*, 165–170.

62. Boylan, C.; Doig, H.R. Effect of saccade size on presaccadic spike potential amplitude. *Investigative Ophthalmol. Vis. Sci.* **1989**, *30*, 2521–2527.

63. Riemslag, F.C.; Van der Heijde, G.L.; Van Dongen, M.M.; Ottenhoff, F. On the origin of the presaccadic spike potential. *Electroencephalogr. Clin. Neurophysiol.* **1988**, *70*, 281–287. [CrossRef]

64. Thickbroom, G.W.; Mastaglia, F.L. Presaccadic spike potential. Relation to eye movement direction. *Electroencephalogr. Clin. Neurophysiol.* **1986**, *64*, 211–214. [CrossRef]

65. Scott, D.F.; Moffett, A.; Bickford, R.G. Comparison of two types of visual evoked potentials: Pattern reversal and eye movement (lambda). *Electroencephalogr. Clin. Neurophysiol.* **1981**, *52*, 102–104. [CrossRef]

66. Yagi, A. Averaged cortical potentials (lambda responses) time-locked to onset and offset of saccades. *Physiol. Psychol.* **1981**, *9*, 318–320. [CrossRef]

67. Thickbroom, G.W.; Knezevic, W.; Carroll, W.M.; Mastaglia, F.L. Saccade onset and offset lambda waves: Relation to pattern movement visually evoked potentials. *Brain Res.* **1991**, *551*, 150–156. [CrossRef]

68. Gaarder, K.; Krauskopf, J.; Graf, V.; Kropfl, W.; Armington, J.C. Averaged brain activity following saccadic eye movement. *Science* **1964**, *146*, 1481–1483. [CrossRef]

69. Kazai, K.; Yagi, A. Integrated effects of stimulation at fixation points on EFRP (eye-fixation related brain potentials). *Int. J. Psychophysiol.* **1999**, *32*, 193–203. [CrossRef]

70. Kazai, K.; Yagi, A. Contrast dependence of lambda response. *Int. Congr. Ser.* **2005**, *1278*, 61–64. [CrossRef]

71. Marton, M.; Szirtes, J.; Donauer, N.; Breuer, P. Sccade-related brain potentials in semantic categorization tasks. *Biol. Psychol.* **1985**, *20*, 163–184. [CrossRef]

72. Barlow, J.S. Brain information processing during reading: Electrophysiological correlates. *Dis. Nerv. Syst.* **1971**, *32*, 668–672.

73. Kutas, M.; Hillyard, S.A. Event-related potentials in cognitive science. In *Handbook of Cognitive Neuroscience*; Gazzaniga, M.S., Ed.; Plenum: New York, NY, USA, 1984.

74. Joyce, C.A.; Gorodnitsky, I.F.; King, J.W.; Kutas, M. Tracking eye fixations with electroocular and electroencephalographic recordings. *Psychophysiology* **2002**, *39*, 607–618. [CrossRef]

75. Takeda, Y.; Sugai, M.; Yagi, A. Eye fixation related potentials in a proof reading task. *Int. J. Psychophysiol.* **2001**, *40*, 181–186. [CrossRef]

76. Reichle, E.D.; Tokowicz, N.; Liu, Y.; Perfetti, C.A. Testing an assumption of the E-Z Reader model of eye-movement control during reading: Using event-related potentials to examine the familiarity check. *Psychophysiology* **2011**, *48*, 993–1003. [CrossRef]

77. Raney, G.E.; Rayner, K. Event-Related Brain Potentials, Eye Movements, and Reading. *Psychol. Sci.* **1993**, *4*, 283–286. [CrossRef]

78. Dambacher, M.; Kliegl, R. Synchronizing timelines: Relations between fixation durations and N400 amplitudes during sentence reading. *Brain Res.* **2007**, *1155*, 147–162. [CrossRef]

79. Ashby, J.; Martin, A.E. Prosodic phonological representations early in visual word recognition. *J. Exp. Psychol.* **2008**, *34*, 224–226. [CrossRef] [PubMed]

80. Sereno, S.C.; Rayner, K.; Posner, M.I. Establishing a time-line of word recognition: Evidence from eye movements and event-related potentials. *Neuroreport* **1998**, *9*, 2195–2200. [CrossRef]

81. Raney, G.E.; Rayner, K. The effects of word frequency during two readings of a text. Presented at the meeting of the Psychonomic Society, San Francisco, CA, USA, November 1991.

82. Raney, G.E. Monitoring changes in cognitive load during reading: An event-related brain potential and reaction time analysis. *J. Exp. Psychol.* **1993**, *19*, 51–69. [CrossRef]

83. Morris, R.K. Sentence context effects on lexical access. In *Eye Movements and Visual Cognition: Scene Perception and Reading*; Rayner, K., Ed.; Springer-Verlag: New York, NY, USA, 1992; pp. 317–332.

84. Raney, G.E.; Fischler, I.; Hardonk, M. ERP evidence of lexical and message level priming in sentences. Presented at the meeting of the Psychonomic Society, St. Louis, MO, USA, November 1992.

85. Kliegl, R.; Nuthmann, A.; Engbert, R. Tracking the mind during reading: The influence of past, present, and future words on fixation durations. *J. Exp. Psychol.* **2006**, *135*, 12–35. [CrossRef]

86. Dambacher, M.; Kliegl, R.; Hofmann, M.; Jacobs, A.M. Frequency and predictability effects on event-related potentials during reading. *Brain Res.* **2006**, *1084*, 89–103. [CrossRef]

87. Balota, D.A.; Chumbley, J.I. Where are the effects of frequency in visual word recognition tasks? Right where we said they were! Comment on Monsell, Doyle, and Haggard (1989). *J. Exp. Psychol.* **1990**, *119*, 231–237. [CrossRef]

88. Baccino, T. Eye movements and concurrent event-related potentials': Eye fixation-related potential investigations in reading. In *The Oxford Handbook of Eye Movements*; Liversedge, S.P., Gilchrist, I., Everling, S., Eds.; Oxford University Press: New York, NY, USA, 2011.

89. Himmelstoss, N.A.; Schuster, S.; Hutzler, F.; Moran, R.; Hawelka, S. Co-registration of eye movements and neuroimaging for studying contextual predictions in natural reading. *Lang. Cogn. Neurosci.* **2019**, 1–18. [CrossRef]

90. Kliegl, R.; Dambacher, M.; Dimigen, O.; Sommer, W. Oculomotor control, brain potentials, and timelines of word recognition during natural reading. In *Current Trends in Eye Tracking Research*; Horsley, M., Eliot, M., Knight, B.A., Reilly, R., Eds.; Springer: Cham, Switzlerland, 2014; pp. 141–155.

91. Nikolaev, A.R.; Meghanathan, R.N.; van Leeuwen, C. Combining EEG and eye movement recording in free viewing: Pitfalls and possibilities. *Brain Cogn.* **2016**, *107*, 55–83. [CrossRef]

92. Ehinger, B.V.; Gross, K.; Ibs, I.; Koenig, P. A new comprehensive Eye-Tracking Test Battery concurrently evaluating the Pupil Labs Glasses and the EyeLink 1000. *bioRxiv* **2019**, 536243. [CrossRef]

93. Holmqvist, K. Common Predictors of Accuracy, Precision and Data Loss in 12 Eye-Trackers. 2017. Available online: https://www.researchgate.net/publication/321678981 (accessed on 11 February 2019).

94. Holmqvist, K.; Nystrom, M.; Andersson, R.; Dewhurst, R.; Jarodzka, H.; van de Weijer, J. *Eye Tracking. A Comprehensive Guide to Methods and Measures*; Oxford University Press: Oxford, UK, 2011.

95. Holmqvist, K.; Nyström, M.; Mulvey, F. Eye tracker data quality: What it is and how to measure it. In Proceedings of the Symposium on Eye Tracking Research and Applications, Santa Barbara, CA, USA, March 2012; Association for Computing Machinery: New York, NY, USA, 2012; pp. 45–52. [CrossRef]

96. Croft, R.J.; Barry, R.J. Removal of ocular artifact from the EEG: A review. *Clin. Neurophysiol.* **2000**, *30*, 5–19. [CrossRef]

97. Delorme, A.; Sejnowski, T.; Makeig, S. Enhanced detection of artifacts in EEG data using higher-order statistics and independent component analysis. *Neuroimage* **2007**, *34*, 1443–1449. [CrossRef] [PubMed]

98. Gratton, G. Dealing with artifacts: The EOG contamination of the event-related brain potential. *Behav. Res. Methods Instrum. Comput.* **1998**, *30*, 44–53. [CrossRef]

99. Plöchl, M.; Ossandón, J.P.; König, P. Combining EEG and eye tracking: Identification, characterization, and correction of eye movement artifacts in electroencephalographic data. *Front. Hum. Neurosci.* **2012**, *6*, 278. [CrossRef] [PubMed]

100. Yagi, A.; Ogata, M. Measurement of work load using brain potentials during VDT tasks. In *Advances in Human Factors/Ergonomics*; Anzai, Y., Ogawa, K., Mori, H., Eds.; Elsevier: Tokyo, Japan, 1995; Volume 20, pp. 823–826. [CrossRef]

101. Frey, A.; Lemaire, B.; Vercueil, L.; Guérin-Dugué, A. An eye fixation-related potential study in two reading tasks: Reading to memorize and reading to make a decision. *Brain Topogr.* **2018**, *31*, 640–660. [CrossRef]

102. Kornrumpf, B.; Dimigen, O.; Sommer, W. Lateralization of posterior alpha EEG reflects the distribution of spatial attention during saccadic reading. *Psychophysiology* **2017**, *54*, 809–823. [CrossRef]

103. Kretzschmar, F.; Pleimling, D.; Hosemann, J.; Füssel, S.; Bornkessel-Schlesewsky, I.; Schlesewsky, M. Subjective Impressions Do Not Mirror Online Reading Effort: Concurrent EEG-Eyetracking Evidence from the Reading of Books and Digital Media. *PLoS ONE* **2013**, *8*, e56178. [CrossRef]

104. Metzner, P.; von der Malsburg, T.; Vasishth, S.; Rösler, F. Brain Responses to World Knowledge Violations: A Comparison of Stimulus- and Fixation-triggered Event-related Potentials and Neural Oscillations. *J. Cogn. Neurosci.* **2015**, *27*, 1017–1028. [CrossRef]

105. Vignali, L.; Himmelstoss, N.A.; Hawelka, S.; Richlan, F.; Hutzler, F. Oscillatory brain dynamics during sentence reading: A fixation-related spectral perturbation analysis. *Front. Hum. Neurosci.* **2016**, *10*, 191. [CrossRef]

106. Baccino, T.; Manunta, Y. Eye-fixation-related potentials: Insight into parafoveal processing. *J. Psychophysiol.* **2005**, *19*, 204–215. [CrossRef]

107. Hutzler, F.; Braun, M.; Võ, M.L.H.; Engl, V.; Hofmann, M.; Dambacher, M.; Jacobs, A.M. Welcome to the real world: Validating fixation-related brain potentials for ecologically valid settings. *Brain Res.* **2007**, *1172*, 124–129. [CrossRef]

108. Kretzschmar, F.; Bornkessel-Schlesewsky, I.; Schlesewsky, M. Parafoveal versus foveal N400s dissociate spreading activation from contextual fit. *NeuroReport* **2009**, *20*, 1613–1618. [CrossRef]

109. Simola, J.; Holmqvist, K.; Lindgren, M. Right visual field advantage in parafoveal processing: Evidence from eye-fixation-related potentials. *Brain Lang.* **2009**, *111*, 101–113. [CrossRef] [PubMed]

110. Dimigen, O.; Kliegl, R.; Sommer, W. Trans-saccadic parafoveal preview benefits in fluent reading: A study with fixation-related brain potentials. *NeuroImage* **2012**, *62*, 381–393. [CrossRef] [PubMed]

111. Henderson, J.M.; Luke, S.G.; Schmidt, J.; Richards, J.E. Co-registration of eye movements and event-related potentials in connected-text paragraph reading. *Front. Syst. Neurosci.* **2013**, *7*, 28. [CrossRef] [PubMed]

112. Hutzler, F.; Fuchs, I.; Gagl, B.; Schuster, S.; Richlan, F.; Braun, M.; Hawelka, S. Parafoveal X-masks interfere with foveal word recognition: Evidence from fixation-related brain potentials. *Front. Syst. Neurosci.* **2013**, *7*, 33. [CrossRef] [PubMed]

113. Kretzschmar, F.; Schlesewsky, M.; Staub, A. Dissociating word frequency and predictability effects in reading: Evidence from coregistration of eye movements and EEG. *J. Exp. Psychol.* **2015**, *41*, 1648–1662. [CrossRef]

114. Kornrumpf, B.; Niefind, F.; Sommer, W.; Dimigen, O. Neural correlates of word recognition: A systematic comparison of natural reading and rapid serial visual presentation. *J. Cogn. Neurosci.* **2016**, *28*, 1374–1391. [CrossRef]

115. López-Peréz, P.J.; Dampuré, J.; Hernández-Cabrera, J.A.; Barber, H.A. Semantic parafoveal-on-foveal effects and preview benefits in reading: Evidence from fixation related potentials. *Brain Lang.* **2016**, *162*, 29–34. [CrossRef]

116. Metzner, P.; Von Der Malsburg, T.; Vasishth, S.; Rösler, F. The importance of reading naturally: Evidence from combined recordings of eye movements and electric brain potentials. *Cogn. Sci.* **2016**, *41*, 1232–1263. [CrossRef]

117. Niefind, F.; Dimigen, O. Dissociating parafoveal preview benefit and parafovea-on-fovea effects during reading: A combined eye tracking and EEG study. *Psychophysiology* **2016**, *53*, 1784–1798. [CrossRef]

118. Weiss, B.; Knakker, B.; Vidnyánszky, Z. Visual processing during natural reading. *Sci. Rep.* **2016**, *6*, 26902. [CrossRef] [PubMed]

119. Loberg, O.; Hautala, J.; Hämäläinen, J.A.; Leppänen, P.H. Semantic anomaly detection in school-aged children during natural sentence reading–A study of fixation-related brain potentials. *PLoS ONE* **2018**, *13*, e0209741. [CrossRef] [PubMed]

120. Degno, F.; Loberg, O.; Zang, C.; Zhang, M.; Donnelly, N.; Liversedge, S.P. Parafoveal previews and lexical frequency in natural reading: Evidence from eye movements and fixation-related potentials. *J. Exp. Psychol.* **2019**, *148*, 453–474. [CrossRef] [PubMed]

121. Degno, F.; Loberg, O.; Zang, C.; Zhang, M.; Donnelly, N.; Liversedge, S.P. A co-registration investigation of inter-word spacing and parafoveal preview: Eye movements and fixation-related potentials. *PLoS ONE* **2019**, *14*, e0225819. [CrossRef]

122. Loberg, O.; Hautala, J.; Hämäläinen, J.A.; Leppänen, P.H. Influence of reading skill and word length on fixation-related brain activity in school-aged children during natural reading. *Vis. Res.* **2019**, *165*, 109–122. [CrossRef]

123. Frey, A.; Ionescu, G.; Lemaire, B.; López-Orozco, F.; Baccino, T.; Guérin-Dugué, A. Decision-making in information seeking on texts: An eye-fixation-related potentials investigation. *Front. Syst. Neurosci.* **2013**, *7*, 39. [CrossRef]

124. Marx, C.; Hawelka, S.; Schuster, S.; Hutzler, F. An incremental boundary study on parafoveal preprocessing in children reading aloud: Parafoveal masks overestimate the preview benefit. *J. Cogn. Psychol.* **2015**, *27*, 549–561. [CrossRef]

125. Hutzler, F.; Schuster, S.; Marx, C.; Hawelka, S. An investigation of parafoveal masks with the incremental boundary paradigm. *PLoS ONE* **2019**, *14*, e0203013. [CrossRef]

126. Kliegl, R.; Hohenstein, S.; Yan, M.; McDonald, S.A. How preview space/time translates into preview cost/benefit for fixation durations during reading. *Q. J. Exp. Psychol.* **2013**, *66*, 581–600. [CrossRef]

127. Kazai, K.; Yagi, A. Comparison between the lambda response of eye-fixation–related potentials and the P100 component of pattern- reversal visual evoked potentials. *Cogn. Affect. Behav. Neurosci.* **2003**, *3*, 46–56. [CrossRef]

128. Hauk, O.; Davis, M.H.; Ford, M.; Pulvermüller, F.; Marslen-Wilson, W.D. The time course of visual word recognition as revealed by linear regression analysis of ERP data. *Neuroimage* **2006**, *30*, 1383–1400. [CrossRef] [PubMed]
129. Sereno, S.C.; Brewer, C.C.; O'Donnell, P.J. Context effects in word recognition: Evidence for early interactive processing. *Psychol. Sci.* **2003**, *14*, 328–333. [CrossRef] [PubMed]
130. Ehinger, B.V.; Dimigen, O. Unfold: An integrated toolbox for overlap correction, non-linear modeling, and regression-based EEG analysis. *PeerJ* **2019**, *7*, e7838. [CrossRef] [PubMed]
131. Hohenstein, S.; Matuschek, H.; Kliegl, R. Linked linear mixed models: A joint analysis of fixation locations and fixation durations in natural reading. *Psychon. Bull. Rev.* **2017**, *24*, 637–651. [CrossRef] [PubMed]
132. Clifton, C., Jr.; Staub, A.; Rayner, K. Eye movements in reading words and sentences. In *Eye Movements: A Window on Mind and Brain*; van Gompel, R., Ed.; Elsevier: Amsterdam, The Netherlands, 2007; pp. 341–372.
133. Amsel, B.D. Tracking real-time neural activation of conceptual knowledge using single-trial event-related potentials. *Neuropsychologia* **2011**, *49*, 970–983. [CrossRef] [PubMed]
134. Cornelissen, T.; Sassenhagen, J.; Võ, M.L.H. Improving free-viewing fixation-related EEG potentials with continuous-time regression. *J. Neurosci. Methods* **2019**, *313*, 77–94. [CrossRef] [PubMed]
135. Kristensen, E.; Guerin-Dugué, A.; Rivet, B. Regularization and a general linear model for event-related potential estimation. *Behav. Res. Methods* **2017**, *49*, 2255–2274. [CrossRef]
136. Kristensen, E.; Rivet, B.; Guérin-Dugué, A. Estimation of overlapped Eye Fixation Related Potentials: The General Linear Model, a more flexible framework than the ADJAR algorithm. *J. Eye Mov. Res.* **2017**, *10*, 1–27. [CrossRef]
137. Smith, N.J.; Kutas, M. Regression-based estimation of ERP waveforms: I. The rERP framework. *Psychophysiology* **2015**, *52*, 157–168. [CrossRef]
138. Smith, N.J.; Kutas, M. Regression-based estimation of ERP waveforms: II. Nonlinear effects, overlap correction, and practical considerations. *Psychophysiology* **2015**, *52*, 169–181. [CrossRef]
139. Woldorff, M.G. Distortion of ERP averages due to overlap from temporally adjacent ERPs: Analysis and correction. *Psychophysiology* **1993**, *30*, 98–119. [CrossRef] [PubMed]
140. Hauk, O.; Giraud, A.L.; Clarke, A. Brain oscillations in language comprehension. *Lang. Cogn. Neurosci.* **2017**, *32*, 533–535. [CrossRef]

MDPI

St. Alban-Anlage 66

4052 Basel

Switzerland

Tel. +41 61 683 77 34

Fax +41 61 302 89 18

www.mdpi.com

Vision Editorial Office

E-mail: vision@mdpi.com

www.mdpi.com/journal/vision